Introduction to
STOCHASTIC PROCESS

Other Books of Interest

A First Course in Asymptotic Theory of Statistics (81-7319-260-X)
T.K. Chandra

A First Course in Probability (81-7319-068-2)
T.K. Chandra and D. Banerjee

A First Course on Parametric Inference (81-7319-196-4)
B.K. Kale

Frontiers in Probability and Statistics (81-7319-115-8)
S.P. Mukherjee, S.K. Basu and B.K. Sinha

Mathematics and Statistics in Engineering and Technology (81-7319-245-6)
A. Chattopadhyay

Recent Developments in Operational Research (81-7319-383-5)
Manju Lata Agarwal and Kanwar Sen

Sample Survey Theory (81-7319-137-9)
Des Raj and P. Chandhok

Sampling Theory and Methods (81-7319-370-3)
S. Sampath

Small Area Estimation in Survey Sampling (81-7319-220-0)
Parimal Mukhopadhyay

Statistical Inference and Design of Experiments (81-7319-281-2)
Ulhas J. Dixit and Meena R. Satam

Introduction to STOCHASTIC PROCESS

A.K. Basu

AMSTERDAM / FEB 2004

Alpha Science International Ltd.
Pangbourne England

Professor A.K. Basu
Editor of Stochastic Modelling and Applications
Department of Statistics, Calcutta University
Kolkata-700 019, India

Alpha Science International Ltd.
P.O. Box 4067, Pangbourne RG8 8UT, UK

ISBN 1-84265-105-6

Printed in India.

Preface

From my thirty years experience of teaching the subject of stochastic process to graduate and undergraduate students I reached the conclusion that there are plenty of books in the market either of an elementary level or of a highly specialized sophisticated level which are very much technical in nature. The adequate demand for intermediate level texts sound in both rigor and application. The present book is an attempt to meet this demand. This text grew out of my lectures on advanced undergraduate courses and graduate level courses at Canadian and U.S. Universities and a postgraduate course at Calcutta University.

Moreover, most of the textbooks available lack enough worked out problems and exercises to illustrate the theory. The present book containing many relevant exercises at the end of each chapter is written in genuine class notebook fashion. All the theorems are proved in sufficient details unlike others which prove only the main steps and leave the details to the readers. The difficult theorems are motivated by applications and supplemented with exercises. For inquisitive readers further reading materials, an appendix and a bibliography have been appended at the end. Large number of worked out exercises, ranging from easy to difficult, make this textbook self-taught for students with some background. For the benefit of students all the exercises are solved and appear as a separate section at the end of the book to help weaker students. The text contain a good number of examples and counter examples illuminating the subtleties and finer points of the general theory. A novel feature is a chapter on 'Time Series and Stochastic Integrals' in the appendix. The book consists of 10 chapters, an appendix and a section of more than 250 exercises.

Of course, in preparing my lectures, I borrowed heavily from the existing books in the field and the final text reflects this. In particular, the books by Kerlin and McGregor, D.R. Cox, Parzen, Chung, Prabhu were significant contributors. The only pre-requisites for the book are a sound knowledge in advanced calculus and elementary probability of the level of Feller Vol. I or my book on "Measure Theorety and Probability" (Prentice-Hall, India).

This book is suitable for one year (or two semesters) courses offered to postgraduate students of Indian universities and advanced undergraduate and first year graduate students of US universities and colleges in statistics, mathematics, operations research, engineering and business administration. The content of this book is based on UGC (University Grant Commission) of India recommended syllabus for postgraduate students in Indian universities.

My warm thanks and appreciation go to M/s DYZICOM, Kolkata, for their expert typing of the manuscript. I am also indebted to my friends and students for their enthusiasm and encouragement while writing this book.

A.K. Basu

Preface

Abbreviations and Notations

$P(A)$	:	Probability of the event A
r.v.	:	Random variable
p.d.f.	:	Probability density function (or density function)
p.g.f.	:	Probability generating function
p.f.	:	Probability mass function
d.f.(c.d.f.)	:	Distribution (cumulative) function
iff	:	If and only if
Ch.f.	:	Characteristic function
u.i.	:	Uniformly integrable
i.o.	:	Infinitely often
a.s.	:	Almost surely
a.e.	:	Almost everywhere
i.i.d.	:	Independent, identically distributed
$E(X)$	:	Expectation of the r.v. X
L.T.	:	Laplace Transform
w.r.t.	:	With respect to
l.h.s.(r.h.s.)	:	Left (right) hand side
$\sim$	:	Asymptotic to
$F * G$	:	Convolution of two distributions F and G
F^{n*}	:	Convolution of n i.i.d. r.vs. having common d.f. F
$\Phi(.)$	:	Distribution function of standard normal r.v.
CLT	:	Central limit theorem
SLLN	:	Strong law of large numbers
M.C.	:	Markov chain
M.P.	:	Markov process
LIL	:	Law of the iterated logarithm
$\mathscr{B}(\mathscr{G})$ or $\sigma(\mathscr{G})$	:	σ-algebra generated by the class $\mathscr{G}$
$\mathscr{B}(X)$	:	σ-algebra generated by the r.v. X
$\xrightarrow{\text{a.s. or a.e.}}$	:	Convergence almost surely or almost everywhere
$\xrightarrow{\text{P or d, or } \pounds}$	:	Convergence in probability or in distribution or in law
$\xrightarrow{L_p}$	:	Convergence in mean of order p
$\mathscr{B}^n$	:	Class of n-dimensional Borel sets
$\wedge$	:	Minimum of
$\vee$	:	Maximum of
$P(s)$	:	Probability generating function (p.g.f.) of a r.v.

S_N	:	Sum of N i.i.d. r.v.s X_i, where N is a r.v.
$G(s)$	:	p.g.f. of $\{g_n\}$
$F^*(s)$	:	$\int_0^\infty e^{-st}\,dF(x)$ (L.T. of r.v. X or d.f. F)
$\{X(t)\}$; $\{X_t\}$	:	Stochastic process in continuous time
$C_{s,t}$; $C(s, t)$	:	Covariance function between $X(t)$ and $X(s)$
p_{ij}	:	Transition probability from state i to j
$N(t)$	:	Number of occurrances of an event by time t (upto epoch t)
$C(z)$	:	Autocavariance generating function of $\{X_t\}$
$\rho(k)$; ρ_k	:	Correlation function of X_t at lag k
ρ	:	Utilization factor
$A \Rightarrow B$	:	A implies B

Contents

Preface *v*
Abbreviations and Notations *vii*

1. Introduction **1**
1.1 Notion of Stochastic Processes *1*
1.2 Different Types of Stochastic Processes *2*
1.3 An Introduction to Stationary Processes *2*
Exercises and Complements *5*

2. Discrete Time Markov Chains **8**
2.1 Definition and Transition Probabilities *8*
2.2 A Few More Examples *10*
2.3 Classification of States *12*
2.4 Limit Theorems for Markov Chains *23*
2.5 Stationary Distribution *27*
2.6 Special Chains and Foster Type Theorems *32*
2.7 Theorems Regarding Finite Markov Chains *37*
2.8 Methods of Evaluation of the n-Step Transition Probability *39*
Exercises and Complements *43*

3. Random Walks **51**
3.1 Different Types of Random Walks *51*
3.2 Sequential Analysis *58*
3.3 Wald's Equation and Wald's Identity *59*
3.4 Fluctuation Theory *62*
Exercises and Complements *62*

4. Renewal Theory **65**
4.1 Introduction *65*
4.2 Renewal Equation *66*
4.3 Renewal Theorems *70*
4.4 Central Limit Theorem for Renewal Theory *74*
4.5 Delayed and Equilibrium Renewal Processes *75*
4.6 Residual and Excess Life Times *78*
4.7 Renewal Reward Process *80*
4.8 Replacement Policies Connected with Renewal Theory *82*
Exercises and Complements *84*

5. Branching Process **87**
5.1 Introduction and History of Branching Process *87*
5.2 Properties of Generating Functions *88*
5.3 Probability of Extinction *91*
5.4 Distribution of Total Number of Progeny *96*
5.5 Continuous Parameter (Markov) Branching Process *98*
Exercises and Complements 105

6. Continuous Time Discrete State Markov Processes **107**
6.0 Axiomatic Definition and Transition Function *107*
6.1 Differentiability of Transition Function *108*
6.2 Kolmogorov Differential Difference Equation *112*
6.3 Infinitesimal Generators and Examples *114*
6.4 Birth and Death Processes *117*
6.5 The Yule Process *118*
6.6 Some Population Models *121*
Exercises and Complements 128

7. Poisson Process **133**
7.1 Different Definitions and Their Equivalence *133*
7.2 Poisson Process and Renewal Theory *136*
7.3 Properties of Poisson Process *138*
7.4 Characterization of Poisson Process *140*
7.5 Generalization of Poisson Process *142*
7.6 Order Statistics and Poisson Process *146*
Exercises and Complements 148

8. Continuous Time and Continuous State Markov Process **150**
8.1 Diffusion Processes *150*
8.2 Kolmogorov's Forward and Backward Diffusion Equations *155*
8.3 Wiener Process (Brownian Motion) *156*
8.4 Simple Properties of Wiener Process and First Passage Time Distribution *158*
8.5 Ornstein-Uhlenbeck Process and Gaussian Processes *162*
Exercises and Complements 172

9. Time Series Analysis **175**
9.1 Introduction and Stationary Time Series *175*
9.2 Different Approaches of Time Series *184*
9.3 Correlogram Analysis *187*
9.4 Wold's Decomposition and Wiener-Khintchine and Bochner's Theorem for Stationary Time Series *194*
9.5 Spectral Densities of some Standard Processes and Autocovariance Generating Function *199*
9.6 Estimation of Spectrum *204*

9.7 Forecasting *212*
9.8 Estimation of Parameter in the First Order Autoregressive Model AR(1) *215*
Exercises and Complements 222

10. Queueing Theory **227**
10.1 Introduction *227*
10.2 Standard Queuing Models *230*
10.3 The Model $M/M/C$ *239*
10.4 The Concept of Imbeded Markov Chains in Queuing Models *243*
10.5 The Queueing Models $M/G/1$ and $GI/M/1$ *246*
10.6 System Where Customers Arrive in Stages: $E_k/M/1$ Model *251*
10.7 Machine Interference Problem *254*
10.8 Bulk Queues: Elementary Results *256*
Exercises and Complements *257*

Appendices
I. Sample Function of Brownian Motion *259*
II. Second Order Stochastic Analysis *271*
III. Stochastic Integral (Ito Integral) *287*
IV. Some Important Martingale Theorems *299*

Solution to Exercises **303**
References **411**
Subject Index **415**

1

Introduction

1.1 Notion of Stochastic Processes

Loosely speaking, the mathematical description of a random phenomenon as it changes in time is a stochastic process. Since the last century there has been greater realisation that stochastic (or non-deterministic) models are more realistic than deterministic models in many situations. Observations taken at different time points rather than those taken at a fixed period of time began to draw the attention of scientists. The physicists and communication engineers played a leading role in the development of dynamic indeterminism. Many a phenomenon occurring in physical and life sciences are studied not only as a random phenomenon but also as one changing with time or space. Similar considerations are also made in other areas such as social sciences, economics and management sciences, and so on. The scope of applications of stochastic processes which are functions of time or space or both is ever increasing.

Definition

A *stochastic process* is a family of random variables $\{X_t\}$, where t takes values in the *index set* T (sometimes called a parameter set or a time set).

The values of X_t are called the state space and will be denoted by S.

If T is countable then the stochastic process is called a *stochastic sequence* (or *discrete parameter stochastic process*). If S is countable then the stochastic process is called a *discrete state (space) process*.

If S is a subset of the real line the stochastic process is called a *real valued process*.

If T takes continuously uncountable number of values like $(0, \infty)$ or $(-\infty, \infty)$ the stochastic process is called a *continuous time process*. To emphasize its dependence on t and sample point w, we shall denote the stochastic process by $X(t, w), t \in T, w \in \Omega$ i.e. for each $w \in \Omega$, $X_t = X(t, w)$ is a function of t.

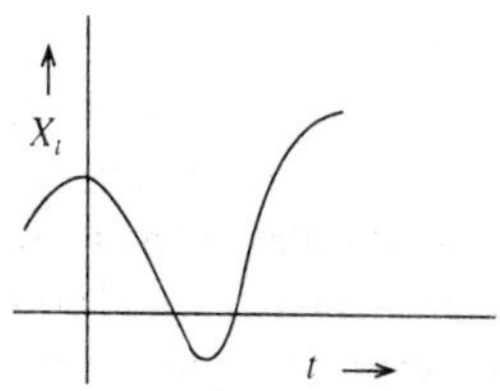

This graph is known as the "typical sample function" or "realization of the stochastic process" $X(t, w)$.

1.2 Different Types of Stochastic Processes

Following are the most important types of stochastic processes we come across:

1. Independent stochastic sequence (Discrete time process)

$T = [1, 2, 3, \ldots]$ and $\{X_t, t \in T\}$ are independent random variables.

2. Renewal process (Discrete time process)

Here $T = [0, 1, 2, 3, \ldots]$, $S = [0, \infty]$.

If X_n are i.i.d. non-negative random variables and $S_n = X_1 + \ldots + X_n$ then $\{S_n\}$ forms a *discrete time* (renewal process).

3. Independent increment process (Continuous time process)

$T = [t_0, \infty]$, where t_0 be any real number (+ or –). For every

$$t_0 < t_1 < \ldots < t_n, \; t_i \in T, \; i = 1, 2, \ldots, n \tag{1.1}$$

if $X_{t_0}, X_{t_1} - X_{t_0}, X_{t_2} - X_{t_1}, \ldots, X_{t_n} - X_{t_{n-1}}$ are independent for all possible choices of (1.1), then the stochastic process $\{X_t, t \in T\}$ is called *independent increment stochastic process.*

4. Markov process

If $P[X_{t_{n+1}} \in A | X_{t_n} = a_n, X_{t_{n-1}} = a_{n-1}, \ldots, X_{t_0} = a_0]$

$= P[X_{t_{n+1}} \in A | X_{t_n} = a_n]$ holds for all choices of

$$t_0 < t_1 < t_2 < \ldots < t_{n+1}, \; t_i \in T \cdot i = 0, 1, 2, \ldots, n + 1 \tag{1.2}$$

and $A \in \mathscr{B}$, the Borel field of the state space S, then $\{X_t, t \in T\}$ is called a *Markov process*.

5. Martingale or fair game process

If
$$E[X_{t_{n+1}} | X_{t_n} = a_n, X_{t_{n-1}} = a_{n-1}, \ldots, X_{t_0} = a_0] = a_n$$

i.e. $E[X_{t_{n+1}} | X_{t_n}, \ldots, X_{t_0}] = X_{t_n}$ a.s. for all choices of the partition (1.1), then $\{X_t, t \in T\}$ is called a *Martingale process.*

6. Stationary process

If the joint distribution of $(X_{t_1+th}, \ldots, X_{t_n+h})$ are the same for all $h > 0$ and

$$t_1 < t_2 < \ldots < t_n, \; t_i \in T, \; t_i + h \in T$$

then $\{X_t, t \in T\}$ is called a *stationary process* (strictly stationary process).

7. Point processes

When a countable set or sets of points randomly distributed on the real line or any arbitrary sets we call the family of random variables governed by the distribution of those random points as point processes. They represent number of occurrences of a particular event as time progresses. Poisson process and Renewal processes are particular cases of point processes. Cox processes (Doubly Poisson processes) are other examples of Point processes.

1.3 An Introduction to Stationary Processes

A stochastic process $\{X_t, t \in T\}$ with $EX_t^2 < \infty$ for all $t \in T$ is called *covariance stationary* or stationary in the wide-sense or *weakly stationary* if its covariance function $C_{s,t} = E(X_t X_s)$ depends only on the difference $|t-s|$ for all $t, s \in T$. Note that in our definition we have taken a zero mean stochastic process.

Examples of stationary processes

(a) Electrical pulses in communication theory are often postulated to describe a stationary process. Of course, in any physical system there is a transient period at the beginning of a signal. Since typically this has a short duration compared to the signal length, a stationary model may be appropriate. In electrical communication theory, often both the electrical potential and the current are represented as complex variables. Here we may encounter complex-valued stationary processes.

(b) The spatial and/or planar distributions of stars of galaxies, plants and animals, are often stationary. Time parameter set T might be Euclidean space, the surface of a sphere or the plane.

A stationary distribution may be postulated for the height of a wave and T is taken to be a set of longitudes and latitudes, again two dimensional.

(c) Economic time series, such as unemployment, gross national product, national income etc., are often assumed to correspond to a stationary process, at least after some correction for long-term growth has been made.

Ergodicity

The behavior in which sample averages formed from a process converge to some underlying parameter of the process is termed ergodic. To make inference about the underlying laws governing an ergodic process, one need not observe separate independent replications of entire processes or sample paths. Instead, one need only observe a single realization of the process, but over a sufficiently long span of time. Thus, it is an important practical problem to determine conditions that lead to a stationary process being ergodic. The theory of stationary processes has a prime goal the clarification of ergodic behavior and the prediction problem for processes falling in the wide range of extremeties.

In covariance stationary process usually the added condition that $E(X_t)$ does not depend on t is imposed. But it should be noted that in order for a stochastic process with $E(X_t^2) < \infty$ to be covariance stationary it is not necessary that its mean function $m(t) = E(X_t)$ be a constant. Consider the example: $X(t) = \cos\left(\frac{2\pi t}{L}\right) + Y(t)$, where $Y(t) = N(t+L) - N(t)$, $\{N(t), t \geq 0\}$ be a Poisson process with intensity parameter λ (to be defined in Chapter 7) and L is a positive constant. Its mean function $m(t) = E(X_t) = \lambda(t + L) - \lambda(t) + \cos\left(\frac{2\pi t}{L}\right)$ is functionally dependent on t. But

$$\text{Cov}\,(X(t), X(s)) = \text{Cov}\,(Y(t), Y(s))$$
$$= \begin{cases} \lambda(L - |t-s|) & \text{if } |t-s| \leq L \\ 0 & \text{if } |t-s| > L \end{cases}$$

depends on $t - s$ only.

Example 1 (*Gaussian process*). If the joint distribution of $(X_{t_1}, \ldots, X_{t_n})$ for all $t_1, \ldots, t_n \in T$ is multivariate normal, then $\{X_t, t \in T\}$ is a Gaussian process. Such type of process appears in e.g. physics and atmospheric turbulence theory. If a *Gaussian process is wide sense stationary then it is strictly stationary.* Now the multivariate normal distribution of $(X_{t_1}, \ldots, X_{t_g})$ is completely determined by its mean vector $(EX_{t_1}, \ldots, EX_{t_n})$ and covarience-matrix $(C_{t_i, t_i}) = (EX_{t_i} X_{t_j})$ $i, j = 1, 2, \ldots, n$. If a Gaussian process is covariance stationary then $E(X_{t_i}^2) < \infty$, $E(X_{t_j}^2) < \infty$ and $C_{t_i, t_j} = C_{ij}$ is a function of $| t_i - t_j |$.

Example 2 Let $\{X_n, n \geq 1\}$ be uncorrelated random variables with mean 0 and variance 1.

Then
$$C_{n,m} = \text{Cov}(X_n, X_m) = E(X_n, X_m)$$
$$= \begin{Bmatrix} 0 \text{ if } n \neq m \\ 1 \text{ if } n = m \end{Bmatrix}$$

So $\{X_n, n \geq 1\}$ is covariance stationary.

If moreover, X_n's are normally identically distributed then $\{X_n, n \geq 1\}$ is strictly stationary.

A process which is not stationary is called *evolutionary.*

Example 3 Poisson process

$P[X_t = k] = \exp(-\lambda t)(\lambda t)^k / k!$, $\lambda > 0$, $k = 0, 1, \ldots$ with the mean function $m(t) = \lambda t$, $\text{Var}(X_t) = \lambda t$ are functions of t and is evolutionary. Also note that $\text{Cov}(X_t, X_s) = \min(t, s)$ in not a function of $(t - s)$ only.

Example 4 Let $X(t) = A \cos(wt) + B \sin(wt)$, $t > 0$, where A and B are uncorrelated random variables each with mean 0 and variance 1 and $w > 0$ (constant). Here $EX(t) = 0$,

$$E(X(t) \cdot X(s)) = \cos(wt) \cdot \cos(ws) + \sin(ws) \cdot \sin(wt)$$
$$= \cos((s - t) \cdot w), \text{ and } E(X^2(t)) = 1.$$

Therefore the process is covariance stationary.

Example 5 Let $Z_t = \sum_{k=1}^{N} A_k e^{i\theta_k t}$, $i = \sqrt{-1}$, where $\theta_1, \theta_2, \ldots, \theta_N$ are real constants, A_k's are uncorrelated with mean 0 and $\text{Var}(A_i) = \sigma_i^2$. Here $E(Z_t) = m(t) = 0$ and

$$C_{s,t} = E(Z_s, \bar{Z}_t) = E\left(\sum_{k=1}^{N} A_k e^{i\theta_k s} \overline{\sum_{j=1}^{N} A_j e^{i\theta_j s}}\right)$$
$$= \sum_{j=1}^{N} \sum_{k=1}^{N} E(A_j A_k) e^{i(\theta_k s - \theta_j t)}$$
$$= \sum_{k=1}^{N} \sigma_k^2 e^{i\theta_k (s-t)}$$

$$\text{Var}(Z_t) = \sum_{k=1}^{N} \sigma_k^2.$$

Therefore, the process is stationary in the wide-sense.

Example 6 Let $X_n = \sum_{k=0}^{\infty} a^k \xi_{n-k}$, where ξ_i's are uncorrelated with mean zero and variance $\sigma^2 < \infty$ and $|a| < 1$.

$$C_{m,n} = E(X_m X_n) = \sum_{j=0}^{\infty} \sum_{k=0}^{\infty} a^{j+k} E(\xi_{m-k} E_{n-j})$$

$$= \begin{cases} \sigma^2 \sum_{k=0}^{\infty} a^{2k+n-m} & \text{if } m - n \leq 0 \\ \sigma^2 \sum_{k=m-n}^{\infty} a^{2k+n-m} & \text{if } m - n > 0 \end{cases}$$

$$= \frac{\sigma^2}{1 - a^2} a^{|m-n|} \text{ if } |a| < 1.$$

Therefore, the process is covariance stationary.

Stationary processes are appropriate for describing many phenomena that occur in communication theory, astronomy, biology and economics.

A stochastic process $\{X_t, t \in T\}$ is called *regular* if $EX_t^2 < \infty$ for all $t \in T$. A regular stochastic process is called *orthogonal* if $E(X_t, X_s) = 0$ for $t, s \in T, t \neq s$.

Exercises and Complements

Exercise 1.1 Let a, b, c be independent r.v.'s uniformly distributed on [0, 1]. What is the probability that $ax^2 + bx + c$ has real roots?

Exercise 1.2 Let X be a Poisson r.v. with parameter $\lambda > 0$. Suppose λ itself is a r.v. following a gamma distribution with density $f(\lambda) = \frac{1}{\sqrt{n}} \lambda^{n-1} e^{-\lambda}, \lambda \geq 0$. Show that

$P(X = k) = \frac{\sqrt{k+n}}{\sqrt{n}\sqrt{k+1}} (1/2)^{k+n}$, $k \geq 0$ (note that when n is a positive integer X is negative binomial with $p = 1/2$).

Exercise 1.3 The following experiment is performed. An observation is made of a Poisson r.v. X with parameter λ. Then a binomial event Y with probability p of success is repeated X number of times and Y successes are observed. What is the distribution of Y?

Exercise 1.4 Let $\{X_t, t \geq 0\}$ be a continuous time stochastic process with independent increments. Also $P(X_0 = 0) = 1$. If $\phi(\theta, t - u)$ is the characteristic function of a single increment i.e.

$$\phi(\theta, t - u) = E[\exp(i\theta(X_t - X_u))],$$

prove that the joint characteristic function of $X_{t_1}, X_{t_2}, \ldots, X_{t_n}$ where $t_1 < t_2 < \ldots < t_n$ is

$$\phi\left(\sum_{j=1}^{n} \theta_j, t_1\right) \phi\left(\sum_{j=2}^{n} \theta_j, t_2 - t_1\right) \ldots \phi(\theta_n, t_n - t_{n-1}).$$

Exercise 1.5 Prove that every continuous parameter Stochastic process with independent increments is a Markov process.

Exercise 1.6 Let T be a nonnegative discrete random variable. Prove that T has a geometric distribution iff

$$P[T > x + y | T > x] = P[T > y] \text{ for all integers } x, y \geq 0.$$

Exercise 1.7 Let T be a non-negative continuous random variable. Prove that T has an exponential distribution iff $P[T > x + y | T > x] = P[T > y]$.

Exercise 1.8 Let T be a nonnegative random variable such that $X(T)$ is a stochastic process and for a fixed value of T, say t, $X(t)$ has a gamma density

$$f_{x(t)}(u) = \frac{1}{\sqrt{\alpha t}} (\lambda u)^{\alpha t - 1} e^{-\lambda u}, \alpha > 0, \lambda > 0.$$

Assume that the distribution of T is $F(t) = P[T \leq t]$.

(a) Derive an expression for $E[e^{-cX(T)}]$

(b) Prove that T has a negative binomial distribution, i.e.

$$P\left[T = \frac{\beta + k}{\alpha}\right] = \binom{\beta + k - 1}{k} p^{\beta} (1 - p)^k, \beta > 0$$

$$1 > p = 1 - q > 0, k = 0, 1, 2, \ldots$$

then $X(T)$ also has a gamma density. Derive its parameters.

(c) Prove that if $X(T)$ has a gamma distribution, i.e.

$E[e^{-sx(T)}] = \left(\frac{\mu}{\mu + s}\right)^{\beta}$ where $\beta > 0$, $\lambda > \mu \geq 0$, then conversely, T has a negative binomial distribution. Determine its parameters.

Exercise 1.9 The random variables X and Y have the following properties: X is positive i.e. $P(X > 0) = 1$, with continuous density function $f(x)$ and $Y | X$ has a uniform distribution on $[0, X]$.

Prove that if Y and $X - Y$ are independently distributed, then

$$f(x) = a^2 x e^{-ax}, x > 0, a > 0.$$

Exercise 1.10 Let U and V be independent $N(0, 1)$ r.vs. Let T be the circumference of the unit circle $T = [0, 2\pi]$ and $Y(t) = U \sin t + V \cos t Z(t) = - U \cos t + V \sin t$. Then show that $X(t) = (Y(t), Z(t))$, $t \in T$ is a strictly stationary process.

Exercise 1.11 Let $X(t) = A \cos \lambda t + B \sin \lambda t$, where A and B are uncorrelated random variables with common mean zero and common variance σ^2. Decide whether $X(t)$ is covariance stationary.

Exercise 1.12 Let $\{X_i\}$ and $\{Y_i\}$ be two sequences of r.vs. on a probability space with $E(X_i) = E(Y_i) = 0$, $\text{Var}(X_i) = \text{Var}(Y_i) = \sigma^2 < \infty$, $E(X_i X_j) = E(Y_i Y_j) = E(X_i Y_j) = 0$ for all $i \neq j$ and $EX_1 Y_1 = 0$ for all i. Let $\{Z(t), t \geq 0\}$ be a stochastic process defined by

$$Z(t) = \sum_{j=1}^{n} \{X_j \cos(\lambda_j \cdot t) + Y_j \sin(\lambda_j \cdot t)\}$$

Determine its covariance function. Is the process second order stationary? Is the process strictly stationary? Justify your answer.

Exercise 1.13 Let $X(t) = \sin wt$, where w is uniformly distributed on $[0, 2\pi]$ (i) Show that $\{X(t), t = 1, 2, \ldots\}$ is covarience stationary but is not strictly stationary (ii) show that $\{X(t), t \geq 0\}$ is neither covariance stationary nor strictly stationary.

Exercise 1.14 Let $X_n = \cos n\, U$ $(n \geq 1)$, where U is uniformly distributed on $[-\pi, \pi]$. Show that $\{X_n\}$ is covariance stationary but not strictly stationary.

Exercise 1.15 *(Random Binary Noise)* A stochastic process $X(t)$ is called random binary noise if $X(t) = Y(t - u)$, where Y is a process taking value ± 1 with probability 1/2 on successive intervals of fixed length T and such that the r.v.'s $Y(t)$ are independent for values of t lying in non overlapping intervals and U is the random shift of time such that U is uniformly distributed on $(0, T)$ and is independent of $Y(t)$. Show that $X(t)$ is wide sense stationary.

Exercise 1.16 *(Random Telegraph Signal Process)*
Let $X(t)$ be a stochastic process defined by $X(t) = Y(-1)^{N(t)}$, $t \geq 0$ where $N(t)$, $t \geq 0$ is a Poisson process with parameter λ and Y is a r.v. independent of the process $N(t)$, $t \geq 0$ with probabilities specified by $P(Y = \pm 1) = 1/2$. Show that $X(t)$, $t \geq 0$ is a covariance stationary process.

Exercise 1.17 *(Model of a Germination Process)*
Let $T_1, T_2, \ldots, T_N$ be the germination times of N seeds planted at time $t = 0$ and they are assumed to be i.i.d. with common unknown d.f. $H(t) = P(T \geq t)$ with $H(0) = 0$ and $H(\infty) = 1 - p$, where $0 \leq p < 1$ be the probability that a seed may fail to germinate. Also assume that the average proportion of germination in a small interval $(S, S + \Delta S)$ is approximately $\lambda > 0$. Find the probability distribution of X_t, the number of seeds that will germinate in $(0, t)$.

Exercise 1.18 *(FKG inequality)*
Let $X_1, X_2, \ldots, X_n$ be i.i.d. r.v.'s having Bernouli distribution with parameter p. Let f be increasing such that $f(\mathbf{x}) \leq f(\mathbf{y})$ whenever $x_i \leq y_i$ for each i. Let $\mathbf{x} = (x_1, \ldots x_n)$.

(a) Show that $E_{p_1}(f(\mathbf{x})) \leq E_{p_2}(f(\mathbf{x})$ if $p_1 \leq p_2$

(b) Let f and g be increasing function on $[0, 1]^n$ to R. Show by induction on n that $\text{cov}\,(f(\mathbf{x}), g(\mathbf{x})) \geq 0$.

Exercise 1.19 *(Antithetic variables)*
Let $g(x_1, \ldots, x_n)$ be an increasing function in all its variables, and let $\{U_r\}$ be i.i.d. uniform r.v.'s. on $[0, 1]$ showthat

$$\text{cov}\,\{g(U_1, \ldots, U_n), g(1 - U_1, \ldots, 1 - U_n)\} \leq 0.$$

[Hint. Use FKG inequality] Explain how this can help in the efficient estimation of $I = \int_0^1 \cdots \int_0^1 g(\mathbf{x})\, dx_1 \ldots dx_n$.

2

Discrete Time Markov Chain

2.1 Definition and Transition Probabilities

Here S = a countable set, $T = \{0, 1, 2, \ldots\}$, $\{X_n, n \geq 0\}$ is a stochastic process satisfying $P[X_{n+1} = j \mid X_0 = i_0, X_1 = i_1, \ldots, X_n = i_n] = P[X_{n+1} = j \mid X_n = i_n]$, the Markov property. Then the stochastic process $\{X_n, n \geq 0\}$ is called a Markov chain (M.C.). We shall assume that the M.C. is stationary i.e. $P[X_{n+1} = j \mid X_n = i] = p_{ij}$ is independent of n for all $i, j \in, S$. Let $P = (P_{ij})$; $i, j \in S$ be a finite or countably infinite dimensional matrix with elements p_{ij}.

The matrix P is called the *one step transition matrix* of the M.C. or simply the Transition matrix or the Probability matrix of the M.C.

Example (Random Walk) A random walk on the (real) line is a Markov chain such that

$$p_{jk} = 0 \text{ if } k \neq j - 1 \text{ or } j + 1.$$

Transition is possible only to neighbouring states (from j to $j - 1$ and $j + 1$). Here state space is

$$S = \{\ldots, -3, -2, -1, 0, 1, 2, 3, \ldots\}.$$

Theorem 2.1 The Markov chain $\{X_n, n \geq 0\}$ is completely determined by the transition matrix P and the initial distribution $\{p_k\}$, defined as $P[X_0 = k] = p_k \geq 0$, $\sum_{k \in s} p_k = 1$.

Proof

$$\begin{aligned}
&P[X_0 = i_0, X_1 = i_i, \ldots, X_n = i_n] \\
&\quad = P[X_n = i_n \mid X_{n-1} = i_{n-1}, X_{n-2} = i_{n-2}, \ldots, X_1 = i_1 \ldots X_0 = i_0] \\
&P[X_{n-1} = i_{n-1}, X_{n-2} = i_{n-2}, \ldots, X_1 = i_1, X_0 = i_0] \\
&\quad = P[X_n = i_n \mid X_{n-1} = i_{n-1}] P[X_{n-1} = i_{n-1}, \ldots, X_0 = i_0] \\
&\quad = p_{i_{n-1} i_n} p_{i_{n-2} i_{n-1}} \; P[X_{n-2} = i_{n-2}, \ldots, X_0 = i_0] \\
&\quad = p_{i_{n-1} i_n} p_{i_{n-2} i_{n-1}} \cdots p_{i_1 i_2} p_{i_0 i_1} p_{i_0} \text{ (by induction).}
\end{aligned}$$

Definition 2.1 A vector $u = (u_1, u_2, \ldots, u_n)$ is called a *probability vector* if the components are non-negative and their sum is one.

Definition 2.2 A square matrix $P = (p_{ij})$ is called a *stochastic matrix* if each of its row is a *probability vector* i.e. if each element of P is non-negative and the sum of the elements in each row is one, e.g.

$$\begin{pmatrix} 0 & 1 & 0 \\ 1/2 & 1/6 & 1/3 \\ 1/3 & 2/3 & 0 \end{pmatrix} \text{ is a stochastic matrix.}$$

A transition matrix P of a M.C. is a stochastic matrix.

Definition 2.3 A stochastic matrix A is said to be *regular* if all the entries of some power A^m is positive (m is a positive integer), e.g.

$$A = \begin{pmatrix} 0 & 1 \\ 1/2 & 1/2 \end{pmatrix} \text{ is regular, since } A^2 = \begin{pmatrix} 1/2 & 1/2 \\ 1/4 & 1/4 \end{pmatrix} > 0.$$

Problems

1. Suppose P is a stochastic matrix, then show that P^n is also a stochastic matrix for all $n > 1$.
2. If P^n is stochastic, is P stochastic?
3. Show that 1 is an eigenvalue if A is a stochastic matrix, i.e.

$$|\lambda I - A| = 0 \Rightarrow \lambda = 1.$$

Consider a sequence of trials with possible outcomes $E_1, E_2, \ldots, E_k \ldots$ To the pairs of outcomes (E_j, E_k) we can associate some numbers (i.e. conditional probabilities) P_{jk}. The $\{E_k\}$ are referred to as the possible states of the system. Instead of saying that the nth trial results in E_k one says that the nth step leads to E_k or that E_k is entered at the nth step.

We shall denote by $P_{jk}^{(n)}$ the probability of transition from E_j to E_k in *exactly n steps* i.e. the conditional probability of entering E_k at the nth step from E_j. This is the *sum of all the probabilities of all possible paths* $E_j \to E_{j_1} \to \ldots E_{j_{n-1}} \to E_k$ of length n starting at E_j and *ending at* E_k

In particular, $$p_{jk}^{(1)} = p_{jk}.$$

Theorem 2.2 *(Chapman-Kolmogorov equation)*

$$p_{ij}^{(n)} = \sum_{k \varepsilon s} p_{ik}^{(n-1)} p_{kj}$$

Proof $$P_{ij}^{(n)} = P[X_n = j \mid X_0 = i]$$

$$= \sum_{k \varepsilon s} P[X_n = j \mid X_{n-1} = k, X_0 = i] P[X_{n-1} = k \mid X_0 = i]$$

$$= \sum_{k \varepsilon s} P[X_n = j \mid X_{n-1} = k] P[X_{n-1} = k \mid X_0 = i] \text{ (by Markov property)}$$

$$= \sum_{k \varepsilon s} P_{kj}^{(1)} P_{ik}^{(n-1)} = \sum_{k \varepsilon s} P_{kj} P_{ik}^{(n-1)}$$

Corollary For $$0 \le m \le n, p_{ij}^{(n)} = \sum_{k \varepsilon s} p_{ik}^{(n-m)} p_{kj}^{(m)}$$

Note Actually $P_{ij}^{(n)}$ are the elements of P^n, where $P = (p_{ij})$ and the corollary express the matrix multiplication rule $P^{m+n} = P^m P^n$.

To be consistant we define

$$P_{ij}^{(0)} = P[X_o = j \mid X_0 = i] = \delta_{ij} \text{ (Kronecker's delta).}$$

Transition probabilities of a non-homogeneous (non-stationary) M.C. $\{X_n\}$ also satisfies Chapman-Kolmogorov equation given by

$$P_{ij}^{(m,n)} = \sum_{k \varepsilon s} p_{ik}^{(m-u)} P_{kj}^{(u,n)} \text{ for } m < u < n$$

and for all values of i, j in S, where $p_{ij}^{(m,n)} = P(X_n = j \mid X_m = i)$.

In terms of transition probability matrices Chapman-Kolmogorov equation may be written as

$$P^{(m,n)} = P^{(m,u)} P^{(u,n)} \quad \text{for all } n > u > m > 0.$$

where $\quad P^{(m,n)} = (p_{ij}^{(m,n)})$.

We also have

$$\begin{aligned} P^{(m,n)} &= P^{(m,n-1)} P^{(n-1,n)} \\ &= P^{(m,n-2)} P^{(n-2,n-1)} P^{(n-1,n)} \ldots \\ &\ldots \\ &= P^{(m,m+1)} P^{(m+1,m+2)} \ldots P^{(n-1,n)} \end{aligned}$$

Proof

$$\begin{aligned} P_{ij}^{(m,n)} &= \sum_{k \varepsilon S} P(X_n = j \mid X_{n-1} = k, X_m = i) P(X_{n-1} = k \mid X_m = i) \\ &= \sum_{k \varepsilon S} P(X_n = j \mid X_{n-1} = k)\, P(X_{n-1} = k \mid X_m = i) \\ &= \sum_{k \varepsilon s} p_{kj}^{(n-1,n)} p_{ik}^{(m,n-1)} = \sum_{k \varepsilon s} p_{ik}^{(m,n-1)} p_{kj}^{(n-1,n)}. \end{aligned}$$

2.2 A Few More Examples

Examples

(a) Independent trials

$P^n = P$ for all $n \geq 1$, where $p_{ij} = p_j$ i.e. all the rows are same.

(b) Success runs

Consider an infinite sequence of Bernoulli trials and at the nth trial the system is in the state E_j if the last failure occurred at the trial number $n - j$, $j = 0, 1, 2, \ldots$ and zero-th trial counts as failure. In other words, the index j equals the length of uninterrupted run of successes ending at nth trial.

Here

$$p_{ij}^{(n)} = \begin{cases} qp^j \text{ for } j = 0, 1, 2, \ldots, i + n - 1 \\ p^j \text{ for } j = j + n \\ 0 \text{ otherwise} \end{cases}$$

This follows either directly or from Chapman-Kolmogorov's equation. It can be shown that P^n converges to a matrix whose all elements in the column j equals qp^j, where the transition matrix P is given by

$$P_{ij} = P(X_n = j \mid X_{n-1} = i) = \begin{cases} p \text{ if } j = i + 1 \\ q \text{ if } j = 0 \\ 0 \text{ otherwise.} \end{cases}$$

(c) Two state M.C.

There are two possible states E_1 and E_2 in which the matrix of transition probability is of the form

$$P = \begin{pmatrix} 1-p & p \\ a & 1-a \end{pmatrix}, 0 < p < 1 \text{ and } 0 < a < 1.$$

The system is said to be in state E_1 if a particle moves in the positive direction and in E_2 if the direction is negative.

(d) Random walk with absorbing barriers

Let the possible states be $E_0, E_1, E_2, \ldots, E_k$. Consider the matrix of transition probabilities

$$P = \begin{pmatrix} 1 & 0 & 0 & 0 & \ldots & 0 & 0 & 0 \\ q & 0 & p & 0 & \ldots & 0 & 0 & 0 \\ 0 & q & 0 & p & \ldots & 0 & 0 & 0 \\ \ldots & \ldots & \ldots & \ldots & \ldots & \ldots & \ldots & \ldots \\ 0 & 0 & 0 & 0 & \ldots & q & 0 & p \\ 0 & 0 & 0 & 0 & \ldots & 0 & 0 & 1 \end{pmatrix}$$

From each of the 'interior' states $E_1, E_2, \ldots, E_{k-1}$ transitions are possible to the right and left neighbour with $p_{i,i+1} = p$, $p_{i,i-1} = q$. However, no transition is possible from either E_0 or E_k to any other state.

The system may move from one state to another, but once E_0 or E_k is reached the system stays there permanently.

(e) Random walk with reflecting barriers

$$P = \begin{pmatrix} q & p & 0 & 0 & \ldots & 0 & 0 & 0 \\ q & 0 & p & 0 & \ldots & 0 & 0 & 0 \\ 0 & q & 0 & p & \ldots & 0 & 0 & 0 \\ \ldots & \ldots & \ldots & \ldots & \ldots & \ldots & \ldots & \ldots \\ 0 & 0 & 0 & 0 & \ldots & q & 0 & p \\ 0 & 0 & 0 & 0 & \ldots & 0 & q & p \end{pmatrix}$$

This chain may be interpreted in Gambling language by considering two players playing for unit stakes with the agreement that every time a player loses all his money his adversary returns it, so that the game can continue forever i.e. a random walk in a finite interval such that whenever the particle is at point 1 it has probability p of moving to position 2 and probability q to stay. The system is in state E_k, if the capitals are k and $R - k + 1$ respectively.

2.3 Classification of States

2.3.1 Decomposition of state space

It may be possible that $p_{ij} = 0, p_{ij}^{(2)} = 0$ but $p_{ij}^{(3)} > 0$. We say that the state j is *accessible* from state i if $p_{ij}^{(n)} > 0$ for some $n > 0$. In notation $i \to j$, i.e. i leads to j. If $i \to j$ and $j \to i$, then i and j *communicate* and we denote this by $i \leftrightarrow j$.

Definition 2.4 The state i is *essential* if $i \to j$ implies $i \leftarrow j$, i.e. if any state j is accessible from i, then i is accessible from that state. We shall let $\mathfrak{I}$ denote the set of all essential states. States that are not *essential* are called *inessential.*

Lemma 2.1 $i \leftrightarrow j$ defines an equivalence relation on $\mathfrak{I}$, the class of essential states.

Proof $i \leftrightarrow i$ (reflexivity)

(i) Since for each i, $\sum_{j\varepsilon s} p_{ij} = 1$ there exists at least one j for which $p_{ij} > 0$. But if i is essential then there exists $m \geq 1$ such that $p_{ji}^{(m)} > 0$. So by Chapmen-Kolmogrov equation $p_{ii}^{(m+1)} \geq p_{ij} p_{ji}^{(m)} > 0$.

(ii) $i \leftrightarrow j \Leftrightarrow j \leftrightarrow i$ (symmetry)

(iii) $i \leftrightarrow j$ and $j \leftrightarrow k \Rightarrow i \leftrightarrow k$ (transitivity)

Proof of (iii)

To prove $i \to k$, since $i \to j$ $p_{ji}^{(n)} > 0$ for some $n \geq 1$ and $j \to k$, $p_{jk}^{(m)} > 0$ for some $m \geq 1$.

Claim: $p_{ik}^{(l)} > 0$ for some $l \geq 1$

$$0 < p_{ij}^{(n)} p_{jk}^{(m)} \leq \sum_{j\varepsilon s} p_{ij}^{(n)} p_{jk}^{(m)} = p_{ik}^{(n+m)} \text{ (Chapman-Kolmogorov)}$$

Taking $l = m + n$,

$$i \to k \text{ and similarly } k \to i \Rightarrow i \leftrightarrow k.$$

By Lemma 2.1, i.e. $\mathfrak{I} = \cup \{C(i)$, where $C(i) = \{j \in \mathfrak{I} \mid i \leftrightarrow j\}$ is called a *communicating class*, i.e. the class of essential states is partitioned into disjoint equivalent classes (communicating classes).

Definition 2.5

A Markov chain is called *irreducible* (or *ergodic*) if there is only one communicating class, i.e. if all states communicate with each other or every state can be reached from every other state.

Definition 2.6

A subset C of S is called *closed* (or *transient*) if it is impossible to leave C in one step i.e. $p_{ij} = 0$ for all $i \in C$ and all $j \notin C$.

Lemma 2.2 If C is closed then
(i) for all $n \geq 1$, $p_{ij}^{(n)} = 0$ if $i \in C, j \notin C$ i.e. no state outside C can be reached from any state in C.
(ii) $\sum_{k \in C} p_{ik}^{(n)} = 1$ for $i \in C$ and all $n \geq 1$.

Proof (i) : $n = 2$

$$p_{ij}^{(2)} = \sum_{k \in C} p_{ik} p_{kj} = \sum_{k \in C} p_{ik} p_{kj} + \sum_{k \notin C} p_{ik} p_{kj} = 0 \text{ for } j \notin C$$

Since $\qquad P_{ik} = 0$ for $k \notin C$ and $i \in C$

$\qquad P_{kj} = 0$ for $k \in C$.

By induction $p_{ij}^{(n)} = 0$ if $i \in C$ and $j \notin C$.

(ii) $$1 = \sum_{k \in S} p_{ik}^{(n)} = \sum_{k \in C} p_{ik}^{(n)} + \sum_{k \notin C} p_{ik}^{(n)} \text{ for all } i \in C$$

But $\quad \sum_{k \in C} p_{ik}^{(n)} = 0 \rightarrow \sum_{k \in C} p_{ik}^{(n)} = 1$

Corollary 2.1 A Markov chain is *irreducible* if the only closed set in S is S itself, i.e. there is no closed set other than the set of all states.

Theorem 2.3 (p_{ik}) $i, k \in C$ is Markov matrix if C is closed, i.e. if in the matrix P^n all rows and all columns corresponding to states outside the closed set C is deleted, there remains stochastic matrices for which the fundamental relations (Chapman-Kolmogorov equation)

$$p_{jk}^{(n+1)} = \sum_{k \in C} p_{ji} p_{jk}^{(n)} \text{ and } p_{jk}^{(m+n)} = \sum_{i \in C} p_{ji}^{(m)} p_{jk}^{(n)} \text{ hold.}$$

(This means that we have a M.C. defined on C and this subchain can be studied independently of all other states). Proof follows from Lemma 2.2 and is omitted.

Definition 2.7 A single-state $\{k\}$ forming a closed set is called *absorbing* (*trapping*) i.e. a state such that the system remains in that state once it enters there. Thus a state k is absorbing if the kth row of the transition matrix P has 1 on the main diagonal and 0 elsewhere.

Example 2.1

$$\begin{array}{cc} & \begin{array}{ccccccccc} 1 & 2 & 3 & 4 & 5 & 6 & 7 & 8 & 9 \end{array} \\ P = & \begin{pmatrix} 0 & 0 & 0 & 1/2 & 0 & 0 & 0 & 0 & 1/2 \\ 0 & 1/4 & 1/4 & 0 & 1/4 & 0 & 0 & 1/4 & 0 \\ 0 & 0 & 0 & 0 & 0 & 0 & 0 & 1 & 0 \\ 1 & 0 & 0 & 0 & 0 & 0 & 0 & 0 & 0 \\ 0 & 0 & 0 & 0 & 1 & 0 & 0 & 0 & 0 \\ 0 & 1/3 & 0 & 0 & 0 & 0 & 1/3 & 0 & 1/3 \\ 0 & 1/3 & 0 & 0 & 0 & 1/3 & 1/3 & 0 & 0 \\ 0 & 0 & 1 & 0 & 0 & 0 & 0 & 0 & 0 \\ 0 & 0 & 0 & 1/2 & 0 & 0 & 0 & 0 & 1/2 \end{pmatrix} \end{array}$$

In this example $p_{55} = 1$, i.e. the state 5 is absorbing.

{3, 8} is a closed set since $p_{83} = p_{38} = 1$. From 1, passages are possible into 4 and 9 and from there only to 1, 4 and 9. So {1, 4, 9} is a closed set.

From 2, transition is possible to 2, 3, 5 and since {3, 8} is closed and 5 is absorbing, the closure of 2 is {2, 3, 5, 8}. But the closure of 6 and 7 are the whole state space {1, 2, 3, 4, 5, 6, 7, 8, 9}. The determination of the closed sets can be simplified by rearranging or renumbering the states in the order 5, 3, 8, 1, 4, 9, 2, 6, 7 and the rearranged matrix is:

$$\begin{array}{c} \begin{matrix} 5 & 3 & 8 & 1 & 4 & 9 & 2 & 6 & 7 \end{matrix} \\ \begin{pmatrix} 1 & 0 & 0 & 0 & 0 & 0 & 0 & 0 & 0 \\ 0 & 0 & 1 & 0 & 0 & 0 & 0 & 0 & 0 \\ 0 & 1 & 0 & 0 & 0 & 0 & 0 & 0 & 0 \\ 0 & 0 & 0 & 0 & 1/2 & 1/2 & 0 & 0 & 0 \\ 0 & 0 & 0 & 1 & 0 & 0 & 0 & 0 & 0 \\ 0 & 0 & 0 & 1/2 & 1/2 & 0 & 0 & 0 & 0 \\ 1/4 & 1/4 & 1/4 & 0 & 0 & 0 & 1/4 & 0 & 0 \\ 0 & 0 & 0 & 0 & 0 & 1/3 & 1/3 & 0 & 1/3 \\ 0 & 0 & 0 & 0 & 0 & 0 & 1/3 & 1/3 & 1/3 \end{pmatrix} \end{array}$$

Hence, the communicating classes are {5}, {3, 8}, {1, 4, 9}, {2} and {6, 7}.

Example 2.2 A man tosses a fair coin until 3 heads occur in a row. Let X_n denote the largest string of heads ending at the nth trial.

Then with positive probabilities $X_n = 0, 1, 2, 3$. This is a Markov chain with state space 0, 1, 2, 3.

The transition matrix is given by

$$\begin{pmatrix} 1/2 & 1/2 & 0 & 0 \\ 1/2 & 0 & 1/2 & 0 \\ 1/2 & 0 & 0 & 1/2 \\ 0 & 0 & 0 & 1 \end{pmatrix}$$

3 is an absorbing state. Therefore, M.C. is not irreducible. The communicating classes are {0, 1, 2} and {3}.

2.3.2 Classification of states

Let $f_{ij}^{(n)} = P[X_n = j, X_{n-1} \neq j, X_{n-2} \neq j, \ldots, X_1 \neq j \mid X_0 = i]$, i.e. probability of arriving at j at time n for the first time, given that the process starts at i.

Define $f_{ij}^{(0)} = 0$. Note that $f_{ij}^{(n)} = P[T_{ij} = n]$, where

$$T_{ij} = \min \{n: X_n = j \mid X_0 = i\}$$

$f_{ij}^{(n)}$ are called the first entrance probability at nth step if $i \neq j$ and recurrence probability at the nth step.

Note $f_{ij}^{(1)} = p_{jj}$ gives the diagonal of the transition matrix.

Theorem 2.4 $p_{ij}^{(n)} = \sum_{m=1}^{n} f_{ij}^{(m)} p_{jj}^{(n-m)}$ for all $m = 1, 2, \ldots n$.

Proof $p_{ij}^{(n)} = P[X_n = j \mid X_0 = i]$

$$= \sum_{m=1}^{n} \underbrace{P[X_n = j}_{A}, \underbrace{X_m = j, X_{m-1} \neq j, \ldots, X_1 \neq j}_{B_m} \underbrace{|X_0 = i]}_{C}$$

$$= \sum_{m=1}^{n} P[AB_m \mid C]$$

where B_m are disjoint (mutually exclusive) and $\bigcup_{m-1}^{n} B_m \supset A$.

Hence
$$P_{ij}^{(n)} = \sum_{m=1}^{n} \frac{P(AB_m C)P(B_m C)}{P(C)P(B_m C)}$$

$$= \sum_{m=1}^{n} P(A \mid B_m C)P(B_m \mid C)$$

$$= \sum_{m=1}^{n} P[X_n = j \mid X_m = j, X_{m-1} \neq j, \ldots, X_1 \neq j, X_0 = i]$$

$$P[X_m = j, X_{m-1} \neq j, \ldots, X_1 \neq j \mid X_0 = i]$$

$$= \sum_{m=1}^{n} P(X_n = j \mid X_m = j)\, f_{ij}^{(m)}$$

$$= \sum_{m=1}^{n} P_{ji}^{(n-m)}\, f_{ij}^{(m)}$$

Corollary 2.2 Let $P_{ij}(s) = \sum_{n=0}^{\infty} P_{ij}^{(n)} s^n$ (for $|s| < 1$) be the G.F. of $\{P_{ij}^{(n)}\}$

and
$$F_{ij}(s) = \sum_{n=0}^{\infty} f_{ji}^{(n)} s^n \text{ (for } |s| < 1\text{) be the G.F. of } \{f_{ij}^{(n)}\}.$$

Then
$$P_{ij}(s) = \delta_{ij} + F_{ij}(s)\, P_{jj}(s)$$

Proof
$$P_{ij}(s) = \sum_{n=1}^{\infty} p_{ji}^{(n)} s^n + p_{ij}^{(0)}$$

$$= \delta_{ij} + \left(\sum_{n=1}^{\infty} p_{jj}^{(n-m)} s^{n-m} \right) f_{ij}^{(m)} s^m$$

$$= \delta_{ij} + \sum_{n=1}^{\infty} P_{jj}(s) f_{ij}^{(m)} s^m$$

$$= \delta_{ij} + P_{jj}(s) \left(\sum_{n=1}^{\infty} f_{ij}^{(m)} s^m \right)$$

$$= \delta_{ij} + P_{jj}(s) F_{ij}(s)$$

Write $f_{ij} = \sum_{n=1}^{\infty} f_{ij}^{(n)}$ = the probability that starting from i the system will ever pass through j. Hence, $0 \le f_{ij} \le 1$.

Definition 2.8 A state $i \in s$ of a Markov chain is said to be *transient* if $f_{ii} < 1$ and *recurrent* (Persistent) if $f_{ii} = 1$. If $f_{ij} = 1$, then $\{f_{ij}^{(n)}\}$ is a proper probability distribution and we shall refer to it as the *first passage* time distribution of j (starting from i). Let $\mu_i = \sum_{n=1}^{\infty} n f_{ii}^{(n)}$.

If $f_{ii} = 1$, then $\mu_i \le \infty$ is the *mean recurrence* time for i.

The first question we can ask concerning any state i is 'whether a return to it is certain or not.' The next question would be 'whether the mean recurrence time μ_i is finite or not.'

Definition 2.9 A state i is called *positive* (or ergodic) if i is recurrent and $\mu_i < \infty$. A state i is called *null* if i is recurrent and $\mu_i = \infty$.

Theorem 2.5 Let $i \in S$ of a Markov chain. Then

(i) i is recurrent iff $\sum_{n=0}^{\infty} p_{ii}^{(n)} = \infty$

(ii) i is transient iff $\sum_{n=0}^{\infty} p_{ii}^{(n)} < \infty$

(iii) if i is recurrent then $\mu_i = \sum_{n=1}^{\infty} n f_{ii}^{(n)} = \lim_{s \to 1-} \dfrac{1}{(1-s)P_{ii}(s)}$

Before proving the theorem let us state the following important theorems from Analysis which will be quoted frequently.

2.3.3 A few important theorems (for countable space)

Let $\{a_n\}$ be a sequence of real numbers such that $0 \le a_n \le 1$, $n = 0, 1, 2, \ldots$

Let $A(z) = \sum_{n=0}^{\infty} a_n z^n$ be the generating function of $\{a_n\}$

1. *Pringsheim* : $A(z)$ converges in a circle $|z| < r$, where $r \le 1$. $Z = r$ is a singularity of $A(z)$.

2. *Abel* : If $\sum_{n=0}^{\infty} a_n = a < \infty$, then $\lim_{z \to 1-} A(Z) = a$.

3. *Tauber* : If $\lim_{z \to 1-} A(Z) = a \le \infty$, then $\sum_{n=0}^{\infty} a_n = a$.

4. *Cesaro-Abel* : If $\lim_{n \to \infty} \dfrac{1}{n} \sum_{k=1}^{\infty} a_k = L < \infty$, then

$$\lim_{z \to 1-} (1-z)A(z) = L.$$

5. *Cesaro-Tauber* : If $\lim_{z \to 1-} (1-z)A(z) = L < \infty$, then

$$\lim_{z \to 1-} \frac{1}{n} \sum_{k=0}^{n} a_k = L.$$

For proofs of 2 and 3 see the book of Karlin (page 46, Introduction of Stochastic process) and proofs of 1, 4 and 5 see the book of Tichmarsh-Theory of Functions.

Let $n = 0, 1, 2, \ldots$.

1. *(Lebesgue) Dominated Convergence Theorem*

If (i) $\lim_{n \to \infty} a_{n,m}$ exists for every m

(ii) $| a_{nm} | \le b_m$ (independent of n) for all $m \ge 0$

(iii) $\sum_{m=0}^{\infty} b_m < \infty$

then $\lim_{n \to \infty} \sum_{m=0}^{\infty} a_{nm} = \sum_{m=0}^{\infty} \lim_{n \to \infty} b_{nm}$.

2. *Fatou's Lemma*

If (i) $a_{nm} \ge 0$ for all m, n,

(ii) $\lim_{n \to \infty} a_{nm}$ exists for all m,

then
$$\lim_{n \to \infty} \left[\sum_{m=0}^{\infty} a_{nm} \right] \ge \sum_{m=0}^{\infty} [\lim_{n \to \infty} a_{nm}].$$

3. *Fubini's Theorem*

In order that
$$\sum_{n=0}^{\infty} \sum_{m=0}^{\infty} a_{nm} = \sum_{m=0}^{\infty} \sum_{n=0}^{\infty} a_{nm}$$

it is sufficient that at least one of the following conditions is satisfied:

(i) $a_{nm} \ge 0$ for all $n. m$

(ii) $\sum_{n=0}^{\infty} \sum_{m=0}^{\infty} | a_{nm} | < \infty$

(iii) $\sum_{m=0}^{\infty} \sum_{n=0}^{\infty} |a_{nm}| < \infty$

Proof of Theorem 2.5

(i) Suppose i is recurrent and that $\Rightarrow f_{ii} = 1$.

By Abel's theorem,

$$\lim_{s \to 1-} F_{ii}(s) = 1 \Rightarrow \lim_{s \to 1-} 1 - F_{ii}(s) = 0$$

By Corollary to Theorem 2.4. $P_{ij}(s) = \delta_{ij} + F_{ij}(s)\, P_{jj}(s)$ and in particular

$$P_{ii}(s) = 1 + F_{ii}(s) P_{ii}(s) \Rightarrow P_{ii}(s) = \frac{1}{1 - F_{ii}(s)} \tag{2.0}$$

Then $\lim_{s \to 1-} P_{ii}(s) = \infty$.

By Tauber's Theorem, $\sum_{n=0}^{\infty} P_{ii}^{(n)} = \infty$.

Conversely, suppose $\sum_{n=0}^{\infty} P_{ii}^{(n)} = \infty$ and also suppose i is not recurrent i.e. it is transient i.e. $f_{ii} < 1$. By Abel's theorem,

$$\lim_{s\to 1-} F_{ii}(s) < 1 \Rightarrow \lim_{s\to 1-} 1 - F_{ii}(s) > 0.$$

Again by Corollary to Theorem 2.4,

$$\lim_{s\to 1-} P_{ii}(s) < \infty \Rightarrow \sum_{n=0}^{\infty} p_{ii}^{(n)} < \infty \text{ (By Tauber's theorem).}$$

Hence, i must be recurrent.

(ii) i is transient i.e. $f_{ii} < 1$ iff $\sum_{n=0}^{\infty} p_{ii}^{(n)} < \infty$

The result follows from the proof of part (i).

(iii) $F_{ii}'(1) = \sum_{n=1}^{\infty} n f_{ii}^{(n)}$, since $F_{ii}(s) = \sum_{n=0}^{\infty} s^n f_{ii}(n)$.

Also $F_{ii}'(1) = \lim_{s\to 1-} \dfrac{1 - F_{ii}(s)}{1-s}$ (if i is recurrent, $F_{ii}(1) = f_{ii} = 1$)

$= \lim_{s\to 1-} \dfrac{1}{(1-s)P_{ii}(s)}$ (by Corollary to Theorem 2.4).

Theorem 2.6 (*Solidarity Theorem*) All states in a communicating class of a Markov chain are of the same type, i.e. states are all positive or all null or all transient.

In particular, in an irreducible Markov chain all states are of the same type.

Proof Suppose $i \in S$ and $j \leftrightarrow i$, i.e. $p_{ij}^{(N)} > 0$, for some N and $p_{ji}^{(M)} > 0$ for some M. Then $\theta = p_{ji}^{(N)} P_{ji}^{(M)} > 0$.

1. *Let i be transient*

By repeated applications of Chapman-Kolmogorov equation

$$p_{ii}^{(N+n+M)} \ge p_{ij}^{(N)} p_{jj}^{(n)} p_{ji}^{(M)} = \theta p_{jj}^{(n)}, \theta > 0$$

$$\therefore \sum_{n=0}^{\infty} p_{ii}^{(N+n+M)} \ge \theta \sum_{n=0}^{\infty} p_{jj}^{(n)} \quad \text{and} \quad \infty > \sum_{n=0}^{\infty} p_{ii}^{(n)} \ge \sum_{n=0}^{\infty} p_{ii}^{(N+n+M)} \ge \theta \sum_{n=0}^{\infty} p_{jj}^{(n)}.$$

Therefore, $\sum_{n=0}^{\infty} p_{jj}^{(n)} < \infty$.

By Theorem 2.5 part(ii), j is transient.

2. *Let i be recurrent*

Then as before $p_{jj}^{(N+n+M)} \ge \theta p_{ii}^{(n)}$ and hence

$$\sum_{n=0}^{\infty} p_{jj}^{(N+n+M)} \ge \theta \sum_{n=0}^{\infty} p_{ii}^{(n)} = \infty \text{ (i is recurrent by Theorem 2.5 part (ii))}$$

Now $\sum_{n=0}^{\infty} p_{jj}^{(n)} \ge \sum_{n=0}^{\infty} p_{jj}^{(N+n+M)} = \infty$.

By Theorem 2.5 part (i), j is recurrent.

3. *Let i be recurrent*
We had proved earlier

$$\left.\begin{aligned} p_{jj}^{(N+M+n)} &\geq \theta p_{ii}^{(n)} \\ \text{and} \quad p_{ii}^{(N+M+n)} &\geq \theta p_{ii}^{(n)} \end{aligned}\right\} \text{if } i \leftrightarrow j.$$

Also $\dfrac{1}{s^{N+M}} \sum_{n=0}^{\infty} p_{jj}^{(n)} s^n \geq \sum_{n=0}^{\infty} p_{jj}^{(N+M+n)} s^n \geq \theta \sum_{n=0}^{\infty} p_{ii}^{(n)} s^n$ for $0 < s < 1$.

Therefore $$\frac{1}{s^{N+M}} P_{jj}(s) \geq \theta P_{ii}(s) \quad \text{for } 0 < s < 1.,$$

where $P_{jj}(s)$ is the G.F. of $\{p_{jj}^{(n)}\}$ and $P_{jj}(s)$ is the G.F. of $\{p_{jj}^{(n)}\}$

$$\therefore \qquad P_{ii}(s) \leq \frac{1}{\theta} \frac{1}{s^{N+M}} P_{jj}(s) \text{ for } 0 < s < 1 \tag{2.1a}$$

Similarly, $$\frac{1}{s^{M+N}} P_{ii}(s) \geq \theta P_{jj}(s) \text{ for } 0 < s < 1 \tag{2.1b}$$

and hence $$(1 - s)\theta s^{M+N} P_{jj}(s) \leq (1 - s) P_{ii}(s) \text{ for } 0 < s < 1 \tag{2.2b}$$

Let $s \to 1-$. Then by Theorem 2.5 (iii) and (2.1a) and (2.1b), we get

$$\frac{\theta}{\sum_{n=0}^{\infty} n f_{jj}^{(n)}} \leq \frac{1}{\sum_{n=0}^{\infty} n f_{jj}^{(n)}} \leq \frac{1}{\theta} \frac{1}{\sum_{n=0}^{\infty} n f_{jj}^{(n)}} \; (0 < \theta < 1) \tag{2.3}$$

Hence $$\sum_{n=0}^{\infty} n f_{ii}^{(n)} < \infty \Rightarrow \sum_{n=0}^{\infty} n f_{jj}^{(n)} < \infty$$

and $$\sum_{n=0}^{\infty} n f_{ii}^{(n)} < \infty \Rightarrow \sum_{n=0}^{\infty} n f_{jj}^{(n)} < \infty$$

Thus, i positive $\Leftrightarrow j$ is positive and i null $\Leftrightarrow j$ null.

Exercise 2.1. Homogeneous Random Walk
Here state space is given by

$$S = \{\ldots, -2, -1, 0, 1, 2 \ldots\}$$

$$p_i = p, \; q_i = q \text{ for all } i \geq 1.$$

This is an irreducible M.C. Hence by solidarity theorem it is enough to consider the state $\{0\}$ only. The n-step recurrence probability is

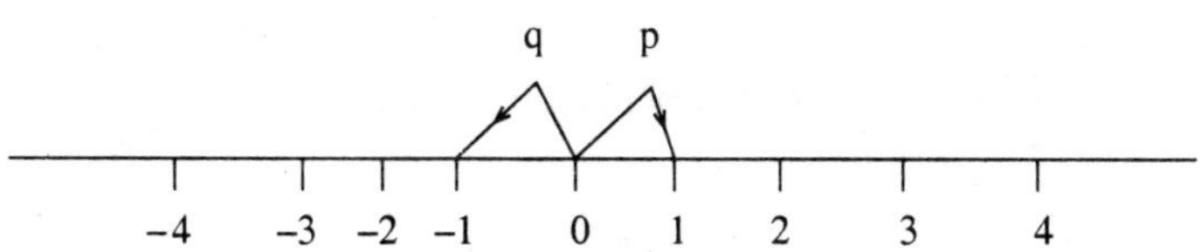

$$P_{00}^{(n)} = \begin{cases} 0 \text{ if } n \text{ is odd} \\ \dbinom{2m}{m} p^m q^m \text{ if } n \text{ is even } (n = 2m) \end{cases}$$

$\{0\}$ is transient iff $\sum_{n=0}^{\infty} P_{00}^{(n)} < \infty$

and recurrent iff $\sum_{n=0}^{\infty} P_{00}^{(n)} = \infty$ (by Theorem 2.5)

$$\sum_{m=1}^{\infty} \binom{2m}{m} p^m q^m \cong \sum_{m=1}^{\infty} \frac{(4pq)^m}{(\pi m)^{1/2}} < \infty \text{ if } 4pq < 1$$
$$= \infty \text{ if } 4pq \geq 1.$$

(using Stirling's approximation for $m! \cong \sqrt{2\pi}\, e^{-m} m^{m+1/2}$) $4pq > 1$ is impossible for if $4pq > (p+q)^2$ then $0 > (p-q)^2$.

Hence
$$4pq < 1 \text{ if } p \neq q$$
$$= 1 \text{ if } p = q = \frac{1}{2}.$$

Therefore $\sum_{m=0}^{\infty} p_{00}^{(2m)}$ converges faster than the geometric series $\sum_m (4pq)^m$ if $p \neq 1/2$.

Hence Random walk is recurrent iff $p = \frac{1}{2}$ and transient iff $p \neq \frac{1}{2}$.

We have shown in Exercise 2.1 that a symmetric Random walk in one dimension is recurrent. Similarly it can be proved that in 2-dimensions a symmetric Random walk is recurrent. But Polya proved that in $k \geq 3$ dimensions a symmetric Random walk is transient.

Exercise 2.2 Show that a symmetric Random walk in three dimensions is transient.

Solution $P_{00}^{(2n+1)} = 0$, $n = 0, 1, 2, \ldots$

and
$$P_{00}^{(2n)} = \sum_{0 \leq i+j \leq n} \frac{(2n)!}{i!\, i!\, j!\, j!\, (n-i-j)!\, (n-i-j)!} (1/6)^{2n},$$

since the probabilities of a transition of one unit to the east, west, north, south, up or down all equal 1/6.

(Consider all paths i units to the east, i units to the west, j units to the south, j units to the north and $k = n - i - j$ units to each of up and down).

$$P_{00}^{(2n)} = \frac{1}{2^{2n}} \binom{2n}{n} \sum_{0 \leq i+j \leq n} \left[\frac{n!}{i!\, j!\, (n-i-j)!}\right]^2 (1/3)^{2n} \tag{2.4}$$

$$\leq C_n \frac{1}{2^{2n}} \binom{2n}{n} (1/3)^n \text{ where } C_n = \max_{0 \leq j \leq n} \left[\frac{n!}{i!\, j!\, (n-i-j)!}\right] \tag{2.5}$$

since
$$\sum_{0 \leq i+j \leq n} \frac{n!}{i!\, j!\, (n-i-j)!} (1/3)^n = 1.$$

We shall show that for large n the value of C_n is attained for $i = j \sim n/3$. Let i_0, j_0 maximize $\dfrac{n!}{i!\, j!\, (n-i-j)!}$ subject to $0 \leq i + j \leq n$.

Then

$$\frac{n!}{j_0!\,(i_0-1)!\,(n-j_0-i_0+1)!} \le \frac{n!}{j_0!\,i_0!\,(n-j_0-i_0)!},$$

$$\frac{n!}{j_0!\,(i_0+1)!\,(n-j_0-i_0-1)!} \le \frac{n!}{j_0!\,i_0!\,(n-j_0-i_0)!},$$

$$\frac{n!}{(j_0-1)!\,i_0!\,(n_0-j_0-i_0+1)!} \le \frac{n!}{j_0!\,i_0!\,(n-j_0-i_0)!}$$

and
$$\frac{n!}{(j_0+1)!\,i_0!\,(n-j_0-i_0-1)!} \le \frac{n!}{j_0!\,i_0!\,(n-j_0-i_0)!}$$

which lead to $n - j_0 - 1 \le 2i_0 \le n - j_0 + 1$ and $n - i_0 - 1 \le 2j_0 \le n - i_0 + 1$. Hence for large n, $i_0 \sim n/3$ and $j_0 \sim n/3$.

Putting $i = j = n/3$ in (2.5), we obtain from (2.4)

$$P_{00}^{(2n)} \le \frac{n!}{(n/3)!(n/3)!(n/3)!2^{2n}3^n}\binom{2n}{n}$$

$$\sim \frac{3\sqrt{3}}{2\pi^{3/2}n^{3/2}}$$

(using Stirling's formula for factorials).

Thus, $\sum\limits_n p_{00}^{(n)} = \sum\limits_n P_{00}^{(2n)} < \infty$. Hence, '0' is a transient state. Since the M.C. is irreducible, the Random walk is always transient.

Example. 2.3 (Success Chains or Runs)

Consider a M.C. with possible states $E_0, E_1, E_2, \ldots$ with transition probability matrix

$$P = \begin{pmatrix} p_0 & q_0 & 0 & & & \ldots \\ p_1 & 0 & q_1 & 0 & & \ldots \\ p_2 & 0 & 0 & q_2 & 0 & \ldots \\ \ldots & & & & & \ldots \end{pmatrix}$$

Let k represent the 'age' of the system. When the system reaches k the aging process continues with probability q_k, but with probability p_k it rejuvenates and starts afresh with age zero. The successive passages through the state E_0 again represent a *recurren event* and the probability that a recurrence time equal k is again given by the product $q_0q_1q_2 \ldots q_{k-1}p_{k-1}$. This is actually a sequence of trials in which at the nth trial the system is in state k if the last failure occurred at the trial number $n - k$ (i.e. index k equals the length of uninterrupted block of successes ending at the nth trial). By convention we take $0 < p_i < 1$ which implies that the chain is irreducible. The nature of the transition matrix P shows that a first return at the nth trial can occur only through the sequence $E_0 \to E_1 \to E_2 \to \ldots \to E_{n-1} \to E_0$, and so for $n \ge 1$,

$$f_{00}^{(n)} = q_0 q_1 q_2 \ldots q_{n-2} p_{n-1}, f_{00}^{(0)} = 1 \text{ and } f_{00}^{(1)} = p.$$

The recurrence probability

$$f_{00} = \sum_{n=1}^{\infty} f_{00}^{(n)} = \sum_{n=1}^{\infty} (q_0 q_1 \ldots q_{n-2} p_{n-1})$$

$$= \sum_{n=1}^{\infty} (q_0 q_1 \ldots q_{n-2} - q_0 \ldots q_{n-1})$$

$$(\text{since } p_{n-1} = 1 - q_{n-1})$$

$$= 1 - \lim_{n\to\infty} (q_0 q_1 \ldots q_n) = 1 - \lim_{n\to\infty} U_n,$$

where $U_n = q_0 q_1 \ldots q_n$, $n > 0$ and $U_{-1} = 1 = q_{-1}$ by convention.

Now $f_{00}^{(n)} = U_{n-2} - U_{n-1}$. Hence, the M.C. is transient iff $f_{00} < 1$

i.e. iff $\lim_{n\to\infty} U_n \neq 0$.

We claim that it is positive recurrent iff $\sum_{n=0}^{\infty} U_n < \infty$,

null recurrent iff $\sum_{n=0}^{\infty} U_n = \infty$, and $\lim_{n\to\infty} U_n = 0$.

The proof of this follows from the following arguments.

In case of recurrent M.C. the mean recurrence time becomes

$$\sum_{n=1}^{\infty} n f_{00}^{(n)} = \sum_{n=1}^{\infty} \left(\sum_{k=1}^{\infty} 1 \right) f_{00}^{(n)} = \sum_{k=1}^{\infty} \sum_{n=k}^{\infty} f_{00}^{(n)}$$

$$= \sum_{k=1}^{\infty} U_{k-2} (\text{since } \lim_{n\to\infty} U_n = 0)$$

$$= U_{-1} + \sum_{k=0}^{\infty} U_k = 1 + \sum_{k=0}^{\infty} U_k = 1 + \sum_{k=0}^{\infty} \prod_{i=0}^{k} q_i$$

Hence, M.C. is positive recurrent iff $\sum_{k=0}^{\infty} U_k < \infty$.

Lemma 2.3 If $0 < p_i < 1$ then $\lim_{n\to\infty} U_n = 0$ iff $\sum_{i=1}^{\infty} p_i = \infty$.

Proof

$$q_i = 1 - p_i \leq e^{-p_i}$$

Hence

$$U_n = \prod_{i=0}^{n} q_i \leq e^{-\sum_{i=0}^{n} p_i}$$

Thus, $\sum_{i=0}^{n} p_i = \infty \Rightarrow U_n \to 0$.

Conversely, if $U_n \to 0$ but $\sum_{i=0}^{n} p_i < \infty$, then

$$\frac{U_n}{U_{k-1}} = \prod_{i=k}^{n} (1 - p_i) \geq 1 - p_k - p_{k+1} - \ldots - p_n > 0 \text{ for large } n \text{ if}$$

$$\sum_{i=0}^{\infty} p_i < \infty \text{ for some } k > 1.$$

Hence $\lim\limits_{n\to\infty} \dfrac{U_n}{U_{k-1}} > 0$ for some $k > 1$.

Therefore $\lim\limits_{n\to\infty} U_n = 0$ is contradicted and hence

$$U_n \to 0 \Rightarrow \sum_{i=0}^{\infty} p_i = \infty.$$

Thus, the M.C. is transient iff $\sum\limits_{i=0}^{\infty} p_i < \infty$

and recurrent iff $\sum\limits_{i=0}^{\infty} p_i = \infty$.

2.4 Limit Theorems for Markov Chains

Definition 2.10 Let d(i) be the greatest common divisor of those $n \geq 1$ for which $p_{ii}^{(n)} > 0$. Then $d(i)$ is called the period of the state i. If $d(i) = 1$, then the state i is called *aperiodic.*

Note $i \leftrightarrow j$, then $d(i) = d(j)$.

There exists n_1 and n_2 such that $p_{ij}^{(n_1)} > 0$ and $p_{ji}^{(n_2)} > 0$.

Now $p_{ii}^{(n_1+n_2)} \geq p_{ij}^{(n_1)} p_{ji}^{(n_2)} > 0$ and hence $d(i)$ is a divisor of $n_1 + n_2$.

If $p_{jj}^{(n)} > 0$, then $p_{ii}^{(n_1+n+n_2)} \geq p_{ij}^{(n_1)} p_{jj}^{(n)} p_{ji}^{(n_2)} > 0$ (by Chapman Kolmogorov equation).

Hence, $d(i)$ is a divisor of $n_1 + n + n_2$. So $d(i)$ must be a divisor of n if $p_{jj}^{(n)} > 0$.

Thus $d(i)$ is a divisor of $\{n \geq 1 : p_{jj}^{(n)} > 0\}$. Since $d(j)$ is the largest of such divisors, $d(i) \leq d(j)$. Hence, by symmetry $d(j) \leq d(i)$.

Hence $d(i) = d(j)$. Therefore having a period d is a class property.

Note If $p_{ii} > 0$, then $d(i) = 1$ and this implies that a sufficient condition for an irreducible M.C. to be aperiodic is that $p_{ii} > 0$ for some $i \in S$. Hence a queueing chain is aperiodic.

Theorem 2.7 *Limit Theorem* (*for diagonal elements*)
Let j be any state in a M.C. As $n \to \infty$.

(i) if j is transient, then $p_{jj}^{(n)} \to 0$

(ii) if j is null recurrent, then $p_{jj}^{(n)} \to 0$

(iii) if j is positive (recurrent) and

(a) aperiodic, then $p_{jj}^{(n)} \to \dfrac{1}{\sum\limits_{n=1}^{\infty} n f_{jj}^{(n)}} = \dfrac{1}{\mu_j}$ (mean recurrence time of j)

(b) periodic with period $d(j)$ then $p_{jj}^{(nd(j))} \to \dfrac{d(j)}{\mu_j}$.

Write $d(j)/\mu_j = \pi_j$.

Corollary 2.3 Let a M.C. be irreducible and aperiodic. Then j is transient iff $\sum_{n=1}^{\infty} p_{jj}^{(n)} < \infty$.

j is null (recurrent) iff $\sum_{n=1}^{\infty} p_{jj}^{(n)} = \infty$ and $p_{jj}^{(n)} \to 0$,

j is positive (recurrent) iff $p_{jj}^{(n)} \to \pi_j > 0$.

Proof of Theorem 2.7 (i) If j is transient, by Theorem 2.5.

$$\sum_n p_{jj}^{(n)} < \infty \Rightarrow p_{jj}^{(n)} \to 0 \text{ as } n \to \infty.$$

As a part of the proof of Theorem 2.7 let us state the following famous lemma without proof.

Lemma 2.4 (Erdös, Pollard and *Feller* in Discrete Renewal Theorem) Let $\{f_n\}$ be a sequence such that $f_0 = 0, f_n \geq 0, \sum_n f_n = 1$ and the greatest common divisor (G.C.D.) of those n for which $f_n > 0$ is $d (\geq 1)$. A second sequence $\{V_n\}$ is defined as follows:

$$v_0 = 1, v_n = \sum_{m=1}^{n} f_m v_{n-m} \ (n \geq 1).$$

Then
$$\lim_{n\to\infty} v_{(nd)} = \begin{cases} d/\mu & \text{if } \mu = \sum_n nf_n < \infty \\ 0 & \text{if } \mu = \infty \end{cases}$$

By Theorem 2.4,
$$p_{jj}^{(n)} = \sum_{m=1}^{n} p_{jj}^{(n-m)} f_{jj}^{(m)} \qquad (n \geq 1)$$

and this can be written as $v_n = \sum_{m=1}^{n} f_m \, v_{n-m}$ (the discrete renewal equation). Thus the lemma gives $\lim_{n\to\infty} p_{jj}^{(n)} = 0$ (since in null recurrent case $\mu = \infty$).

(iii) By definition of positive recurrence, $\mu_j = \sum_n n f_{jj}^{(n)} < \infty$.

(a) So by the lemma $p_{jj}^{(n)} \to \dfrac{1}{\sum_n n f_{jj}^{(n)}}$ if the M.C. is aperiodic i.e. $d = 1$.

(b) $p_{jj}^{(nd(j))} \to \dfrac{d(j)}{\mu_j}$ if j has period $d(j)$.

Proof of Corollary 2.3 follows from Theorem 2.5 and Theorem 2.7.

(i) j is transient iff $\sum_{n=1}^{\infty} p_{jj}^{(n)} < \infty$ (from Theorem 2.5)

(ii) j is recurrent iff $\sum_{n=0}^{n} p_{jj}^{(n)} = \infty$ (from Theorem 2.5)

$$p_{jj}^{(n)} \to 0 \Rightarrow \frac{1}{\mu_j} = 0 \Rightarrow \mu_j = \infty, \text{ i.e. } j \text{ is null recurrent iff } p_{jj}^{(n)} \to 0$$

$$p_{jj}^{(n)} \to \pi_j > 0 \Rightarrow \mu_j < \infty \text{ i.e. positive recurrent iff } p_{jj}^{(n)} \to \pi_j > 0.$$

Proof i is recurrent $\Rightarrow \sum_{n=1}^{n} p_{ii}^{(n)} = \infty$.

(i) Now i is positive iff $\mu_i = \sum_{n=1}^{n} n f_{ii}^{(n)} = \lim_{s \to 1-} \frac{1}{(1-s)P_{ii}(s)} < \infty$

i.e. if $\frac{1}{\mu_i} = \frac{1}{\sum_{n=1}^{n} n f_{ii}^{(n)}} = \lim_{s \to 1-} (1-s)P_{ii}(s) > 0$ (by Theorem 2.5)

i.e. if $\lim_{n \to \infty} \frac{1}{n.} \sum_{k=1}^{n} p_{ii}^{(k)} > 0$ (by Cesaro-Tauber Theorem).

Conversely by Cesaro-Abel Theorem,

if $\lim_{n \to \infty} \frac{1}{n} \sum_{k=0}^{n} p_{ii}^{(k)} > 0$

then $\lim_{n \to 1-} (1-s)P_{ii}(s) > 0$

i.e. $\mu_i < \infty$ and hence i is positive recurrent.

(ii) Similarly i is null if

$$\mu_i = \lim_{e \to 1-} \frac{1}{(1-s)P_{ii}(s)} = \infty$$

i.e. $$\lim_{s \to 1-} (1-s)P_{ii}(s) = 0$$

$$\Rightarrow \lim_{n \to \infty} \frac{1}{n} \sum_{k=0}^{n} p_{ii}^{(k)} = 0 \text{ (by Cesaro-Tauber Theorem)}$$

Conversely, $\lim_{n \to \infty} \frac{1}{n} \sum_{k=0}^{n} p_{ii}^{(k)} = 0$

$$\Rightarrow \lim_{s \to 1-} (1-s)P_{ii}(s) = 0 \text{ (by Cesaro-Abel Theorem)}$$

$\Rightarrow \mu_i = \infty \Rightarrow i$ is null recurrent.

Exercise 2.4. Homogeneous Random Walk (Continued from Exercise 2.1) is null recurrent if $p = q$.

This M.C. is transient if $p \neq q$ (proved in Exercise 2.1 before) and recurrent if $p = q = \frac{1}{2}$.

Proof Now $p_{00}^{(2n)} \cong \frac{(4pq)^n}{\sqrt{\pi n}} \to 0$ as $n \to \infty$ if $p = q = \frac{1}{2}$

Also $$p_{00}^{(2n+1)} = 0 \text{ for all } n \geq 1.$$

Hence $$p_{00}^{(n)} \to 0 \text{ as } n \to \infty$$

and
$$\sum_n p_{00}^{(n)} \geq \sum_n p_{00}^{(n)} \cong \sum_n \frac{1}{\sqrt{\pi n}} = \infty.$$

(Here the chain is not aperiodic, the period is 2).
By Theorem 2.7. the M.C. is null recurrent.
Therefore, M.C. can never by positive recurrent.

Theorem 2.8 *Limit Theorem (for off diagonal elements).*
Let j be a fixed state in a M.C. and i be an arbitrary state. Then as $n \to \infty$

(i) If j is transient, then $p_{ij}^{(n)} \to 0$ as $n \to \infty$
(ii) If j is null recurrent, then $p_{ij}^{(n)} \to 0$
(iii) If j is positive recurrent and the M.C. is aperiodic,

then
$$p_{ij}^{(n)} \to \frac{f_{ij}}{\mu_j}$$

Corollary 2.4 Let a M.C. be irreducible and aperiodic then $\lim_{n\to\infty} p_{ij}^{(n)}$ exists for all i and j, but is *independent* of i.

Proof If j is transient or null, the result is obvious from Theorem 2.8. If j is positive, then this can be proved if $f_{ij} = 1$.

Lemma 2.5 If j is recurrent and $j \to i$, then i is recurrent and
$$f_{ji} = f_{ij} = 1.$$

Proof Assume $j \neq i$, for otherwise there is nothing to prove.
Since $f_{ji} > 0$, there exists n_0 such that $p_{ji}^{(n_0)} > 0$

and
$$p_{ji}^{(m)} = 0 \text{ for } 0 < m < n_0 \tag{2.6}$$

Since $p_{ji}^{(n_0)} > 0$, we can find states $i_1, \ldots, i_{n_0-1}$ such that $p_{ji_1} \ldots p_{ji_{n_0-1}} > 0$ and none of the states $i_1, \ldots, i_{n_0-1}$ equal j or i, for if one of them did equal j or i, it would be possible to go from j to i with positive probability in fewer then n_0 steps in contradiction to (2.6).

Suppose $f_{ij} < 1$. Then a M.C. starting from i has positive probability $1 - f_{ij}$ of never hitting j and that implies it has positive probability $p_{i_{j_1}} \ldots p_{i_{n_0-1}i}$ $(1 - f_{ij})$ of visiting the states $i_1, \ldots, i_{n_0-1}, i$ successively in the first n_0 steps and never returning to j after n_0 steps. But if this happens then the M.C. never returns to j at any time $n > 1$ and that contradicts the fact that j is recurrent. So $f_{ij} = 1$. Since $f_{ij} = 1$, there exists n_1 such that $p_{ij}^{(n_1)} > 0$. Now $p_{ii}^{(n_1+n+n_0)} \geq p_{ij}^{(n_1)} p_{jj}^{(n)} p_{ji}^{(n_0)}$ and hence $\sum_{n=1}^{\infty} p_{ii}^{(n)} \geq \sum_{n=1}^{\infty} p_{ii}^{(n_1+n+n_0)} \geq p_{ij}^{(n_1)} p_{ji}^{(n_0)} \sum_{n=1}^{\infty} p_{jj}^{(n)} = \infty$ (since j is recurrent).

Hence, i is recurrent. Since i is recurrent and $i \to j$ $(f_{ij} = 1)$ from the first part of proof it follows that $f_{ji} = 1$.

Proof of Theorem 2.8 By Theorem 2.4,
$$p_{ij}^{(n)} = \sum_{m=1}^{n} f_{ij}^{(m)} p_{jj}^{(n-m)} = \sum_{m=1}^{n'} f_{ij}^{(m)} p_{jj}^{(n-m)} + \sum_{m=n'+1}^{n} f_{ij}^{(m)} p_{jj}^{(n-m)} \tag{2.7}$$

where $0 < n' < n (n \geq 1)$.

For $\varepsilon > 0$, take n' and n so large that

$$\sum_{m=n'+1}^{n} f_{ij}^{(m)} < \varepsilon. \tag{2.8}$$

For (i) and (ii) when j is *transient* or *null recurrent* take n so large that $p_{jj}^{(n-m)} < \varepsilon$ for all $0 \le m < n' < n$ (by Theorem 2.7). Now from (2.7) and (2.8), we have

$$0 \le p_{ij}^{(n)} - \sum_{m=1}^{n'} f_{ij}^{(m)} p_{jj}^{(n-m)} = \sum_{m=n'+1}^{n} f_{ij}^{(m)} p_{jj}^{(n-m)} \le \sum_{m=n'+1}^{n} f_{ij}^{(m)}$$

$$< \varepsilon \text{ (since } p_{jj}^{(n-m)} \le 1) \tag{2.9}$$

$$0 \le \lim_{n\to\infty} p_{ij}^{(n)} \le \varepsilon + \varepsilon \sum_{m=1}^{n'} f_{ij}^{(m)} \text{ (from equation (2.9) and } \sum_{m=1}^{n'} f_{ij}^{(m)} \le 1)$$

$$\le \varepsilon + \varepsilon = 2\varepsilon, \text{ for all } \varepsilon > 0.$$

Therefore $p_{ij}^{(n)} \to 0$ as $n \to \infty$.

For (iii) (j is positive recurrent and the M.C. is aperiodic) taken $n \to \infty$ and n' fixed. Then

$$0 \le \lim_{n\to\infty} p_{ij}^{(n)} - \lim_{n\to\infty} \sum_{m=1}^{n'} f_{ij}^{(n)} p_{jj}^{(n-m)} < \varepsilon \text{ (by (2.9))}$$

$$= \lim_{n\to\infty} p_{ij}^{(n)} - \sum_{m=1}^{n'} f_{ij}^{(m)} \frac{1}{\mu_j} \text{ (by Theorem 2.7 (iii))}$$

$$= \lim_{n\to\infty} p_{ij}^{(n)} - \frac{1}{\mu_j} \sum_{m=1}^{n'} f_{ij}^{(m)} < \varepsilon.$$

Now take then $n' \to \infty$, then

$$0 \le \lim_{n\to\infty} p_{ij}^{(n)} - \frac{1}{\mu_j} f_{ij} < \varepsilon \text{ for } \varepsilon > 0. \qquad (f_{ij} = \sum_{m=1}^{\infty} f_{ij}^{(m)})$$

Therefore $p_{ij}^{(n)} \to f_{ij}/\mu_j$.

2.5 Stationary Distribution

Definition 2.10 A probability distribution is $\{v_j\}$ (i.e. $v_j \ge 0$, $\sum_j v_j = 1$) is called a *stationary distribution for a Markov chain* with transition matrix (p_{ij}) if

$$v_j = \sum_i v_i p_{ij} \text{ for all } j = 1, 2, \ldots$$

$$= \sum_i \left(\sum_k v_k p_{ki}\right) p_{ij}$$

$$= \sum_k v_k \sum_i p_{ki} p_{ij} \qquad \text{(by Fubini's Theorem)}$$

$$= \sum_k v_k \, p_{kj}^{(2)} \qquad \text{(by Chapman-Kolmogorov)}$$

$$\ldots = \sum_k v_k \, p_{kj}^{(n)} \qquad \text{(by induction)} \tag{2.10}$$

Suppose a stationary distribution $\pi = (\pi_1, \pi_2, \ldots)$ exists. Also suppose

$$\lim_{n\to\infty} p_{ij}^{(n)} = \pi_j \geq 0 \text{ for all } i \geq 1 \tag{2.11}$$

Then π is called the *steady state* distribution of the M.C. with transition matrix (p_{ij}).

If the initial distribution $\{a_j^{(0)}\}(a_j^{(0)} = P(X_0 = j))$ is stationary, we have the marginal distribution of X_n given by $a_j^{(n)}$ (i.e. $a_j^{(n)} = P(X_n = j)) = \sum_i a_j^{(0)} a_{ij}^{(n)} = a_j^{(0)}$ (using (2.10)).

Thus, the unconditional (or marginal) distribution of X_n is independent of n and we may therefore say that the system (or the process) is in *statistical equilibrium*. Suppose conversely, that the distribution of X_n *is independent* of n. Then the initial distribution $a_0 = (a_1^{(0)}, a_2^{(0)}, \ldots)$ i.e. $a_j^{(0)} = P(X_0 = j) = P(X_1 = j) = \sum_i a_i^{(0)} p_{ij} = a_j^{(1)} = \ldots$ and consequently, a_0 is a stationary distribution. Therefore the distribution of X_n is independent of n iff the initial distribution is a stationary distribution. Suppose (2.11) holds. Since $a_j^{(n)} = \sum_i a_i^{(0)} p_{ij}^{(n)} \to \pi_j$, we see that the limiting distribution of X_n is given by π.

In other words, we can say that if (2.11) holds and a stationary distribution π as $n \to \infty$. Denote $\dfrac{1}{\mu_j} = \pi_j$.

Theorem 2.9 (*Limit theorem on stationary distribution*)
Let a M.C. be irreducible, aperiodic and positive (i.e. the M.C. is ergodic).

Then (i) $\lim_{n\to\infty} p_{ij}^{(n)} = \pi_j$

(ii) $\pi_j > 0, \sum_{j\varepsilon S} \pi_j = 1$

(iii) $\pi_j = \sum_{j\varepsilon S} \pi_k p_{kj}$

Moreover, (ii) and (iii) determine $\{\pi_i, i \in S\}$ completely.

(Note that in this case the steady state probabilities or the long run distribution is identical with the stationary distribution of the chain).

Proof (i) follows from Theorem 2.8 and the lemma thereafter.

(ii) $\pi_j = \dfrac{1}{\mu_j} > 0$ (since j is positive recurrent $\Rightarrow \mu_j < \infty$). Suppose S_M is a subset of state space S with exactly M states.

Now
$$\sum_{j\in S_M} p_{ij}^{(n)} \leq \sum_{j\in S} p_{ij}^{(n)} = 1$$

Let
$$n \to \infty, \text{ then } \sum_{j\in S_M} \pi_j \leq 1.$$

Then taking limit
$$M \to \infty, \sum_{k\in S} \pi_j \leq 1. \tag{2.12}$$

$$\sum_{j\in S_M} p_{ik}^{(n)} p_{kj} \leq \sum_{j\in S} p_{ik}^{(n)} p_{kj} = p_{ij}^{(n+1)} \quad \text{(Chapman-Kolmogorov)}$$

Let
$$n \to \infty, \text{ then } \sum_{k\in S_M} \pi_k p_{kj} \leq \pi_j.$$

Then letting

$$M \to \infty, \text{ we get } \sum_{j\in S} \pi_k p_{kj} \le \pi_j. \tag{2.13}$$

$$\sum_{j\in S} \pi_j p_{ji}^{(2)} = \sum_{j\in S} \pi_j \sum_{k\in S} p_{jk} p_{ki} = \sum_{j\in S} \sum_{k\in S} \pi_j p_{jk} p_{ki}$$

$$= \sum_{k\in S} \left(\sum_{j\in S} \pi_j p_{jk} \right) p_{ki} \le \sum_{k\in S} \pi_k p_{ki} \le \pi_i$$

By induction, $$\sum_{j\in S} \pi_j p_{ji}^{(n)} \le \pi_i \text{ for all } n \ge 1,\ i \in S. \tag{2.14}$$

Now $$\pi_k = \pi_k \left(\sum_{j\in S} p_{kj}^{(n)} \right) \quad \left(\text{Since } \sum_{j\in S} p_{kj}^{(n)} = 1\right)$$

$$\Rightarrow \quad \sum_{k\in S} \pi_k = \sum_{k\in S} \sum_{j\in S} \pi_k p_{kj}^{(n)}$$

$$= \sum_{j\in S} \sum_{k\in S} \pi_k p_{kj}^{(n)} \text{ (by Fubini's Theorem)} \tag{2.15}$$

Suppose $\sum_{k\in S} \pi_k p_{kj}^{(n)} < \pi_j$. Then, $\sum_{j\in S} \sum_{k\in S} \pi_k p_{kj}^{(n)} < \sum_{j\in S} \pi_j$ and

$$\sum_{k\in S} \pi_k < \sum_{j\in S} \pi_j \text{ which is a contradiction.}$$

Thus $$\sum_{k\in S} \pi_k p_{kj}^{(n)} = \pi_j \text{ for } n \ge 1. \tag{2.16}$$

In particular, for $n \ge 1$, $\sum_{j\in S} \pi_j p_{ji} = \pi_i$. This proves (iii).

Moreover, by Lebesgue Dominated Covergence Theorem and part (i), letting $n \to \infty$ in (2.16), $\sum_{j\in S} \pi_j \pi_i = \pi_i$.

Now $\pi_i > 0$, that gives $\sum_{j\in S} \pi_j = 1$.

To show that the solution given by (ii) and (iii) is unique, suppose that $\{x_i, i, \in S\}$ is another such solution satisfying $x_i > 0$, $\sum_{j\in S} \pi_j = 1$ and $x_i = \sum_{j\in S} x_j p_{ji}$ for all $i \in S$.

$$x_i = \sum_{j\in S} x_j p_{ji} = \sum_{j\in S} \left(\sum_{k\in S} x_k p_{kj} \right) p_{ji}$$

$$= \sum_{k\in S} x_k \left(\sum_{j\in S} p_{kj} p_{ji} \right) \quad \text{(by Fubini's Theorem)}$$

$$= \sum_{k\in S} x_k p_{ki}^{(2)} = \dots$$

$$= \sum_{k\in S} x_k p_{ki}^{(n)} \text{ for all } n = 1, 2, \dots$$

By the Lebesgue Dominated Convergence Theorem, letting $n \to \infty$, $x_i = \sum_{k\in S} x_k \pi_i = \pi_i \sum_{k\in S} x_k = \pi_i$ for all $i \in S$.

Thus, the solution $\{\pi_i, i \in S\}$ is unique.

Note that condition (iii) of Theorem 2.9 can be written in a matrix equation $\pi(P-I)=0$ where $\pi=(\pi_1, \pi_2, \ldots)$. This matrix P–I is such that its off-diagonal elements are non-negative and the diagonal elements are strictly negative. In case of finite chain with k-states, from the irreducibility condition it follows that the matrix P–1 is of rank (k–1). In this case $\pi=(\pi_1, \pi_2, \ldots \pi_k)$ can be easily obtained from the homogeneous equations $\pi(P-1)=0$ and the normalizing condition $\sum_{i=1}^{k} \pi_i = 1$.

We state the following theorem without proof.

Theorem 2.10 (*Limit theorem for periodic chain*)
Let the M.C. be irreducible, positive recurrent having a stationary distribution. If the chain is periodic with period d, then for each pair i, j of states in S there is an integer r, $0 \le r < d$ such that $p_{ij}^{(n)}=0$ unless $md+r=n$ for some non-negative integer m

$$\lim_{m\to\infty} p_{ij}^{(md+r)} = \pi_j,\ \pi_j = \frac{d}{\mu_j}.$$

Note

(i) if $i=j$, $r=0$, Theorem 2.10 is proved before (see Theorem 2.7).

(ii) if $r=0$, $d=1$, Theorem 2.10 is proved before (see Theorem 2.8).

It is extremely useful to calculate higher order transition probabilities. We shall illustrate this with a simple example. In general, P^n is difficult to compute except for 2-state M.C.

Example 2.4.

Consider the M.C. with $P = \begin{matrix} & \begin{matrix} 0 & \quad 1 \end{matrix} \\ \begin{matrix} 0 \\ 1 \end{matrix} & \begin{pmatrix} 1-p & p \\ q & 1-q \end{pmatrix} \end{matrix}$

Denote the initial distribution by $\pi_0=(\pi_0(0), \pi_0(1))$, row vector $\pi_n = (P(X_n = 0), P(X_n = 1))$ represents the marginal distribution of X_n.

We compute $p_{00}^{(n)} = P_0(X_n=0)$, the conditional probability of $X_n = 0$ given that $X_0 = 0$. Let $\pi_0(0) = P(X_0 = 0)$.

For a stationary chain

$$p_{01} = P(X_{n+1} = 1 \mid X_n = 0) = p,\ p_{10} = P(X_{n+1} = 0 \mid X_n = 1) = q,$$

$$p_{00} = 1 - p,\ p_{11} = 1 - q,\ p_{10} = \pi_0(1) = P(X_0 = 1) = 1 - \pi_0(0).$$

$$P_0(X_{n+1} = 0) = P_0(X_n = 0)p_{00} + P_0(X_n = 1)p_{10} \tag{2.17}$$

$$P_0(X_n = 0)(1-p) + P_0(X_n = 1)q = (1-p-q)P_0(X_n = 0) + q$$

(2.17) is a first order difference equation. The solution is

$$P_{00}^{(n)} = P_0(X_n = 1) = \frac{q}{p+q} + (1-p-q)^n\left(1 - \frac{q}{p+q}\right)\ (p_{00}(0) = 1),$$

Similarly, $$p_{01}^{(n)} = P_0(X_n = 1) = \frac{p}{p+q} - (1-p-q)^n\frac{p}{p+q},$$

$$p_{10}^{(n)} = \frac{q}{p+q} - (1-p-q)^n \frac{q}{p+q},$$

and
$$p_{11}^{(n)} = \frac{p}{p+q} + (1-p-q)^n \frac{q}{p+q},$$

Thus,
$$P^n = \frac{1}{p+q}\begin{bmatrix} q & p \\ q & p \end{bmatrix} + \frac{(1-p-q)^n}{p+q}\begin{bmatrix} p & -p \\ -q & q \end{bmatrix}$$

$$\to \frac{1}{p+q}\begin{bmatrix} q & p \\ q & p \end{bmatrix} \text{ as } n \to \infty \text{ if } \begin{Bmatrix} 0 < p+q < 2 \\ p < 1, q < 1 \end{Bmatrix}$$

This is the steady state distribution and also a stationary distribution. Hence the marginal distribution of X_n is given by

$$P(X_n = 0) = \frac{q}{p+q} + (1-p-q)^n \frac{p}{p+q} \to \frac{q}{p+q}$$

$$P(X_n = 1) = \frac{q}{p+q} - (1-p-q)^n \frac{p}{p+q} \to \frac{p}{p+q}$$

Moreover $0 < p + q < 2$ i.e. $|1 - p - q| < 1$.

So we have a unique stationary distribution π determined by,

$$\pi_0 = \frac{q}{p+q}, \pi_1 = \frac{p}{p+q} \text{ and } p_{ij}^{(n)} \to \pi_j \text{ if } 0 < p+q < 2.$$

π does not depend on initial distribution.

Alternativcly we may find the stationary distribution from Theorem 2.9. This M.C. is obviously irreducible and aperiodic $(0 < p, q < 1)$. Stationary distribution is the solution of $x_j = \sum_{i \in S} x_i p_{ij}$. Therefore

$$(x_1, x_2) = (x_1, x_2)\begin{bmatrix} 1-p & p \\ q & 1-q \end{bmatrix}$$

$$= (x_1(1-p) + x_2 q, x_1 p + x_2(1-q))$$

or $x_1 = x_1(1-p) + x_2 q$ and $x_2 = x_1 p + x_2(1-q)$

or $x_2 q - x_1 p = 0$ and $x_1 p - x_2 q = 0$. Hence, we get

$$x_2 / p = x_1 / q = \frac{x_1 + x_2}{p+q} = \frac{1}{p+q} \text{ (since } x_1 + x_2 = 1)$$

Therefore
$$x_2 = \frac{p}{p+q}, x_1 = \frac{q}{p+q}.$$

Let us make a few commets about existence of steady state distribution of M.C.

1. Even if a M.C. is irreducible the steady state distribution may not always exist (since the irreducible chain may be periodic).
2. If the M.C. has more than one closed sets then the steady state probabilities (i.e. the fixed probability vector π) do not exist.
3. A M.C. is called regular if there exists a positive integer m such that every element of P^m except those relating to transient states is strictly positive.

Exercise 2.5. $P = \begin{pmatrix} 0.6 & 0 & 0.4 \\ 0.3 & 0.5 & 0.2 \\ 0.7 & 0 & 0.3 \end{pmatrix}$. Here state 2 is transient.

and $$P^n \to \begin{pmatrix} \frac{7}{11} & 0 & \frac{4}{11} \\ \frac{7}{11} & 0 & \frac{4}{11} \\ \frac{7}{11} & 0 & \frac{4}{11} \end{pmatrix} \text{ as } n \to \infty$$

and fixed probability vector $\pi = \left(\frac{7}{11}, 0, \frac{4}{11}\right)$ satisfies the equation $\pi P = \pi$.

Proposition 2.1 Let P be a regular stochastic matrix. Then

(i) P has a unique fixed probability vector π are all positive (those elements corresponding to non-transient states) and $\pi P = \pi$.

(ii) $P^n \to T$ whose rows are each fixed points.

(iii) Also $\pi P^n \to \pi$ for any probability vector π.
(Proof follows from Theorem 2.8 and Theorem 2.9).

2.6 Special Chains and Foster Type Theorems

2.6.1 Special chains

If the Markov Chain is infinite, the number of equations given by $\pi(P - I) = 0$ will be infinite involving an infinite number of unknowns. In some particular cases we can solve these equations. The following examples will illustrate this point.

Example 2.5 Birth and Death Chain (Non-Homogeneous Random Walk)
Consider a birth and death chain on $\{0, 1, 2, \ldots, d\}$ or a set of non-negative integers i.e. where $d = \infty$. Assume that the chain is irreducible i.e. $p_j > 0$ and $q_j > 0$ in case $0 \le j \le d$ (i.e. when d is finite) $p_j > 0$ for $0 \le j < \infty$ and $q_j > 0$ for $0 < j < \infty$ if d is infinite. Consider the transition matrix

$$\begin{pmatrix} r_0 & p_0 & 0 & \cdot\cdot & \cdot & \\ q_1 & r_1 & p_1 & 0\,. & \cdot\cdot & \\ 0 & q_2 & r_2 & p_2 & 0.\,. & \cdot \\ 0 & 0 & q_3 & r_3 & p_3 & 0 \\ \cdot & \cdot & \cdot & \cdot & \cdots & \end{pmatrix}$$

when $d < \infty$ we assume that $r_i = 0$ for $i \ge 0$ and $p_0 = 1$.

Particular Case: First consider that d is still infinite and $r_1 = 0$ for $i \ge 0$, $p_0 = 1$. The stationary distribution is given by

$$X = (x_0, x_1, x_2, \ldots) = (x_0, x_1, x_2, \ldots)\begin{pmatrix} 0 & 1 & 0 & 0 & \ldots \\ q_1 & 0 & p_1 & 0 & \ldots \\ 0 & q_2 & 0 & p_2 & \ldots \\ & \ldots & \ldots & & \end{pmatrix} \quad (2.18)$$

or $X = XP$. Let $x_0 \neq 0$. Then

$$\begin{aligned} x_0 &= x_1 q_1, \\ x_1 &= x_0 + x_2 q_2, \\ x_3 &= x_2 p_2 + x_4 q_4, \\ x_4 &= \ldots \\ &\ldots \end{aligned}$$

Define $$y_i = \frac{x_i}{x_0}, y_0 = 1, i = 1, 2, 3, \ldots$$

Then $$y_1 = 1/q_1, y_1 = 1 + y_2 q_2 \text{ or } y_2 = \frac{y_1 - 1}{q_2} = \frac{1 - q_1}{q_1 q_2} = \frac{p_1}{q_1 q_2}$$

$$y_3 = \frac{p_1 p_2}{q_1 q_2 q_3}, \ldots, y_n = \frac{p_1 p_2 \cdots p_{n-1}}{q_1 q_2 \cdots q_n} > 0 \quad \text{for all } n = 1, 2, \ldots$$

(by assumption that all p, q's are > 0).

By Theorems 2.9 and 2.11, the non-homogeneous random walk is positive recurrent if $0 < \sum_0^\infty x_i < \infty$ i.e. $\sum_1^\infty y_i < \infty$ i.e. iff

$$\sum_{n=1}^{\infty} \frac{p_1 p_2 \cdots p_{n-1}}{q_1 q_2 \cdots q_n} < \infty.$$

Note that $\sum_0^\infty y_i = \sum_0^\infty x_i / x_0 = 1/x_0$, since $\sum_0^\infty y_i = 1$.

Therefore $x_n = y_n x_0 = y_n / \sum_0^\infty y_i$ gives the stationary distribution provided $\sum_0^\infty y_i < \infty \cdot x_0$ still has to be determined.

Now $$1 = x_0 + \sum_{i=1}^{\infty} x_i = x_0 + \sum_{i=1}^{\infty} x_0 y_i \Rightarrow x_0 = \frac{1}{1 + \sum_{i=1}^{\infty} y_i}$$

and so $x_0 > 0$, iff $\sum_{i=1}^{\infty} y_i < \infty$ i.e. iff $\sum_{i=1}^{\infty} y_i < \infty$ (since $y_0 = 1$).

In fact if $\sum_i y_i = \infty$, the solution to (2.18) is either identically zero or has infinite sum ($\sum_i x_i = \infty$) and hence has no stationary distribution.

In general case also the stationary distribution exists iff $\sum y_i < \infty$. The system of equations $\sum_{i \in S} \pi_i p_{ij} = \pi_j$ becomes $\pi_0 r_0 + \pi_1 q_1 = \pi_0$, $\pi_{j-1} p_{j-1} + \pi_j r_j + \pi_{j+1} q_{j+1} = \pi_j$ for $j \geq 1$. Since $p_j + q_j + r_j = 1$, these equations reduce to $q_1 \pi_1 - p_0 \pi_0 = 0$ (since $q_0 = 0$),

$$q_{j+1}\pi_{j+1} - p_j\pi_j = q_j\pi_j - p_{j-1}\pi_{j-1}, j \geq 1.$$

By induction, $q_{j+1}\pi_{j+1} - p_j\pi_j = 0$, $j \geq 1$ and hence $\pi_{j+1} = \dfrac{p_j}{q_{j+1}}\pi_0, j \geq 1.$

Consequently, $$\pi_j = \frac{p_0 p_1 \cdots p_{j-1}}{q_1 q_2 \cdots q_j}\pi_0, j \geq 1.$$

Set $$y_j = \begin{cases} 1 \text{ if } j = 0 \\ \dfrac{p_0 p_1 \cdots p_{j-1}}{q_1 q_2 \cdots q_j} \text{ if } j \geq 1. \end{cases}$$

Then we have $\pi_j = y_j\,\pi_0$ for $j \geq 0$.

Thus, the M.C. has unique stationary distribution given by

$$\pi_j = \frac{y_j}{\sum_{j=0}^{\infty} y_j}, \quad j > 0.$$

Similarly, when $d < \infty$, then the unique stationary distribution is given by $\pi_j = \dfrac{y_j}{\sum_{j=0}^{d} y_j}, 0 \leq j \leq d.$

As a particular case of the last example consider the *Ehrenfest chain.* The transition matrix is given by

$$p_{ij} = \begin{cases} i/d & \text{if } j = i - 1 \\ 1 - i/d & \text{if } j = i + 1 \\ 0 & \text{otherwise} \end{cases} \text{ and } S = \{0, 1, 2, \ldots, d\}$$

such chain arises in exchange of heat or gas molecules between two isolated bodies. Consider two boxes labelled 1 and 2 and d balls labelled 1, 2, 3, . . . d. An integer is chosen at random. Then ball labelled by the chosen integer is removed from a box and put in the other box. Trials are repeated. Let X_n be the number of balls in box 1 after n trials.

Then X_n is a M.C. on $S = \{0, 1, 2, 3\}$, $d = 3$.

$$(p_{ij}) = \begin{pmatrix} 0 & 1 & 0 & 0 \\ 1/3 & 0 & 2/3 & 0 \\ 0 & 2/3 & 0 & 1/3 \\ 0 & 0 & 1 & 0 \end{pmatrix}$$

This is an irreducible birth and death chain with

$$y_0 = 1, y_1 = \frac{1}{\frac{1}{3}} = 3, y_2 = \frac{1 \cdot \frac{2}{3}}{\frac{1}{3} \cdot \frac{2}{3}} = 3, y_3 = \frac{1 \cdot \frac{2}{3} \cdot \frac{1}{3}}{\frac{1}{3} \cdot \frac{2}{3} \cdot 1} = 1.$$

Hence, the unique stationary distribution is given by

$$\pi = \left(\frac{1}{8}, \frac{3}{8}, \frac{3}{8}, \frac{1}{8}\right)$$

No steady state probability exists because $p_{ii}^{(n)} = 0$ for all odd values of n. It has periodic behaviour

$$P^n \simeq \begin{pmatrix} \frac{1}{4} & 0 & \frac{3}{4} & 0 \\ 0 & \frac{3}{4} & 0 & \frac{1}{4} \\ \frac{1}{4} & 0 & \frac{3}{4} & 0 \\ 0 & \frac{3}{4} & 0 & \frac{1}{4} \end{pmatrix} \text{ when } n \text{ is even and}$$

$$P^n \simeq \begin{pmatrix} 0 & \frac{3}{4} & 0 & \frac{1}{4} \\ \frac{1}{4} & 0 & \frac{3}{4} & 0 \\ 0 & \frac{3}{4} & 0 & \frac{1}{4} \\ \frac{1}{4} & 0 & \frac{3}{4} & 0 \end{pmatrix} \text{ when } n \text{ is odd and } n \text{ is large.}$$

Modified Ehrenfest chain

Here in addition we select at random one of the two boxes and put the removed balls into this box.

P_{10} = Probability that the ball selected is from box 1 and the box selected is box $2 = \frac{1}{3} \cdot \frac{1}{2} = \frac{1}{6}$.

Similarly, $p_{12} = \frac{2}{3} \cdot \frac{1}{2} = \frac{1}{3}, p_{13} = 0, p_{11} = 1 - p_{10} - p_{12} - p_{13} = \frac{1}{2}, d = 3$.

$$(p_{ij}) = \begin{pmatrix} 1/2 & 1/2 & 0 & 0 \\ 1/6 & 1/2 & 1/3 & 0 \\ 0 & 1/3 & 1/2 & 1/6 \\ 0 & 0 & 1/2 & 1/2 \end{pmatrix}$$

Again from Theorems 2.9 and 2.11 the unique stationary distribution is given by

$$\pi = \left(\frac{1}{4}, \frac{3}{8}, \frac{3}{8}, \frac{1}{8}\right)$$

Here $p_{ii} = \frac{1}{2} > 0$. The M.C. is aperiodic.

Note If some $r_i > 0$, then $p_{ii} = r_i > 0$ and the chain is aperiodic. In particular Modified Ehrenfest chain is aperiodic. Suppose $r_i = 0$ for all i, i.e. we have a non-homogeneous random walk. Then in one transition the state of the chain changes either from an odd numbered state to an even numbered or from an even numbered state to an odd numbered. In particular, the chain can return to its initial state only after an even number of transitions. Hence, the period of the chain is 2 or multiple of 2. Since $p_{00}^{(2)} = p_0 q_1 > 0$, the chain is periodic with period 2 and so the Ehrenfest chain is periodic with period 2.

Example 2.6 Consider the Markov chain with transition matrix

$$P = \begin{pmatrix} 0 & 1 & 0 \\ 1/2 & 0 & 1/2 \\ 0 & 1 & 0 \end{pmatrix}$$

This chain is irreducible. We have

$$P^2 = \begin{pmatrix} 1/2 & 0 & 1/2 \\ 0 & 1 & 0 \\ 1/2 & 0 & 1/2 \end{pmatrix}, P^3 = P \ldots$$

so that $p_{ii}^{(2n)} > 0, p_{ii}^{(2n+1)} = 0$ for each i.

The states are periodic with period 2. It follows that $\sum_n p_{ii}^{(n)}$ diverges so that the state i is recurrent. (Note that a periodic state must necessarily be recurrent, though a recurrent state may or may not be periodic).

Now $$f_{11}^{(1)} = 0, f_{11}^{(2)} = 1 \text{ so that } f_{11} = \sum_1^{\infty} f_{11}^{(n)} = 1$$

i.e. State 1 is recurrent and by solidarity theorem the other two states 0 and 2 are also recurrent.

Now $\mu_1 = \sum_n n f_{11}^{(n)} = 1 \times 0 + 2 + 1 = 2$. State 1 is positive recurrent. Thus the states of the chain are periodic (each with period 2) and positive recurrent. By Theorem 2.7, $p_{ii}^{(2n)} \to d/\mu = 2/2 = 1$ for all i.

2.6.2 Foster type theorems

The following theorems, associated with Foster, give criteria for transient and recurrent chains in terms of solution of certain equations. Assume that the M.C. is *irreducible.*

Theorem 2.11 (Foster, 1953) Let the Markov chain be irreducible. Assume that there exists x_k, $k \in S$ such that $x_k = \sum_{k \in S} x_i p_{ik}$ and $0 < \sum_{k \in S} |x_k| < \infty$. Then the Markov Chain is positive recurrent (this is a sort of converse of Theorem 2.9).

Proof Since $y_k = \dfrac{1}{\sum_{k \in S} |x_k|} > 0$, $\sum_{k \in S} y_k = 1$.

Without loss of generality $\{x_k, k \in S\}$ is a stationary distribution of a M.C. Then

$$x_k = \sum_{k \in S} x_i p_{ik}^{(n)} \text{ for all } n = 1, 2, \ldots \tag{2.19}$$

Suppose that there is no positive state.

Since the M.C. is irreducible, then all the states are either transient or null. In that case $p_{ik}^{(n)} \to 0$ as $n \to \infty$ for all $i, k \in S$. By Lebesgue Dominated Convergence Theorem, taking $n \to \infty$ in (2.19)

$$x_k = \sum_{i \in S} (x_i).\, 0 = 0 \text{ for all } k \in S \tag{2.20}$$

But $0 < \sum_{k \in S} x_k < \infty$ is a contradiction to (2.20).

Hence, there is at least one positive recurrent state. Since M.C. is irreducible, by Solidarity Theorem the M.C. must be positive recurrent.

Conclusion An ireducible aperiodic M.C. has a stationary distribution iff all states are positive recurrent.

Theorem 2.11(a) If the M.C. is positive recurrent the system of equations $x_i = \sum_{j=0}^{\infty} x_j p_{ji}$ has a solution such that $0 < \sum_{j=0}^{\infty} x_j < \infty$.

(Proof may be found in Karlin and Taylor's book.)

Theorem 2.12 The M.C. is transient iff $x_i = \sum_{j=0}^{\infty} p_{ij} x_j$ has a solution for $i \neq 0$, which is bounded and non-constant i.e. all x_i's are not equal.

Theorem 2.13 The M.C. is positive recurrent if $x_i \geq \sum_{j=0}^{\infty} p_{ij} x_j$ has a solution such that $x_i \to \infty$ as $i \to \infty$ (see Chung's book on Markov Chains with Stationary Transition Probabilities).

2.7 Theorems Regarding Finite Markov Chain

Theorem 2(a). In a M.C. with a finite number of states, there is no null state and not all states can be transient.

Proof Suppose the chain has $N < \infty$ states. If all states are transient, then letting $n \to \infty$ in the relation $\sum_{j=0}^{N} p_{ij}^{(n)} = 1$ we get $0 = 1$ (since by Theorem 2.8, $\lim_{n \to \infty} p_{ij}^{(n)} = 0$ for each j), which is absured and hence not all states in a finite M.C. are transient. Consider the subchain C_1 formed by a closed set of null recurrent states. Then $\sum_{j \in C_1} p_{ij}^{(n)} = \alpha$ (say) > 0. Letting $n \to \infty$, $0 = \alpha > 0$ which is also absurd. So there cannot be any null recurrent state in a finite M.C.

Theorem 2(b). An irreducible M.C. having a finite number of states is positive recurrent.

Proof By previous theorem, there is no null recurrent state and not all states are transient. Suppose there is one transient state. Then all states are transient by Solidarity Theorem. Hence, all states are positive recurrent.

Exercise 2.6 If a finite M.C. is irreducible, aperiodic and has doubly stochastic transition matrix, then show that $\lim_{n\to\infty} p_{ij}^{(n)} = 1/k$, where k is the number of states in the chain.

Solution If j is a positive recurrent state in an aperiodic irreducible chain then

$$p_{ij}^{(n)} \to \pi_j > 0 \text{ (by Theorem 2.9).}$$

Hence $1 = \sum_{i=1}^{k} p_{ij}^{(n)}$ for all j and $n \ge 1$,

$$\begin{pmatrix} p_{11} & p_{12} \cdots p_{1k} \\ p_{21} & p_{22} \cdots p_{2k} \\ \cdots & \\ p_{k1} & p_{k2} \cdots p_{kk} \end{pmatrix} \begin{matrix} = 1 \\ = 1 \\ = 1 \end{matrix}, \quad \sum_{i=1}^{k} \lim_{n\to\infty} p_{ij}^{(n)} = 1$$

$$\begin{matrix} 1 & 1 \ldots 1 \end{matrix}$$

Therefore $k\,\pi_j = 1 \Rightarrow \pi_j = \dfrac{1}{k}$.

Exercise 2.7. Find the unique fixed probability vector of the regular stochastic matrix

$$P = \begin{bmatrix} 0 & 1 & 0 \\ 1/6 & 1/2 & 1/3 \\ 0 & 2/3 & 1/3 \end{bmatrix}$$

What matrix does P^n approach?

Solution

$$P^2 = \begin{bmatrix} \frac{1}{6} & \frac{1}{2} & \frac{1}{3} \\ \frac{1}{12} & \frac{23}{36} & \frac{5}{18} \\ \frac{1}{9} & \frac{5}{9} & \frac{1}{3} \end{bmatrix} > 0, \text{ so } P \text{ is regular.}$$

Fixed vector $\pi = (x, y, z)$ satisfies the equations

$$(x, y, z) \begin{bmatrix} 0 & 1 & 0 \\ \frac{1}{6} & \frac{1}{2} & \frac{1}{3} \\ 0 & \frac{2}{3} & \frac{1}{3} \end{bmatrix} = (x, y, z)$$

or
$$\left.\begin{aligned} \frac{1}{6}y &= x \\ x + \frac{1}{2}y + \frac{2}{3}z &= y \\ \frac{1}{3}y + \frac{1}{3}z &= z \end{aligned}\right\} \text{ or } \begin{aligned} y &= 6x \\ 6x + 3y &= 4z = 6y \\ y + z &= 3z \end{aligned}$$

or
$$\left.\begin{aligned} y &= 6x \\ 6x + 4z &= 3y \\ y &= 2z \end{aligned}\right\}$$
Hence on zero solution is feasible.
Setting $x = 1$, we get
$y = 6$ and $z = 3$.

Since $1 + 6 + 3 = 10$, the probability vector or stationary distribution is given by $\pi = \left(\frac{1}{10}, \frac{6}{10}, \frac{3}{10}\right)$. Obviously

$$P^n \to T = \begin{bmatrix} \frac{1}{10} & \frac{6}{10} & \frac{3}{10} \\ \frac{1}{10} & \frac{6}{10} & \frac{3}{10} \\ \frac{1}{10} & \frac{6}{10} & \frac{3}{10} \end{bmatrix}.$$

Exercise 2.8 Suppose a machine can be in one of three states:
(1) broken beyond repair, (2) in need of adjustment or
(2) working properly, with the following transition probabilities,

$$P = \begin{bmatrix} 1 & 0 & 0 \\ \frac{1}{4} & \frac{1}{2} & \frac{1}{4} \\ \frac{1}{18} & \frac{8}{18} & \frac{9}{18} \end{bmatrix}.$$

Under these conditions : (1) what is the probability that the machine will be in state i ($i = 1, 2, 3$) after n time period? (2) What is the probability that the machine is in state i when n becomes infinitely large (i.e. the steady state probability or long run behaviour of the machine)? (It is given that the machine is initially in need of adjustment).

2.8 Methods of Evaluation of the *n*-Step Transition Probability

(a) Method of Spectral Decomposition

Let P be a NXN matrix with latent roots $\lambda_1, \ldots, \lambda_N$ all distinct and simple. Then
$(P - \lambda_i I)\, U_i = 0$ for the column latent vector U_i and
$V_i'(P - \lambda_i I) = 0$ for the row latent vector V_i.

$A_i = U_i V_i'$ are called latent or spectral matrix associated with λ_i, $i = 1, \ldots, N$.
The following properties of A_i's are well known:

(i) A_i's are idempotent, i.e. $A_i^2 = A_i$,

(ii) they are orthogonal, i.e. $A_i A_j = 0$ $(i \neq j)$,

(iii) they give spectral decomposition $P = \sum_{i=1}^{N} \lambda_i A_i$. It follows from (i) to (iii), that

$$P^k = \left(\sum_{i=1}^{N} \lambda_i A_i\right)^k = \sum_{i=1}^{N} \lambda_i^k A_i = \sum_{i=1}^{N} \lambda_i^k U_i V_i'. \tag{2.21}$$

Also we know that $P^k = UD^k U^{-1}$ (by Diagonalisation Theorem) where

$$U = (U_1, U_2, \ldots, U_N)$$

and

$$D = \begin{bmatrix} \lambda_1 & 0\ldots & 0 \\ 0 & \lambda_2 & \vdots \\ 0 & \ldots & \lambda_N \end{bmatrix}$$

Since the latent vectors are determined uniquely only upto a multiplicative constant, we have chosen them such that $U_i' V_i = 1$. From (2.21) one can get any power of P knowing λ_i's and A_i's.

Example 2.7 We shall illustrate the last method with the help of Exercise 2.8 of Section 2.7.

In our problem, $P = \begin{bmatrix} 1 & 0 & 0 \\ \frac{1}{4} & \frac{1}{2} & \frac{1}{4} \\ \frac{1}{18} & \frac{8}{18} & \frac{9}{18} \end{bmatrix}$ with characteristic equation

$$\begin{bmatrix} 1-\lambda & 0 & 0 \\ \frac{1}{4} & \frac{1}{2}\lambda & \frac{1}{4} \\ \frac{1}{18} & \frac{8}{18} & \frac{9}{18}\lambda \end{bmatrix} = 0 \text{ or } (1-\lambda)\left(\lambda^2 - \lambda + \frac{5}{36}\right) = 0.$$

So the eigenvalues are $\lambda_1 = 1, \lambda_2 = \frac{1}{6}, \lambda_3 = \frac{5}{6}$.

Corresponding to $\lambda_1 = 1$, taking $a_1 = 1$ in the matrix equation

$$\begin{bmatrix} 1-\lambda_1 & 0 & 0 \\ \frac{1}{4} & \frac{1}{2}\lambda_1 & \frac{1}{4} \\ \frac{1}{18} & \frac{8}{18} & \frac{9}{18}\lambda_1 \end{bmatrix} \begin{bmatrix} a_1 \\ b_1 \\ c_1 \end{bmatrix} = 0 \text{ we get the characteristic vector } U_1 = \begin{bmatrix} 1 \\ 1 \\ 1 \end{bmatrix}.$$

Corresponding to $\lambda_2 = \frac{1}{6}$, taking $a_2 = 0$, the characteristic vector is $U_2 = \begin{bmatrix} 0 \\ 3 \\ -4 \end{bmatrix}$.

Corresponding to $\lambda_3 = \frac{5}{6}$, taking $a_3 = 0$, the characteristic vector is $U_3 = \begin{bmatrix} 0 \\ 3 \\ 4 \end{bmatrix}$.

Hence, the matrix whose columns are characteristic vectors is given by

$$U = \begin{bmatrix} 1 & 0 & 0 \\ 1 & 3 & 3 \\ 1 & -4 & 4 \end{bmatrix} \text{ and } U^{-1} = \frac{1}{24}\begin{bmatrix} 24 & 0 & 0 \\ -1 & 4 & -3 \\ -7 & 4 & 3 \end{bmatrix}.$$

So

$$P^k = UD^kU^{-1} = \frac{1}{24}\begin{bmatrix} 1 & 0 & 0 \\ 1 & 3 & 3 \\ 1 & -4 & 4 \end{bmatrix}\begin{bmatrix} 1 & 0 & 0 \\ 0 & \left(\frac{1}{6}\right)^k & 0 \\ 0 & 0 & \left(\frac{5}{6}\right)^k \end{bmatrix}\begin{bmatrix} 24 & 0 & 0 \\ -1 & 4 & -3 \\ -7 & 4 & 3 \end{bmatrix}$$

$$\text{or}\quad P^k = \begin{bmatrix} 1 & 0 & 0 \\ 1 - \frac{r_k}{2} - \frac{3s_k}{8} & \frac{r_k}{2} & \frac{3s_k}{8} \\ 1 - \frac{r_k}{2} - \frac{2s_k}{8} & \frac{2s_k}{8} & \frac{r_k}{2} \end{bmatrix}$$

where $r_k = \left(\frac{5}{6}\right)^k + \left(\frac{1}{6}\right)^k$, $s_k = \left(\frac{5}{6}\right)^k - \left(\frac{1}{6}\right)^k$.

Now $r_k \to 0$ and $s_k \to 0$ as $k \to \infty$.

$$\text{Therefore } P^k \to \begin{bmatrix} 1 & 0 & 0 \\ 1 & 0 & 0 \\ 1 & 0 & 0 \end{bmatrix}.$$

This shows that irrespective of the initial distribution (state), the machine will be in state 1 (irrepairably broken). So state 1 is an absorbing or a trapping state.

Exercise 2.9. Derman (1963) designed a model for making the Maintenance or Replacement decisions.

At the beginning of time period	State of equipment at the end of time period	
	Operating properly	Need of Adjustment
Operating Properly	0.90	0.10
Need of Adjustment	0.01	0.99

$$\text{Here } P = \begin{bmatrix} 0.90 & 0.10 \\ 0.10 & 0.99 \end{bmatrix}.$$

Suppose the equipment starts out working properly in period zero. Calculate the unconditional probabilities of the state of the equipments and the conditional probabilities after time period n.

Solution The initial distribution $a^{(0)} = (1, 0)$,

$$a^{(1)} = a^{(0)}P = (1, 0)\begin{pmatrix} .90 & .10 \\ .01 & .99 \end{pmatrix} = (0.90, 0.10) \text{ and}$$

$$a^{(2)} = a^{(1)}P = (a^{(0)}P^2) = (1, 0)\begin{pmatrix} 0.8110 & 0.1890 \\ 0.0189 & 0.9811 \end{pmatrix} = (0.8110, 0.1890)$$

In general $a^{(n)} = a^{(0)}P^n$. We shall calculate P^n by another method.

(b) Method of Caley-Hamilton

Caley-Hamilton Theorem Every square matrix satisfies its own characteristic equation.

The characteristic equation of P is given by $|P - \lambda I| = 0$. In the last example

$$\begin{vmatrix} 0.9 - \lambda & 0.10 \\ 0.01 & 0.99 \end{vmatrix} = 0 \Rightarrow \lambda^2 - 1.89\lambda + 0.89 = 0.$$

By Caley-Hamilton Theorem,

$$P^2 - 1.89\,P + 0.89\,I = 0 \Rightarrow P^2 = 1.89\,P - 0.89\,I$$

$$\Rightarrow \quad P^3 = 1.89\,P^2 - 0.89\,P = 1.89(1.89\,P - 0.89\,I) - 0.89\,P$$

$$= 2.6821\,P - 1.6821\,I.$$

Similarly, any power of P can be calculated in this manner.

(c) Method of Generating Functions

Since 1 is always a latent root of a stochastic matrix P. $(I - tP)$ is non-singular for $0 < t < 1$.

Hence $(I - AP)^{-1}$ exists. But since $(tP)^n \to 0$,

$I + tP + t^2P^2 + \ldots = (I - tP)^{-1}$. Therefore, $p_{ij}^{(n)}$ is the coefficient of t^n in the (i,j)th element of $(I - tP)^{-1}$, which is the generating function of $\{p_{ij}^{(n)}\}$. Suppose $S = \{0, 1, 2, 3\}$ and

$$P = \begin{pmatrix} 1 & 0 & 0 & 0 \\ q & p & 0 & 0 \\ 0 & q & p & 0 \\ 0 & 0 & q & p \end{pmatrix}, p + q = 1, p, q \geq 0.$$

Then $|I - tP| = (1 - t)(1 - tP)^2$ and

$$(I - tP)^{-1} = \begin{bmatrix} (1-t)^{-1} & 0 & 0 & 0 \\ \dfrac{tq}{(1-t)(1-tp)} & \dfrac{1}{1-tp} & 0 & 0 \\ \dfrac{(tq)^2}{(1-t)(1-tp)^2} & \dfrac{tq}{(1-tp)^2} & \dfrac{1}{1-tp} & 0 \\ \dfrac{(tq)^2}{(1-t)(1-tp)^3} & \dfrac{(tq)^2}{(1-tp)^3} & \dfrac{tq}{(1-tp)^2} & \dfrac{1}{1-tp} \end{bmatrix}$$

Expanding the (i, j)th element of $(I - tP)^{-1}$ in powers of t, we get $p_{ij}^{(n)}$ as the coefficient of t^n.

(d) Method of First Entrance

If $f_{jk}^{(m)}$ be the probability that a M.C starting with state j will reach k for the first time at the mth step, we obtain from Theorem 2.4 the generating functions

$$\left.\begin{aligned} P_{jk}(s) &= F_{jk}(s)\, P_{kk}(s)\ (j \neq k) \\ P_{jj}(s) &= F_{jj}(s) P_{jj}(s) + 1 \end{aligned}\right\}$$

and

Hence we can determine $P_{jk}(s)$ and $P_{kk}(s)$ if we know $F_{jk}(s)$ and $F_{jj}(s)$.

Exercises and Complements

Exercise 2.1 Find the closed sets in the following Markov chain with transition matrix:

(a)

$$\begin{array}{c} \\ 1\\2\\3\\4\\5 \end{array}\begin{array}{c} \begin{array}{ccccc} 1 & 2 & 3 & 4 & 5 \end{array} \\ \begin{bmatrix} 0.2 & 0.3 & 0.5 & 0.0 & 0.0 \\ 0.7 & 0.3 & 0.0 & 0.0 & 0.0 \\ 0.0 & 1.0 & 0.0 & 00 & 0.0 \\ 0.0 & 0.0 & 0.0 & 0.4 & 0.6 \\ 0.0 & 0.0 & 0.0 & 1.0 & 0.0 \end{bmatrix} \end{array}$$

(b)

$$\begin{array}{c} \\ 1\\2\\3\\4 \end{array}\begin{array}{c} \begin{array}{cccc} 1 & 2 & 3 & 4 \end{array} \\ \begin{bmatrix} 0.0 & 0.0 & 1.0 & 0.0 \\ 1.0 & 0.0 & 0.0 & 0.0 \\ 0.3 & 0.7 & 0.0 & 0.0 \\ 0.6 & 0.2 & 0.2 & 0.0 \end{bmatrix} \end{array}$$

Exercise 2.2 A service agency assigns its jobs to a particular worker in the following way. The maximum number of jobs assigned to him, in addition to one he is working on at any time is $N(\geq 1)$. If he can not finish the assigned jobs on a given day, he starts with the remaining ones the following day. However, if any time of the day he finishes all the jobs assigned to him, he returns to his own work and becomes unavailable for any more agency jobs for that day. Let p_j be the probability that $j(\geq 0)$ new jobs arrive during a service period. Let X_n be the number of jobs assigned to him at the end of the nth service. Under what conditions is $\{X_n\}$ a Markov chain? Determine its transition probability martix and classify its states.

Exercise 2.3 (*Bartky's sampling inspection scheme*). In a sampling inspection procedure successive sampling of size N is taken. If in the initial sample the number of defective is zero, the lot is accepted. If the number of defective exceeds a predetermined number a, the lot is rejected. From the second sample onward one defective per sample is allowed. Thus after n such samples, the lot will be accepted if the total number of defectives is $\leq n$ and rejected if the number of defectives is $> a + n$. Let X_k = number of defectives out of N – One at the kth sample. Let S_n = Total excess number of defectives in the lot. Find the distribution of X_k. Show that $\{S_n\}$ is a Markov chain. Find the transition matrix for $\{S_n\}$ in terms of distribution of X_k. Also classify the states of $\{S_n\}$.

Exercise 2.4 *A Simple Waiting Model (Queueing).* In a simple queueing model a server serves one customer (if any) at time instant 0, 1, 2, ... Let ξ_n be the number of customers arrive in the time interval $(n, n + 1)$ and we assume $\{\xi_n, n \geq 0\}$ is a sequence of i.i.d. nonnegative integer valued r.v.'s with $P(\xi_0 = k) = p_k$, $\sum_{k=0}^{\infty} p_k = 1$ and there is a waiting room for at most m customers (including the customer being served). Let X_n be the number of customers present at time n, including the one being served. Show that X_n is a Markov chain with states 0, 1, . . ., m. Find its transition matrix in terms of $\{p_k\}_0^{\infty}$.

Exercise 2.5 (*Storage Model*). Consider a dam which can hold at most m units of water. Assume that during day n a quantity η_n of units of water flow into the dam, $n \geq 0$, where $\{\eta_n, n \geq 0\}$ is a sequence of i.i.d. nonnegative integer valued r.vs. with $P(\eta_0 = k) = p_k$, $\sum_{k=0}^{\infty} p_k = 1$. Any overflow is lost with no possibility of recovery. At the end of any day when the dam is not dry, one unit of water is released. Show that $\{X_n'\}$, the content of the dam at the beginning of day n and $\{X_n''\}$, the content of the dam at the end of day n are Markov chains on appropriate state spaces. Find the transition matrix of $\{X_n'', n \geq 0\}$ in terms of $\{p_k\}_0^{\infty}$.

(Note: $\{X_n', n \geq 0\}$ is a M.C. identical to that with the preceding Queueing example with states space $(0, 1, \ldots, m)$ and hence it is omitted for any further discussion.)

Exercise 2.6 A Markov chain remains Markov if the time is reversed, i.e.

$$P(X_n = i_n \mid X_{n+1} = i_{n+1}, \ldots, X_{n+k} = i_{n+k}) = P(X_n = i_n \mid X_{n+1} = i_{n+1})$$

Exercise 2.7 Show that the eigenvalues of a stochastic matrix are in absolute value at most equal to 1.

Exercise 2.8 (*A model for the spread of a contagious disease in a small community*). Let ξ_m, $m \geq 0$ be the number of individuales getting infected at time m (i.e. the number of infectious individuals at time $m + 1$) and X_n, $n \geq 0$, the number of susceptibles still uninfected at time n. Let us assume that the chance of contact between two individuals be $0 < p = 1 - q < 1$ and the contacts to be independent. Show that $\{X_n\}$ is a Markov chain with states space $\{0, \ldots, r\}$ when the initial number of uninfected susceptibles is r. Find its transition matrix. If τ be the duration of epidemic defined as smallest integer τ such that $\xi_\tau = 0$, express the number of cases in the epidemic in terms of $X_\tau (X_{\tau-1} = X_\tau)$.

Exercise 2.9 Show that in a Markov chain, if the present is specified, then the past is independent of the future in the following sense. If $k < m < n$ then

$$P(X_n = i_n, X_k = i_k \mid X_m = i_m) = P(X_n = i_n \mid X_m = i_m) P(X_k = i_k \mid X_m = i_m)$$

Exercise 2.10 A Markov chain is of order $m < k$ if

$$P(X_{n_{k+m}} = x_{n_{k+m}}, \ldots, X_{n_{k+1}} = x_{n_{k+1}} \mid X_{n_j} = x_{n_j}, 1 \leq j \leq k)$$

$$= P(X_{n_{k+m}} = x_{n_{k+m}}, \ldots, X_{n_{k+1}} = x_{n_{k+1}} \mid X_{n_k} = x_{n_k})$$

for integers $n_1 < n_2 < \ldots < n_{k+m}$.

Show that if $\{X_n, n \geq 1\}$ is a Markov chain, then it is a Markov chain of order $m < k$.

Exercise 2.11 If $P = (p_{ij})^{N \times N}$ stochastic matrix, then show that there exists a stochastic matrix $Q = (q_{ij})^{N \times N}$ such that $\lim_{n \to \infty} \sum_{m=1}^{n} p_{ij}^{(m)} = q_{ij}$, $i, j = 1, 2, \ldots, N$. Moreover,

$$QP = PQ = Q \text{ and } Q^2 = Q.$$

Exercise 2.12 Prove that in a finite Markov chain a class C of states is recurrent iff it is closed.

Exercise 2.13 (a) If a Markov chain starts in state i then show that the probability of returning to i at least r times equals f_{ii}^r.

(b) If a Markov Chain starts in state i then show that the probability that state j occurs at least r times equals $f_{ij}[f_{jj}]^{r-1}$.

(c) Calculate the probabilities of returning to i infinitely often when $f_{ii} < 1$ and when $f_{ii} = 1$.

Exercise 2.14 Let $\upsilon(i)$ be the number of visits to the state i and $P_i[A]$ is the conditional probability of A starting from the state i. Then show that

(a) $P_i[\upsilon(j) = n] = f_{ij}[f_{ij}]^{n-1}[1 - f_{jj}],\ n \geq 1$

where f_{ij} is the conditional probability of ever visiting j starting from i.

(b) $P_k[\upsilon(j) < \infty] = 1$ and $\mu(k, i) = \dfrac{f_{kj}}{1 - f_{ii}} < \infty,\ k \in S$

where $\mu(k, i)$ is the expected number of visit to i starting from k when i is transient.

(c) Hence show that if i is a transient state, then the M.C. visits i only finite number of times and the expected number of visits is also finite.

Exercise 2.15 In a Markov chain whatever be the states i and j, prove the *Doeblin's formula*

$$f_{ij} = \lim_{m \to \infty} \frac{\sum_{n=1}^{m} p_{ij}^{(n)}}{1 + \sum_{n=1}^{m} p_{jj}^{(n)}} \tag{1}$$

In particular show that for

$$j = i \quad f_{ii} = 1 - \lim_{m \to \infty} \frac{1}{1 + \sum_{n=1}^{m} p_{ii}^{(n)}} \tag{2}$$

Exercise 2.16 Let $\{Y_n, n \geq 1\}$ be a sequence of i.i.d. r.v.s having discrete uniform distribution on $\{0, 1, 2, \ldots k\}$. Let $X_n = \max\{Y_1, \ldots, Y_n\}, n \geq 1$. Show that $\{X_n\}$ is a Markov chain (M.C.) and obtain its one-step transition probability matrix. Discuss about the states of this M.C.

Exercise 2.17 Consider the *polya urn scheme* with a white and b black balls in a box. Each time one ball is drawn and is replaced together with an additional ball of the same colour. The procedure is repeated several times. Let S_n equals the total number of white balls in the box at the end of the nth step. Show that $S_0, S_1, \ldots, S_n, \ldots$ is a nonhomogeneous Markov chain with transition probabilities given by

$${}_n p_{i,j+1} = \frac{i}{a+b+n},\ {}_n p_{i,i} = \frac{a+b+n-i}{a+b+n},\ {}_n p_{i,j} = 0 \text{ if } j \neq i, i+1.$$

Exercise 2.18 If $X_n = \max\{Y_1, Y_2, \ldots Y_n\}$ where Y_i denotes the number on the face turning up in the ith toss of a die with faces $1, 2, \ldots, 6$. Show that $\{X_n\}$ is a Markov chain. Obtain its transition matrix.

Exercise 2.19 Inputs at the jth period, X_j, into a dam of finite capacity k follows a geometric distribution $P(X_j = k) = qp^k (k = 0, 1, \ldots)$.

There is unit release unless the dam is empty. Show that the contents of the dam $\{Y_n\}$ at any time n form a Markov chain. Write down its transition probability matrix.

Exercise 2.20 Let $\{X_n\}$ be a M.C. on $\{1, 2, \ldots M\}$. The conditional distribution of X_{n+1} given $X_n = j$ is the discrete uniform distribution on $\{M - j + 1, \ldots, M\}$. Obtain the stationary distribution.

Exercise 2.21 Consider a Markov chain having state space $\{0, 1, 2\}$ and the following transition probability matrix P.

$$P = \begin{bmatrix} 0.0 & 1.0 & 0.0 \\ 0.5 & 0.0 & 0.5 \\ 0.0 & 1.0 & 0.0 \end{bmatrix}$$

Identify the recurrent states of the chain.

Exercise 2.22 Consider a communication system which transmit the two digits 0 and 1 through several stages. Let $\{X_n, n \geq 1\}$ be the digit leaving the nth stage of the system and X_0, the digit entering the first stage leaving the 0th stage. At each stage there is a constant probability q that the digit which enters will be transmitted unchanged, and probability p otherwise, $p + q = 1$.

Find the n-step transition matrix of the two-state Markov chain and the steady state probabilities if there is any. Also find the probability that the digit entering the first stage is 0 given the digit leaving the mth stage is 0 when the initial distribution is given by $P(X_0 = 0) = a = 1 - p(X_0 = 1)$.

Exercise 2.23 Let ρ_{xy} denotes the probability that the Markov chain starting at state x will be in state y at some positive time; also $N(y)$ denotes the number of times $n \geq 1$ that the chain is in state y. Show that $P[N(y) \geq 2 \mid X_0 = x] = \rho_{xy}\rho_{yy}$, X_0 being the initial state. Extend your arguments to establish that in general $P(N(y) \geq m \mid X_0 = x) = \rho_{xy}\, \rho_{yy}^{m-1}$, $m \geq 1$. Hence show that if y is a transient state, then for every initial state x, $E[N(y) \mid X_0 = x] < \infty$. Now argue to deduce that a Markov chain having a finite state space must have atleast one recurrent state.

Exercise 2.24 Deduce that (with usual notation)

$$p_{jj}^{(m)} p_{jj}^{(n)} \leq p_{jj}^{(n+m)} \leq p_{jj}^{(m)} p_{jj}^{(n)} + (1 - p_{jj}^{(m)})$$

Exercise 2.25 On a genetic inbreading experiment with single pair of Allelomorphos A and a, the progeny of a mating $Aa \times Aa$ results in 50% heteroztgotes Aa where the proportion of hetrozygotes being reduced by one-half at each generation. Find the transition matrix of various genotypes and the proportion of hetrozygotes at rth generation and in the long run.

Exercise 2.26 Consider a dam for water storage with Capacity k units, k being a positive integer. Let there be independently identically distributed integer-valued daily inputs to the dam with probability generating function $\sum_{j=0}^{\infty} g_j Z^j$. Any overflow is lost. If the dam is not empty, $m \geq 1$ units is released at the end of each day. The dam contains Z_n units at the end of nth day. Show that Z_n is a Markov chain and write down the transition matrix. When is the chain irreducible? Find also the transition matrix when the input has a geometric distribution.

Exercise 2.27 *Modelling of soil erosion effect and modelling of a rock crunching process (Kolmogorov 1941).*

Let $\{Y_n\}_1^{\infty}$ be the sequence of r.v.s representing annual yields of a given crop in a crop-producing area A when it is not affected by erosion. Let R_n be the percentage reduction of annual yield due to soil erosion in the nth year and $U_n = R_n/100$. Then show that when

A is subject to soil erosion the resulting annual yield in the nth year is given by $X_n = Y_n \pi_{i=1}^{n} Z_i, Z_i = 1 - U_i$. In case $\{Z_n\}_1^\infty$ are i.i.d. and independent of $\{Y_n\}_1^\infty$, $L_n = \pi_{i=1}^{n} Z_i$ is called "loss-rate function". Show that $\{L_n\}_1^\infty$ is a Markov chain.

Exercise 2.28 N black balls and N white balls are placed in two urns so that each urn contains N balls. At each step one ball is selected at random from each urn and the two balls interchange. The state of the system is the number of white balls in the urns. Determine the transition matrix of the Markov chain. Find its steady state probabilities and show that these do not depend on the initial distribution of black and white balls in the two urns.

Exercise 2.29 Consider two urns A and B containing a total of N balls. Assume that at time t there are exactly k balls in A.

(a) A time $t + 1$ urn A is chosen with probability k/N or B is chosen with probability $\frac{N-k}{N}$. Then a ball is selected from A with probability p or from B with probability q and placed in the previously chosen urn. Determine the transition matrix for this Markov chain.

(b) Now assume that at time $t + 1$ a ball and an urn are chosen with probability depending on the contents of the urn, i.e. both a ball and the urn A are chosen with probability k/N or both a ball and urn B are chosen with probability $\frac{N-k}{N}$. Determine the transition matrix of the Markov chain with states represented by the contents of A.

(c) Determine the equivalence class in (a) and (b).

Exercise 2.30 Every stochastic $n \times n$ matrix corresponds to a Markov chain for which it is the one step transition matrix. Similarly every $n \times n$ two-step transition matrix of a Markov chain is a stochastic matrix but the converse is not true. Give a counter example.

Exercise 2.31 If P is the one-step transition matrix of a finite aperiodic irreducible Markov chain, prove that for some positive integer n all terms of P^n are positive.

Exercise 2.32 If j is a transient state prove that for all $i \in S$, $\sum_{n=1}^{\infty} p_{ij}^{(n)} < \infty$.

Exercise 2.33 If an infinite irreducible Markov chain has a doubly stochastic transition matrix, prove that the chain can not be positive recurrent.

Exercise 2.34 Suppose the quantities $\{\pi_{ij}, i \cdot j \cdot \in S\}$ are defined as the "Cesaro means" $\pi_{ij} = \lim_{n\to\infty} \frac{1}{n} \sum_{m=1}^{n} p_{ij}^{(m)}$.

(a) Show that π_{ij} always exists for all $i, j \in S$.

(b) Show that $\sum_{j\in S} \pi_{ij} \le 1$ for all $i \in S$.

(c) Show that $\pi_{ij} = \sum_k \pi_{ik} p_{kj} = \sum_k p_{ik}\pi_{kj} = \sum_k \pi_{ik}\pi_{kj}$.

(d) If j is a positive recurrent state, show that

$$\pi_{ij} = \frac{f_{ij}}{\sum_{n=1}^{\infty} n f_{jj}^{(n)}} = f_{ij}\pi_j \text{ for all } i \in S.$$

Exercise 2.35 Show that if a finite irreducible Markov chain has a doubly stochastic transition matrix, then all the stationary probabilities are equal.

Exercise 2.36 Let N_i be the number of returns to state i, starting at i. If i is transient, then prove that $P(N_i = r) = (1 - f_{ii})(f_{ii})^r$. Also find $E(N_i)$.

Exercise 2.37 A positive recurrent irreducible Markov chain is given.

Let $\mu_{ij} = \sum_{n=1}^{\infty} n f_{ij}^{(n)}$. Show that

$$\mu_{ij} = p_{ij} + \sum_{k \neq j} p_{ik}[1 + \mu_{kj}].$$

Exercise 2.38 Let j and k be arbitrary states in a (arbitrary) Markov chain, prove the following statements:

(a) $\sup_n p_{jk}^{(n)} \leq f_{jk} \leq \sum_{n=1}^{\infty} p_{jk}^{(n)}$.

(b) $j \rightarrow k$ iff $f_{jk} > 0$.

(c) $j \leftrightarrow k$ iff $f_{jk} f_{kj} > 0$.

(d) $\sum_{n=1}^{\infty} p_{jk}^{(n)} = f_{jk}\left[\sum_{n=1}^{\infty} p_{kk}^{(n)} + 1\right]$.

(e) If $f_{kk} < 1$ then $\sum_{n=1}^{\infty} p_{kk}^{(n)} = \dfrac{f_{kk}}{1 - f_{kk}}$.

Exercise 2.39 Consider the unconditional (marginal) probabilities in a Markov chain $p_j^{(n)} = P(X_n = j), n \geq 1, \; p_j^{(0)} = P(X_0 = j) = \alpha_j$ and $\sum_{j \in S} \alpha_j = 1$.

(a) Show that $p_j^{(n)} = \sum_{i \in S} p_{ij}^{(n)} \alpha_i$ and that

$$\sum_{j \in S} p_j^{(n)} = 1 \text{ for all } n \geq 0.$$

(b) If the Markov chain is irreducible and aperiodic then show that $\pi_j = \lim_{n \to \infty} p_j^{(n)}$.

(c) Assume that the Markov chain has a stationary distribution $\{\pi_j\}$. Evaluate $p_j^{(n)}$ if $\alpha_k = \pi_k$, Explain.

Exercise 2.40 Consider a Markov chain on $S = \{0, 1, 2, \ldots\}$ with transition matrix

$$p_{jk} = e^{-\lambda} \sum_{n=0}^{k} \binom{j}{n} p^n q^{j-n} \lambda^{k-n}/(k-n)!, \text{ where } p + q = 1, p, q > 0.$$

Show that the Markov chain is irreducible, aperiodic and positive recurrent. Also prove that $\pi_j = e^{-\lambda/q}(\lambda/q)^j/j!$ is the stationary distribution.

Exercise 2.41 Assume state $\{0\}$ of a Markov chain is positive recurrent. Take $\{0\}$ to be the initial state of the Markov chain. Let $\{W_n, n \geq 1\}$ denote successive recurrence times which are independent and identically distributed r.vs. with finite mean and with a generating function $G(t) = \sum_{k=1}^{\infty} t^k P(W_1 = k), |t| < 1$. Define Y_n as the time of the last visit to state $\{0\}$ before time n.

Show that

$$\sum_{n=0}^{\infty} t^n \sum_{j=0}^{n} x^j P(Y_n = j) = \frac{1 - G(t)}{(1-t)(1 - G(xt))}$$

Exercise 2.42 Consider a discrete time Markov chain with states $0, 1, \ldots, N$ whose transition matrix has elements

$$p_{ij} = \begin{cases} \mu_i \text{ if } j = i - 1 & i, j = 0, 1, 2, \ldots, N \\ \lambda_i \text{ if } j = i + 1 & \\ 1 - \mu_i - \lambda_i \text{ if } j = i & \\ 0 & \text{otherwise} \end{cases}$$

Suppose that $\mu_0 = \lambda_0 = \mu_N = \lambda_N = 0$, and all other μ_i's and λ_i's are positive and that initial state of the process is k. Determine the absorption probabilities at 0 and N.

Exercise 2.43 Consider an irreducible positive recurrent Markov chain with initial state $X_0 = i$. Let $N_n(i)$ be the number of visits to state i in the first n trials. Prove that

$$\lim_{n\to\infty} \frac{E(N_n(i))}{n} = \frac{1}{\mu_i},$$

where μ_i is the mean recurrence time to state i, i.e. $\mu_i = \sum_{n=1}^{\infty} n f_{ii}^{(n)}$.

Exercise 2.44 Consider an irreducible Markov chain. For some state; call it 0. Prove that if $f_{i0} \geq \alpha > 0$ for every $i \neq 0$ then the chain is recurrent, where f_{i0} is the probability of reaching state 0 from state i.

Exercise 2.45

(a) Let a finite state Markov chain be given. Prove that $j \in S$ is transient iff there exists a state $k \in S$ such that $j \to k$ but $k \nrightarrow j$.

(b) Give a counter example to (a) if we drop the "finiteness" of the Markov chain.

Exercise 2.46 Let a Markov chain be given with only absorbing states (E) and transient states ($E^c = S - E$).

(a) Show that transition matrix P can be put in the form

$$P = \begin{matrix} & E & E^c \\ E & I & O \\ E^c & R & Q \end{matrix}$$

(b) If $i, j \in E^c$, let N_{ij} be the mean number of times the chain is in state j, starting at i. Prove that $N = \sum_{k=0}^{\infty} Q^{(k)}$ where $Q = Q(0)$ is defined in (a).

(c) Define $B_{ij} = P$ {The Markov chain is absorbed at $j \mid X_0 = i$} for $j \in E$, $i \in E^c$. If $E \neq \phi$ (null set), show that $B = NR$ (with $B = (B_{ij})$).

(d) Derive from this $\lim_{n\to\infty} P^n = \begin{pmatrix} I & O \\ B & O \end{pmatrix}$.

Exercise 2.47 Prove that if the one-step transition matrix P of a discrete time Markov chain is symmetric, so is the n-step transition matrix $P^{(n)}$ for all $n \geq 1$. Moreover this Markov chain cannot be positive recurrent (except for finite chain).

Exercise 2.48 Let M be an irreducible, aperiodic positive recurrent discrete time Markov chain. Assume that the initial distribution coincides with stationary distribution. For all i, $j \in S$ and integer $m \geq 1$, define $(R^{(m)})_{ji} = R_{ji}^{(m)} = P(X_n = i \mid X_m = j)$.

(a) Show that this chain is also positive recurrent, and determine its stationary distribution.

Exercise 2.49 (*Markov-Bernoulli chain, Wang (1981)*)
Consider a sequence of dependent trials such that each trial has only two outcomes denoted by '1' and '0' and dependency is connected by a simple Markov chain model having transition probabilities

$$\begin{matrix} & 0 & 1 \\ \begin{matrix} 0 \\ 1 \end{matrix} & \begin{pmatrix} 1-(1-c)p & (1-c)p \\ (1-c)(1-p) & (1-c)p+c \end{pmatrix} \end{matrix}, 0 < p < 1, 0 < c < 1$$

with initial distribution $p_1 = P(X_0 = 1) = p = 1 - P(X_0 = 0)$.

Find the marginal probabilities $p_n = P(X_n = 1)$. If $S_n = S_1 + \ldots + X_n$, show that $E(S_n) = np$, $\text{var}(S_n) = np(1-p) + 2p(1-p)\left(\frac{c(n-1)}{1-c} - c^2\frac{1-c^{n-1}}{(1-c)^2}\right)$. If as $n \to \infty$ and $p \to 0$ such that $np \to \lambda$, then show that $E(S_n) \to \lambda$ and var $(S_n) \to \lambda + \frac{2\lambda c}{1-c}$.

Exercise 2.50 Let $P = (p_{ij})$ be a transition matrix of an irreducible Markov chain and suppose that P is idempotent Markov matrix, i.e. $P^2 = P$. Prove that $p_{ij} = p_{jj}$ for all $i, j, \in S$ and that the chain is aperiodic.

Exercise 2.51 (Ergodic theorem in discrete time). Let X_n be an irreducible M.C. Let μi be the mean recurrence time of state i. Let $V_i(n) = \sum_{k=0}^{n-1} I_{[x_k=i]}$ be the number of visits to i upto $n-1$ and f be any bounded function on S. Show that:

(a) $n^{-1}V_i(n) \xrightarrow{a.s.} \mu_i^{-1}$ as $n \to \infty$

(b) if $\mu_i < \infty$ for all i, then $\frac{1}{n} = \sum_{k=0}^{n-1} f(x_k) \to \sum_{i\in s} f(i)/\mu_i$ as $\to \infty$.

Exercise 2.52 (Estimating transition matrix). Let $\{X_k, 0 \le k \le n\}$ be a M.C. with initial distribution $\alpha_i = P(X_0 = i)$. The log-likelihood function $L(P) = \log\ (\alpha_{x0}\, p_{x_0x_1} \cdots p_{x_{n-1}x_n})$. Show that:

(a) $L(P) = \log\ \alpha_{x_0} + \sum_{i|j} N_{ij} \log p_{ij}$ where N_{ij} is the number of transitions from i to j,

(b) Viewed as a function of p_{ij}, $L(P)$ is maximal when $p_{ij} = \hat{p}_{ij} = N_{ij}/\sum_k N_{ik}$,

(c) If X is irreducible and ergodic then $\hat{p}_{ij} \to p_{ij}$ a.s. as $n \to \infty$.

3

Random Walks

3.0 Introduction

Let $\{X_n, n = 0, 1, 2, 3, \ldots\}$ be a sequence of independent discrete random variables taking integral values only and $S_n = X_1 + X_2 \ldots + X_n (n = 0, 1, 2, \ldots)$. Then the sequence $\{S_n\}$ is a M.C. whose transition probabilities are given by,

$^{(m)}p_{ij} = P(S_{m+1} = j \mid S_m = i) = P(X_{m+1} = j - i),\ i, j = \ldots, -2, -1, 0, 1, 2, \ldots$ (non-homogeneous random walk).

The chain represents a Random walk of a particle along a straight line, the magnitude of 'jump' at time n being given by the random variable X_n. If X_0 is denotes the initial position of a particle then its position after n jumps (at time n) is given by S_n. When X_n's are also indentically distributed, $^{(n)}p_{ij} = p_{j-i}$ where $p_j = P(X_n = j)$. We have then a homogeneous Random walk (RW). Such Random walks occur in fluctuation theory (sums of discrete or continuous random variables). In classical RW, $P(X_n = +1) = p$, $p(X_n = -1) = q = 1 - p$.

3.1 Different Types of Random Walks

(a) Unrestricted Random Walk

In this the elements of transition matrix is given by $p_{i,i+1} = p$, $p_{i,i-1} = q$, for all integer $i(\ldots, -1, 0, 1, 2, \ldots)$.

If $0 < p < 1$, the chain is irreducible. Then we have

$$p_{ij}^{(n)} = P(S_n = j - i) = \binom{n}{(n-j+i)/2} p^{\frac{n+j-i}{2}} q^{\frac{n-j+i}{2}} \quad \text{if } n \text{ is even}$$

$$= 0 \text{ if } n \text{ is odd.}$$

and
$$p_{00}^{(n)} = \binom{n}{\frac{n}{2}} (pq)^{n/2}$$

The period of the chain is 2.

It is transient if $p \neq \frac{1}{2}$ and null recurrent if $p = \frac{1}{2}$.

(b) Random Walk with an Absorbing Barrier

In this walk the elements of transition matrix are given by $p_{i,i+1} = p$, $p_{i,i-1} = q$, $(p + q = 1)$, $p_{00} = 1$ for all $i \geq 1$.

'0' is an absorbing state and the remaining states are all transient. 0, –1, –2, –3, . . . are condensed into a single absorbing state '0'.

Let $f_{i0}^{(n)}$ = Probability of visiting '0' from i, first time in n steps

$$= \binom{i}{n}\binom{n}{(n-1)/2} p^{(n-i)/2} q^{(n+i)/2}$$

Probability of visiting '0' from i ever,

$$f_{i0} = \sum_n f_{i0}^{(n)} \text{ satisfies difference equations}$$

$$f_{i0} = pf_{i+1,0} + qf_{i-1,0} \text{ for } i > 1, f_{10} = pf_{20} + q.$$

Hence solving we get

$$f_{i0} = \begin{cases} 1 \text{ if } p \leq q \\ (q/p)^i \text{ if } p \geq q. \end{cases}$$

(c) Random Walk with Two Absorbing Barries

Here the elements of transition matrix is given by

$$p_{i,i+1} = p,\ p_{i,\,i-1} = q \text{ for } 1 \leq i \leq a-1,\ p_{00} = 1,\ p_{aa} = 1.$$

'0' and 'a' are absorbing and remaining states are transient.

Examples of Random Walks with Absorbing Barriers

Gambler's Ultimate Ruin Problem

The fortune of a gambler forms a M.C. with transition matrix

$$p_{ij} = \begin{cases} p \text{ if } j = i+1 \\ q \text{ if } j = i-1 \\ 0, \text{ otherwise} \end{cases} \quad \text{and } i = 2, 3, \ldots, s$$

$$p_{ij} = \begin{cases} 1 \text{ if } j = 1 \\ 0 \text{ if } j \neq 1 \end{cases} \quad \text{and } i = 1 \text{ and } s.$$

More explicitly the transition matrix is given by

$$P = \begin{bmatrix} 1 & 0 & 0 & & 0 \\ q & 0 & p & 0 & 0 \\ 0 & q & 0 & p & 0 \\ \cdot\cdot & \cdot\cdot & \cdot\cdot & \cdot\cdot & \cdot\cdot \\ 0 & 0 & 0 & 0 & 1 \end{bmatrix}$$

In terms of gambling this can be described as follows:

If two gamblers play a series of games in which the probability of a particular player winning is p for each game ($q = 1 - p$ is the probability of losing a game). If the player loses he gives one unit of money to his opponent and if he wins he receives one unit from his opponent. If this particular player starts with x units of money and his opponent with $s - x$ units, what is the probability of the player losing all his money? The absorbing barriers are '0' and 's'. When barrier '0' is reached the gambler is ruined.

Solution Let $p(x)$ be the probability of the particular player losing all his money if he now has x units. Then we have the difference equation

$$p(x) = p \cdot p(x+1) + q \cdot p(x-1) \text{ if } 1 < x < s-1$$

$$p(1) = p \cdot p(2) = q,\ p(s-1) = q \cdot p(s-2)$$

Boundary conditions are: $p(0) = 1$, $p(s) = 0$

Auxiliary equation is $px^2 - x + q = 0$ or $(x-1)(x - q/p) = 0$

Solutions are $x = 1$ and q/p.

General solution is $p(x) = A + B(q/p)^x$

From the boundary conditions $1 = p(0) = A + B$

$$0 = p(s) = A + B(q/p)^s$$

$$B = \frac{1}{1-(q/p)^s},\ A = \frac{-(q/p)^s}{1-(q/p)^s}$$

Hence
$$p(x) = \begin{cases} \dfrac{(q/p)^s - (q/p)^x}{(q/p)^s - 1} & \text{if } p \neq q \\ 1 - \dfrac{x}{s} \text{ if } p = q \end{cases} \tag{3.1}$$

The last expression follows from the fact that if $\frac{q}{p} = r \to 1$, then

$\lim p(x) = 1 - \frac{x}{s}$ (by L' Hospital's Rule).

Now suppose the gambler is playing against a person with infinite capital, i.e. $s \to \infty$ (a random walk with one absorbing barrier) then from (3.1) we get

$$\lim_{s\to\infty} p(x) = \begin{cases} 1 \text{ if } q > p \\ (q/p)^x \text{ if } q < p \end{cases}$$

Let us investigate the effect of changing stakes.

If the amount of money held by two players are doubled, then

$$p_2(x) = \frac{(q/p)^{2s} - (q/p)^{2x}}{(q/p)^{2s} - 1} = p(x).\ \frac{(q/p)^s + (q/p)^x}{(q/p)^s + 1}$$

depends only on the ratio (q/p).

Let $p(s)$ be the Gambler's ultimate winning probability.

Then $p(s) = \dfrac{(p/q)^s - (p/q)^{s-x}}{(p/q)^s - 1}$ (replacing p in (3.1) by q and x by $s - x$)

Evidently, $p(0) + p(s) = 1$.

Let D_x be the expected number of steps (time) required for absorption to one of the two absorbing states (i.e. the expected duration of the game). Then D_x satisfies the difference equation

$$D_x = pD_{x+1} + qD_{x-1} + 1 \quad \text{if } 0 < x < s \tag{3.2}$$

with boundary conditions $D_0 = 0$, $D_s = 0$.

This follows from the fact that if the first trial results in success the game continues as if the initial position has been $x + 1$. Conditional expectation of the duration assuming success at the first trial is $D_{x+1} + 1$. More formally

$$E(D_x \mid \text{first trial is a success}) = D_{x+1} + 1.$$

Then $D_x = (D_{x+1} + 1)\ p + (D_{x-1} + 1)q = pD_{x+1} + qD_{x-1} + p + q$ with boundary conditions: $D_0 = 0 = D_s$.

A formal solution is

$$D_x = \frac{x}{q-p}\ (p \neq q) \text{ with } \Delta_x = p\ \Delta_{x+1} + q\ \Delta_{x-1}$$

(where Δ_x is the difference of two solutions of (3.2)).

To get the general solution we proceed as follows.

Boundary conditions give $A + B = 0$ and $A + B(q/p)^s = -\dfrac{s}{q-p}$

Solving A and B we get, $D_x = \dfrac{x}{q-p} + A + B(q/p)^x$

$$= \begin{cases} \dfrac{x}{q-p} - \dfrac{s}{q-p}\dfrac{1-(q/p)^x}{1-(q/p)^s}, & p \neq q \\ x(s-x) \text{ if } p = q = \dfrac{1}{2} \end{cases}$$

which is the expected duration of the game in the gambler's ruin problem.

(d) Random Walk with a Reflecting Barrier

$$P_{i,i+1} = p,\ P_{i,i-1} = q \text{ for } i \geq 1,\ p_{00} = q,\ p_{01} = p.$$

Here we imagine a barrier placed at $-1/2$ such that every time the particle moves to the left from 0, it is reflected at the barrier and returns to '0'. The chain is irreducible if $0 < p < 1$. To classify its states, consider the system of equations $y_i = \sum_{j=1}^{\infty} P_{ij} y_j$. Then we get

$$y_i = py_{i+1} + qy_{i-1} \quad (i \geq 1)$$

i.e. $$p(y_{i+1} - y_i) = q(y_i - y_{i-1})\ (i \geq 2),\ y_1 = py_2.$$

Therefore by iteration we obtain

$$y_{i+1} - y_i = y_1(q/p)^i,\ i \geq 1$$

and $$y_i - y_1 = y_1\{(q/p) + (q/p)^2 + \ldots + (q/p)^{i-1}\}$$

Hence $y_i = \dfrac{1-(q/p)^i}{1-(q/p)} y_1, i \geq 1$, so that y_i is bounded if $p > q$. Thus by **Theorem 2.13** (Foster-type theorem) The states are all transient if $p > q$ and recurrent if $p \leq q$, then the stationary distribution is given by

$$\pi_j = \sum_{i=0}^{\infty} \pi_i p_{ij} = p\pi_{j-1} + q\pi_{j+1} \quad (j \geq 1)$$

$$\pi_0 = q\pi_0 + q\pi_1 \Rightarrow \pi_1 = \pi_0 \frac{(1-q)}{q} = \pi_0 \frac{p}{q}.$$

Proceeding successively $\pi_j = (p/q)^j \pi_0 (j \geq 0)$,

where $$\pi_0 \{1 + (p/q) + (p/q)^2 + \ldots\} = 1. \tag{3.3}$$

If $p = q$, the series diverges and consequently $\pi_0 = 0$ and $\pi_j = 0$ $(j \geq 0)$ so that stationary distribution does not exist. Thus, if $p = q$, the states are null recurrent. If $p < q$, by (3.3) $\pi_0 = 1 - p/q > 0$, and is the stationary distribution (the states are positive recurrent).

(e) Random Walk with two Reflecting Barriers

If in addition to the barrier at –1/2, there is a second reflecting barrier at $a + \frac{1}{2}$, so that $p_{aa} = p$, $p_{a,a-1} = q$, $(p_{i,\,i+1} = p, p_{i,i-1} = q, i = 1, \ldots, a-1, p_{10} = q, p_{01} = p)$. We have a finite chain on $\{0, 1, 2, \ldots, a\}$. In this case stationary distribution exists and is in fact the geometric distribution of the previous case truncated at a, i.e.

$$\pi_0 = \frac{1 - p/q}{1 - (p/q)^{a+1}} \text{ and } \pi_j = \frac{(1 - p/q)(p/q)^j}{1 - (p/q)^{a+1}}, \quad j = 1, \ldots, a$$

and this is irreducible $(0 < p < 1)$ and positive recurrent.

3.1.1 Waiting time for a gain

Let $\{X_i\}$ be sequence of i.i.d r.v.s with common distribution

$P(X_k = 1) = p$, $p(X_k = -1) = q$, $p + q = 1$ and $S_n = X_1 + \ldots + X_n$, $S_0 = 0$.

In gambling terminology S_n, $n \geq 1$ is the Peter's accumulated gain at the end of the nth trial if Peter and Paul play for unit stakes. Now consider the event $A_n = \{S_1 \leq 0, S_2 \leq 0, \ldots, S_{n-1} \leq 0, S_n = 1\}$.

Thus, the nth trial is the first to render Peter's accumulated gain positive. The event A_n is called first visit to +1 or the index n is the passage time through 1 in random walk terminology.

Let $\phi_n = P(A_n)$. Define $\phi_0 = 0$, $\phi_1 = p$.

If the event holds for $n > 1$, then $S_1 = -1$ and there exists a smallest integer $\upsilon < n$ such that $S_\upsilon = 0$. The outcome of the first n trials may be described as follows: (1) At the first trial Peter looses an unit amount. (2) It takes $\upsilon - 1$ further trials for Peter to reestablish the initial situation. (3) It takes exactly $n - \upsilon$ further trials for Peter to attain a positive net gain. These events depend on non-overlapping blocks of trials and are therefore mutually independent, (2) and (3) have probabilities $\phi_{\upsilon-1}$ and $\phi_{n-\upsilon}$.

Now the event A_n occurs iff the events (1) to (3) occcur for some $\upsilon < n$. Summing over all possible $1 \le \upsilon \le n - 1$, we get

$$\phi_n = q(\phi_1 \phi_{n-2} + \ldots + \phi_{n-2} \phi_1) = q \sum_{\upsilon=2}^{n-1} \phi_{\upsilon-1} \phi_{n-\upsilon}$$

Multiplying both sides by s^n and summing $n = 2, 3, \ldots$, we get

$$\sum_{n=2}^{\infty} \phi_n s^n = q s \left\{ \sum_{n=2}^{\infty} \phi_{n-1} s^{n-1} \right\}^2$$

$$= q s \left\{ \sum_{n=0}^{\infty} \phi_n s^n \right\}^2 = qs\, \Phi^2(s)$$

(since $\phi_0 = 0$) (3.4)

Also $$\sum_{n=2}^{\infty} \phi_n s^n = \Phi(s) - \phi_0 - \phi_1 s = \Phi(s) - ps \tag{3.5}$$

where $\Phi(s)$ is the G.F. of $\{\phi_n\}$.

Equation (3.3) follows from the fact that

$$\phi_1 \phi_{n-2} + \ldots + \phi_{n-2} \phi_1 = (n-1) \text{ th term of } \{\phi_n\} * \{\phi_n\}.$$

and $$\sum_{n=2}^{\infty} q\{\phi_1 \phi_{n-2} + \ldots + \phi_{n-2} \phi_1\} s^n$$

$$= qs \sum_{n=2}^{\infty} \{\phi_1 \phi_{n-2} + \ldots + \phi_{n-2} \phi_1\} s^{n-1}$$

Hence from (3.3) and (3.4), $\Phi(s) - ps = qs\Phi^2(s)$.

Therefore, $\Phi(s) = \dfrac{1 - \sqrt{1 - 4pqs^2}}{2qs}$ is the unique, bounded solution near $s = 0$.

$$\Phi(s) = \frac{1}{2qs}(-1) \sum_{k=1}^{\infty} \binom{\frac{1}{2}}{k} (-1)^k (4pq)^k s^{2k}$$

$$= \frac{1}{2q} \sum_{k=1}^{\infty} \binom{\frac{1}{2}}{k} (-1)^{k-1} (4pq)^k s^{2k-1}$$

Therefore $$\phi_{2k-1} = \binom{\frac{1}{2}}{k} \frac{(-1)^{k-1}}{2q} (4pq)^k \text{ and } \phi_{2k} = 0,\ k \ge 1.$$

Now $$\sum_{n=0}^{\infty} \phi_n = \Phi(1) = \frac{1 - |p - q|}{2q} = \begin{cases} p/q & \text{if } p < q \\ 1 & \text{if } p \ge q \end{cases}$$

$P(S_n$ remains negative forever)

$$= 1 - \sum_{n=1}^{\infty} \phi_n = \begin{cases} \dfrac{q-p}{q} & \text{if } p < q \\ 0 & \text{if } p \ge q \end{cases}$$

Therefore, $P(S_n$ will sooner or later become positive$) = 1$ if $p \geq q$ and E (number of trials preceding the first positive sum S_n) $= \Phi'(1) = \begin{cases} \infty & \text{if } p = q = \frac{1}{2} \\ \frac{1}{p-q} & \text{if } p > q \end{cases}$

3.1.2 Returns to equilibrium

Let A_k be the event of equalization of the accumulated number of successes and failures occurs at the kth trial if $S_k = 0$. Let $u_k = P(S_k = 0)$. The number of trials is necessarily even and the probability of a return to the origin at the $2n$th trial is given by

$$U_{2n} = \binom{2n}{n} p^n q^n = (-1)^n \binom{-\frac{1}{2}}{n} (4pq)^n$$

The G.F. of $\{U_{2n}\}$ is $U(s) = \sum_{n=0}^{\infty} U_{2n}\, s^{2n}$

$$= \sum_{n=0}^{\infty} (-1)^n \binom{-\frac{1}{2}}{n} (4pqs^2)^n = (1 - 4pqs^2)^{-\frac{1}{2}}$$

The first return to (origin) equilibrium,

$$B_{2n} = [S_k \neq 0, \text{ for } k = 1, \ldots, 2n-1, S_{2n} = 0].$$

Let $P(B_{2n}) = f_{2n}$.

Consider two sub-events with $X_1 = 1$, $X_1 = -1$ and denote their probabilities by f_{2n}^+ and f_{2n}^-, i.e.

$$f_{2n}^+ = P(B_n \cap (X_1 = 1)) \text{ and } f_{2n}^- = P(B_n \cap (X_1 = -1)).$$

Now $f_{2n}^- = q\, \phi_{2n-1}$ (because first $2n-2$ partial sums $X_2 + X_3 + \ldots + X_n \leq 0$, but the next one is positive)

As before let $\phi_n = P[S_1 \leq 0, S_2 \leq 0, \ldots, S_n = 1]$

Then the G.F. of $\{f_{2n}^-\}$ is

$$F^-(s) = \sum_{n=1}^{\infty} f_{2n}^-\, s^{2n} = sq \sum_{n=1}^{\infty} \phi_{2n-1}\, s^{2n-1}$$

$$= qs\, \Phi(s) = qs \frac{1 - (1 - 4pqs^2)^{\frac{1}{2}}}{2qs}$$

By symmetry, $F^+(s) = F^-(s)$ and hence

$$\sum_{n=1}^{\infty} f_{2n} s^{2n} = F(s) = F^+(s) = F^-(s) = 1 - (1 - 4pqs^2)^{\frac{1}{2}}$$

$$\left(= 1 - \frac{1}{U(S)} \text{ in general} \right)$$

Hence probability that a return to equilibrium occurs sooner or later will be

$$F(1) = \sum_{n=1}^{\infty} f_{2n}$$

$$= 1 - (1 - 4pq)^{\frac{1}{2}} = 1 - |p - q|$$

$$= \begin{cases} 2q & \text{if } p > q \\ 2p & \text{if } p < q \\ 1 & \text{if } p = q = \frac{1}{2} \end{cases}$$

Hence, if $p = q = \frac{1}{2}$ a return to equilibrium is certain.

Differentiating $F(s) = 1 - (1 - s^2)^{1/2}$, we get

E(waiting time for the first return to origin) $= F'(1) = \infty$ if $p = q = \frac{1}{2}$

if $$p = q = \frac{1}{2},\ U(s) = \frac{1 - F(s)}{1 - s^2}$$

As both U and F are power series in s^2 this relation differs only notationally from

$$Q(s) = \frac{1 - P(s)}{1 - s},$$

where $p(s) = \sum_{j=0}^{\infty} p_j s^j$, $Q(s) = \sum_{j=0}^{\infty} q_j s^j$ and $q_j = p_{j+1} + p_{j+2} + \ldots$.

Hence, by uniqueness theorem for G.F.

$$U_{2n} = f_{2n+2} + f_{2n+4} + \ldots$$

that is when $p = \frac{1}{2}$, the probability that $S_{2n} = 0$ equals the probability that the $2n$ sums $S_1, S_2, \ldots S_{2n}$ are different from zero.

3.2 Sequential Analysis

An important problem arising in Wald's sequential analysis is concerned with the random variable $N = N(a, b)$, where $N = \min\{n \mid S_n \le -b \text{ or } S_n \ge a\}$ is the first exist time from the interval $(-b, a)$.

We ignore the trivial case $P(X_i = 0) = 1$.

Let X_i are i.i.d. r.v.s and $S_n = X_1 + \ldots + X_n$.

Theorem 3.1 (*C. Stein 1947*)

N is a proper random variable with finite moments of all order, i.e.

(i) $P(N < \infty) = 1$ and (ii) $E(N)^k < \infty$ for all $k = 1, 2, \ldots$

Proof (i) We shall show, more specifically that there exists $A > 0$ and

$$0 < \delta < 1 \text{ independent of } n \text{ and } P[N \ge n] \le A\,\delta^n \tag{3.5}$$

Let $C = a + b$ and r be a positive integer.

Let $S_1^* = X_1 + \ldots + X_r$, $S_2^* = X_{r+1} + X_{r+2} + \ldots + X_{2r}, \ldots,$

$$S_k^* = X_{(k-1)r+1} + \ldots + X_{kr}$$

We have, $P[N \geq kr] \leq P[\mid S_1^* \mid < C, \mid S_2^* \mid < C, \ldots, \mid S_{k-1}^* \mid < C] = p^{k-1}$, where $p = P[\mid S_k^* \mid < C]$ is independent of k and $0 < p < 1$.

Obviously, $p > 0$.

If $p = 1$, then $E(S_k^*)^2 = rEX_i^2 + r(r-1)(EX_i)^2$ (since X_i's are i.i.d.)

Since $E(X_i^2) > 0, E(S_k^*)^2 > C^2$ by choosing r large enough. But $p = 1 \Rightarrow E(S_k^*)^2 \leq C^2$, which is a contradiction. Therefore $p \neq 1$ and $P(N < \infty) = 1$.

(ii) For $t > 0$ and positive integer k, $n^k < e^{tn}$ for large n,

$$\sum_{n=m}^{\infty} n^k P[N = n] \leq \sum_{n=m}^{\infty} e^{tn} P[N \geq n] \leq A \sum_{n=m}^{\infty} (\delta e^t)^n < \infty \text{ if } \delta e^t < 1.$$

Hence
$$E(N^k) = \sum_{n=1}^{\infty} n^k P[N = n]$$
$$= \sum_{n=1}^{m-1} n^k P[N = n] + \sum_{n=m}^{\infty} n^k P[N = n]$$

Definition 3.1 N is called a stopping rule if N is a non-negative integer-valued random variable and the event $[N \geq n]$ depends on $X_1, X_2, \ldots X_{n-1}$ only, i.e. $[N = n]$ is measurable with respect to $\mathscr{B}(X_1, \ldots, X_{n-1})$ ($X_1, \ldots, X_{n-1}$, need not be i.i.d. r.v.s).

3.3 Wald's Equation and Wald's Identity

Theorem 3.2 (*Wald's equation*) Let $\{X_i\}$ be a sequence of i.i.d. r.v.s with $E(N) < \infty$. If $E \mid X_1 \mid < \infty$ then $E(S_N) = (EX_1)\, EN$.
If moreover, $\sigma^2 = \text{var}(X_1) < \infty$, then $E(S_N - N\mu)^2 = \sigma^2 E(N)$, where $\mu = E(X_1)$.

Proof
$$E(S_N) = \sum_{n=1}^{\infty} E(S_N \mid N = n)\, P[N = n]$$
$$= \sum_{n=1}^{\infty} \sum_{i=1}^{n} P[N = n]\, E(X_i \mid N = n)$$
$$= \sum_{i=1}^{\infty} \sum_{n=i}^{\infty} P[N = n]\, E(X_i \mid N = n)$$
(interchanging the order of summation)

$$\left|\sum_{i=1}^{\infty} \sum_{n=i}^{\infty} E(X_i \mid N = n)\, P(N = n)\right| \leq \sum_{i=1}^{\infty} \sum_{n=i}^{\infty} E(|X_i| \mid N = n)\, P(N = n)$$
$$= E \mid X_t \mid E(N) < \infty \qquad \text{(Fubini condition is satisfied)}$$

Therefore

$$E(S_N) = \sum_{i=1}^{\infty} P[N \geq i]\, E(X_i \mid N \geq i) \text{ (since } N \geq i \text{ depends on } X_1, \ldots, X_{i-1} \text{ only)}$$
$$= \sum_{i=1}^{\infty} P[N \geq i]\, E(X_i) = E(X_i)\, E(N).$$

Let $N_n = \min(N, n)$. Now let $N_n \to N$ monotonically, it follows from the Monotone convergence theorem that

$$EN_n \to E(N) \text{ as } n \to \infty \tag{3.7}$$

Since $\{(S_n - n\mu)^2 - n\,\sigma^2, \mathcal{F}_n), n \geq 1\}$ is a martingale (prove it).
We can apply optional sampling theorem to obtain (see Appendix iv)

$$E(S_{N_n} - n\mu)^2 = \sigma^2\, EN_n \tag{3.8}$$

Now let $m \geq n$. Since martingales have orthogonal increments we have, by (3.7) and (3.8),

$$E(S_{N_m} - \mu N_m - (S_{N_n} - \mu N_n))^2 = E(S_{N_m} - \mu N_m)^2 - E(S_{N_n} - \mu N_n)^2$$
$$= \sigma^2 (EN_m - EN_n) \to 0 \text{ as } n, m \to \infty,$$

that is $S_{N_n} - \mu N_n$ converges in L_2 as $n \to \infty$.

However, since we already know that $S_{N_n} - \mu N_n \to S_N - \mu N$ as $n \to \infty$, it follows that

$$E(S_{N_n} - \mu N_n)^2 \to E(S_N - \mu N)^2 \text{ as } n \to \infty,$$

which together with (3.7) and (3.8), completes the proof.

3.3.1 Wald's fundamental identity

Let $X_1, X_2, \ldots$ are i.i.d. r.v.s with $S_n = X_1 + X_2 + \ldots + X_n$ and N is a stopping rule.

Let $F_n(x) = P[S_n \leq x]$, $F_1(x) = F(x) = P[X_1 \leq x]$ and m.g.f. of X_1 is given by

$$\phi(\theta) = \int_{-\infty}^{\infty} e^{\theta x} dF(x) < \infty \text{ if } \phi(\sigma) < \infty, \text{ where } \sigma = \text{Re}(\theta).$$

We also assume that

$$\phi(\sigma) < \infty \text{ for all } \sigma, -\beta < \sigma < \alpha < \infty,\ \alpha, \beta > 0.$$

Under these conditions, $P[e^X < 1 - \delta] > 0$ and $P[e^X > 1 + \delta] > 0$, $\delta > 0$.

$\phi(\theta)$ has a minimum at $\theta = \theta_0 \neq 0$, where θ_0 is the root of the equation $\phi(\theta) = 1$.

Wald's Sequential Analysis presented the so-called Wald's identify

$$E(e^{\theta S_N} / [\phi(\theta)]^N) = 1 \text{ for } \phi(\theta) < \infty \text{ and } |\phi(\theta)| \geq 1.$$

Actually we shall give the proof of a more general theorem in Random walk due to Miller and Kemperman (1961).

Define $F_n(x) = P[S_n \leq x;\ N \geq n]$, $N = \min\{n \mid S_n \notin (-b, a), 0 < a, b < \infty\}$ and

the series $F(z, \theta) = \sum_{n=0}^{\infty} z^n \int_{-b}^{a} e^{\theta x} dF_n(x)$.

Then

$$E(e^{\theta S_N} z^N) = 1 + [z\phi(\theta) - 1]\, F(z, \theta) \text{ for all } \theta \tag{3.9}$$

which is known as *Miller and Kemperman's Identity*.

If $\phi(q) = 1/z$ we get Wald's Identify.

Proof Let $F_0(x) = \begin{cases} 0 \text{ if } x \le 0 \\ 1 \text{ if } x \ge 0 \end{cases}$

and $F_n(x) = P[S_n \le x; N \ge n],\ n \ge 1$

$= P[-b < S_k < a,\ k = 1, \ldots n-1,\ S_n \le x]$

Then $dF_n(x) = f_n(x)\,dx = P[-b < S_1, S_2, \ldots, S_{n-1} < a,\ x < S_n < x + dx]$

$= P[N = n,\ x < S_n < x + dx]$ (if $x < -b$ or $x > a$)

is the joint probability that the time N for absorption is n and that the position reached when absorption occurs between x and $x + dx$. Hence if we take Laplace transform with respect to n and with respect to x over absorbing states we have

$$E(e^{\theta S_N} z^N) = \sum_{n=1}^{\infty} z^n \left(\int_{-\infty}^{-b} e^{\theta x} dF_n(x) + \int_{a}^{\infty} e^{\theta x} dF_n(x) \right)$$

$$= \sum_{n=1}^{\infty} z^n \left(\int_{-\infty}^{\infty} - \int_{-b}^{a} \right) e^{\theta x} dF_n(x)$$

$$= \sum_{n=1}^{\infty} z^n \int_{-\infty}^{\infty} e^{\theta x} dF_n(x) - F(z, \theta) + 1$$

where $F(z, \theta) = \sum_{n=0}^{\infty} z^n \int_{-b}^{a} e^{\theta x} dF_n(x)$.

This series converges for all θ for which $-\beta < \sigma < \alpha$ and $|z| < [\phi(\sigma)]^{-1}$, since

$$\left| z^n \int_{..}^{\infty} e^{\theta x} dF_n(x) \right| \le |z|^n \int e^{\sigma x}\, dF_n(x) = |z|^n\ [\phi(\sigma)]^n$$

Now, $F_n(x) = \int_{-b}^{u} F(x - y) dF_{n-1}(x)$ and $S_n = S_{n-1} + X_n$

$$\int_{-\infty}^{\infty} e^{\theta x} dF_n(x) = \int_{-\infty}^{\infty} e^{\theta x} dF(x) \int_{-b}^{a} e^{\theta x} dF_{n-1}(x)$$

$$= \phi(\theta) \int_{-b}^{a} e^{\theta y} dF_{n-1}(y)$$

Therefore $E(e^{\theta s_N} z^N) = \sum_{n=1}^{\infty} z^n \phi(\theta) \int_{-b}^{a} e^{\theta y} dF_{n-1}(y) - F(z, \theta) + 1$

$$= z\phi(\theta) \sum_{n=1}^{\infty} z^{n-1} \int_{-b}^{a} e^{\theta y} dF_{n-1}(y) - F(z, \theta) + 1$$

$$= 1 + [z\phi(\theta) - 1]\ F(z, \theta) \tag{3.10}$$

To get Wald's identity, we let $z \to [\phi(\theta)]^{-1}$ in equation (3.9)

$$E(e^{\theta S} N[\phi(\theta)]^{-N}) = 1 + \lim_{z \to [\phi(\theta)]^{-1}} [z\phi(\theta) - 1]\, F(z, \theta)$$

$$= 1.$$

The Case of Single Barrier

If either a or $b \to \infty$, then we have a single barrier. If $a = \infty$, $b < \infty$, $N = N(b) = \min\{n \mid S_n \le -b\}$. In this case the previous Theorem 3.1 (Stein) does not apply and we have $P[N < \infty] \le 1$ but Wald's Identity still holds as the Identity (3.9), can be regarded as analytic continuation of the other beyond the common range of validity.

3.4 Fluctuation Theory

In this section $X_1, X_2, \ldots, X_n, \ldots$ are i.i.d. r.v.s.

Theorem 3.3 If $E \mid X_i \mid < \infty$, then

$$P[N(b) < \infty] = 1 \text{ if } E\, X_i \le 0$$

$$< 1 \text{ if } E\, X_i > 0$$

For Proof see Chung and Fuchs (1951) and Chung and Ornstein (1962), Memoirs of American Math. Society.

Definition 3.2 If S is uncountable, and $S_n = X_1 + \ldots + X_n$ are Markov, X_i's being independent, then x is called a possible value of the state space S of the Markov chain if there exits an n such that

$P[\mid S_n - x \mid < \delta] > 0$ for all $\delta > 0$. A state x is called recurrent if

$P[\mid S_n - X \mid < \delta$ i.o.$] = 1$ i.e. $S_n\, \varepsilon (x - \delta, x + \delta)$ i.o. with probability one.

We shall conclude this section by stating two very important and famous theorems whose proofs are beyond the scope of this book.

Theorem 3.4 (*Chung and Fuchs*)

Either every state is recurrent or no state is recurrent. (ref. Spitzer-Random Walk (1962)).

Theorem 3.5 (*Chung and Ornstein*)

If $E \mid X_i \mid < \infty$, then recurrent values exist iff $E(X_i) = 0$.

Exercises and Complements

Exercise 3.1 In a simple random walk with two absorbing barriers at 0 and a let the position X_n at the nth step be given by $X_n = X_{n-1} + Z_n$ where Z_n's are i.i.d. r.vs. taking values 1 and -1 with corresponding probabilities p and $q = 1 - p$. Let $\pi_k(n)$ be the probability of absorption at 0 of the random walk in n-steps starting from position k.

Show that the generating function $G_k(s) = \sum_{n=0}^{\infty} \pi_k(n) s^n$, $\mid s \mid > 1$ is given by

$$(q/p)^k \frac{\lambda_1^{a-k}(s) - \lambda_2^{a-k}(s)}{\lambda_1^a(s) - \lambda_2^a(s)},$$

where $\lambda_1(s)$ and $\lambda_2(s)$ are given by

$$\lambda_1(s) = \frac{1 + (1 - 4pqs^2)^{1/2}}{2ps}, \lambda_2(s) = \frac{1 - (1 - 4pqs^2)^{1/2}}{2ps}.$$

Also show that

$$\pi_k(n) = 2^n p^{(n-k)/2} q^{(n+k)/2} \int_0^1 \cos^{n-1}(\pi x)\sin(\pi x)\sin(k\pi x)dx.$$

What will be the value of $\pi_k(n)$ in case of simple absorbing barrier at 0 when playing against an infinitely rich opponent?

Exercise 3.2 In a random walk with two absorbing barriers at $-n$ and a, let the position X_n at the nth step be given by $X_n = X_{n-1} + Z_n$, where Z_n's are i.i.d. r.v.s taking values 1, −1, 0 with corresponding probabilities p, q, $1 - p - q$.

If $f_{ja}^{(n)} = P(-b < X_1, X_2, \ldots, X_{n-1} < a, X_n = a \mid X_0 = j)$,

Show that the generating function of $\{ f_{ja}^{(n)} \}$ is given by

$$F_{ja}(s) = \frac{[\lambda_1(s)]^{j+b} - [\lambda_2(s)]^{j+b}}{[\lambda_1(s)]^{a+b} - [\lambda_2(s)]^{a+b}}$$

where $\lambda_1(s)$ and $\lambda_2(s)$ are the roots of the equation

$$ps\lambda^2 - \lambda[1 - s(1 - pq)] + qs = 0.$$

If the random walk starts from the origin, what will be the expression of the generating function.

Exercise 3.3 Show that in a simple Random walk with two reflecting barriers at 0 and a, the stationary distribution π_j is given by

$$\pi_j = \frac{1 - p/q}{1 - (p/q)^{a+1}}(p/q)^j \quad (j = 0, 1, \ldots, a),$$

where $\pi_j = \lim_{n\to\infty} p_{ij}^{(n)}$; and $p_{ij}^{(n)}$ has the usual interpretation that it represents the probability that the particle occupies the state j at time n having started in the state i; what will be the value of π_j when the Random walk is homogeneous (i.e. $p = q$)?

Exercise 3.4 (*Wald's Lemma*) Let X be a r.v. such that (a) $E(X)$ exists and $\neq 0$, (b) the m.g.f. $\phi(\theta) = E(e^{\theta X}) < \infty$ for all real θ and (c) there exists a $\delta > 0$ such that $P(e^X < 1 - \delta) > 0$ and $P(e^X > 1 + \delta) > 0$. Show that the equation $\phi(\theta) = 1$ has a unique root $\theta_0 \neq 0$ (i.e. $\phi(\theta)$ has a unique minimum at $\theta = \theta_0$).

Exercise 3.5 (*Continuation of Exercise 3.4*). Let $X_1, \ldots, X_n, \ldots$ are i.i.d. r.v.s satisfying the conditions of Exercise 3.4. Define $S_N = X_1 + \ldots + X_N$.

Let $E_0 = E(e^{\theta_0 S_N} \mid S_N \leq -b)$, $E_1 = E(e^{\theta_0 S_N} \mid S_N \geq a)$

Prove that $P(S_N \leq -b) = \dfrac{E_1 - 1}{E_1 - E_0}$, where $N = \min\{n: S_n \notin (-b, a), 0 < a, b < \infty\}$.

Exercise 3.6 (*Continuation of Exercises 3.4 and 3.5*)
If $\phi(\theta)$ exists for all real values of θ at any values θ in the domain $|\phi(\theta)| \geq 1$. Give an alternative proof (by differentiation and putting $\theta = 0$) of

$$E(N) = E(S_N)/E(X_n) \text{ if } EX_n \neq 0$$

$$= E(S_N^2)/E(X_n^2) \text{ if } EX_n = 0.$$

Exercise 3.7 Let $X_1, X_2, \ldots$ be i.i.d. rvs taking values in the integres $Z = \{\ldots, -2, -1, 0, 1, 2, \ldots\}$ and having finite mean. Show that the M.C. $\{S_n, n \geq 1\}$ given by $S_n = \sum_1^n X_i$ is transient if $E(X_1) \neq 0$.

Exercise 3.8 Consider a symmetric random walk S_n with $S_0 = 0$. Let $T = \min\{n \geq 1 : S_n = 0\}$ be the time of first return of the walk to its starting point. Show that

$$P(T = 2n) = \frac{1}{2n-1}\binom{2n}{n} 2^{-2n}$$

and deduce that $E(T^\alpha) < \infty$ iff $\alpha < 1/2$.

[Hint. Use Stirling's formula $n! \sim n^{n+1/2} e^{-n} \sqrt{2\pi}$.]

4

Renewal Theory

4.1 Introduction

Let X_n, $n = 1, 2, \ldots$, be the nonnegative i.i.d r.v.s with $S_n = X_1 + \ldots + X_n$, $n \geq 1$, $S_0 = 0$. F is the common d.f. of X and assume $P(X_n = 0) < 1$. Define $N(t) = \sup \{n \mid S_n \leq t\}$. The process $\{N(t), t \geq 0\}$ is called the *Renewal Process*.

To fix our ideas X_i can be taken to represent the life time of the machines being replaced. The first machine is installed at time $t = 0$ and is replaced instantaneously at time $t = X_1$. The replaced machine is again replaced at time $t = X_1 + X_2$, and so on. If we write $S_n = X_1 + \ldots + X_n$, the partial sum S_n can be interpreted to be the time at which the nth replacement is made. $N(t)$ is the largest value of n for which $S_n \leq t$. In other words $N(t)$ is the number of renewals that would have occurred at time t. The Renewal Theory, in a sense, is a special case of a Random Walk with absorbing barrier. We are sampling the X_i until S_n shoots the barrier at time t and $N(t) + 1$ is the sample size when we stop. Hence the Renewal Theory is also linked with Sequential Analysis in statistics.

$\{N(t), t \in (0, \infty)\}$ is called the *Renewal Counting Process*. We can also write $N(t) = \max\{n \mid S_n \leq t\}$.

We want to find $P[N(t) = n]$ given F. To compute this we proceed as follows:

$$P[S_2 \leq t] = \int_0^{\infty} F(t-u)dF(u)$$

$$= \int_0^{t} F(t-u)dF(u)$$

$$= F * F(t) = F^{(2)}(t), \ldots$$

$$P[S_n \leq t] = F^{(n)}(t) = \int_0^t F^{(n-1)}(t-u)dF(u), n \geq 1 \tag{4.1}$$

Define $$F^{(0)}(t) = \begin{cases} 0 \text{ if } t < 0 \\ 1 \text{ if } t \geq 0. \end{cases}$$

Now $$P[N(t) = n] = P[S_1 \leq t, S_2 \leq t, \ldots, S_n \leq t, S_{n+1} > t]$$

$$= P[S_n \leq t, S_{n+1} > t] \text{ (by nonnegativeness of } X_1)$$

$$= P[S_n \le t, S_n + X_{n+1} > t]$$

$$= P[t - X_{n+1} < S_n \le t]$$

$$= \int_0^\infty P[t - X_{n+1} < S_n \le t \mid u < X_{n+1} \le u + du]dF(u)$$

$$= \int_0^\infty P[t - u < S_n \le t \mid u < X_{n+1} \le u + du]dF(u)$$

$$= \int_0^\infty P[t - u < S_n \le t]dF(u) \text{ (since } S_n \text{ is independent of } X_{n+1})$$

$$= \int_0^\infty \{F^{(n)}(t) - F^{(n)}(t-u)\}dF(u) \qquad \text{(by (4.1))}$$

$$= \int_0^\infty F^{(n)}(t)\,dF(u) - \int_0^\infty F^{(n)}(t-u)dF(u)$$

$$= F^{(n)}(t) - \int_0^t F^{(n)}(t-u)dF(u)$$

Definition 4.1 $H(t) = E(N(t))$ is called the *Renewal Function.*

4.2 Renewal Equation

Theorem 4.1

(a) $P[N(t) = n] = F^{(n)}(t) - F^{(n+1)}(t)$

(b) $H(t) = \sum_{n=1}^{\infty} F^{(n)}(t)$

(c) $H(t) = F(t) + \int_0^t H(t-u)dF(u)$, the so-called integral equation of Renewal Theory (Renewal equation).

(d) $\{N(t), t \in [0, \infty)\}$ is completely determined by $H(t)$.

Proof (b) $H(t) = \sum_{n=0}^{\infty} nP[N(t) = n]$

$$= P[N(t) = 1] + 2P[N(t) = 2] + \dots$$

$$= F^{(1)}(t) - F^{(2)}(t) + 2F^{(2)}(t) - 2F^{(3)}(t) + \dots$$

$$= F^{(1)}(t) + F^{(2)}(t) + F^{(3)}(t) + \dots$$

$$= \sum_{n=1}^{\infty} F^{(n)}(t) \text{ provided the series is convergent.}$$

(convergence of the series will be proved in Exercise 4.5)

(c) $H(t) = \sum_{n=1}^{\infty} F^{(n)}(t) = F^{(1)}(t) + \sum_{n=2}^{\infty} F^{(n)}(t)$

$$= F(t) + \sum_{n=1}^{\infty} F^{(n+1)}(t) = F(t) + \sum_{n=1}^{\infty} \int_0^t F^{(n)}(t-u)dF(u)$$

$$= F(t) + \int_0^t \sum_{n=1}^{\infty} F^{(n)}(t-u)dF(u) \text{ (by Fubini Theorem)}$$

$$= F(t) + \int_0^t H(t-u)dF(u)$$

(d) $H(t) = F(t) + \int_0^t H(t-u)dF(u) = F(t) + H * F(u)$ where $*$ is the convolution operator.

Taking Laplace transform on both sides

$$\mathscr{H}(s) = \int_0^{\infty} e^{-st} dH(t) = \mathscr{F}(s) + \mathscr{H}(s)\mathscr{F}(s)$$

or
$$\mathscr{F}(s) = \frac{\mathscr{H}(s)}{1 + \mathscr{H}(s)}, \text{ where } \mathscr{F}(s) = \int_0^{\infty} e^{-st} dF(t)$$

and $\mathscr{H}(s) = \dfrac{\mathscr{F}(s)}{1 - \mathscr{F}(s)}$ (Re(s) > 0). This shows that $H(t)$ and $F(x)$ can be determined uniquely one from the other, since Laplace transform determines a non-decreasing (specially a d.f.) function uniquely. Hence $N(t)$ is completely determined by $H(t)$.

Now $N(t) = \max\{n \mid S_n \le t\}$ and $EN(t) = \sum_{n=1}^{\infty} F^{(n)}(t)$ if $EN(t) < \infty$.
The next theorem will prove that all moments of $N(t)$ is finite.

Theorem 4.2 $N(t)$ is a proper random variable with finite moments of all orders i.e. $P(N(t) < \infty) = 1$ and for $EN^k(t) < \infty$ for $k = 1, 2, 3, \ldots$

Proof We assume $P(X_k = 0) < 1$. Since X_k are i.i.d. and $X_k \ge 0$ a.s., there exists $\alpha > 0$ such that $P[X_k \ge \alpha] > 0$. Let us take without loss of generality $\alpha = 1$. Then $P[X_k \ge 1] = p > 0$.

Define a modified Renewal process with

$$\overline{X}_k = \begin{cases} 0 \text{ if } X_k < 1 \\ 1 \text{ if } X_k \ge 1 \end{cases}$$

Let $\overline{S}_n = \overline{X}_1 + \ldots + \overline{X}_n$ and $\overline{N}(t) = \max\{n \mid \overline{S}_n \le t\}$.

Then $\overline{X}_n$ is a Binomial Process. Since $\overline{S}_n \le S_n$, $N(t) \le \overline{N}(t)$.

$N(t)$ is a proper random variable, since $\overline{N}(t)$ is. Also

$$E(e^{\theta_0 N(t)}) \le E(e^{\theta_0 \overline{N}(t)}) < \infty \text{ for some } 0 < \theta_0 < \log \frac{1}{1-p}.$$

Since $n^k < e^{\theta_0 n}$ for large n and all $k = 1, 2, 3, \ldots$, $EN^k(t) < \infty$ for all $k = 1, 2, \ldots$

For *Binomial Process* renewal occurs at $t = 0, 1, 2, 3, \ldots$

Let N_j be the number of renewals in $(j-1, j]$, $j = 1, 2, 3, \ldots$.
Then $\overline{N}(t) = N_1 + N_2 \ldots + N_T$, where $T = [t]$.
Now $P[N_j = n] = pq^{n-1}$ $(n \geq 1)$ and

$$EN_j = \frac{1}{p}, \text{ Var } N_j = q/p^2 (q = 1-p).$$

$$E(e^{\theta N_j}) = \sum_{n=1}^{\infty} e^{n\theta} pq^{n-1} = pe^{\theta} \sum_{n=1}^{\infty} (qe^{\theta})^{n-1} = pe^{\theta} / (1 - qe^{\theta})$$

$$\left(\text{Re}(\theta) < \log \frac{1}{q}\right).$$

Since $N_1 \ldots, N_T$ are i.i.d, $\overline{N}(t)$ is a T-fold convolution of N_1 and hence

$E(e^{\theta \overline{N}(t)}) = \left[\dfrac{pe^{\theta}}{1 - qe^{\theta}}\right]^T$, the m.g.f. of a negative binomial distribution whose distribution is given by $\begin{bmatrix} T+k-1 \\ k \end{bmatrix} p^T q^k$, $k = 0, 1, 2, \ldots$

$$\overline{H}(t) = E\overline{N}(t) = \sum_{j=1}^{T} E(N_j) = \frac{T}{p}$$

and
$$\text{Var }(\overline{N}(t)) = \sum_{j=1}^{T} \text{Var}(N_j) = \frac{Tq}{p^2}.$$

We have derived before the integral equation and showed that the $H(t) = \sum_{n=1}^{\infty} F^{(n)}(t)$ is a solution of (4.1). Now we have to show that the solution is unique.

Let $V(t)$ be a solution to (4.1), which is *bounded for all finite t*. Then,

$$V(t) = F(t) + \int_0^t V(t-u)dF(u) \text{ for all } t$$

$$= F^{(1)}(t) + \int_0^t dF(u)[F(t-u) + \int_0^{t-u} V(t-u-s)dF(s)]$$

$$= F^{(1)}(t) + F^{(2)}(t) + \int_0^t dF(u) \int_u^t V(t-\theta)\, dF(\theta - u), (u + s = \theta)$$

$$= F^{(1)}(t) + F^{(2)}(t) + \int_0^t V(t-\theta)\, d\int_0^{\theta} F(\theta - u)\, dF(u)$$

(interchanging the order of integration and making the change of limits of integration)

$$\left.\begin{matrix} u \leq \theta \leq t \\ 0 \leq u \leq t \end{matrix}\right\} \Rightarrow \left\{\begin{matrix} 0 \leq u \leq \theta \\ 0 \leq \theta \leq t \end{matrix}\right.$$

$$= F^{(1)}(t) + F^{(2)}(t) + \int_0^t V(t-\theta)\, dF^{(2)}(\theta)$$

and proceeding in similar manner, we find that

$$V(t) = F^{(1)}(t) + F^{(2)}(t) + \ldots + F^{(n)}(t) + \int_0^t V(t-\theta)\, dF^{(n)}(\theta) (n \geq 1) \quad (4.2)$$

Now since the series $\sum_{n=1}^{\infty} F^{(n)}(t)$ converges,

$F^{(n)}(t) \to 0$ as $n \to \infty$ and therefore.

$\int_0^t V(t-\theta) dF^{(n)}(\theta) \leq \sup_{u \leq t} V(u)\, F^{(n)}(t) \to 0$ as $n \to \infty$ and (4.2) gives

$V(t) = \sum_{n=1}^{\infty} F^{(n)}(t) = H(t).$

This proves the uniqueness of $H(t) = \sum_{n=1}^{\infty} F^{(n)}(t)$ as a solution of (4.1). The Renewal (Integral) equation can be generalized as follows:

$$V(t) = G(t) + \int_0^t V(t-x) dF(x), t \geq 0 \quad (4.3)$$

where G and F are known and V is unknown. Equation (4.3) is called the *Renewal Type* equation. An unique solution of $V(t)$ exists in terms of G and F as can be seen from the following.

Theorem 4.3 If $V(t) = G(t) + \int_0^t V(t-x) dF(x), t \geq 0$

$$= G(t) + (V^* F)(t) \quad (4.4)$$

then $$V(t) = G(t) + \int_0^t G(t-x) dH(x). \quad (4.5)$$

Proof Taking Laplace-Stieltjes Transform of (4.4) we get

$$V^*(s) = G^*(s) + V^*(s)\, F^*(s)$$

or $$V^*(s) = \frac{G^*(s)}{1 - F^*(s)}, \text{ where } V^*(s) = \int_0^\infty e^{-st} dV(t)$$

and $$F^*(s) = \int_0^\infty e^{-st} dF(t)$$

$$= G^*(s) \left[1 + \frac{F^*(s)}{1 - F^*(s)} \right] = G^*(s) + G^*(s) \frac{F^*(s)}{1 - F^*(s)}$$

$$= G^*(s) + G^*(s)\, H^*(s),$$

since $$H^*(s) = \sum_{n=1}^{\infty} [F^*(s)]^n$$

$$= F^*(s) \sum_{n=1}^{\infty} [F^*(s)]^{n-1}$$

$$= F^*(s)/[1 - F^*(s)] \text{ if } F^*(s) < 1.$$

Inverting the Laplace transform, we get

$$V(t) = G(t) + \int_0^t G(t-x)dH(x).$$

Example Let $X_1, X_2, \ldots$ be i.i.d. r.v.s with

$$P(X_k \le t) = F(t) = 1 - e^{-\lambda t}, \lambda > 0, t > 0.$$

Then $$F^{(n)}(t) = P[S_n \le t] = 1 - e^{-\lambda t}\left\{\sum_{k=0}^{n-1} \frac{(\lambda t)^k}{k!}\right\}$$

(can be proved by induction on n. Prove first for $n = 2$).

Hence $$F^{(n)}(t) - F^{(n+1)}(t) = P[N(t) = n] = \frac{e^{-\lambda t}(\lambda t)^n}{n!}$$

i.e. the Renewal process generated by exponential random variables is a Poisson process, or in other words, Poisson process is a Renewal process with Exponential Interarrival times.

In this case, $H(t) = E(N(t)) = \lambda t = \dfrac{t}{E(X)}$ for all $t > 0$, i.e.

$$\frac{H(t)}{t} = \lambda, \text{ where } \lambda^{-1} = E(X).$$

4.3 Renewal Theorems

1. *Elementary Renewal Theorem* (Feller 1941)

$$\lim_{t\to\infty} \frac{H(t)}{t} = \begin{cases} 1/\mu & \text{if } 0 < \mu = E(x) < \infty \\ 0 & \text{if } \mu = \infty \end{cases}$$

2. *Blackwell's Renewal Theorem* (1948)

$\lim\limits_{t\to\infty} \dfrac{H(t+h) - H(t)}{h} = 1/\mu$ for fixed $h > 0$ and X_1 is a continuous random variable.

$\lim\limits_{t\to\infty} P$ [renewal at nd] $\to \dfrac{d}{\mu}$ where X_i is a lattice type discrete r.v. and d is the period of the lattice.

Definition 4.2 A random variable X is said to have lattice distribution if $P[X = c + nd] > 0$ where c and d (> 0) are real constants and $n = \pm 1, \pm 2, \ldots$.

3. *Key Renewal Theorem* (W.L. Smith, 1953)

If $Q(t) \geq 0$ and non increasing and $\int_0^\infty Q(t)dt < \infty$ then

$$\lim_{t\to\infty} \int_0^t Q(t-x)dH(x) = \frac{1}{\mu}\int_0^\infty Q(t)dt \text{ whenever } X \text{ is not arithmatic.}$$

A weaker version of the Elementary Renewal Theorem is the following due to J.L. Doob (1948) and is known as *Doob's Renewal Theorem*

$$\lim_{t\to\infty} \frac{N(t)}{t} = \frac{1}{\mu} \text{ a.s.}$$

Proof Assume that $X_1, X_2, \ldots$ and i.i.d. with $0 < \mu < \infty$. and for all $0 < \varepsilon < \mu$, $(\mu - \varepsilon)n \leq S_n \leq (\mu + \varepsilon)n$ for all large n.

$$nt < S_{N(tn)+1} \leq (\mu + \varepsilon)\,(N(tn) + 1). \tag{4.6}$$

Now $N(t) = \max\{k: S_k \leq t\} \Rightarrow S_{N(t)} \leq t$, $S_{N(t)+1} > 1$. Replace t by nt to get

$$nt \geq S_{N(tn)} \geq (\mu - \varepsilon)\, N(tn) \text{ for all large } n. \tag{4.7}$$

From (4.6) and (4.7)

$$\frac{t}{\mu+\varepsilon} < \frac{N(tn)+1}{n} \leq \frac{t}{\mu-\varepsilon} + \frac{1}{n} \text{ for } t > 0 \text{ and large } n. \text{ Hence}$$

$$\frac{N(tn)}{n} \to \frac{t}{\mu} \text{ a.s. as } n \to \infty \text{ and } t > 0,$$

i.e. $\frac{N(tn)}{tn} \to \frac{t}{\mu}$ as $t \to \infty$. Putting $nt = t^*$ we get the result.

Proof of Elementary Renewal Theorem

Assume $0 < \mu < \infty$.

Hence, there exists $\lambda > 0$ such that $P(X_n \geq \lambda) > 0$ for all $n = 1, 2, \ldots$

Define
$$X'_n = \begin{cases} \lambda \text{ if } x_n \geq \lambda \\ 0 \text{ otherwise} \end{cases} \qquad \text{(truncated r.v.)}$$

Then X'_n are i.i.d. r.v.s and $X'_n \leq X_n$ a.s. for all $n = 1, 2, \ldots$.

Define $S'_n = \sum_{k=1}^{n} X'_k$ and $N'(t) = \max\ \{K: S'_k \leq t\}$.

Then $S'_n \leq S_n$ and $N'(t) \geq N(t)$.

Let $P[X_n \geq \lambda] = p > 0$. Then $\frac{X'_n}{\lambda}$ is a Bernoulli random variable with probability of success p. If t is an integer then $N'(t+1) - N'(t)$ is the number of times, the partial sums take on the value $t + 1$.

Also
$$N'(t) = \sum_{j=0}^{[t]} \{N'(j)\}$$

$$E\left(\frac{N(t)}{t}\right)^2 \leq E\left(\frac{N'(t)}{t}\right)^2 = 0(1) \text{ as } t \to \infty, \text{ because}$$

$$E(N'(t)) = T/p, \text{ where } T = [t] \text{ and}$$
$$\text{Var } (N'(t)) = Tq/p^2.$$

Therefore $E(N'(t))^2 = \dfrac{T^2}{p^2} + \dfrac{Tq}{p^2} \simeq \dfrac{t^2}{p^2} + \dfrac{tq}{p^2} = \dfrac{t^2 + tq}{p^2}$ and

$$E\left(\frac{N'(t)}{t}\right)^2 \simeq \frac{1}{p^2} + \frac{q}{p^2 t} = 0(1) \text{ as } t \to \infty, 0 < p < 1. \text{ This implies}$$

$$E\left(\frac{N(t)}{t}\right)^2 \leq c \text{ (constant) for all large } t.$$

Therefore $\left\{\dfrac{N(t)}{t}, t \geq 1\right\}$ is (uniformly integrable) u.i. by the following result.

Exercise If $E X_n^2 \leq c$ for all large n, show that $\{X_n\}$ is u.i.

Proof $$\int_{|X_n| > k} |X_n| \, dP \leq \frac{1}{k} \int X_n^2 \, dP \leq \frac{1}{k} E(X_n^2) \leq \frac{c}{k} \text{ if } n \geq n_0$$

$$\to 0 \text{ as } k \to \infty \textit{ if } n \geq n_0 \tag{4.8}$$

But $$\max_{1 \leq n \leq n_0} \int_{|X_n| > k} | X_n | \, dP \leq c \text{ if } k \to \infty \tag{4.9}$$

Hence, $$\sup_n \int_{|X_n| > k} | X_n | \, dP \to 0 \text{ as } k \to \infty$$

i.e. $\{X_n\}$ is u.i.

By Doob's Renewal theorem $\dfrac{N(t)}{t} \to \dfrac{1}{\mu}$ a.s. and $\left\{\dfrac{N(t)}{t}, t \geq 1\right\}$ is u.i. Hence $\dfrac{N(t)}{t} \xrightarrow{L_1} \dfrac{1}{\mu} \Rightarrow \dfrac{H(t)}{t} \to \dfrac{1}{\mu}$ as $t \to \infty$.

Proof of Blackwell's Theorem which is a special case of Key Renewal Theorem.

Take $$Q(t) = \begin{cases} 1/h \text{ if } 0 \leq t \leq h \\ 0, \quad \text{otherwise} \end{cases}$$

Then $$Q(t - x) = \begin{cases} 1/h \text{ if } t \geq x \geq t - h \\ 0, \quad \text{otherwise} \end{cases}$$

$$\lim_{t \to \infty} \int_0^\infty Q(t - x) dH(x) = \lim_{t \to \infty} \frac{1}{h} \int_{t-h}^t dh(x) = \lim_{t \to \infty} \frac{H(t) - H(t - h)}{h} = \frac{1}{\mu}$$

$$\left(\text{since} \int_0^\infty Q(t) dt = \int_0^h 1/h \, dt = 1\right)$$

i.e. $\lim_{t\to\infty} \dfrac{H(t+h)-H(t)}{h} = \dfrac{1}{\mu} \Rightarrow$ Blackwell's Theorem.

Proof of Key Renewal Theorem

Let
$$\int_0^t Q(t-x)dH(x) = I_1 + I_2,$$

where $I_1 = \int_0^{t/2} Q(t-x)dH(x)$ and $I_2 = \int_{t/2}^{t} Q(t-x)dH(x)$.

Since $Q(t)$ is non-increasing and non-negative, we have

$$0 \le I_1 \le Q(t/2)\int_0^{t/2} dH(x) = Q(t/2)\,H(t/2) \text{ (since } H(0)=0)$$

$$= \left(\frac{t}{2}\,Q(t/2)\right)\left(\frac{H(t/2)}{t/2}\right). \tag{4.10}$$

By Elementary Renewal Theorem $\dfrac{H(t/2)}{t/2} \to 1/\mu$ as $t \to \infty$.
[Similarly for the continuous case using intergrals, the result follows.]

Now let
$$Q\int_0^\infty Q(t)dt = \sum_{n=0}^{\infty}\int_{nh}^{nh+h} Q(t)dt. \tag{4.11}$$

Since $Q(nh+h) \le Q(t) \le Q(nh)$ for $nh \le t \le nh+h$, (4.11) gives,

$$h\sum_{n=1}^{\infty} Q(nh) = h\sum_{n=0}^{\infty} Q(nh+h) \le Q \le h\sum_{n=0}^{\infty} Q(nh) < \infty$$

and $\quad 0 \le Q - h\sum_{n=1}^{\infty} Q(nh) \le hQ(0) < \varepsilon$ if we choose $0 < h < \varepsilon/Q(0)$ (4.12)

If $[t/2h] = N_0$ then

$$I_2 = \int_{t/2}^{t} Q(t-x)dH(x) - \int_0^{t/2} Q(s)[-dH(t-s)], \text{ putting } x = t-s$$

$$= \sum_{n=0}^{N_0-1}\int_{nh}^{nh+h} Q(s)[-dH(t-s)]$$

Hence $\sum_{n=0}^{N_0-1} Q(nh+h[H(t-nh)-H(t-nh-h)]$ (since $H(t) = \sum_n F^{(n)}(t)$ is non-decreasing)

$$\le I_2 \le \sum_{n=0}^{N_0-1} Q(nh)[H(t-nh)-H(t-nh-h)]. \tag{4.13}$$

Now we can choose t so large that (by Blackwell's Theorem)

$$\left|\frac{H(t-nh)-H(t-nh-h)}{h} - \frac{1}{\mu}\right| < \varepsilon \tag{4.14}$$

and
$$h \sum_{n=N_0+1}^{\infty} Q(nh) < \varepsilon \tag{4.15}$$

(Note that t large $\Rightarrow N_0$ is large).
Using (4.12) to (4.15), we have

$$\left(\frac{1}{\mu} - \varepsilon\right)(Q - 2\varepsilon) < I_2 < \left(\frac{1}{\mu} + \varepsilon\right)(Q - \varepsilon) \text{ for arbitrarily small } \varepsilon > 0.$$

Hence, $I_2 \to \dfrac{Q}{\mu}$ as $t \to \infty$.

Therefore $\displaystyle\int_0^t Q(t-x)dH(x) \to \frac{1}{\mu}\int_0^\infty Q(t)dt$ as $t \to \infty$.

4.4 Central Limit Theorem for Renewal Theory

4.4.1 Renewal density and integral equation

If $F(t)$ is absolutely continuous with $F'(t) = f(t)$, then $H'(t) = h(t)$ exists and is given by

$$h(t) = \sum_{n=1}^{\infty} f^{(n)}(t) \text{ where } f^{(n)}(t) = [F^{(n)}(t)]'$$

(since the terms of $H(t)$ are non-negative and bounded we can carry out term-by-term differentiation and obtain the result). $h(t)$ is called the *Renewal density*.

If we "Confuse" expectation with probability then $h(t)dt$ can be interpreted as the probability that a renewal occurs during $(t, t + dt)$. Now if a renewal occurs during $(t, t + dt)$, then the dying individual is either the initial one itself or the one which entered the population at some prior time, say $t - \tau (0 < \tau < t)$. This leads to

$$h(t) = f(t) + \int_0^t h(t-\tau)f(\tau)d\tau$$

which again is a Renewal type integral equation. Such an argument is called the *Renewal argument*. We have seen $\dfrac{N(t)}{t} \to \dfrac{1}{\mu}$ a.s. if $\mu > 0$ (Doob's Renewal Theorem) and

$$\lim \frac{H(t)}{t} = \lim \frac{E(N(t))}{t} = \frac{1}{\mu} \text{ as } t \to \infty$$

(from Elementary Renewal Theorem).

We can prove that as $t \to \infty$, $\begin{cases} E(N(t)) = 0\left(\dfrac{t}{\mu}\right) \\ V(N(t)) \cong \dfrac{t\sigma^2}{\mu^2} \text{ where } \sigma^2 = \text{Var } X_1. \end{cases}$

(see Ex. 4.12)

Central limit Theorem 4.4 $\displaystyle\lim_{t\to\infty} P\left[\frac{N(t) - t/\mu}{\sqrt{(t\sigma^2/\mu^3)}} \le X\right] = \frac{1}{\sqrt{2\pi}}\int_{-\infty}^{x} e^{-\frac{1}{2}u^2} du.$

Proof $P[N(t) < n] = P[S_n > t]$, since

$$P[N(t) < n] = P[N(t) = n-1] + P[N(t) = n-2] + \ldots + P[N(t) = 0]$$
$$= F^{(n-1)}(t) - F^{(n)}(t) + F^{(n-2)}(t) - F^{(n-1)}(t) + \ldots + F^{(0)}(t) - F^{(1)}(t)$$
$$= F^{(0)}(t) - F^{(n)}(t)$$
$$= 1 - F^{(n)}(t)$$
$$= 1 - P[S_n \le t] = P[S_n > t].$$

By CLT of i.i.d. random variables,

$$\lim_{t\to\infty} P\left[\frac{S_n - n/\mu}{\sqrt{n}\sigma} \le x\right] = \frac{1}{\sqrt{2\pi}}\int_{-\infty}^{x} e^{-\frac{1}{2}u^2}\,du$$

Let $n \to \infty$ and $t \to \infty$ in such a way that

$$\frac{t - n\mu}{\sqrt{n}\sigma} = -x\,(x \text{ fixed}) \tag{4.16}$$

Solving (4.16) for n as a function of t and x, we get

$$n = t/\mu \pm x(\sigma^2 t)^{1/2}\,\mu^{-3/2} + o(t) \tag{4.17}$$

because we note that $t > n\mu$ for $x < 0$ and $t < n\mu$ for $x > 0$ in the two solutions of the above equation.

From CLT of i.i.d r.v.s,

$$\lim_{n\to\infty} P[S_n > t] = \lim_{n\to\infty} P\left[\frac{S_n - n\mu}{\sqrt{n}\sigma} > \frac{t - n\mu}{\sqrt{n}\sigma} = -x\right]$$
$$= \frac{1}{\sqrt{2\pi}}\int_{-x}^{\infty} e^{-\frac{1}{2}u^2}\,du = \frac{1}{\sqrt{2\pi}}\int_{-x}^{\infty} e^{-\frac{1}{2}u^2}\,du$$

Now equation (4.17) shows $\dfrac{n - t/\mu}{\sqrt{(t\sigma^2/\mu^3)}} \simeq x$. Therefore, $\lim_{n\to\infty} P[N(t) \le n]$

$$\lim_{n\to\infty} P\left[\frac{N(t) - t/\mu}{\sqrt{(t\sigma^2/\mu^3)}} \le X\right] = \frac{1}{\sqrt{2\pi}}\int_{-\infty}^{x} e^{-\frac{1}{2}u^2}\,du.$$

4.5 Delayed and Equilibrium Renewal Processes

There are two types of generalizations of ordinary Renewal Process. When the first interarrival time X_1 has a distribution function G different from the common d.f.F of $X_2, X_3, \ldots$, we call such a Renewal process a *Modified* or *Delayed Renewal Process*. A Modified Renewal Process in which

$$G(x) = \int_0^x \frac{1 - F(t)}{\mu}\,dt,$$

is called the *Equilibrium* (Stationary) *Renewal Process*.

Let $H_D(t) = \sum_{n=1}^{\infty} G^* F^{(n-1)}(t)$ be the Renewal function of delayed Renewal process. Earlier we have seen for ordinary Renewal process

$$\frac{H(t)}{t} \to \frac{1}{\mu} \text{ as } t \to \infty, \text{ where } \mu = \int_0^{\infty} x\, dF(x).$$

Similarly, $\dfrac{H_D(t)}{t} \to \dfrac{1}{\mu}$ as $t \to \infty$ (by Strong Law).

Hence, the question is whether there is any Delayed Renewal Process for which $H_D(t) = t/\mu$? Also we know that for Poisson process, where the underlying variables are exponential, $H(t) = t/\mu$.

Let
$$F_e(x) = \int_0^x \frac{1 - F(t)}{\mu}\, dt \tag{4.18}$$

The following proposition is an answer in affirmative to our last question.

Proposition 4.1 There exists a modified Renewal process where the initial distribution $G = F_e$ exists such that $H_e(t) = t/\mu$.

Proof If $H_e(t = t/\mu$, then Laplace transform

$$H_e^*(s) = \int_0^{\infty} e^{-st}\left(\frac{t}{\mu}\right) dt = \frac{1}{\mu s^2} \tag{4.19}$$

Also
$$H_e^*(s) = \frac{f_e^*(s)}{s[1 - f^*(s)]} \tag{4.20}$$

where $f_e^*(s)$ is the Laplace Transform of f_e and $f^*(s)$ is the Laplace Transform of f where $F'(x) = f(x)$.

Equation (4.20) follows from the following argument: Integrating by parts Laplace transform of d.f. $F(x)$ is $(F(x))^* = f^*(s)/s$. By multiplicative property of Laplace transform

$$\sum_{n=1}^{\infty} \{F^{(n-1)}(t)\}^* = 1/s \sum_{n=1}^{\infty} [f^*(s)]^{n-1} = 1/s\left(\frac{1}{1 - f^*(s)}\right) \text{ if } f^*(s) < 1$$

and F is absolutely continuous with density f). Now $H_e(t) = \sum_{n=1}^{\infty} F_e * F^{(n-1)}(t)$.

Taking Laplace Transform on both sides

$$H_e^*(t) = \sum_{n=1}^{\infty}\{F_e * F^{(n-1)}(t)\}^* = \frac{1}{s}\sum_{n=1}^{\infty} f_e^*(s)\,[f^*(s)]^{n-1} = \frac{f_e^*(s)}{s[1 - f^*(s)]}.$$

Equating (4.19) and (4.20) $\dfrac{1}{\mu s^2} = \dfrac{f_e^*(s)}{s[1 - f^*(s)]}$

$$\Rightarrow f_e^*(s) = [1 - f^*(s)]/(\mu s). \tag{4.21}$$

Inverting the Laplace Transform, we get

$$f_e(t) = \frac{1 - F(t)}{\mu} \text{ i.e. } F_e(X) = \int_0^x \left(\frac{1 - F(t)}{\mu}\right) dt.$$

As a particular case, a Poisson process is an Equilibrium Renewal process. Stationary or Equilibrium Renewal Process has a simple physical interpretation. Suppose that the Renewal process started at time $t = -\infty$, i.e. at remote past but the observation of the process begins at $t = 0$. Then X_1 has distribution function F_e given by equation (4.18) and $H_e(t) = t/\mu$. (It can be shown that $H(t) \geq \frac{t}{\mu} - 1$ for $t \geq 0$. Refer Statistical Theory of Reliability and Life Testing by Barlow and Proschan. Also AMS, 1964 Paper on Replacement Policy by Barlow and Proschan).

Let $N_e(t)$ be the number of Renewal in an Equilibrium Renewal process.

Result 4.1 We have for all $t \geq 0$ and $n = 0, 1, 2, \ldots$

$$P[N_e(t) \geq n] \geq P[N(t) \geq n] \tag{4.22}$$

iff

$$\int_t^\infty \frac{1 - F(x)}{1 - F(t)} dx \leq \mu \tag{4.23}$$

Proof Assume (4.23) holds.

Since $\frac{1}{\mu}\int_t^\infty (1 - F(x))dx = \frac{1}{\mu}\left[\int_0^\infty \{1-F(x)\}dx - \int_0^t \{1-F(x)\}dx\right]$

$$= 1 - F_e(t)$$

note that for a non-negative r.v., the mean $\mu = \int_t^\infty (1 - F(x))dx$.

Hence, by (4.23), $1 - F_e(t) \leq 1 - F(t)$ for $t \geq 0$

or $$F_e(t) \geq F(t) \text{ for } t \geq 0.$$

If S_n and S'_n denote the nth renewal epochs of $N(t)$ and $N_e(t)$, respectively, then for $n \geq 1$

$$P[N_e(t) \geq n] = P[S'_n \leq t] = \int_0^t F^{(n-1)}(t - x)dF_e(x)$$

$$\geq \int_0^t F^{(n-1)}(t - x)dF(x) = P[S_n \leq t] = P[N(t) \geq n] \text{ for } n \geq 1.$$

(It also holds for $n = 0$).

Further, $P[N_e(t) \geq 0] = P[N(t) \geq 0] = 1$ for all $t > 0$. Thus (4.22) holds if (4.23) holds.

To prove the converse, assume that

$$P[N_e(t) \geq n] = P[N(t) \geq n] \text{ holds for } n = 0, 1, \ldots$$

Since equality holds for $n = 0$, we must have

$$P[N_e(t) = 0] \leq P[N(t) = 0]$$

(since $[N(t) \geq 0] = [N(t) = 0] \cup [N(t) \geq 1]$) or $1 - F_e(t) \leq 1 - F(t)$

as $$P[N(t) = 0] = F^{(0)}(t) - F^{(1)}(t) = 1 - F(t).$$

Corollary 4.1 $H(t) \le t/\mu$ iff $F_e(t) \ge F(t)$

Proof $H(t) = E(N(t)) = \sum_n P[N(t) \ge n] \le \sum_n P[N_e(t) \ge n]$

$$= E(N_e(t)) = \frac{t}{\mu} \text{ iff } F_e(t) \ge F(t).$$

Definition 4.3 A unit (or equivalently its life-time distribution F) is said to be *New Better (Worse) than used NBU (NWU)* if

$$1 - F(x + y) \le (\ge) [1 - F(x)][1 - F(y)]$$

i.e.
$$\frac{\bar{F}(x+y)}{\bar{F}(x)} \le (\ge) \bar{F}(y) \text{ for } x \ge 0, y \ge 0.$$

In other words, a unit is NBU (NWU) if the conditional survival probability of a unit at age x is less (greater) than the corresponding survival probability of a new unit.

Definition 4.4 A unit is said to be *New Better (Worse) than used in Expectation NBUE (NWUE)* iff

$$\int_t^{\infty} \frac{1 - F(x)}{1 - F(t)} dx \le (\ge) \mu.$$

In other words, a unit is NBUE (NWUE) if the conditional expectation of the residual life of a unit at age t is less (greater) than the expectation of the life-time of a new unit.

The above result may be restated as

$$P[N_e(t) \ge n] \ge P[N(t) \ge n], \; t \ge 0, \; n = 0, 1, 2, \ldots$$

as well as $H(t) \le t/\mu$ iff distribution F involving the Renewal process is NBUE.

Note If F belongs to NWUE class then $H(t) \ge \frac{t}{\mu} - 1$ for $t \ge 0$.

4.6 Residual and Excess Life-Times

To a given $t \ge 0$, there corresponds uniquely a Renewal process $N(t)$ such that $S_{N(t)} \le t \le S_{N(t)+1}$, i.e. t falls in the interval of length $X_{N(t)+1}$.

(i) The *residual life-time* of the individual alive at age t is given by the time $Y(t)$ from t to the next Renewal epoch, i.e. $Y_t = S_{N(t)+1} - t$. It is also called *Forward Recurrence time* at t, or Excess life time at t.

(ii) The *Spent life-time* of the individual alive at age t is given by the time upto t since the last renewal epoch, i.e. $Z(t) = t - S_{N(t)}$. It is also called *Backward Recurrence* time at t, or current life time.

(iii) The length of the life-time containing t is given by

$$Y(t) + Z(t) = S_{N(t)+1} - S_{N(t)} = X_{N(t)+1}.$$

Theorem 4.5 $P[Y(t) \le X] = F(t + x) - \int_0^t [1 - F(t + x - y)]dH(y)$ (4.24)

If in addition, F is not lattice, then

$$\lim_{t\to\infty} P[Y(t) \le x] = \frac{1}{\mu}\int_0^x [1 - F(y)]dy \text{ if } \mu < \infty \quad (4.25)$$

$$= 0 \text{ if } \mu = \infty$$

Proof It is clear that $Y(t) > x$ iff no renewals occur in $[t, t + x]$.

Let $P(t) = P[Y(t) > x]$. By conditioning on X_1, we get

$$P(t) = \int_0^\infty P[Y(t) > x \mid X_1 = Y]dF(y).$$

Now following three situations may arise:

(i) $y > t + x$, in which case it is certain that no renewal occurs in $[t, t + x]$ so that $P[Y(t) > x \mid X_1 = y] = P[\text{No renewal in } [t, t + x]] = 1$.

(ii) $t < y < t + x$ in which case one renewal (the one corresponding to $X_1 = y$) occurs in $[t, t + x]$ so that

$$P[Y(t) > x \mid X_1 = y] = P[\text{No renewal in } [t, t + x] \mid X_1 = y] = 0.$$

(iii) $0 < y < t$, in which case the first renewal occurs in $[y, y + dy)$ and then the process restarts itself and

$$P[Y(t) > x \mid X_1 = y] = P[Y(t - y) > x] = P(t - y).$$

considering the above three cases, we get

$$P(t) = \int_0^t P(t - y)dF(y) + \int_{t+x}^\infty 1.dF(y)$$

$$= 1 - F(t + x) + \int_0^t P(t - y)dF(y),$$

which is a Renewal type equation and the solution is given by

$$P(t) = 1 - F(t + x) + \int_0^t [1 - F(t + x - y)]dH(y)$$

which is equivalent to (4.24).

Therefore

$$P(t) = 1 - F(t + x) + \int_0^t [1 - F(t + x - y)]dH(y) \quad (4.26)$$

Now suppose that F is not lattice. Then applying the key Renewal theorem with $Q(t) = 1 - F(t + x)$, we get as $t \to \infty$

$$\int_0^t [1 - F(t + x - y)]dH(y) \to \frac{1}{\mu}\int_0^\infty [1 - F(t + x)]dt$$

$$= \frac{1}{\mu}\int_x^\infty [1 - F(y)]dy \quad (4.27)$$

Hence from (4.26) and (4.27), we have

$$\lim_{t\to\infty} P[Y(t) > x] = \lim_{t\to\infty} P(t) = \frac{1}{\mu}\int_x^\infty [1 - F(y)]dy$$

or $\lim_{t\to\infty} P[Y(t) \le x] = 1 - \frac{1}{\mu}\int_x^{\infty} [1 - F(y)]dy = \frac{1}{\mu}\int_0^{x} [1 - F(y)]dy.$

Theorem 4.6 $P[Z(t) \le x] = \begin{cases} F(t) - \int_0^{t-x} [1 - F(t-y)]dH(y) \text{ if } x \le t \\ 1 \text{ if } x > t \end{cases}$

$$\lim_{t\to\infty} P[Z(t) \le x] = \frac{1}{\mu}\int_x^{\infty} [1 - F(y)]dy \text{ if } F \text{ is not lattice.}$$

Proof The distribution of $Z(t)$ can be obtained immediately by noting that $Z(t) > x$ iff no renewals occur in $[t - x, t]$ i.e. iff $Y(t - x) > x$.

By previous theorem,

$$P[Z(t) \le x] = \begin{cases} F(t) - \int_0^{t-x} [1 - F(t-y)]dH(y) \text{ if } x \le t \\ 1 \text{ if } x > t \end{cases}$$

(since $Z(t) = t - S_{N(t)} \le x \leftrightarrow S_{N(t)} \ge t - x < 0$ if $x > t$, so $Z(t) \le x$ is a sure event) and if F is not lattice then,

$$\lim_{t\to\infty} P[Z(t) \le x] = \int_0^{x} [1 - F(y)]dy/\mu.$$

Note It is to be noted that the result

$$\lim_{t\to\infty} P[Z(t) \le x] = \int_0^{x} [1 - F(y)]dy/\mu$$

also holds for the modified Renewal process regardless of initial distribution G.

4.7 Renewal Reward Process

Consider an ordinary renewal process generated by the sequence $[X_n, n \ge 1\}$ of interarrival times. Suppose that each time a renewal occurs, a reward is received or a cost is incurred (e.g. for replacement of a failed component). Denote the reward or cost associated with nth renewal by Y_n, $n \ge 1$. Now Y_n will usually depend on X_n, the duration of the interarrival time between $(n - 1)^{\text{th}}$ and nth renewal. In such a case we assume that the sequence of pairs $[X_n, Y_n, n \ge 1\}$ are i.i.d. The sequence of pairs of i.i.d. random variables $\{X_n, Y_n\}$ is said to generate a *Renewal Reward process*. Now $Y(t) = \sum_{n=1}^{N(t)} Y_n$ gives the total reward carried (or cost incurred) by time t and $\{Y(t), t \ge 0\}$ is called the renewal reward process generated by $\{X_n, Y_n\}$.

Theorem 4.7 Suppose $E(X) < \infty$, $E \mid Y \mid < \infty$, then

(a) $Y(t)/t \to E(Y)/E(X)$ as $t \to \infty$ a.s.

(b) $E(Y(t)/t) \to E(Y)/E(X)$ *as* $t \to \infty$.

Proof

(a) $$\frac{Y(t)}{t} = \frac{Y(t)}{N(t)} \cdot \frac{N(t)}{t} = \frac{\sum_{n=1}^{N(t)} Y_n}{N(t)} \cdot \frac{N(t)}{t}$$

By Kolmogorov's SILN for i.i.d. r.v.s $S_n/n \to E(X)$ a.s. as $n \to \infty$.
Let $M(t) = \inf \{n: S_n > t\} = N(t) + 1$.

Since $N(t)/t$ or $\frac{M(t)}{t} = (N(t) + 1)/t \to 1/E(X) > 0$ a.s. as $t \to \infty$

$M(t)$ or $N(t) \to \infty$ *as* $t \to \infty$ if $0 < E(X) < \infty$.

Let $S_n^* = \sum_{i=1}^{n} Y_i$, $A = \{\omega: S_n^*(\omega)/n \to E(Y)\}$, $B = \{\omega: M(t, \omega) \to \infty\}$ and $C = \{\omega: S_{M(t)}^*(\omega)/M(t) \to E(Y)\}$.

Then $C \subset AUB$. But $P(A) = 0$ and $P(B) = 0$ and hence $P(C) = 0$.

This is same as $S_{M(t)}^*/M(t) \xrightarrow{\text{a.s.}} E(Y)$ as $t \to \infty$ which implies $\sum_{n=1}^{N(t)} Y_n/N(t)$ $\to EY$ as $t \to \infty$ and also $\frac{N(t)}{t} \to \frac{1}{E(X)}$ as $t \to \infty$ a.s.

Hence $\frac{Y(t)}{t} \to \frac{E(Y)}{E(X)}$ as $t \to \infty$.

[Note that $N(t)$ is not a stopping time since $[N(t) = n]$ is not independent of $X_{n+1}, X_{n+2}, \ldots$ but $N(t) + 1$ is a stopping time since $[N(t)+1 = n]$ is independent of $X_{n+1}, X_{n+2}, \ldots$].

(b) We first note that $N(t) + 1$ is a stopping time for $Y_1, Y_2, \ldots$ since the independence of $N(t) + 1$ and $\{X_{n+1}, X_{n+2}, \ldots\}$ implies the independence of $N(t) + 1$ and $\{Y_{n+1}, Y_{n+2}, \ldots\}$. By Wald's equation,

$$\begin{aligned} E(Y(t)) &= E\left(\sum_{n=1}^{N(t)} Y_n\right) = E\left[\sum_{n=1}^{N(t)+1} Y_n - Y_{N(t)+1}\right] \\ &= E(N(t) + 1)\, E(Y) - E(Y_{N(t)+1}) \\ &= (H(t) + 1)\, E(Y) - E(Y_{N(t)+1}). \end{aligned}$$

Hence

$$\frac{EY(t)}{t} = \frac{H(t) + 1}{t}(EY) - \frac{E(Y_{N(t)+1})}{t}.$$

Thus, it is enough to prove $E(Y_{N(t)+1}/t) \to 0$ as $t\ \infty$.

Let $$g(t) = E(Y_{N(t)+1}) = \int_0^{\infty} E(Y_{N(t)+1} \mid X_1 = x) dF(x)$$

Now $$E(Y_{N(t)+1} \mid X_1 = x) = \begin{cases} g(t - x) \text{ if } x \le t \\ E(Y_1 \mid X_1 = x) \text{ if } x > t. \end{cases}$$

Hence, $$g(t) = h(t) + \int_0^t g(t - x) dF(x) \tag{4.28}$$

where $$h(t) = \int_t^\infty E(Y_1 \mid X_1 = x)dF(x).$$

Also note that, since

$$E \mid Y_1 \mid = \int_0^\infty E[\mid Y_1 \mid |X_1 = x]dF(x) < \infty,$$

$$h(t) \to 0 \text{ as } t \to \infty \text{ and } h(t) \le E \mid Y_1 \mid < \infty \text{ for all } t > 0.$$

The solution of the Renewal type of equation (4.28) yields

$$g(t) = h(t) + \int_0^t h(t-x)dH(x) \text{ where } H(x) = \sum_{n=1}^\infty F^{(n)}(x).$$

Now $\mid h(t) \mid < \varepsilon$ if $t \ge T$.

Hence $$\frac{\mid g(t) \mid}{t} \le \frac{\mid h(t) \mid}{t} + \int_0^{t-T} \frac{\mid h(t-x) \mid}{t} dH(x) + \int_{t-T}^t \frac{\mid h(t-x) \mid}{t} dH(x)$$

$$\le \frac{\varepsilon}{t} + \varepsilon \frac{H(t-T)}{t} + E \mid Y_1 \mid \frac{H(t) - H(t-T)}{t} \quad (\because H(0) = 0)$$

$$\to \varepsilon / \mu \text{ as } t \to \infty \ (\mu > 0) \quad \left[\begin{array}{l} \text{since } \dfrac{H(t)}{t} \to \dfrac{1}{\mu} \text{ as } t \to \infty \\ \dfrac{H(t-T)}{t} \to \dfrac{1}{\mu} \text{ as } t \to \infty \end{array} \right]$$

Since $\varepsilon > 0$ is arbitrarily small, $\dfrac{g(t)}{t} \to 0$ as $t \to \infty$.

Denote $EY/EX = \xi$ and $EY(t) = c(t)$, the average expected reward (or cost) in $[0, t]$. Then from Theorem 4.7 (b) we get for large t, $c(t) = t\xi$.

A more refined asymptotic form of $c(t)$ has been given by Christer (1978) $c(t) = \xi t + \eta$, where ξ and η are functions of parameters and decision variables of renewal process.

4.8 Replacement Policies Connected with Renewal Theory

4.8.1 Age and Block Replacement Policies

The usual replacement policy implies replacement of a component as and when it fails, by a similar new one. The most important replacement policies general use are *age and block replacement policies.*

Definition 4.5 In *age replacement policy* a unit is replaced upon failure or at age T, which ever comes first.

Definition 4.6 In a *block replacement policy* the unit in operation is replaced upon failure and at times $T, 2T, 3T, \ldots$.

The age replacement policy is administratively more difficult to implement, as it requires the age of the unit to be recorded. Whereas block replacement policy, though simpler to administer as the age of the unit need not be recorded, leads to more frequent replacements of relatively new items. The word "block" is used to imply that in practice a block or set of components will be replaced at times $T, 2T, 3T, \ldots$ regardless of which have failed. Suppose that the successive life times of the units are r.v.s with a common d.f. $F(x)$ having mean μ.

T is called the replacement interval under these policies.

Let $N(t)$ = Number of renewals in $[0, t]$ for ordinary renewal process,

$N_A(t, T)$ = Number of renewals in $[0, t]$ under age replacement policy,

$N_D(t, T)$ = Number of renewals in $[0, t]$ under block replacement policy,

Denote by

$H(t) = EN(t)$, $H_A(t, T) = EN_A(t, T)$, $H_D(t, T) = EN_B(t, T)$ the corresponding renewal functions.

Theorem 4.8 (a) $$\lim_{t\to\infty} \frac{N_A(t, T)}{t} = \lim_{t\to\infty} \frac{H_A(t, T)}{t} = \frac{F(T)}{\int_0^T (1 - F(x)dx} \text{ (a.s.)}$$

(b) $$\lim_{t\to\infty} \frac{N_D(t, T)}{t} = \lim_{t\to\infty} \frac{N_D(t, T)}{t} = \frac{H(t)}{T} \text{ (a.s.)}.$$

Proof (a) $\{N_A(t, T)\}$ is a renewal process with interarrival times $\{Y_i\}$ and $F_A(t) = P[Y_i \le t]$. Since t lies between one of the intervals $nT \le t \le (n + 1)T, n = 0, 1, 2, \ldots$,

$$1 - F_A(t) = P[Y_i > t] = [1 - F(T)]^n [1 - F(t - nT)].$$

Hence $$\mu_A = ET_i = \int_0^\infty P[Y_i > t]dt$$

$$= \sum_{n=0}^{\infty} \int_{nT}^{(n+1)T} [1 - F(T)]^n [1 - F(t - nT)]dt$$

$$= \sum_{n=0}^{\infty} [1 - F(T)]^n \left[\int_0^T (1 - F(x))dt\right] \quad \text{(putting } x = t - nT\text{)}$$

$$- \int_0^T [1 - F(x)]dx / F(T), \text{ since}$$

$$= \sum_{n=0}^{\infty} [1 - F(T)]^n = \frac{1}{1 - (1 - F(T))}, F(T) < 1.$$

Then applying corresponding limit theorems for ordinary renewal process we get the desired result.

(b) Note that after successive block replacements at $T, 2T, 3T, \ldots$ the process starts anew, and if $N_{B_r}(T)$ is the number of failures in the interval $[(r - 1) T, rT]$, $r = 1, 2, 3, \ldots$ following block replacements, then $N_{B_r}(T) = N(T)$.

Suppose that for some specified integral value of n, $nT \le t \le (n + 1)T$, i.e. $t = nT + \tau, 0 \le \tau < T$ then $N_B(t, T) = \sum_{r=1}^{n} N_{B_r}(T) + N(\tau)$.

We then get, $$\lim_{t\to\infty} \frac{N_D(t, T)}{t} = \lim_{n\to\infty} \frac{\sum_{r=1}^{n} N_{D_r}(T)}{nT} + \frac{N(\tau)}{\lim_{n\to\infty} nT + \tau}.$$ By Strong's law the last expression is equal to

$$E\left(\frac{N_{D_r}(T)}{T}\right) = E\left(\frac{N(T)}{T}\right) = \frac{H(T)}{T}, \text{ since } \frac{N(\tau)}{\lim nT + \tau} = 0.$$

Exercises and Complements

Exercise 4.1 Assume that customers arrive at a store according to a renewal process and that, independently of the process and of each other, every customer makes a purchase with the same probability p. Show that those customers who make a purchase also form the "renewal points" of a renewal process. Let the time scale be expanded by a factor of $1/p$. Find the interarrival distribution of the new process in terms of the original interarrival distribution.

Exercise 4.2 Let the Laplacc transforms of $F(x)$ (the interarrival distribution) and $W_t(x) = P(Y(t) \le x)$, the d.f. of Residual life time $Y(t)$ in a renewal process be $L_F(s)$ and $L_W(s)$ respectively. Show that

$$L_w(s)e^{-st} = [1 - L_F(s)] \int_t^{\infty} e^{-sy}\,dH(y), \text{ where } H(y) \text{ is the renewal function.}$$

Exercise 4.3 (a) If the random lifetime of an item has d.f. $F(x)$, what is the mean remaining life of an item of age x?

(b) Find the mean total life time of an item when the d.f. $F(x) = 1 - e^{-\lambda x}$, $x \ge 0$, $\lambda > 0$.

Hence show that the mean total life is approximately 2 times the mean life when the renewal process has been in operation for a long time.

Exercise 4.4 Let $\{X_k, k \ge 1\}$ be a renewal process, X_k has distribution $F(t)$, $N(t)$ is the renewal counting function, and $H(t) = E(N(t))$ is the renewal function.

1. Show that
$$\begin{cases} \text{(a) } \lim_{t\to\infty} H(t) = \infty \\ \text{(b) for every } t \ge 0,\ 0 \le H(t) \le \dfrac{F(t)}{1 - F(t)} \le \dfrac{1}{1 - F(t)} \end{cases}$$

2. Let $H_2(t) = E(N^2(t))$

Show that
$$\begin{cases} \text{(a) } H_2(t) = \sum_{n=1}^{\infty} (2n - 1)\, F_{(t)}^{(n)} \\ \text{(b) } H_2(t) = H(t) + 2\displaystyle\int_0^t H(t - u)\,dH(u) \end{cases}$$

3. Assume that X_k has finite first moment $\mu > 0$.

Define $F^*(t) = \dfrac{1}{\mu}\displaystyle\int_0^t [1 - F(x)]dx,\ t \ge 0.$

Show that (a) $F^*(t)$ is a probability distribution of a nonnegative r.v. X^*

(b) If the Laplace-Stieltjes transform of $F(t)$ is $f(s)$, then the Laplace-Stieltjes transform of $F^*(t)$ is $\dfrac{1 - f(s)}{\mu}$.

(c) Use (b) to show that X^* has finite first moment iff X has finite second moment μ_2.

4. Assume that X_k has non-lattice distribution. Define $Q(t) = 1 - F^*(t)$ where $F^*(t)$ is defined in Problem 3.

(a) Show that $\displaystyle\int_0^t Q(t - u)\,dH(u) = H(t) - t/\mu + F^*(t).$

(b) Apply Smith's key renewal theorem to the left hand side of this equality to prove that if X_k has finite second moment μ_2 then

$$\lim_{t\to\infty}\left\{H(t)-\frac{t}{\mu}\right\}=\frac{\mu_2-2\mu^2}{2\mu^2}.$$

(c) Interpret this result graphically.

Exercise 4.5 Show that $EN^k(t) < \infty$ is finite for all $k \geq 1$ and that

$$E[N^k(t)] = \sum_{n=0}^{\infty}[(n+1)^k - n^k]\,F^{(n+1)}(t)$$

Exercise 4.6 Derive the alternative formula for Residual life time $Y(t)$ in a renewal process

$$P(Y(t) \leq x) = \int_t^{t+x}[1-F(t+x-y)]dH(y),\ t \geq 0,\ x \geq 0$$

by arguing that $Y(t) \leq x$ means that there is at least one renewal point in the interval $(t, t+x]$. If the largest such point is S_n, then

$$P(Y(t) \leq x) = \sum_{n=1}^{\infty} P(t \leq S_n \leq t+x \leq S_{n+1}).$$

Exercise 4.7 Prove that Residual life time $Y(t)$ and spent life time $Z(t)$ have joint distribution given by

$$P(Y(t) > x,\ Z(t) \leq z) = \int_{t-z}^{t}[1-F(t+x-y)]dH(y) \text{ for } t \geq 0 \text{ and } 0 < z < t.$$

Exercise 4.8 (Uniform Renewal Process) Let $V_1, V_2, \ldots$ be a sequence of i.i.d. r.v.s with $EV_i = 0,\ EV_i^2 = \sigma^2$.

Assume also that $\{T_n\}$ and $\{V_n\}$ are independent and F_1 is the distribution function of T_1. Define a stochastic process $\{X(t), t \geq 0\}$ as follows.

If $0 \leq t \leq T_1$ let $X(t) = V_1$, if $T_1 + \ldots + T_{n-1} \leq t \leq T_1 + \ldots + T_n$ let $X(t) = V_{n+1}$ $(n = 1, 2, \ldots)$. Show that $\{X(t)\}$ is stationary with covariance function $K(t) = \sigma^2(1 - F_1(t))$.

Also assume that $F_1 = \frac{1}{m}\int_0^t (1-F(x)dx$ and m is the expectation of F, the common d.f. of $T_2, T_3, \ldots$.

Exercise 4.9 Suppose that $EV_1^r < \infty$ for some $r > 1$ and the renewal process is arithmatic. Let $Y(t) = S_{N(t+1)} - t$ be the residual lifetime. Show that

(a) $E(Y(t))^{r-1} \to \frac{1}{r\mu}EX_1^r$ as $t \to \infty$

(b) $E(Y(t)) \to \frac{\sigma^2+\mu^2}{2\mu}$ as $t \to \infty$, where $EX_1 = \mu$ and var $(X_1) = \sigma^2 < \infty$.

Exercise 4.10 We say that the random walk (or renewal process) is *arithmetic* if the common distribution function F has its support on $\{0, \pm d, \pm 2d, \ldots\}$ (or $\{0, d, 2d, \ldots\}$) for some $d > 0$. The largest d with this property is called the span. A random walk or renewal process which is arithmetic with span d is also called *d-arithmetic*.

If the renewal process is d-arithmetic, then show that $u_n = f_n + \sum_{k=0}^{n} u_{n-k} f_k$ where

$$u_n = \sum_{j=1}^{\infty} P(S_j = kd)\ (k \geq 0) \text{ and } f_k = P(X_1 = kd).$$

Exercise 4.11 Let $0 < \mu = EX_1 \le \infty$. Then show that for renewal counting process $N(t)$,

$$E(N(t)/t)^r \to \frac{1}{\mu^r} \text{ as } t \to \infty \text{ all } r > 0.$$

Exercise 4.12 Suppose that $0 < \mu = EX_1$ and $\sigma^2 = Var(X_1) < \infty$. Show that

(a) $E(N(t)) = t/\mu + 0(\sqrt{t})$ as $t \to \infty$

(b) $\text{Var}(N(t)) = \dfrac{t\sigma^2}{\mu^3} + o(t)$ as $t \to \infty$.

Exercise 4.13 Let $Y_t = S_{N(t)+1} - t$ be the excess life at time t in a renewal process and $Z_t = t - S_{N(t)}$ be the spent life.

Then show that

$$\lim_{t\to\infty} P(Y_t > z) = \mu^{-1} \int_z^\infty \{1 - F(y)\}dy,\ z > 0, \text{ where } \mu = E(X_1)$$

and

$$\lim_{t\to\infty} P(Z_t > y, Y_t \ge x) = \mu^{-1} \int_{x+y}^\infty \{1 - F(z)\}d(z).$$

Also show that the total life $S_t = Y_t + Z_t$ has limiting distribution

$$\lim_{t\to\infty} P(S_t \le x) = \mu^{-1} \int_0^x y\, dF(y).$$

Exercise 4.14 (Replacement policy) If $f(x)$ is a p.d.f. associated with a lifetime d.f. $F(x)$, the *hazard rate* is $r(x) = f(x)/[1-F(x)]$. Suppose Y_i is the cost of the ith replacement under age replacement policy where a planned replacement at age T costs C_1 dollars, while a failure replaced at time $x < T$ costs C_2 dollars. Show that the long run mean cost per unit time is given by

$\theta(T) = C_1[1 - F(T)] + C_2F(T) \,/ \displaystyle\int_0^\infty [1 - F(x)]dx$ and also show that the replacement age T^* that minimizes $\theta(T)$ must satisfy

$$r(T^*) \int_0^{T^*} [1 - F(x)]dx - F(T^*) = \frac{C_1}{C_1 - C_2}.$$

Exercise 4.15 Let $N_1(t)$ and $(N_2(t)$ be two independent renewal processes with the same interoccurrence distribution F with mean μ. Let $N(t) = N_1(t) + N_2(t)$. If $N(t)$ is also a renewal process, then show that $N_1(t)$, $N_2(t)$ and $N(t)$ are all Poisson.

5

Branching Process

5.1 Introduction and History

Branching process (popularly known as Galton-Watson process) dates back to 1874 when a mathematical model was formulated by Galton and Watson for the problem of "Extinction of families". The model did not attract for a long time but during the last 40 years much attention has been devoted to it.

An important class of Markov processes with countably many states is the class of branching processes. Let us suppose that we are observing a physical system consisting of a finite number of particles either of the same type or of several different types. With the passage of time each particle can disappear or turn into a group of new particles, independently of other particles. Phenomena described by such a scheme are frequently encountered in natural science and technology, sociology and demography, for example, showers of cosmic rays, growth of large organic cells, the development of biological populations and the spread of epidemics. All of these processes are characterized by the same property, namely, their development has a branching form. A precise definition of processes of this type within the frame-work of the theory of Markov process brings us to the concept of a branching process. In fact, the mathematical model of such kind of emperical processes lead to the idea of a branching process.

The theory of branching random processes is a rapidly developing area of general stochastic processes. A great number of papers and the books of Harris (1963), Jagers (1975), Athreya and Ney (1972), Assmussen and Hering (1983) and other books are some of the results of this growth and they show the development of the theory. The interest in this theory is connected with its applications to a wide spectrum of practical problems. They include the description of various biological populations, the investigation of the transformation processes of particles in nuclear reactors, the investigation of cascade processes, of chemical processes, problems of queueing theory, the theory of graphs and other problems.

Suppose we start with an initial set of objects (or individuals) which form the 0th generation of these objects called *ancestors*. The offsprings reproduced or the objects generated by the objects of the 0th generation are the "direct descendents" of the ancestors and are said to form the first generation; the objects generated by those of the first generation form the second generation, and so no.

5.2 Properties of Generating Functions

Mathematically, we have a sequence of i.i.d. random variables $\{\xi_i\}$ with

$$P[\xi = k] = p_k \geq 0, \sum_{k=0}^{\infty} p_k = 1 \text{ and } X_{n+1} = \sum_{i=1}^{X_n} \xi_i, n = 0, 1, 2, \ldots$$

Here X_n is the number of elements in the nth generation and $\{X_n, n = 0, 1, 2, \ldots\}$ is a discrete parameter integer valued Markov chain with transition probability $p_{ik} = P[X_{n+1} = k | X_n = i]$.

We also assume $P[X_0 = 1] = 1$.

Let $\varnothing(z) = \sum_{k=1}^{\infty} p_k z^k$ be the probability generating function of ξ_i, $\varnothing_0(z) = z$ and $\varnothing_1(z) = \varnothing(z)$.

Define $$\varnothing_n(z) = \sum_{k=1}^{\infty} P[X_n = k] z^k, n = 0, 1, 2, 3, \ldots$$

The nth iterate of $\varnothing(z)$ is $\underbrace{\varnothing[\varnothing[\ldots[\varnothing(z)]\ldots]}_{n\text{-times}}$

Theorem 5.1

(i) $\varnothing_{n+1}(z) = (\varnothing_n(\varnothing(z))$ (5.1)

and (ii) $\varnothing_{n+1}(z) = \varnothing(\varnothing_n(z))$ (5.2)

Proof

(i) $$\varnothing_{n+1}(z) = \sum_{k=0}^{\infty} P[X_{n+1} = k] z^k$$

$$= \sum_{k=0}^{\infty} \left(\sum_{l=0}^{\infty} P[X_{n+1} = k | X_n = l] P[X_n = l] \right) z^k$$

$$= \sum_{l=0}^{\infty} P[X_n = l] \sum_{k=0}^{\infty} P[\xi_1 + \xi_2 + \ldots + \xi_1 = k] z^k$$

(interchanging summations by Fubini Theorem)

$$= \sum_{l=0}^{\infty} P[X_n = l] \left(\sum_{k=0}^{\infty} P[\xi = k] z^k \right)^l \quad (\xi_i\text{'s are i.i.d. r.v.s})$$

$$= \sum_{l=0}^{\infty} P[X_n = l][\varnothing(z)]^l$$

$$= \varnothing_n(\varnothing(z)).$$

Also $$\varnothing_0(z) = P[X_0 = 1]z = z \text{ for } n = 0$$

$$\varnothing_1(z) = \sum_{l=0}^{\infty} P[X_0 = l] \varnothing(z) = \varnothing(z)$$

(ii) By (5.1), $\varnothing_{n+1}(z) = \varnothing_n(\varnothing(z)) = \varnothing_{n-1}[\varnothing(\varnothing(z))]$

$= \varnothing_{n-1}(\varnothing_2(z))$

$= \ldots = \varnothing[\varnothing_n(z)]$, by induction on n.

Note In fact for any $k = 0, 1, 2, \ldots, n$

$$\varnothing_{n+1}(z) = \varnothing_{n-k}(\varnothing_{k+1}(z)).$$

Let $E(\xi_1) = EX_1 = m.$

Corollary 5.1

$$E(X_n) = m^n \tag{5.3}$$

$$\text{Var}(X_n) = \begin{cases} \sigma^2 m^{n-1} \dfrac{m^n - 1}{m - 1} & \text{if } m \neq 1 \\ n\sigma^2 & \text{if } m = 1, \end{cases} \tag{5.4}$$

where $\sigma^2 = \text{Var}(\xi) = \text{Var}(X_1)$.

Proof From $\varnothing_{n+1}(z) = \varnothing_n(\varnothing(z))$, we get

$$\varnothing'_{n+1}(1) = E(X_{n+1}) = \varnothing'_n(\varnothing(1))\varnothing'(1) \tag{5.5}$$

But $\varnothing(1) = \sum_{k=0}^{\infty} p_k = 1$ and $\varnothing'(1) = m = E\xi = E(X_1)$.

Hence $$E(X_{n+1}) = \varnothing'_n(1)m = E(X_n)m = \ldots$$

$$= m^{n+1}E(X_0) = m^{n+1}, \text{ since } P(X_0 = 1) = 1.$$

Note If we assume $P(X_0 = n_0) = 1$, then $EX_n = n_0 m^n$.

We also get from $\varnothing_{n+1}(z) = \varnothing(\varnothing_n(z))$,

$$\varnothing'_{n+1}(1) = \varnothing'(\varnothing_n(1))\varnothing'_n(1)$$

$$\varnothing''_{n+1}(1) = \varnothing''(\varnothing_n(1))[\varnothing'_n(1)]^2 + \varnothing'(\varnothing_n(1))\varnothing''_n(1)$$

$$= \varnothing''(1)(m^n)^2 + m\varnothing''_n(1) \quad (\because \varnothing_n(1) = 1, \varnothing'_n(1) = m^n)$$

$$= m\varnothing''_n(1) + Km^{2n}, \quad \text{where } K = \varnothing''(1).$$

Thus, we have $$Q''_{n+1}(1) = m\varnothing''_n(1) + Km^{2n} \tag{5.6}$$

$$m\varnothing''_n(1) = m^2\varnothing''_{n-1}(1) + Km^{(2n-1)}$$

$$\ldots$$

$$m^{n-1}\varnothing''_n(1) = m^n\varnothing''(1) + Km^{2n-(n-1)}$$

Adding both sides of the above equations, we get

$$\varnothing''_{n+1}(1) = m^n K + Km^{n+1} + \ldots + Km^{2n}$$

$$= m^n K(1 + m + \ldots + m^n)$$

$$= m^n K \frac{m^{n+1} - 1}{m - 1} \text{ if } m \neq 1.$$

Now $\quad \text{Var}(X_{n+1}) = \varnothing''_{n+1}(1) + \varnothing'_{n+1}(1) - (\varnothing'_{n+1}(1))^2$

$$= m^n K \frac{m^{n+1} - 1}{m - 1} + m^{n+1} - m^{2(n+1)} \text{ if } m \neq 1$$

$$= \frac{m^n}{m-1}[K m^{n+1} - K + m^2 - m - m^{n+3} + m^{n+2}]$$

$$= \sigma^2 \frac{m^{n+1} - 1}{m - 1} m^n \text{ if } m \neq 1.$$

$(\because \varnothing''(1) = K = \text{Var}(X) - m + m^2 = \sigma^2 - m + m^2)$

$$\varnothing''_{n+1}(1) = \varnothing''_n(1) + K \quad \text{(from (5.6), putting } m = 1)$$

If $m = 1$, $\varnothing''_{n+1}(1) = (n+1)K$.

Also
$$\text{Var}(X_n) = \varnothing''_n(1) + \varnothing'_n(1) - (\varnothing'_n(1)^2)$$
$$= nK + m^n - m^{2n} = nK \text{ if } m = 1$$

Hence $\text{Var}(X_n) = n\sigma^2$ if $m = 1$.

Note If $m > 1$, the population increases (i.e. explodes). This case is called supercritical case.

If $m < 1$, the population dies out and the case is called subcritical.

If $m = 1$, the case is called critical.

Also note that

if $m < 1$, $\text{Var}(X_n) \to 0$ as $n \to \infty$ and if $m > 1$, $\text{Var}(X_n) \to \infty$ as $n \to \infty$.

Alternative Proof of (5.3) *and* (5.4)

Now the number of members in $(n+1)^{\text{th}}$ generation can be written as $X_{n+1} = \sum_{i=1}^{X_n} \xi_i$, where ξ_i's are i.i.d r.vs. By Wald's Identity $E(X_{n+1}) = E(\xi_1)(EX_n) = mEX_n$ and the rest is the same as before.

Exercise Show that (i) $\text{Var}(X) = E(\text{Var}(X|Y)) + \text{Var}(E(X|Y))$ and

(ii) $\text{Var}\left(\sum_{i=1}^{N} X_i\right) = (EN)\,\text{Var}(X) + (EX)^2\,\text{Var}(N)$ if N is independent of X_{n+1}, $X_{n+2}, \ldots$.

This is a particular case of Wald's Identity for second moment and follows from (i)

Now
$$\text{Var}(X_{n+1}) = E(X_n)\,\text{Var}(\xi_1) + (E\xi_1)^2\,\text{Var}(X_n)$$
$$= m^n\sigma^2 + m^2\,\text{Var}(X_n) \tag{5.7}$$

This is a first order difference equation.

A particular solution of (5.7) is

$$\text{Var}(X_n) = \frac{\sigma^2 m^n}{m - m^2} \text{ if } m \neq 1.$$

A general solution of the difference equation (5.7) is given by $\mathrm{Var}(X_n) = A(m^2)^n$ and a complete solution is given by

$$\mathrm{Var}\,(X_n) = A(m^2)^n + \frac{\sigma^2 m^n}{m - m^2} \quad \text{for } n \geq 1 \text{ and } \mathrm{Var}(X_1) = \sigma^2$$

gives $A = \sigma^2/m(m-1)$.
Therefore

$$\mathrm{Var}(X_n) = \frac{m^{n-1}(m^n - 1)}{m - 1}\sigma^2 \quad \text{for} \quad n \geq 1 \text{ if } m \neq 1$$

and $\qquad \mathrm{Var}(X_n) = n\sigma^2$ if $m = 1$ (from (5.7)).

5.3 Probability of Extinction

Let $P[X_n = 0] = q_n$. Assume $p_0 > 0$.

We are interested to know when the population dies out, i.e. to know $\lim_{n\to\infty} q_n$.

Lemma 5.1 q_n is non-decreasing, $\lim_{n\to\infty} q_n = \pi$ exists and $0 < \pi \leq 1$. Moreover π is the smallest positive root of the equation $\varnothing(z) = z\ (0 < z \leq 1)$.

Proof $q_0 = 0$, since $P[X_0 = 1] = 1$ (by assumption).

$$\varnothing_n(z) = \sum_{k=0}^{\infty} P[X_n = k]z^k \quad \text{and} \quad \varnothing_n(0) = P[X_n = 0] = q_n.$$

Since $\qquad \varnothing_{n+1}(0) = \varnothing(\varnothing_n(0)),\ q_{n+1} = \varnothing(q_n),\ n = 0, 1, \ldots \qquad (5.8)$

Claim

$$\begin{aligned} q_1 &= \varnothing(q_0) > 0 \\ &= \varnothing(0) = p_0 > 0 \quad \text{(by assumption)} \end{aligned}$$

Now $q_0 = 0$ and hence $q_1 > q_0$.

Suppose $q_{n-1} \leq q_n$. Then since $\varnothing(z)$ is increasing in $z (0 \leq z \leq 1)$, we have

$$\varnothing(q_{n-1}) \leq \phi(q_n) \qquad (5.9)$$

or $q_n \leq q_{n+1}$ (by (5.8) and (5.9)) Hence q_n is nondecreasing in n.

$$q_{n+1} = \varnothing(q_n) \leq \varnothing(1) = 1 \text{ for all } n = 1, 2, 3, \ldots$$

Therefore $\lim_{n\to\infty} q_n = \pi$ (say) exists.

Now $q_1 > 0$ and hence $q_n > 0$ for all n and $\{q_n\}$ is nondecreasing and $\pi > 0$. We have $q_{n+1} = \varnothing(q_n)$. Taking limit $n \to \infty$ on both sides, we get

$$\begin{aligned} \pi = \lim_{n\to\infty} q_{n+1} &= \lim_{n\to\infty} \varnothing(q_n) \quad (\because \varnothing(z) \text{ is continuous}) \\ &= \varnothing(\pi). \end{aligned}$$

Now $\varnothing(z) = z \Rightarrow \pi$ is positive and hence π is the positive root of $\varnothing(z) = z$. Assume that S_0 is another positive root of $\varnothing(z) = z$. Then $q_0 = 0 < s_0$. Assume $q_n < s_0$ and which implies

$$q_{n+1} \leq s_0 \quad (\because q_{n+1} = \varnothing(q_n) \leq \varnothing(s_0) = s_0)$$

Now $\varnothing_n(0) = P[X_n = 0] = q_n < 1$.

By induction on n, $q_n < s_0$ for all $n = 0, 1, 2, \ldots$. Hence, $\lim_{n\to\infty} q_n = \pi \le s_0$. Therefore π is the smallest positive root of $\varnothing(z) = z$.

Lemma 5.2 $\varnothing(z) = z$ has a real root in $0 < z < 1$ iff $m > 1$.
$\varnothing(z) = z$ has only $z = 1$ as root in $0 < z \le 1$ iff $m \le 1$.

Before giving the proof of lemma 5.2 let us give an interesting example illustrating lemma 5.2.

Example 5.1 $P[\xi = 0] = \mu > 0, P[\xi = 2] = \lambda > 0, P[\xi = 1] = 1 - \mu - \lambda > 0.$

Then $$\varnothing(z) = \mu + (1 - \mu - \lambda)z = \lambda z^2$$

To find the solution of $\varnothing(z) = z$.

$$\mu + (1 - \mu - \lambda)z + \lambda z^2 = z$$

$$\Rightarrow \quad \mu - (\mu + \lambda)z + \lambda z^2 = 0$$

$$\Rightarrow \quad (z - 1)(\lambda z - \mu) = 0 \Rightarrow z = 1 \text{ or } \mu/\lambda.$$

Then $m = \varnothing'(1) = 1 - \mu - \lambda + 2\lambda = 1 - \mu + \lambda$.
Therefore $m > 1 \Rightarrow 1 - \mu + \lambda > 1$ or $-\mu + \lambda > 0$ or $\lambda > \mu$.

Therefore, if $m > 1$, i.e. $\lambda > \mu$, the population dies out with probability $\pi = \mu/\lambda < 1$ and if $m \le 1$, i.e. $\lambda \le \mu$ i.e. $\mu/\lambda \ge 1$, then $\pi = 1$ i.e. the population is sure to die out (by Lemma 5.2).

Proof of Lemma 5.2 Now G.F. $\varnothing(z) = \sum_{k=0}^{\infty} p_k z^k$.

Then $f(z) = \dfrac{\varnothing(z)}{z} = \dfrac{p_0}{z} + p_1 + p_2 z + p_3 z^2 + \ldots (0 < z < 1)$.

Since this is a power series which is uniformly convergent

$$f'(z) = -\frac{p_0}{z^2} + p_2 + 2p_3 z + \ldots$$

and $$f''(z) = \frac{2p_0}{z^3} + 2p_3 + \ldots > 0 (0 < z \le 1).$$

This $$\Rightarrow f'(z) \uparrow z \text{ and } f'(0+) = -\infty.$$

Case 1 $m > 1$

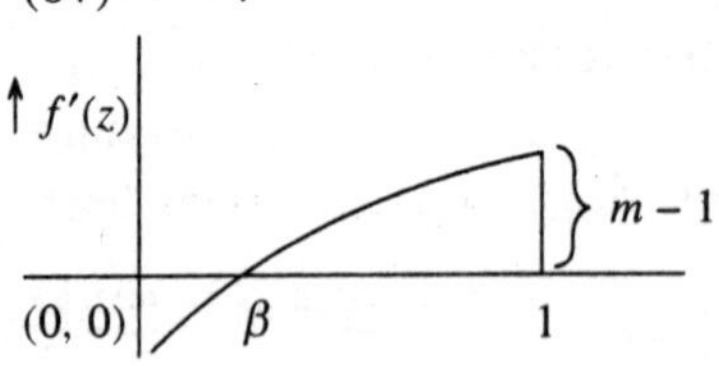

$$f'(1) = -p_0 + p_2 + 2p_3 + 3p_4 + \ldots$$

$$0 = p_0 + p_1 + p_2 + \ldots -1$$

$$f'(1) = p_1 + 2p_2 + 3p_3 + 4p_4 + \ldots -1$$

$$= m - 1 > 0.$$

Thus, there exists a $0 < \beta < 1$ such that $f'(\beta) = 0$ and which $\Rightarrow f'(z) < 0$ for $0 < z < \beta$. Hence $f(z) \downarrow$ in z in $0 < z < \beta$ and $f(z) \uparrow$ in z for $\beta < z < 1$.

Now $$f(1) = \frac{\varnothing(1)}{1} = \frac{1}{1} = 1.$$

$f(0+) = \infty$.

Therefore there exists $0 < \alpha < 1$ such that $f(\alpha) = 1$. i.e. $\varnothing(\alpha) = \alpha$.

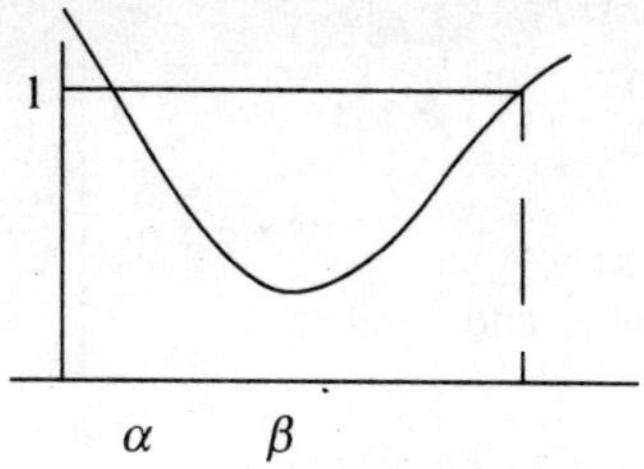

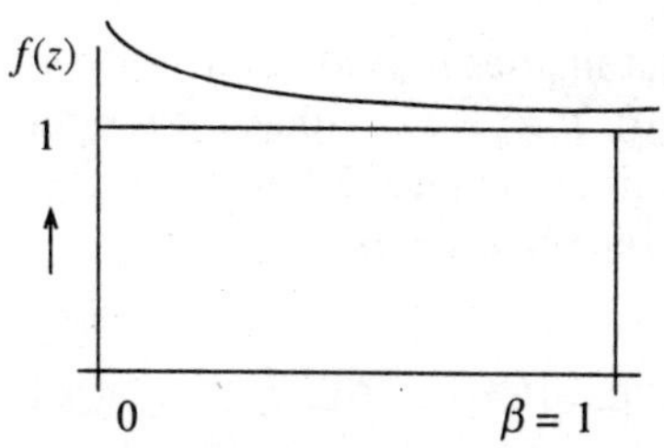

Case 2 $m \le 1$
Then $\beta = 1$. Hence
$f'(z) < 0$ in z for $0 < z < 1$ or $f(z) \downarrow$.
$f(0+) = \infty$ $f(1) = 1$.

Hence, there does not exist $0 < \alpha < 1$ such that $f(\alpha) = 1$ and hence $z = 1$ is the only root of $\varnothing(z) = z$.

Theorem 5.2 If $m > 1$, then with probability π, the population become extinct and with probability $1 - \pi$, the population explodes. If $m \le 1$, then with probability 1, the population becomes extinct. [Regardless of the actual value of the mean $EX_1 = m > 1$, the probability that the nth generation will consist of any positive finite number of individuals tends to zero as $n \to \infty$, with probability of extinction tending to π. In these circumstances, we say that $X_n \to 0$ as $n \to \infty$ with probability $1 - \pi$].

Proof $q_n = P[X_n = 0]$,

$$\varnothing_n(z) = q_n + \sum_{k=1}^{\infty} P(X_n = k) z^k, \; |z| \le 1.$$

Then
$$\lim_{n\to\infty} \varnothing_n(z) = \pi + \sum_{k=1}^{\infty} P[X_\infty = k] z^k \tag{5.10}$$

(by Lebesgue dominated convergence theorem)

Now
$$P(X_n = k) = E(I_{[X_n = k](\omega)}) = \int_{[X_n = k]} dP$$

Claim $\lim_{n\to\infty} \varnothing_n(z) = \pi$.

Case (1) Suppose $0 < z < \pi$. Then by Lemma 5.1. $\varnothing(\pi) = \pi$ and therefore, $\varnothing_n(\pi) = \varnothing(\varnothing_{n-1}(\pi)] = \pi$ (by Theorem 5.1).

Now $\pi = \lim_{n\to\infty} q_n = \lim_{n\to\infty} \varnothing_n(0) \le \lim_{n\to\infty} \varnothing_n(z)$ $(\because \varnothing_n(z) \uparrow \text{ in } z)$

$$\le \lim_{n\to\infty} \varnothing_n(\pi) = \pi.$$

Therefore, if $0 < z < \pi$, $\lim_{n\to\infty} \varnothing_n(z) = \pi$.

Case (2) $\pi < z < 1$. Suppose $\varnothing_n(z) \to \gamma \ne \pi$ as $n \to \infty$.

Therefore, $\lim_{n\to\infty} \varnothing_{n+1}(z) = \lim_{n\to\infty} \varnothing(\varnothing_n(z)) = \varnothing(\lim \varnothing_n(z))$ and hence $\gamma = \varnothing(\gamma)$.

If $\pi < \gamma < 1$, then $f(\gamma) < 1$

i.e. $\varnothing(\gamma) < \gamma \left(\therefore \dfrac{\varnothing(\gamma)}{\gamma} = f(\gamma) < 1 \right)$

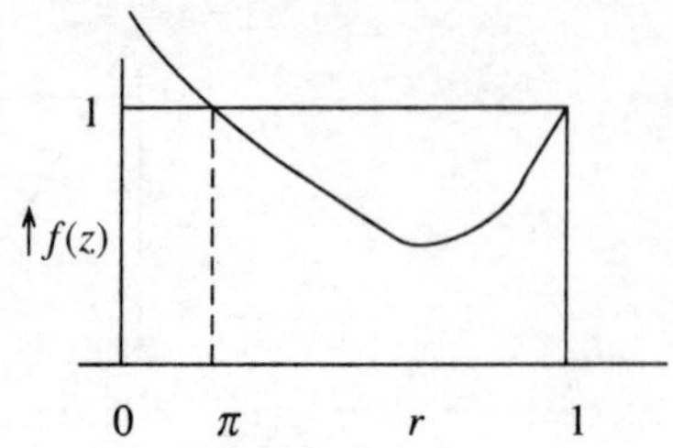

is a contradiction to $\varnothing(\gamma) = \gamma$.

If $0 < \gamma < \pi$, $\varnothing(\gamma) > \gamma$ and again a contradiction arises.

Therefore $\lim_{n\to\infty} \varnothing_n(z) = \pi$, i.e. $\varnothing_n(z)$ converges to a constant function, viz. π and that implies $q_n = P(X_n = 0) \to \pi$ as $n \to \infty$ and all other coefficients of z^k in (5.10), $P(X_n = k) \to 0$ as $n \to \infty$ for $k = 1, 2, \ldots$ and the rest of the proof follows from Lemmas 5.1 and 5.2. Let T be the time till extinction (first passage time to state 0).

Then $[T = n] = [X_n = 0, X_{n-1} \neq 0]$.

Hence $$P[T = n] = q_n - q_{n-1} = P[X_n = 0] - P[X_{n-1} = 0]$$

$(\because X_{n-1} = 0 \Rightarrow X_n = 0$, i.e. $[X_{n-1} = 0] \subset [X_n = 0])$

Hence $$E(T) = \sum_{n=1}^{\infty} n[q_n - q_{n-1}]$$

$$= q_1 - q_0 + 2q_2 - 2q_1 + 3q_3 - 3q_2 + \ldots \tag{5.11}$$

$$= -q_0 - q_1 - q_2 - q_3 - \ldots < 0 \quad \text{which is absurd.}$$

(This is because the series $\sum_{n=1}^{\infty} n(q_n - q_{n-1})$ is convergent but not absolutely convergent so that a rearrangement of the series, as done in (5.11), is not possible.)

Now $\lim_{n\to\infty} q_n = \pi > 0 \Rightarrow q_n = d > 0$. Therefore, $\sum_n q_n$ is divergent.

Example 5.2 $\varnothing(z) = p_0 + p_1 z$, $0 < p_0 < 1$, and $p_0 + p_1 = 1$.

Then $\varnothing_2(z) = p_0 + p_1(p_0 + p_1 z) = p_0 + p_0 p_1 + p_1^2 z$.

Thus $\varnothing_n(z) = p_0 + p_0 p_1 + p_0 p_1^2 + \ldots + p_0 p_1^{n-1} + p_1^n z$ (by iteration)

$$= p_0 \frac{1 - p_1^n}{1 - p_1} + p_1^n z$$

$$= (1 - p_1^n) + p_1^n z$$

Hence $q_n = P(X_n = 0) = 1 - p_1^n \to 1$ as $n \to \infty$.

Also $$\varnothing(z) = z = p_0 + p_1 z \Rightarrow z(1 - p_1) = p_0 \Rightarrow z = 1$$

i.e. the population is sure to die out. Here

$$P(T = n) = q_n - q_{n-1} = 1 - p_1^n - 1 + p_1^{n-1} = p_1^{n-1}(1 - p_1) = p_0 p_1^{n-1}$$

i.e. the population is alive with geometric distribution.

$$E(T) = \sum_{n=1}^{\infty} n p_1^{n-1} p_0 = p_0 \sum_{n=1}^{\infty} n p_1^{n-1}$$

$$= p_0 \frac{1}{(1-p_1)^2} = \frac{p_0}{p_0^2} = \frac{1}{p_0}.$$

Example 5.3 It is frequently assumed that the problem of a mutant gene having k distinct discordants ($k = 0, 1, 2, \ldots$) is governed by a Poisson distribution with mean $\lambda = 1$.

Then $\varnothing(z) = e^{z-1}$ and $\pi = 1$, because

$$\varnothing(z) = \sum_{k=0}^{\infty} \frac{e^{-\lambda}\lambda^k}{k!} z^k = e^{-\lambda(1-z)}$$

$\varnothing(z) = z$ has the explicit solution for $\lambda < 1$ though for $\lambda > 1$, the extinction is proved with probability $0 < \pi < 1$ (iff $\lambda > 1$). If $\lambda = 2$, then $z = e^{2(z-1)}$ has a root $0.2 < 1$ and $\pi = 0, 2$.

The graph of $\varnothing(z)$ in $0 \le z < 1$ between $(0, e^{-\lambda})$ and $(1, 1)$ $\varnothing(z)$ is convex and that the curve $y = \varnothing(z)$ always lies above $y = z$ when $\lambda = 1$. There is no other root of $z = \varnothing(z)$ except 1 in [0, 1].

Hence the probability of extinction is 1.

Theorem 5.3 $E(X_{n+r}|X_n) = m^r X_n$. (5.12)

Proof We first prove the case of $r = 1$.

$$E(X_{n+1}|X_n) = E[\sum_{j=1}^{x_n} \xi_j | x_n] = (X_n)E(\xi_j) = mX_n \quad (5.13)$$

($\because$ ξ_1 are i.i.d. with $E(\xi_1) = m$).

Assume (5.12) is truc for r. We shall prove that it holds for $r + 1$.

Now $E(X_{n+r+1}|X_n) = E(E[X_{n+r+1}, |X_{n+r}, X_{n+r-1} \ldots X_n]|X_n)$

$= E(E[X_{n+r+1}|X_{n+r}]|X_n)$ ($\because$ X_n is Markov)

$= E(X_{n+r}m|X_n) = mE(X_{n+r}|X_n)$ ($\because$ X_n is stationary, apply (5.13))

$= m^{r+1}X_n$.

Therefore, by induction on r we have $E(X_{n+r}|X_n) = m^r X_n$ for $r = 0, 1, 2, \ldots$.

Now consider the new process

$$W_n = \frac{X_n}{m^n}, \; n = 0, 1, 2, \ldots$$

Claim W_n is a Martingale, i.e.

$$E(W_{n+r}|W_n, W_{n-1}, \ldots, W_1, W_0) = W_n \text{ a.s. for } r = 0, 1, 2, \ldots$$

Proof

$$E(W_{n+r}|W_n) = \frac{1}{m^{n+r}} E(X_{n+r}|X_n)$$

$$= \frac{1}{m^{n+r}} X_n m^r = W_n \text{ a.s.}$$

and by Markov property $E(W_{n+r}|W_n, \ldots, W_1, W_0) = E(W_{n+r}|W_n)$ a.s. (for Martingale or fair game or expectation constant process, see author's book "An introduction to measure theoretic probability," Prentice Hall, New Delhi).

Now $\quad EW_0 = EW_r = 1$ and $E(W_n) = 1$ for all $n = 0, 1, 2, \ldots$

Also $\quad EW_n^2 = \dfrac{EX_n^2}{m^{2n}} = \dfrac{1}{m^{2n}}\left[m^{2n} + \dfrac{m^{n-1}(m^n - 1)}{m - 1}\sigma^2\right]$

$$= 1 + \frac{\sigma^2}{m(m-1)}(1 - m^{-n}) \to 1 + \frac{\sigma^2}{m(m-1)} \text{ if } m \neq 1$$

and $V(W_n) = \dfrac{\sigma^2}{m(m-1)}(1 - m^{-n}) \to \dfrac{\sigma^2}{m(m-1)}$ if $m \neq 1$.

If $m > 1$, the limiting variance of $W_n < \sigma^2$.

5.4 Distribution of Total Number of Progeny

Let X_n denote the size of the nth generation $n = 0, 1, 2, \ldots$ and $X_0 = 1$.

Then the random variable $\quad Y_n = \sum_{k=0}^{n} X_k = 1 + X_1 + \ldots + X_n$

denotes the total number of progeny, i.e. the total number of descendents upto and including the nth generation and also including the ancestors.

Theorem 5.4 The G.F. $R_n(s)$ of Y_n satisfies the recursion relation

$$R_n(s) = s\varnothing(R_{n-1}(s)), \tag{5.14}$$

$\varnothing$ being the G.F. of the offspring distribution.

Proof Let $Z_n = X_1 + X_2 + \ldots + X_n$ and $G_n(s)$ be its G.F.
Then $R_n(s) = sG_n(s)$ ($Y_n = 1 + Z_n$ and G.F. of 1 is s).

We have $G_n(s) = \sum_{k=0}^{\infty} P(Z_n = k)s^k$.

$P(Z_n = k) = \sum_{i=0}^{\infty} P$ [total number of descendents in the succeeding $(n - 1)$ generations following the first generation is $k-i$ given $X_1 = i$] $P[X_1 = i]$.

If the process starts with one ancestor then the probability of having r descendents in succeeding m generations is the coefficient of s^r in $G_m(s)$ and if it starts with i ancestors then the probability of having r descendents in the succeeding m generations will be the coefficient of s^r in $[G_m(s)]^i$.

Thus $\quad P(Z_n = k) = \sum_{i=0}^{\infty}$ coefficient of s^{k-i} in $\{G_{n-1}(s)\}^i p_i$

$$= \sum_{i=0}^{\infty} p_i \text{ (coefficient of } s^k \text{ in } \{sG_{n-1}(s)\}^i)$$

$$= \text{coefficient of } s^k \text{ in } \sum_{i=0}^{\infty} p_i \{sG_{n-1}(s)\}^i$$

$$= \text{coefficient of } s^k \text{ in } \varnothing(sG_{n-1}(s)).$$

Therefore $\quad G_n(s) = \sum_{k=0}^{\infty} P(Z_n = k)s^k = \varnothing(sG_{n-1}(s))$

and $\quad R_n(s) = s\varnothing(sG_{n-1}(s)) = s\varnothing(R_{n-1}(s)).$

Theorem 5.5 $\quad \lim_{n\to\infty} R_n(s) = H(s) = \sum_{k=0}^{\infty} p_k s^k$, where $p_k \ge 0$ (5.15)

and $H(s)$ satisfies the relation $H(s) = s\varnothing(H(s))$, $0 < s < 1$.

Further $H(s)$ is the unique root of the equation $t = s\varnothing(t)$ such that $H(s) \le y$, where y is the smallest positive root of $x = \varnothing(x)$ and $H(1) = \sum_{k=0}^{\infty} p_k = y$.

Proof For $0 < s < 1$, $R_1(s) = s\varnothing(s) < s$ ($\because \varnothing(s) < 1$).
$R_2(s) = s\varnothing(R_1(s)) < s\varnothing(s) = R_1(s)$. Assuming $R_m(s) < R_{m-1}(s)$, we get

$$R_{m+1}(s) = s\varnothing(R_m(s)) < R_m(s) \text{ if } 0 < s < 1.$$

Hence by induction on n, $R_n(s) < R_{n-1}(s)$ for all $n = 1, 2, \ldots$.

Thus for $0 < s < 1$, $\{R_n(s)\}$ is a monotone decreasing sequence of functions, and bounded below, and $\lim_{n\to\infty} R_n(s) = H(s)$, say, exists. From Continuity Theorem of G.F.s, it follows that $H(s)$, being the limit of a sequence of G.F.s is the G.F. of a sequence of non-negative numbers p_k such that $H(1) = \sum_{k=0}^{\infty} p_k \le 1$. Taking limit $n \to \infty$ in equation (5.14) of Theorem 5.4 we get $H(s) = s\varnothing(H(s))$. Since $\varnothing$ is continuous, $0 < s < 1$ for some fixed $s(0 < s < 1)$, $H(s)$ is a root of the equation $t = s\,\varnothing(t)$. For fixed $s < 1$, $y = s\varnothing(t)$ is a convex function of t and the graph of $y = s\varnothing(t)$ intersect the line $y = t$ in at most two points. Let θ be the smallest positive root of the equation $x = \varnothing(x)$. Then clearly $\theta \le 1$. The function $t - s\varnothing(t)$ is negative for $t = 0$ and positive for values of t bewteen θ and 1. Thus $t = s\varnothing(t)$ has exactly one root between θ and 1. The unique root of $t = s\varnothing(t)$ equals $H(s)$ and $H(s) \le \theta$.

Clearly $H(1)$ is a root of $x = \varnothing(x)$ and since θ is the smallest root of this equation, $H(1) = \theta$.

Note $H(1) = 1$ or < 1 depending on whether $1 + X_1 + \ldots + X_n + \ldots$ is finite or not with probability 1 and $H(1) = 1$ whenever $m \le 1$ (i.e. the population is sure to die out).

Some well known branching models

1. Lotka Model (1939): Here

$$p_k = bc^{k-1},\ k = 1, 2, \ldots,\ 0 < b, c;\ b + c < 1$$

and $p_0 = 1 - \sum_{k=1}^{\infty} p_k$.

Then $$m = \frac{b}{(1-c)^2} \text{ and P.G.F. } \varnothing(z) = \sum_{k=0}^{\infty} p_k z^k$$

$$= 1 - \frac{b}{1-c} + \frac{bz}{1-cz}$$

The quadratic equation $z = \varnothing(z)$ has roots 1 and $\dfrac{1-(b+c)}{c(1-c)} = z_0$ (say).

If $m = 1$, $z_0 = 1$, and the extinction probability is 1.
If $m > 1$, $z_0 < 1$, and the extinction probability is $\pi = z_0 < 1$.
The 1920 US census figures give $b = 0.2126$, $c = 0.5893$, $(m \simeq 1.25 > 1)$, $\pi = z_0 = 0.819$.

2. Suppose $\varnothing(z)$ is of the form $\varnothing(z) = \dfrac{\alpha + \beta z}{\gamma + \delta z}$, $\alpha\delta - \beta\gamma = 0$ (functional linear transformation or bilinear form).

In that case $\varnothing_n(z) = \dfrac{\alpha_n + \beta_n z}{r_n + \delta_n z}$ (Lotka's model is a particular bilinear form).

5.5 Continuous Parameter Branching Process (Markov Branching Process)

Let $X(t)$ denote the size of the population at time t so that $X(t)$ takes only non-negative integral values. We consider only homogeneous branching process, that is, processes for which $p_{ij}(t_1, t_2) = p_{ij}(t_2 - t_1)$. Let us consider at present, processes with a single type of object. Since in branching processes every object evolves independently of the others we may assume that at the initial instant only one object exists. With the passage of time t it either disappears or turns into k individuals that are all of the same type and constitute the first generation.

Every object of the first generation "lives" independently of the other, and the same probability laws obtain for it as for the initial object "the ancestor". At some instant of time the object either disappears or turns into second generation objects and so forth. The entire process is described by a single integer valued stochastic process $X(t)$ that is equal to the number of objects existing at the instant t. By assumption $X(0) = 1$ a.s.

The set of possible states of the process is the sequence of natural numbers 0, 1, 2, 3, ... Here 0 be an absorbing state. If $X(t) = 0$ then $X(t) = 0$ for all $t > t_0$. If $X(t) = 0$ a.s. for some t, then the process is said to be degenerate process. In other case $X(t)$ vanishes after a finite interval of time with probability < 1. This probability is called the probability of degeneration of the process. It is possible that $X(t)$ increases without bound in time. In case of nuclear reaction this situation can be interpreted as explosive. Thus we can ask the questions: What is the probability of degeneration and what is the $\lim\limits_{t \to \infty} X(t)$?

We state the formal definition of a continuous parameter Markov Branching process as follows.

Definition 5.1 A Stochastic process $\{X(t), t \geq 0\}$ with values on the set of non-negative integers is called a continuous parameter Markov branching process if

(i) it is a stationary Markov chain

(ii) its transition probabilities $p_{ij}(t) = P(X(t) = j | X(0) = i)$ satisfy the relation $p_{ij}(t) = p_{1k}^{(i)}(t)$, the i fold convolution of the probabilities.

Let $p_{ij}(t)$ denote the conditional probability that the system consists of j objects at the instant $t + \tau$ if there were i objects at the instant τ. To solve the problems that arises in the theory of branching processes, we shall find it convenient to use the method of generating functions. We, therefore, introduce the generating function $G_i(z, t)$ of the distributions $\{p_{ij}(t)\}$, for $j = 0, 1, 2, \ldots$

$$\left.\begin{aligned} G_i(z,t) &= \sum_{k=0}^{\infty} z^k p_{ik}(t),\ |z| \le 1,\ i = 1,2,\ldots \\ &= E(Z^{k(t)}) \\ G(z,t) &= G_1(z,t) \text{ and } G(z,0) = z \end{aligned}\right\} \tag{5.16}$$

Since the individuals act independently, the probabilities $p_{ik}(t)$ for the fixed i correspond to the distribution of i independent r.v.s all having the same distribution determined by $G_1(z, t)$.

Thus, we have

$$G_i(z, t) = [G_1(z, t)]^i \tag{5.17}$$

The formula (5.17) characterizes and distinguishes branching process from other continuous time Markov chains. In view of the time homogenity, the Chapman-Kolmogorov equation takes the form

$p_{ik}(t + \tau) = \sum_{j=0}^{\infty} p_{ij}(t)\, p_{jk}(\tau)$. We shall now obtain a continuous time analogue of the relation (5.1) satisfied by a Galton-Watson process.

Lemma 5.3 $G_i(z, t + \tau) = [G(z, t + \tau)]^i = [G(G(z, t), \tau)]^i$, $i \ge 0$, $|z| \le 1$ and in particular,

$$G(z, t + \tau) = G(G(z, t), \tau). \tag{5.18}$$

Proof Using (5.16) and Chapman-Kolmogorov equation, we get

$$\begin{aligned} [G(z, t + \tau]^i &= \sum_{t=0}^{\infty} p_{ij}(t + \tau) z^j,\ i \ge 0 \\ &= \sum_{j=0}^{\infty} z^j \left(\sum_{k=0}^{\infty} p_{ik}(t)\, p_{kj}(\tau) \right) \\ &= \sum_{k=0}^{\infty} p_{ik}(t) \left(\sum_{j=0}^{\infty} p_{kj}(\tau) z^j \right) \end{aligned}$$

(interchanging the order of summation by Fubini's Theorem)

$$\sum_{k=0}^{\infty} p_{ik}(t)\{G(z, \tau)\}^k = [G(G(z, \tau), t)]^i$$

and in particular $G(z, t + \tau) = G(G(z, \tau), t)$.

We shall now proceed to obtain two differential equations (Backward and Forward) satisfied by $G_1(z, \tau)$; these can be obtained from the system of Kolmogorov Backward and Forward differential equation for discrete state continuous parameter Markov chain (Chapter 6). According to general theory (Theorem 6.2), the derivatives

$$\lim_{t\to 0} \frac{p_{ij}(t)}{t} = \lambda_j, j \neq 1 \text{ and } \lim_{t\to 0} \frac{1 - p_{11}(t)}{t} = \lambda_1$$

exist for a regular Markov process. Let us also suppose that the process is conservative so that

$$\lambda_1 = \lambda_0 + \sum_{j=2}^{\infty} \lambda_j < \infty \tag{5.19}$$

Now (5.18) implies $p_{ij}(t) = \lambda_j t + o(t), j = 0, 2, 3, \ldots$

and
$$p_{11}(t) = 1 - \lambda_1 t + o(t) \tag{5.20}$$

With the given initial conditions, the differential equations can be solved in terms of the infinitesemal probabilities λ_k's.

Let $u(z) = \sum_{j=0}^{\infty} \lambda_j z^j$ be the generating function of the sequence $\{\lambda_k\}$.

Theorem 5.6 Let $X(t)$ be a Markov branching process with infinitesimal transition probabilities defined by (5.19) and (5.20). Then the generating function of the probabilities obeys the equations

$$\frac{\partial G}{\partial t} = u(G) \tag{5.21}$$

and
$$\frac{\partial G}{\partial t} = u(z) \frac{\partial G}{\partial z}, \tag{5.22}$$

where $G = G(z, t)$ is the generating function of $\{p_{1k}(t)\}$.

Proof By (5.19), (5.20) and the same argument used to derive Backward Kolmogorov equation in Chapter 6 and time homogenity of branching process $X(t)$, we obtain

$$\frac{dp_{1j}(t)}{dt} = -\lambda_1 p_{1j}(t) + \sum_{\substack{k=0 \\ k\neq 1}}^{\infty} \lambda_k p_{kj}(t) \tag{5.23}$$

Multiplying both sides of equation (5.23) by z^j and summing over j from 0 to ∞, we obtain

$$\frac{\partial G_1(z,t)}{\partial t} = -\lambda_1 G_1(z,t) + \sum_{\substack{k=0 \\ k\neq 1}}^{\infty} \lambda_k G_k(z,t)(|z|\leq 1), \tag{5.24}$$

or on the basis of (5.16),

$$\frac{\partial G(z,t)}{\partial t} = -\lambda_1 G(z,t) + \sum_{\substack{k=0 \\ k\neq 1}}^{\infty} \lambda_k G_k(z,t), \tag{5.25}$$

where we set $G(z, t) = G_1(z, t)$. Finally we obtain the following non-linear equation

$$\frac{\partial G(z,t)}{\partial t} = u(G(z,t)) \text{ with initial condition } G(z, 0) = z. \tag{5.26}$$

Suppose that the conditions under which the second system of Kolmogorov forward differential equations may be applied are satisfied. From the definition of the branching process it follows :

$$p_{kk}(t) = (1 - \lambda_1 t)^k + o(t) = 1 - k\lambda_1 t + o(t),$$

$$p_{kk-1}(t) = k(1 - \lambda_1 t)^{k-1}\lambda_0 t + o(t) = k\lambda_0 t + o(t),$$

$$p_{k,k-j}(t) = o(t), j \geq 2$$

$$p_{k,k+j}(t) = k(1 - \lambda_1 t)\lambda_{j+1} t + o(t) = k\lambda_{j+1} t + o(t), j \geq 1.$$

From which it follows (in the notation of Section 6.1) that

$$q_{jj} = j\lambda_j q_{j,j-1} = j\lambda_0,\ q_{j,j-k} = 0,\ k \geq 2;$$

$$q_{j,j+k} = j\lambda_{k+1},\ j \geq 1.$$

Thus Kolmogorov's forward equation (Section 6.2) takes the form

$$\frac{dp_{1j}(t)}{dt} = -j\lambda_1 p_{1j}(t) + \sum_{\substack{k=0 \\ k\neq 0}}^{j+1} p_{kj}(t) k\lambda_{j-k+1} \tag{5.27}$$

Multiplying (5.27) by z^j and summing with respect to j from 0 to ∞ we obtain

$$\frac{\partial G(z,t)}{\partial t} = u(z)\frac{\partial G(z,t)}{\partial z} \quad (|z| < 1) \tag{5.28}$$

with initial condition (5.26).

Solution of equations (5.22) *and* (5.21)

$G(z, t)$ can be obtained by solving either the backward equation (5.21) or the forward equation (5.22) with the help of initial condition (5.26).

Noting that $\frac{d}{dz}\int_0^z f(x)dx = f(z)$ and using $G(z, 0) = z$, equation (5.21) is equivalent to

$$\int_s^{c(x,t)} \frac{dx}{u(x)} = t \tag{5.29}$$

provided $u(x) \neq 0$ in $G(z, 0) = z \leq x \leq G(z, t)$. Thus $G(z, t)$ can be obtained theoretically from (5.21). Equation (5.29) can be rewritten as

$$\theta(G) - \phi(z) = t,\ \phi(z) = \int_0^z \frac{dz}{u(z)} = t \tag{5.30}$$

Similarly, solution of (5.22) and (5.26) takes the form

$$G(z,t) = \psi\left(t + \int_0^x \frac{du}{u(z)}\right), \tag{5.31}$$

where $\psi(t)$ is the inverse of the function $t = \phi(z) = \int_0^x \frac{dz}{u(z)}$.

This solution coincides with (5.30).

Example 5.4 (Linear Birth and Death Process)
In this special Markov branching process if we put $\lambda_2 = q$, $\lambda_0 = p$, $\lambda_1 = p + q$ and $\lambda_k = 0$ we get a generating function $u(z) = p - (p + q)\, z = qz^2$. In this case a particle disappears with probability $pt + o(t)$ after an interval of time t, divides into two particles with probability $qt + o(t)$ at the end of the same interval, or keep its original state with probability $1-(p + q)t + o(t)$ for that interval. The probability of dividing into more than two particles is equal to $o(t)$. The process in question coincides with a linear birth and death process. We have

$$t = \phi(z) = \int_0^z \frac{dz}{u(z)} = \int \frac{dz}{p-(p+q)z+z^2} = \frac{1}{q-p}\ln\frac{z-1}{z-p}\ (p \neq q)$$

For the inverse function $z = \psi(t)$ we obtain the expression

$$z = \psi(t) = \frac{1-\beta e^{qt(t-\beta)}}{1-e^{qt(1-\beta)}} \quad \text{and} \quad \beta = p/q \tag{5.32}$$

Then from equation (5.31), we get

$$G(z,t) = \psi[1+\phi(z)] = \frac{z-\beta-\beta(z-1)e^{qt(1-\beta)}}{z-\beta-(z-1)e^{qt(1-\beta)}}.$$

Expanding $G(z, t)$ in a series of powers of z, we get

$$G(z,t) = \frac{1-e^{t(q-p)}}{1-\beta^{-1}e^{t(q-p)}} + \sum_{n=1}^{\infty}\frac{(1-\beta)^2[1-e^{t(q-p)}]^{n-1}e^{t(q-p)}}{[\beta-e^{t(q-p)}]^{n+1}}z^n \tag{5.33}$$

which leads us to the formulae

$$p_{10}(t) = \frac{1-e^{t(q-p)}}{1-\beta^{-1}e^{t(q-p)}} \tag{5.34}$$

$$p_{1n}(t) = (1-\beta)^2\frac{[1-e^{t(q-p)}]^{n-1}e^{t(p-q)}}{[\beta-e^{t(q-p)}]^{n+1}},\ n \geq 1 \tag{5.35}$$

Special case: If $p = q$, then

$$\phi(z) = \int_0^z \frac{dz}{p(1-z)^2} = \frac{1}{p}\frac{z}{1-z},\ z = \psi(t) = 1-\frac{1}{1+pt}, \tag{5.36}$$

Using equations (5.36) in (5.31), we get

$$G(z,t) = \frac{pt}{1+pt} + \sum_{n=1}^{\infty}\frac{(pt)^{n-1}}{(1+pt)^{n+1}}z^n \tag{5.37}$$

Consequently, powers of z give

$$p_{10}(t) = \frac{pt}{1+pt},\ p_{1n}(t) = \frac{(pt)^{n-1}}{(1+pt)^{n+1}},\ n = 1, 2, \ldots \tag{5.38}$$

The following asymptotic relationships as $t \to \infty$ are immediate consequences of formulae (5.34) to (5.38): If $q \leq p$, then $p_{10}(t) \to 1$. $p_{1n}(t) \to 0$ for $n \geq 1$ as

$t \to \infty$. If $q \le p$, the branching process degenerates with probability 1, i.e. all objects eventually disappear. If $q > p$, the probability of degeneration of the process is equal to $\beta = p/q < 1$. However, if the objects do not disappear the number of them increases without bound in time, because

$$P(X(t) > N \mid X(t) > 0) = 1 - \frac{1}{p_{10}(t)} \sum_{k=1}^{N} p_{1k}(t) \to 1 \quad \text{for every } N.$$

Let us look at the question of asymptotic behavior as $t \to \infty$ of a branching process in the general case. In what follows we shall need the moments of $X(t)$.

5.5.1 Moments of the process $X(t)$

Since we are using generating functions it will be convenient to introduce factorial moments. We define

$M_k(t) = E[X(t)(X(t) - 1) \ldots (X(t) - k + 1)]$. It is not difficult to set up linear differential equations that are satisfied by $M_k(t)$'s.

Differentiating equation (5.32), we get

$$\frac{\partial \phi_1(z,t)}{\partial t} = u'(G)\phi_1(z,t) \quad \text{and} \quad \phi_1(z,0) = 1, \tag{5.39}$$

where $$\phi_1(z,t) = \frac{\partial}{\partial z} G(z,t) = E[X(t) z^{X(t)-1}]$$

and $$u'(z) = -\lambda_1 + \sum_{k=2}^{\infty} k \lambda_k z^{k-1} \tag{5.40}$$

It follows from (5.39) that $\phi_1(z,t) = \exp\left\{\int_0^t u'(G)\,dt\right\}(|z| < 1)$.

As $z \to 1$, $u'(z) \to \lambda_1 + \sum_{k=2}^{\infty} k\lambda_k = m_1 < \infty$ (assume) and $G(z,t) \to 1$, increasing monotonically and hence uniformly with respect to t.

Therefore $\lim_{z \to 1} \phi_1(z,t) = \lim_{z \uparrow 1} E[X(t) z^{X(t)-1}] = EX(t) = \sum_{k=1}^{\infty} k p_k(t)$.

$$\text{Thus } M_1(t) = \lim_{z \to 1} \phi_1(z,t) = \lim_{z \to 1} \exp\left\{\int_0^t u'(G)\,dt\right\} = e^{m_1 t} \tag{5.41}$$

This result is immediately generalized to factorial moments of higher order.

Lemma 5.4 Suppose that

$$m_r = \sum_{n=1}^{\infty} n(n-1)\ldots(n-r+1)\lambda_n < \infty, r = 1, 2, \ldots k, \ldots \tag{5.42}$$

Then the factorial moments $M_r(t)$, for $r = 1, 2, \ldots k$ of the process $X(t)$ are finite and they satisfy the first order linear differential equation with the constant coefficients.

Proof Let us differentiate equation (5.22) successively with respect to z (for $|z| < 1$). Setting

$$\phi_k(z,t) = \frac{\partial^k G(z,t)}{\partial z^k} = E[X(t)(X(t)-1)\ldots(X(t)-k+1)z^{X(t)-k}],$$

we obtain

$$\left.\begin{aligned} \frac{\partial \phi_2(z,t)}{\partial t} &= u'(G)\phi_2(z,t) + u''(G)\phi_1^2(z,t), \\ \frac{\partial \phi_k(zt)}{\partial t} &= u'(G)\phi_k(z,t) + \psi_k(z,t), \end{aligned}\right\} \tag{5.43}$$

where $\psi_k(z, t)$ is a polynomial in $\phi_1(z, t), \phi_2(z, t), \ldots, \phi_{k-1}(z, t)$ and the coefficients depend on $u''(G), \ldots, u^{(k)}(G)$. The initial conditions are given by $\phi_k(z, 0) = 0$ for $k \geq 2$. The solution of equation (5.43) is of the form

$$\phi_k(z,t) = \exp\left[\int_0^t u'(G)\,dt\right]\int_0^t \psi_k(z,s)\exp\left[-\int_0^t u'(G)\,d\theta\right]ds.$$

Using the same consideration as in the case of $k = 1$, we obtain by induction

$$M_k(t) = \lim_{z\to 1}\phi_k(z,t) = \theta^{m_1 t}\int_0^t \theta^{-m_1\theta}\psi_k(1,s)\,ds.$$

Here $M_k(t)$ obviously satisfies equation (5.43), in which we have set $z = 1$. In particular

$$M_2(t) = \begin{cases} \dfrac{m_2}{m_1}(e^{m_1 t}-1)e^{m_1 t} & \text{if } m_1 \neq 0 \\ m_2 t & \text{if } m_1 = 0 \end{cases} \tag{5.44}$$

and variance of $X(t)$ is given by $\mathrm{Var}(X(t)) = \left(\dfrac{m_2}{m_1} - 1\right)e^{m_1 t}(e^{m_1 t}-1), (m_1 \neq 0).$

Let us proceed to the discussion on probability of extinction. Since the events $X(t) = 0$ constitute an increasing sequence of events, the probability of extinction is given by

$$q = P(\lim_{t\to\infty} X(t) = 0) = \lim_{t\to\infty} P(X(t) = 0) = \lim_{t\to\infty} p_{10}(t).$$

The following theorem illustrates under what conditions the branching process become extinct.

Theorem 5.7 The probability of degeneration of a branching process coincides with the smallest non-negative root of the equation $u(x) = 0$. If $u'(1) = m_1 = -\lambda_1 + \sum\limits_{k=2}^{\infty} k\lambda_k < \infty$, then for $u'(1) \leq 0$ the probability of degeneration is $q = 1$. If $u'(1) > 0$, then $q < 1$.

Proof Since $p_{10}(1) = G(0, t)$, it follows from (5.22) that

$$\frac{dp_{10}(t)}{dt} = u(p_{10}(t)), p_{10}(t) = \int_0^t u(p_{10}(y))dy, p_{10}(0) = 0 \quad (5.45)$$

we have $u(1) = 0$ and $u(0) \geq 0$.

If $\lambda_0 = 0$, we get a trivial solution $p_{10}(t) = 0$ for equation (5.45) and the theorem is trivial (note $u(0) = \lambda_0$). Suppose that $u(0) = \lambda_0 > 0$. We note that x_0 is the smallest positive root of the equation $u(x) = 0$ in $0 \leq x \leq 1$. In other words, $u(x) = 0$ for $x = x_0$, while $u(x) > 0$ in $0 \leq x \leq x_0$. Now $p_{10}(t)$ is non-decreasing function of t and so from (5.45) we find $u(p_{10}(t)) \geq 0$ for $t > 0$, i.e. $u(p_{10}(t))$ is non-negative for all $t > 0$. Hence $p_{10}(t) \leq x_0$ for all $t > 0$.

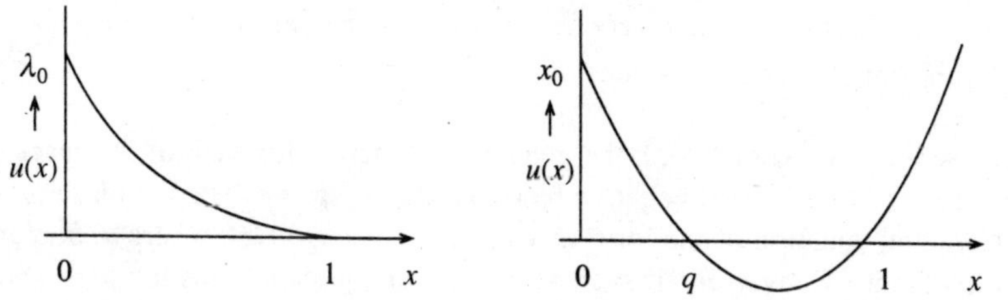

Now $p_{10}(t)$ cannot be equal to x_0 for all $t > 0$, because if $p_{10}(t_0)$ were equal to x_0 for $t_0 > 0$, then by virtue of uniqueness of the solution of equation (5.45) we would have $p_{10}(t) = x_0$, which is impossible. Since the limit $q = \lim_{t\to\infty} p_{10}(t) \leq 1$ exists, it follows from equation (5.45) that $\lim_{t\to\infty} p'_{10}(t) = u(q)$ also exists by continuity of u and this implies $u(q) = 0$. Otherwise the quantity

$$p_{10}(t) = \int_{t_0}^{t} p'_{10}(t)dt + p_{10}(t_0) \to \infty \text{ as } t \to \infty.$$

Thus $q = x_0$. If $x_0 < 1$, $u(x)$ is an increasing function at the point $x_0 = 1$ and $u'(1) > 0$ provided it exists. If $x_0 = 1$, then $u'(1) \leq 0$. In case of linear birth and death process

$$u(x) = \lambda(x - 1)(x - \mu/\lambda). \text{ Thus } q = 1 \text{ iff } \lambda \leq \mu \text{ iff } u'(1) \leq 0.$$

Exercises and Complements

Exercise 5.1 Let $f(s) = 1 - p(1 - s)^\beta$, where p and β are constants and $0 < p < 1$, $0 < \beta < 1$. Prove that $f(s)$ is a probability generating function and its iterates are

$$f_n(s) = 1 - p^{1+\beta+\ldots+\beta^{n-1}}(1 - s)\beta^n \text{ for } n = 1, 2, \ldots$$

Exercise 5.2 Show that $E\left(\sum_{n=1}^{\infty} X_n\right) = m/(1 - m)$ when $m = E(X_1) < 1$ in a branching process.

Exercise 5.3 Let $f(s) = as^2 + bs + c$, where a, b, c are positive and $f(1) = 1$. Assume that the probability of extinction is $d (0 < d < 1)$, prove that $d = c/a$.

Exercise 5.4 Let $q_n = P(X_n = 0) = \phi_n(0)$, where ϕ_n is the nth iterate of the probability generating function $\phi(s) = \sum_{k=0}^{\infty} p_k s^k$. Prove that $q_n = \phi_{n-1}(p_0)$, $n > 1$.

Exercise 5.5 Let X_n denote the number of offsprings in the nth generation of a discrete time branching process. The function $\phi_n(s)$ is the nth iterate of the probability generating function $\phi(s)$, $m = EX_1$, $\sigma^2 = \text{Var}(X_1)$ and $P(X_0 = 1) = 1$.

1. Assume $n > l$. Show that the bivariate generating function of X_l and X_n is
$$\sum_{k=0}^{\infty}\sum_{j=0}^{\infty} P[X_l = k, X_n = j]\, u^k x^j = \phi_l\,[u\,\phi_{n-l}(x)].$$
2. Prove that $E(X_n X_l) = m^{n-l} E(X_l^2)$ if $n > l$.
3. Find R_{nl}, the correlation coefficient for X_n and X_l, and discuss the behavior of $\lim_{n\to\infty} R_{nl}$ for fixed values of l.

Exercise 5.6 Show that $\phi(z)$, the generating function for each of the cases (i) Poisson, (ii) binomial (iii) negative binomial the offspring distribution satisfies a differential equation of the form $(c + dz)\,\phi'(z) = a + b\phi(z)$, where a, b, c and d are constants. Conversely, if such a differential equation holds for $\phi(z)$, show that the following recursive formula applies:

if $P(X_n = k + 1) > 0$, then, for $k \geq 0$ and $n \geq 1$

$$(c + dq_{n-1})P(X_n = k+1) = aP(X_{n-1} = k-1) + \frac{1}{k+1}\sum_{j=0}^{k}[b(k+1-j) - d_j]$$

$$P(X_n = j)P(X_{n-1} = k+1-j), \quad \text{where} \quad q_n = P(X_n = 0).$$

Exercise 5.7 Let $m > 1$. Assume that $X_0 = 1$ and $0 < EX_1^2 < \infty$. Show that the sequence X_n / EX_n is an a.s. convergent martingale.

Exercise 5.8 At time 0 a blood culture starts with one red cell. At the end of one minute, the red cell dies and is replaced by one of the combinations with probabilities as indicated: 2 red cells with probability 1/4, 1 red cell and 1 white cell with probability 2/3, 2 white cells with probability 1/12. Each red cell lives for one minute and gives birth to off springs in the same way as the parent cell. Each white cell lives for one minute and dies without reproducing. Assume that individual cells behave independently. What is the probability that the entire culture dies out eventually?

Exercise 5.9 Let E_n be the event that extinction occurs at sometime after n and $Y_n = X_n/(n\sigma^2)$, $\sigma^2 > 0$. Show that if $m = 1$ and $\phi''(z) < \infty$ then

$$P(Y_n \leq y \mid E_n) \to 1 - e^{-2y}, \text{ as } n \to \infty$$

Exercise 5.10 (Branching immigration) Each generation of a branching process (single ancestor) is augmented by a random number of immigrations who are indistinguishable from the other member of the population. Suppose that number of immigrants are independent of each other and of the past history of the process, each such number having p.g.f. $H(z)$. Show that $\phi_{n+1}(z) = \phi_n(\phi(z))\, H(z)$.

6

Continuous Time Discrete State Markov Processes

6.0 Axiomatic Definition and Transition Function

Here $T = [t: 0 \leq t \leq \infty]$, S = denumerable (countable) set. Recall that in discrete parameter Markov chain transition probabilities satisfy the following conditions:

(1) $0 \leq p_{ij} \leq 1$ (2) $\sum_j p_{ij} = 1$

(3) $p_{ij}^{(n+1)} = \sum_{k \in S} p_{ik}^{(n)} p_{kj}$ (4) $p_{ij}^{(0)} = \delta_{ij}$ (Kronecker's delta).

Now define
$$p_{ij}^{(t)} = P[X(t) = j \mid X(0) = i]$$
$$= P[X(t+s) = j \mid X(s) = i]$$

(by stationarity) as the transition function of a continuous parameter Markov process. Let $\{X(t), t \in R^+\}$ be a Markov process with countable state space S and tanstition probability function $p_{ij}(t)$, $i, j \in S$, $t \in R^+$.

We assume the following axioms about transition function $p_{ij}(t)$:

1. $0 \leq p_{ij}(t) \leq 1$
2. $\sum_{j \in S} p_{ij}(t) = 1$ for all $i = 1, 2, \ldots$ and $t \geq 0$.
3. Chapman-Kolmogorov equation (or semigroup property)
   $p_{ij}(t+h) = \sum_{k \in S} p_{ik}(t) p_{kj}(h), t \geq 0, h \geq 0$
4. $p_{ij}(0) = \delta_{ij}$
5. $\lim_{t \downarrow 0} p_{ij}(t) = \delta_{ij}$ (makes Markov process standard or regular)

Note that by axioms 1 to 5 $\{p_{ij}(t)\}$ are right continuous functions at $t = 0$ and in terms of Transition (Markov) matrix, $P(t) = (p_{ij}(t))$, the above axioms can be rewritten as

(1) $\mathbf{1} \geq P(t) \geq \mathbf{0}$

where **1** and **0** are matrices with all elements 1 and 0, respectively

(2) $P(t)\begin{pmatrix}1\\1\\\vdots\\1\end{pmatrix} = \begin{pmatrix}1\\1\\\vdots\\1\end{pmatrix}$

(3) $P(t + h) = P(t)P(h)$

(4) $P(0) = I$

(5) $\lim_{t \downarrow 0} P(t) = I.$

Suppose now $P(s)$ for $0 \leq s \leq h$ is known. For each t we can find an integer n such that $nh \leq t \leq (n + 1)h$. Then

$$P(t) = P(nh + t - nh) = P(nh)\, P(t - nh) \text{ (From (3))}$$

$$= P(\underbrace{h + h + \ldots + h}_{n \text{ times}})\, P(t - nh)$$

$$= P(h)\, P(h) \ldots P(h)\, P(t - nh) = P^n(h)\, P(t - nh).$$

We want to define $p'_{ij}(t)$ at least for $t = 0$. It will actually be shown that owing to the rich additional structure reflected in (3), continuity implies that $P(t)$ is in fact differentiable in t, i.e. $p'_{ij}(t) = dp_{ij}(t)/dt$ exists for all pair $(i, j) \in S$ and all $t \geq 0$. At $t = 0$, of course "derivative" refers to right hand derivative. Let us first prove continuity of $p_{ij}(t)$ for all $t \geq 0$ and $i, j \in S$.

Lamma 6.1 $P(t)$ is continuous for $t \geq 0$ (6.1)

Proof From (3) $P(t + s) = P(t)\, P(s)$, $t, s \geq 0$.
If $s = h > 0$, then we have

$$\lim_{h \to 0+} P(t + h) = P(t) \lim_{h \to 0+} P(h) = P(t)I = P(t)$$

i.e. $P(t)$ is right continuous.
If $t > 0$ and $0 < h < t$

$$P(t) = P(t - h)\, P(h) \tag{6.2}$$

But $\| P(h) - I \| < \varepsilon$ if $| h | < \delta$ [by axiom (5)].
Hence for $| h | < \delta$, $P^{-1}(h)$ exists and tends to I as $h \to 0$.

Therefore $$P(t) = P(t) \lim_{h \to 0+} P^{-1}(h) = \lim_{h \to 0+} (P(t)P^{-1}(h))$$

$$= \lim_{h \to 0+} (P(t - h)P(h)P^{-1}(h)) \quad \text{[by (6.2)]}$$

$$= \lim_{h \to 0+} P(t - h).$$

Thus, $P(t)$ is left continuous and hence $P(t)$ is continuous.

6.1 Differentiability of Transition Function

From Chapman-Kolmogorov's equation

$$P(t+s) = P(t).\ P(s)$$

or $$P\ (t+s) - P(t) = P(t)\ P(s) - P(t)$$

or $$P\ (t+s) - P(t) = P(t)[P(s) - I]$$

$$\lim_{s \downarrow 0} \frac{P(t+s) - P(t)}{s} = \lim_{s \downarrow 0} \frac{P(t)[P(s) - I]}{s}$$

$$\Rightarrow \quad P'(t) = P(t) \lim_{s \downarrow 0} \frac{P(s) - I}{s} \text{ if } \lim_{s \downarrow 0} \frac{P(s) - I}{s} \text{ exists}$$

Lemma 6.2 $P_{ii}(t) > 0$ for all $t > 0$ and $i \in S$.

Proof $p_{ii}(0) = 1 > 0$.
If $t \downarrow 0$, $p_{ii}(t) \to 1$ continuously by (5) and (6.1).
For some

$$t_0 > 0,\ p_{ii}\ (t) > 0 \text{ for } 0 \le t \le t_0 \tag{a}$$

(by continuity of $P_{ii}(t)$).
Choose n such that

$$\left(t = \frac{t}{n} n\right) \frac{t}{n} \le t_0 \tag{b}$$

$$p_{ii}(t) = \sum_{j \in S} p_{ij}(t/n) P_{ji}\left(\frac{n-1}{n} t\right) \text{ (by (3))}$$

$$\ge p_{ii}\left(\frac{t}{n}\right) p_{ii}\left(\frac{n-1}{n} t\right)$$

By induction (or repetitions) on $n \ge \left(p_{ii}\left(\frac{t}{n}\right)\right)^n > 0$ by (a) and (b).

Theorem 6.1 (a) $\lim_{t \downarrow 0} \frac{1 - p_{ii}(t)}{t} = q_{ii}$ exists but may be ∞

$$= -p'_{ii}(0).$$

(b) $\lim_{t \downarrow 0} \frac{p_{ij}(t)}{t} = q_{ij} < \infty$ if $i \ne j$.

(c) For every state i of a finite state Markov process, q_{ii} is finite and is equal to $\sum_{j \ne i} q_{ij}$.

(d) For all $i \in S$, $\sum_{j \ne i} q_{ij} \le q_{ii}$.

Proof (a) We shall first see that for every $i \in S$,

$$-p'_{ii}(0) = \lim_{t \to 0} \frac{1 - p_{ii}(t)}{t}.$$

By semigroup property $p_{ii}(t+s) \ge p_{ii}(t)\ p_{ii}\ (s)$.
Taking log on both sides and writing $-\log p_{ii}(t) = \varnothing(t)$

(this is well defined since $p_{ii}(t) > 0$ for $t > 0$), we get

$$\varnothing(t+s) \le \varnothing(t) + \varnothing(s) \tag{6.3}$$

Since $$0 < p_{ii}(t) \le 1,\ \varnothing(t) \ge 0.$$

Let $q_{ii} = \sup_{t>0} \dfrac{\varnothing(t)}{t}$. Then $0 \le q_{ii} \le \infty$.

If $q_{ii} < \infty$, there exists $t_0 > 0$ such that $\varnothing(t_0)/t_0 \ge q_{ii} - \varepsilon$.

By (6.3)

$$\varnothing(t_0) \le \varnothing(nt) + \varnothing(\delta) \le \varnothing((n-1)t) + \varnothing(t) + \varnothing(\delta) \le \ldots \le n\varnothing(t) + \varnothing(\delta)$$

if $t_0 = nt + \delta$, $0 \le \delta < t$, $n = 1, 2, \ldots$

Hence $$q_{ii} - \varepsilon \le \frac{\varnothing(t_0)}{t_0} \le \frac{n\,\varnothing(t) + \varnothing(\delta)}{t_0}$$

$$= \frac{nt}{t_0}\varnothing(t)/t + \frac{\varnothing(\delta)}{t_0}.$$

Therefore $$q_{ii} - \varepsilon \le \underline{\lim}_{t\to 0}\left[\frac{nt}{t_0}\cdot\frac{\varnothing(t)}{t} + \frac{\varnothing(\delta)}{t_0}\right] \tag{6.4}$$

But $t \to 0 \Rightarrow \dfrac{nt}{t_0} \to 1$ and $\varnothing(\delta) \to 0$ (by axiom (5) and $p_{ii}(\delta) \to 1$ as $\delta \to 0$).

Therefore $$\underline{\lim}_{t\to 0} \varnothing(t)/t = \underline{\lim}_{t\to 0}\left[\frac{nt}{t_0}\cdot\frac{\varnothing(t)}{t} + \frac{\varnothing(\delta)}{t_0}\right] \tag{6.5}$$

Hence by definition of q_{ii},

$$\overline{\lim}_{t\to 0} \frac{\varnothing(t)}{t} \le q_{ii} \tag{6.6}$$

By (6.4), (6.5) and (6.6), we have

$$q_{ii} - \varepsilon \le \underline{\lim}_{t\to 0} \varnothing(t)/t \le \overline{\lim}_{t\to 0} \frac{\varnothing(t)}{t} = q_{ii}$$

Since $\varepsilon > 0$ is arbitrary, we have

$$\overline{\lim}_{t\to 0} \frac{\varnothing(t)}{t} = \underline{\lim}_{t\to 0} \frac{\varnothing(t)}{t} = q_{ii}.$$

If $q_{ii} = \infty$, we can replace $q_{ii} - \varepsilon$ by an arbitrary large constant M and then obtain $M \le \underline{\lim}_{t\to 0} \dfrac{\varnothing(t)}{t}$; thus $\infty = \underline{\lim}_{t\to 0} (\varnothing(t)/t)$.

In either case we have $\lim_{t\to 0} \dfrac{\varnothing(t)}{t} \le q_{ii}$.

Now $$\lim_{t\to 0} \frac{1 - p_{ii}(t)}{t} = \lim_{t\to 0} \frac{1 - \exp(-\varnothing(t))}{\varnothing(t)} \cdot \frac{\varnothing(t)}{t} = q_{ii}$$

(since $\varnothing(t) \to 0$ as $t \to 0$, by L-Hospital's rule we get the last equation).

(b) Let $c \in (1/2, 1)$ be fixed. Then by lemma (6.2), there exists an $s > 0$ sufficiently small so that $p_{ii}(s) > c$ and $p_{ij}(s) > c$. Let $0 < s \leq nt < \delta$. Consider a stationary discrete parameter M.C. $\{X_n\}_0^\infty$ with state space S such that

$$P(X_{n+1} = j \mid X_n = i) = p_{ij}(t).$$

Then for all $n > 1$, we have

$$p_{ij}(nt) = P(X_n = j \mid X_0 = i) \geq P(X_1 = j, X_n = j \mid X_0 = i)$$

$$+ \sum_{r=1}^{n-1} P(X_1 \neq j, \ldots X_{r-1} \neq j, X_r = i, X_{r+1} = j, X_n = j \mid X_0 = i)$$

$$\geq cp_{ij}(t)\left(1 + \sum_{r=1}^{n-1} P(X_1 \neq j, \ldots, X_{r-1} \neq j, X_r = i \mid X_0 = i)\right)$$

Since each term on the right corresponds to a possible way of going from i to j in n-steps (relative to $(p_{ij}(t))$) and these paths are mutually exculsive but not necessarily exhaustive.

On the other hand,

$$P(X_1 \neq j, \ldots, X_{r-1} \neq j, X_r = i \mid X_0 = i)$$

$$= P(X_1 \neq j, \ldots, X_{r-2} \neq j, X_r = i \mid X_0 = i) - P(X_1 \neq j, \ldots, X_{r-2}$$

$$\neq j, X_{r-1} = j, X_r = i \mid X_0 = i)$$

$$= P(X_1 \neq j, \ldots, X_{r-3} \neq j, X_r = i \mid X_0 = i) - P(X_1 \neq j, \ldots, X_{r-3}$$

$$\neq \quad j, X_{r-2} = j, X_r = i \mid X_0 = i) - P(X_1 \neq j, \ldots, X_{r-2} \neq j, X_{r-1} = j,$$

$$X_r = i \mid X_0 = i),$$

and so on. Continuing in similar way we obtain

$$P(X_1 \neq j, \ldots, X_{r-1} \neq j, X_r = i \mid X_0 = 1)$$

$$P(X_r = i \mid X_0 = i) - \sum_{k<r} P(X_1 \neq j, \ldots, X_{k-1} \neq j, X_k = j) P(X_r)$$

$$= i \mid X_k = j) \geq c - (1 - c) \sum_{k<r} P(X_1 \neq j, \ldots X_{k-1} \neq j, X_k = j \mid X_0$$

$$= i \mid \geq 2c - 1.$$

Consequently,

$p_{ij}(nt) \geq (2c - 1)np_{ij}(t)$. Let $h < \delta$, $t < \delta$, and $n = [h/t]$,

where $[x]$ is the interger part of x; then, $p_{ij}(t)/t \leq p_{ij}([h/t]t)/\{[h/t]t_0(2c - 1)\}$.

Letting, $t \to 0+$, we obtain

$$\lim_{t \to 0+} \sup (p_{ij}(t)/t) \leq p_{ij}(h)\ \{hc(2c - 1)\} < \infty \text{ for all } t > 0.$$

Therefore,

$$\lim_{t \to 0+} \sup (p_{ij}(t)/t) \leq \lim_{t \to 0+} \inf (p_{ij}(h)/\{h(2c - 1)c\},$$

which means that the limit $q_{ij} = \lim_{t \to 0+} (p_{ij}(t)/t) < \infty$ exists.

(c) We shall prove $q_{ii} < \infty$ for a finite state process. Now $1 = p_{ij}(h) + \sum_{\substack{j=1 \\ j \neq i}}^{N} p_{ij}(h)$, where N is the number of states, or $1 - p_{ii}(h) = \sum_{j \neq i}^{N} p_{ij}(h)$.

Dividing by h and letting $h \to 0$ and by part (a)

$$q_{ij} = \sum_{\substack{j=1 \\ j \neq i}}^{N} q_{ij} < \infty.$$

(d) Since $\sum_{j \neq i}^{\infty} p_{ij}(h) = 1 - p_{ii}(h)$, $\sum_{j \neq i}^{N} P_{ij}(h) \leq 1 - p_{ii}(h)$ for every finite number N of states.

Dividing both sides by h and leting $h \to 0$ leads to $\sum_{j \neq i}^{N} q_{ij} \leq q_{ii}$.

Since N is arbitrary and all terms are positive, the result follows. The matrix $Q = (q_{ij})$ is called the generator of the Markov process.

Example 6.1 A birth and death precess has state space $S = \{0, 1, 2, \ldots\}$ and its generater Q (for definition see Section 6.3) has elements

$$q_{ii} = \lambda_i + \mu_i \text{ and } q_{ij} = \begin{cases} \lambda_i \text{ if } j = i + 1 \\ 0 \text{ if } j \neq i - 1 \text{ or } i + 1, i = 0, 1, 2, \ldots \\ \mu_i \text{ if } j = i - 1 \end{cases}$$

λ_i and μ_i are called birth and death rates, respectively.

Definition 6.1 A state $i \in S$ is called *instantaneous* if $q_{ii} = \infty$ and is called *stable* if $0 \leq q_{ii} < \infty$. A state $i \in S$ is called *absorbing* if $q_{ii} = 0$.

Exercise 6.1 Show that $q_{ii} = 0$ iff $p_{ii}(t) = 1$.
We shall see that Kolmogorov backward equation $P'(t) = AP(t)$ gives a solution $p(t) = e^{At}$ (Section 6.3) from which it follows that $P\{X(\tau) = i, 0 \leq \tau \leq t \mid X(0) = i\} = e^{-q_{ii}t}$ and is equal to 1 for all $t > 0$ if $q_{ii} = 0$ and this shows that $p_{ii}(t) = 1$. We shall also derive this as distribution of waiting time of the initial state i.

Definition 6.2 A continuous time standard Markov process is called *conservative (nonexplosive)* if $\sum_{j \neq i} q_{ij} = q_{ii} < \infty$ for all $i \in S$.

Note that a birth and death process is conservative, since $\sum_{j \neq i} q_{ij} = \lambda_i + \mu_i = q_{ii} < \infty$.
Being a pure birth process a Poisson process is also conservative.

6.2 Kolmogorov Differential Difference Equation

Theorem 6.2 In a conservative Markov process, the transition probability functions $p_{ij}(t)$ are differentiable for all $i, j, \in S$ and $t \geq 0$ and they satisfy Kolmogorov's differential equations :

(i) *Backward Equation*

$$p'_{ij}(t) = \sum_{k \neq i} q_{ik} p_{kj}(t) - q_{ii} p_{ij}(t) \text{ for all } i, j \in S, t \geq 0.$$

(ii) *Forword Equation*

$$p_{ij}(t) = \sum_{k \neq i} q_{kj} p_{ik}(t) - q_{jj} p_{ij}(t) \text{ for all } i, j \in S, t \geq 0$$

Proof (i) $p_{ij}(t + s) - p_{ij}(t) = \sum_{k \in S} p_{ik}(s) p_{kj}(t) - p_{ij}(t)$

$$= \sum_{k \neq i} p_{ik}(s) p_{kj}(t) + p_{ii}(s) p_{ij}(t) - p_{ij}(t)$$

$$= \sum_{k \neq i} p_{ik}(s) p_{kj}(t) - [1 - p_{ii}(s)] p_{ij}(t)$$

Claim

$$\lim_{s \downarrow 0} \sum_{k \neq i}^{\infty} \frac{p_{ik}(s) p_{kj}(t)}{s} = \sum_{k \neq i}^{\infty} q_{ik} p_{kj}(t) \tag{6.7}$$

$$\liminf_{s \downarrow 0} \sum_{k \neq i}^{\infty} \frac{p_{ik}(s)}{s} p_{kj}(t) \geq \sum_{k \neq i}^{N} q_{ik} q_{kj}(t) \tag{6.8}$$

for every positive integer N (by Theorem 6.1 (b))

and hence

$$\sum_{k \neq i}^{\infty} q_{ik} p_{kj}(t) \leq \liminf_{s \downarrow 0} \sum_{k \neq i}^{\infty} \frac{p_{ik}(s) p_{kj}(t)}{s} \tag{6.9}$$

Take $N > k$, then $\sum_{k \in S} p_{ik}(s) p_{kj}(t) = \sum_{k=0}^{N} p_{ik}(s) p_{kj}(t) + \sum_{k=N+1}^{\infty} p_{ik}(s) p_{kj}(t)$

$$\leq \sum_{k=0}^{N} p_{ik}(s) p_{kj}(t) + \sum_{N+1}^{\infty} p_{ik}(s)$$

$$= \sum_{k=0}^{N} p_{ik}(s) p_{kj}(t) + \left(1 - \sum_{k=1}^{N} p_{ik}(s)\right)$$

$$= \sum_{k=0}^{N} p_{ik}(s) p_{kj}(t) + 1 - p_{ii}(s) - \sum_{k \neq i}^{N} p_{ik}(s).$$

Hence $\sum_{k \neq i}^{\infty} \frac{p_{ik}(s) p_{kj}(t)}{s} \leq \sum_{k \neq i}^{N} \frac{p_{ik}(s) p_{kj}(t)}{s} + \frac{1 - p_{ii}(s)}{s} - \sum_{k \neq i}^{N} \frac{p_{ik}(s)}{s}$

Therefore $\limsup_{s \downarrow 0} \sum_{k \neq i}^{\infty} \frac{p_{ik}(s) p_{kj}(t)}{s} \leq \sum_{k \neq i}^{N} q_{ik} p_{kj}(t) + q_{ii} - \sum_{k \neq i}^{N} q_{ik}$

(by Theorems 6.1(a), and 6.1(b)).

Now letting $N \to \infty$ (since the M.P. is conservative),

$$\limsup_{s \downarrow 0} \sum_{k \neq i}^{\infty} \frac{p_{ik}(s)}{s} p_{kj}(t) \leq \sum_{k \neq i}^{\infty} q_{ik} p_{kj}(t) \tag{6.10}$$

(the last two terms cancel).

Combining (6.9) and (6.10), the backward equation follows from

$$\limsup_{s\downarrow 0} \sum_{k\neq i}^{\infty} \frac{p_{ik}(s)}{s} p_{kj}(t) \leq \sum_{k\neq i}^{\infty} q_{ik} p_{kj}(t) \leq \liminf_{s\downarrow 0} \sum_{k\neq i}^{\infty} \frac{p_{ij}(s)}{s} p_{kj}(t),$$

since $\lim_{s\downarrow 0} \frac{1}{s} \sum_{k\neq i}^{\infty} p_{ik}(s) p_{kj}(t)$ exists and is equal to $\sum_{k\neq i}^{\infty} q_{ik} p_{kj}(t)$.

(ii) $$p_{ij}(s+t) - p_{ij}(s) = \sum_{k\in S} p_{ik}(s) p_{kj}(t) - p_{ij}(s)$$

$$= \sum_{k\in S} p_{ik}(s)[p_{kj}(t) - \delta_{kj}]$$

and hence the result (ii) can be proved similarly as (i) after dividing by t and letting $t \to 0$.

Since the scope and validity of the forward equation is rather involved we are not discussing any more details of the proof.

6.3 Infinitesimal Generators and Examples

Definition 6.3 The infinitesimal generator or matrix or transition rates of a Markov processes is the matrix $A = (a_{ij})$ defined by

$$a_{ij} = \begin{pmatrix} q_{ij} \text{ if } i \neq j \\ -q_{ii} \text{ if } i = j \end{pmatrix}$$

The important problem is, however, to construct transition probabilities $p_{ij}(t)$ having prescribed infinitesimal matrix A satisfying $a_{ij} \geq 0$ for all $i, j \in S$, $i = j$ and

$$-a_{ii} = \sum_{j\neq i} a_{ij} \tag{6.11}$$

Now Kolmogorov's backward and forward equations can be written in the matrix form as

$$P'(t) = AP(t) \text{ and } P'(t) = P(t)\, A \text{ respectively} \tag{6.12}$$

In case that state space S is finite it is known from the theory of ordinary differential equations that, subject to the initial condition $P(0) = I$, the unique solution of (6.12) is given by $P(t) = e^{tA}$, $t \geq 0$, where the matrix e^{tA} is defined by

$$e^{tA} = \sum_{n=0}^{\infty} (tA)^n / n! = I + \sum_{n=1}^{\infty} t^n A^n / n! \tag{6.13}$$

Suppose that A is an $(N + 1) \times (N + 1)$ matrix with eigenvalues $\lambda_0, \lambda_1, \lambda_N$, possibly with repetitions, and corresponding eigen vectors $U^{(0)}, U^{(1)}, \ldots, U^{(N)}$ which are linearly independent. Let

$U = (U^{(0)}, U^{(1)}, \ldots, U^{(N)})$. Then by orthogonality

$$AU = (\lambda_0 U^{(0)}, \lambda_1 U^{(1)}, \ldots, \lambda_N U^{(N)})$$

$$A^2 U = A(\lambda_0 U^{(0)}, \lambda_1 U^{(1)}, \ldots \lambda_N U^{(N)}) = (\lambda_0^2 U^{(0)}, \ldots, \lambda_N^2 U^{(N)})$$

$$A^n U = (\lambda_0^n U^{(0)}, \lambda_1^n U^{(1)}, \ldots, \lambda_N^n U^{(N)}).$$

Hence $e^{tA}U = \left(\sum_{n=0}^{\infty} t^n/n!\, A^n \right) U = \sum_{n=0}^{\infty} t^n/n!(\lambda_0^n U^{(0)}, \ldots, \lambda_N^n U^{(N)})$

$$= (e^{t\lambda_0} U^{(0)}, e^{t\lambda_1} U^{(1)}, \ldots, e^{t\lambda_N} U^{(N)})$$

$$= U \wedge (t),$$

where $\wedge(t) = \begin{pmatrix} e^{t\lambda_0} & 0 & \cdots \\ 0 & e^{t\lambda_1} & \cdots \\ \vdots & \vdots\ \vdots & e^{t\lambda_N} \end{pmatrix}.$

Therefore $e^{tA} = U \wedge (t)\, U^{-1}$.

Hence the method of spectral decomposition gives the unique solution of (6.12) as $P(t) = \sum_{j=0}^{N} \exp(\lambda_j t) U^{(j)} V^{(j)\prime}$), where $U^{(j)}$ and $V^{(j)}$ are the corresponding right and left eigen vectors of A.

In case the state space S is countably infinite, results analogous to those for the finite case can be obtained under the following restrictive condition:

$$a = \sup_{i,j} |\, a_{ij} \,| < \infty \qquad (6.14)$$

In view of condition (6.11), condition (6.14) is equivalent to

$$\sup_i |\, a_{ii} \,| < \infty \qquad (6.15)$$

The solution $P(t)$ is still given by (6.13), i.e. for each $j \in S$,

$$p_{ij}(t) = \delta_{ij} + ta_{ij} + t^2/2!\, a_{ij}^{(2)} + \ldots + t^n/n!\, a_{ij}^{(n)} + \ldots \qquad (6.16)$$

for all $i \in S$, where $a_{ij}^{(n)}$ are the elements of A^n, and $A = (a_{ij})$.

Proof of (6.16) By (6.14), (6.11) and by induction on n, we have

$$|\, a_{ij}^{(n)} \,| \leq a\,(2a)^{n-1}, \text{ for } i, j \in S, n \geq 1 \qquad (6.17)$$

Thus, the series on right side of (6.16) converges to an analytic function $h_{ij}(t)$, say, absolutely for all t. Hence term-by-term differentiation of this series for $h_{ij}(t)$ one can easily verifies Kolmogorov's backward equation with initial conditions $p_{ij}(0) = 0$ for $i \neq j$ and $p_{ii}(0) = 1$.

To verify the Chapman-Kolmogorov equations, note that

$$P(t+s) = e^{(t+s)A} = I + (t+s)A + \frac{(t+s)^2}{2!} A^2 + \ldots + \frac{(t+s)^n}{n!} A^n + \ldots$$

$$= \left[I + tA + \frac{t^2}{2!} A^2 + \ldots + t^2/n! A^n + \ldots \right]$$

$$\times \left[I + sA + \frac{s^2}{2!} A^2 + \ldots + \frac{s^m}{m!} A^m + \ldots \right]$$

$$= e^{tA} e^{sA} = P(t)P(s)$$

which follows from the identity

$$\sum_{m+n=\upsilon} \frac{t^m}{m!} A^m \frac{s^n}{n!} A^n = \left(\sum_{m+n=\upsilon} \frac{t^m s^n}{m!n!} \right) A^\upsilon = \frac{(t+s)^\upsilon}{\upsilon!} A^\upsilon$$

Also, the functions $P_i(t) = \sum_{j \in S} p_{ij}(t)$ satisfy

$$P_i'(t) = \sum_{j \in S} p_{ij}'(t) = \sum_{j \in S} \sum_{k \in S} a_{ik} p_{kj}(t)$$

$$= \sum_{k \in S} a_{ik} P_k(t), t \geq 0$$

with initial conditions $P_i(0) = \sum_{j \in S} \delta_{ij} = 1$ for $i \in S$. Since $P_i(t) = 1$ for, all $t \geq 0$, all $i \in S$ clearly satisfy these equations, we have $P_i(t) = 1$ for all t by uniqueness of such solutions. Solution (6.17) have been shown to satisfy all conditions for being transition functions except for non-negativity which will follow from the following more general result.

Theorem 6.3 Given any infinitesimal transition matrix

$$A = (a_{ij}) \text{ satisfying } -a_{ij} = \sum_{j \neq i} a_{ij}, a_{ij} \geq 0 \text{ for } i \neq j$$

there exists a smallest nonnegative solution $P(t)$ of the backward equations

$$P_{ij}'(t) = \sum_k a_{ik} P_{kj}(t), i, j \in S \tag{6.18}$$

with initial conditions $p_{ij}(0) = \delta_{ij}$, $i, j \in S$.

This solution satisfies $\sum_{j \in S} p_{ij}(t) \leq 1$ for all $i, \in S, t \geq 0$.

Moreover, when the equality holds in the last inequality for all $i \in S$ and $t \geq 0$, the solution is unique.

Proof Write $\lambda_i = -a_{ii}$ for all $i \in S$.

Multiplying both sides of backward equations $p_{ij}'(t) = \sum_{k \in S} a_{ik} p_{kj}(t)$ by the "integrating factor" $e^{\lambda_i s}$ we get

$$e^{\lambda_i s} p_{ik}'(s) = e^{\lambda_i s} \sum_{j \in S} a_{ij} p_{jk}(s)$$

or $$\frac{d}{ds}(e^{\lambda_i s} p_{ik}(s)) = \lambda_i e^{\lambda_i s} p_{ik}(s) + e^{\lambda_i s} \sum_{j \in S} a_{ij} p_{jk}(s) = \sum_{j \neq i} e^{\lambda_i s} a_{ij} p_{ik}(s)$$

On integrating between 0 and t we have, remembering that $p_{ik}(0) = \delta_{ik}$,

$$e^{\lambda_i t} p_{ik}(t) = \delta_{ik} + \sum_{j \neq i} \int_0^t e^{\lambda_i s} a_{ij} p_{ik}(s) ds,$$

or $$p_{ik}(t) = \delta_{ik} e^{-\lambda_i t} + \sum_{j \neq i} \int_0^t e^{-\lambda_i (t-s)} a_{ij} p_{jk}(s) ds \text{ for } t \geq 0, i, k \in S \tag{6.20}$$

Reversing the steps shows that (6.18) together with (6.19) follow from (6.20). Thus (6.18) and (6.19) are equivalent to the system of integral equations (6.20).

To solve (6.20) start with the first approximation

$$p_{ik}(0) = \delta_{ik} e^{-\lambda_i t} \text{ for } i, k \in S \text{ and } t \geq 0$$

and compute successive approximations, recursively, by

$$p_{ik}^{(n)}(t) = \delta_{ik} e^{-\lambda_i t} + \sum_{j \neq i} \int_0^t e^{-\lambda_i (t-s)} a_{ij} p_{jk}^{(n-1)}(s) ds \tag{6.21}$$

Since $a_{ij} \geq 0$, $i \neq j$, it is clear that $p_{ik}^{(1)}(t) \geq p_{ik}^{(0)}(t)$.

Then from (6.21) by induction it follows that $p_{ik}^{(n+1)}(t) \geq p_{ik}^{(n)}(t)$ for all $n \geq 0$.

Thus $$\bar{p}_{ik}(t) = \lim_{n \to \infty} p_{ik}^{(n)}(t) \text{ exists.}$$

6.4 Birth and Death Processes

Definition 1 A birth and death process is a conservative M.P. with infinitesimal generator

$$A = \begin{pmatrix} -\lambda_0 & \lambda_0 & 0 & 0 & 0 & \cdots \\ \mu_1 & -(\lambda_1 + \mu_1) & \lambda_1 & 0 & 0 & \cdots \\ 0 & \mu_2 & -(\mu_2 + \lambda_2) & \lambda_2 & 0 & \cdots \\ \vdots & \vdots & \vdots & \vdots & & \end{pmatrix}$$

where λ_i and μ_j, $i = 0, 1, 2, \ldots, j = 1, 2, 3 \ldots$ are the birth and death rates, respectively.

Definition 2 (Karlin) A birth and death process is a continuous time Markov process on the state space $S = \{0, 1, 2, \ldots\}$ with stationary transition functions $p_{ij}(t) = P(X(t+s)) = j \mid X(s) = i)$ and such that

(i) $p_{ij}(0) = \delta_{ij}$

(ii) $P_{ii+1}(h) = \lambda_i h + o(h)$ for $i \geq 0$ and $h \downarrow 0$

(iii) $p_{ii-1}(h) = \mu_i h + o(h)$ for $i \geq 0$ and $h \downarrow 0$

(iv) $p_{ii}(h) = 1 - (\lambda_i + \mu_i) h + o(h)$ for $i \geq 0$ and $h \downarrow 0$ and
$p_{ij}(h) = o(h)$ for $j \neq i+1, i-1, i$ as $h \downarrow 0$ and all $i \geq 1$.

(v) $\lambda_i > 0$ for $i \geq 0$ and $\mu_i > 0$ for $i \geq 1$ and $\mu_0 = 0$ are the birth and death rates, respectively. λ_i and μ_i are also called parameters of birth and death processes.

Note These two definitions can be proved to be identical.
The backward Kolmogorov equation will be given by

$$p'_{ij}(t) = \mu_i p_{i-1,j}(t) - (\lambda_j + \mu_i) p_{ij}(t) + \lambda_i p_{i+1,j}(t), \, i \geq 1$$

and $p'_{oj}(t) = -\lambda_0 p_{oj}(t) + \lambda_0 p_{1j}(t)$ with initial conditions $p_{ij}(0) = \delta_{ij}$.
The forward Kolmogorov equation for birth and death processes are

$$p'_{ij}(t) = \lambda_{j-1} p_{i,j-1}(t) - (\lambda_j + \mu_j) p_{ij}(t) + \mu_{j+1} p_{i,j+1}(t), \, j \geq 1$$

and $p'_{io}(t) = -\lambda_0 p_{io}(t) + \mu_1 p_{i1}(t)$ with initial conditions $p_{ij}(0) = \delta_{ij}$.

Proof For birth and death processes, Chapman-Kolmogorov equation gives

$$p_{ij}(t+h) = \sum_{k=0}^{\infty} p_{ik}(h)\,p_{kj}(t) = p_{i,\,i-1}(h)p_{i-1,j}(t)$$

$$+\ p_{ii}(h)\,p_{ij}(t) + p_{i,j+1}(h)p_{i+1,j}(t) + \sum_k^* p_{ik}(h)p_{kj}(t)$$

where $\sum_k^*$ is over all $k \neq i-1, i, i+1$. Using axioms (ii), (iii) and (iv) we obtain

$$\sum_k^* p_{ik}(h)p_{kj}(t) \le \sum_k^* p_{ik}(h) = 1 - [p_{ii}(h) + p_{i,i-1}(h) + p_{i,i+1}(h)]$$

$$= 1 - [1 - (\lambda_i + \mu_i)h + o(h) + \mu_i h + o(h) + \lambda_i h + o(h)]$$

$$= o(h) \text{ as } h \to 0.$$

Hence $p_{ij}(t+h) = \mu_i\, hp_{i-1,j}(t) + (1 - (\lambda_i + \mu_i)h)\, p_{ij}(t) + \lambda_i h\, p_{i+1,j}(t) + o(h)$.

Transposing the term $p_{ij}(t)$ to the left hand side and dividing the equation by h, we obtain (after letting $h \to 0$ the backward Kolmogorov's equation)

$$p_{ij}(t) = \mu_i p_{i-1,j}(t) - (\lambda_i + \mu_i)p_{ij}(t) + \lambda_i p_{i+1,j}(t).$$

6.5 The Yule Process

This a pure birth process (that arises in physics and biology) with parameters $\lambda_i = \beta(N+i)$ and $\mu_i = 0$, $\beta > 0$. Kolmogorov's backward equation becomes (putting $i = n$)

$$p'_{N+n}(t) = -\beta(N+n)p_{N+n}(t) + \beta(N+n-1)p_{N+n-1}(t),\ n \ge 1 \quad (6.22)$$

and

$$p'_N(t) = -\beta N p_N(t)$$

where $p_{N+n}(t) = p_{N,N+n}(t) = P(X(t) = N + n / x(0) = N)$.

If $N = 1$, (6.22) becomes

$$p_{n+1}(t) - \beta(n+1)p_{n+1}(t) + \beta n p_n(t) \quad (6.23)$$

with initial conditions $p_1(0) = 1$, $p_n(0) = 0$, $n = 2, 3, \ldots$.

By induction, $p'_1(t) = -\beta p_1(t)$ implies $p_1(t) = ce^{-\beta t}$ and initial conditions give the solution of (2) as

$$p_n(t) = e^{-\beta t}(1 - e^{-\beta t})^{n-1},\ n \ge 0.$$

To get a general solution of (6.22) (i.e. when $N > 1$) we shall first express the Kolmogorov's differential equation for the Yule process as a partial differential equation in terms of the G.F. of the process.

Let $G(t, z) = \sum_{n=0}^{\infty} p_{N+n}(t)\, z^{N+n}$, $|z| < 1$, $n \ge 1$. Then it is uniformly convergent and

$$\frac{\delta G}{\delta t} = \sum_{n=0}^{\infty} p'_{N+n}(t)z^{N+n} = \sum_{n=0}^{\infty} [-\beta(N+n)p_{N+n}(t) + \beta(N+n-1)p_{N+n-1}(t)]z^{N+n}$$

$$= -\beta \sum_{n=0}^{\infty} (N+n)p_{N+n}(t)z^{N+n} + \beta \sum_{n=0}^{\infty} (N+n-1)p_{N+n-1}(t)z^{N+n}$$

$$=-\beta z \sum_{n=0}^{\infty} (N+n) p_{N+n}(t) z^{N+n-1} + \beta z^2 \sum_{m=0}^{\infty} (N+m) p_{N+m}(t) z^{N+m-1} (m=n-1)$$

$$= -\beta \frac{\delta G}{\delta z} z + \beta z^2 \frac{\delta G}{\delta z}$$

Therefore, $\frac{\delta G}{\delta t} = -\beta z(1-z)\frac{\delta G}{\delta z}$ or $\frac{\delta G}{\delta t} + \beta z(1-z)\frac{\delta G}{\delta z} = 0$ with boundary conditions (1) $P_{N+n}(0) = \delta_{N,N+n}$.

$$(2) \quad \sum_{n=0}^{\infty} P_{N+n}(0) z^{N+n} = z^N = G(0, z). \tag{6.24}$$

Now $$u(x, y, z)\frac{\delta z}{\delta x} + v(x, y, z)\frac{\delta z}{\delta y} = \varnothing(x, y, z)$$

represents a Recatti type of partial differential equation (p.d.e.).

To solve Recatti type p.d.e. we find the so called ch. equations (susidiary equations) given by

$$\frac{dx}{u(x, y, z)} = \frac{dy}{v(x, y, z)} = \frac{dz}{\varnothing(x, y, z)}.$$

The fundamental property of Recatti equations is that the general solution is homographic function of constant of integration. Hence

$$z = f_1 + cf.$$

In fact their solution comprises two one-parameter families of surfaces which can be denoted by

$$P(x, y, z) = c_1 \text{ and } Q(x, y, z) = c_2.$$

Here in our Yule process

$$z = G, x = t, y = z, \varnothing(x, y, z) = 0; \begin{cases} u(x, y, z) = 1 \text{ and} \\ v(x, y, z) = \beta z(1 - z). \end{cases}$$

Thereofore

$$\frac{dt}{1} = \frac{dz}{\beta z(1-z)} = \frac{dG}{0} \Rightarrow \frac{dG}{dt} = \frac{0}{1} = 0 \Rightarrow G(t, z) = c_1, \beta\, dt = \frac{dz}{z(1-z)}$$

or $$\frac{dz}{dt} = \beta z(1-z) \Rightarrow \frac{z}{z-1} e^{-\beta t} = c_2.$$

Elemenating one of the constants c_1 and c_2 we get the general solution given by $G(t, z) = \psi\left(\frac{z}{z-1} e^{-\beta t}\right)$, where ψ is an unknown function to be determined from initial conditions.

The initial condition gives $G(0, z) = z^N = \psi\left(\frac{z}{z-1}\right)$.

Put $z = \frac{u}{u-1}$ or $u = \frac{z}{z-1}$. Then

$\psi(u) = \left(\dfrac{u}{u-1}\right)^N$. Substituting $u = \dfrac{z}{z-1}e^{-\beta t}$ we get

$$G(t, z) = \left[\frac{ze^{-\beta t}}{1 - z(1 - e^{-\beta t})}\right]^N \tag{6.25}$$

For $N = 1$, coefficient of z^n in (6.25) gives $p_n(t) = e^{-\beta t}(1 - e^{-\beta t})^{n-1}$.

Theorem 6.4 Let T_i be the time that the conservative M.P. stays in the state i. Then $F(t) = P[T_i \le t] = 1 - e^{-q_{ii}t}$. Note T_i is called holding time at state i.

Proof $\overline{F}(t) = P[T_i > t] = P[X(s) = i, 0 \le s \le t \,|\, X(0) = i]$

$$= \lim_{n\to\infty} P\left[X(t)=i, X\left(\frac{t}{n}\right)=i, X\left(\frac{2t}{n}\right)=i, X\left(\frac{t(n-1)}{n}\right)=i \,\Big|\, X(0)=i\right]$$

(by separability of stochastic process, since a conservative continuous parameter M.P. is separable; see Doob (or appendix I).

To see why the last equality holds, define $A_n = \left\{X\left(\dfrac{k}{n}t\right) = i, k = 0, 1, \dots n\right\}$.

Then $A_n \downarrow$ in n and $A_n = \lim\limits_{n\to\infty} A = \bigcap\limits_{n=1}^{\infty} A_n = \{X(s) = i$ for all s in $[0, t]$ which is a rational multiple of $t\} = \{T_i > t\} \subset A$.

Since the sample paths are step functions, if a sample path is not in $\{T_i > t\}$ then there occurs a jump to state j, different from i, at some $t_0 (0 < t_0 \le t)$. The case $t_0 = t$ may be excluded, since it is not in A. Because each sample path is right-continuous step function, there is a time point $t_1 > t_0$ such that $X(s) = j$ for $t_0 \le s < t_1$. Since there is some s of the form $s = tk/n \le t$ in every nondegenerate interval, it follows that $X(s) = j$ for some s of the form $(k/n)t < t$; this implies that this sample path is not in A_n and hence, not in A. Therefore, $\{T_i > t\} \supset A$.

Hence $\{T_i > t\} = A$.

$$\overline{F}(t) = \lim_{n\to\infty} P\left[X(t) = i \,\Big|\, X\left(\frac{n-t}{n}t\right) = i, \dots, X\left(\frac{t}{n}\right) = i, X(0) = i\right]$$

$$\times P\left[X\left(\frac{n-1}{n}t\right) = i, \dots, X\left(\frac{t}{n}\right) = i \,\Big|\, X(0) = i\right]$$

$$= \lim_{n\to\infty} P\left[X(t) = i \,|\, X\left(\frac{n-t}{n}t\right) = i\right] P\left[X\left(\frac{n-1}{n}t\right) = i, \dots \,|\, X(0) = i\right]$$

(by Markov property)

$$= \lim_{n\to\infty} \left[p_{ii}\left(\frac{t}{n}\right)\right]^n \quad \text{(by induction)}$$

$$= \lim_{n\to\infty} \left[1 - \frac{t}{n} q_{ii} + o\left(\frac{t}{n}\right)\right]^n$$

$=e^{-tq_{ii}}$

$$\left(\lim_{t\downarrow\infty}\frac{1-p_{ii}(t)}{t}=q_{ii} \text{ (by Theorem 6.1)} \Rightarrow p_{ii}(t)=1-tq_{ii}+o(t) \text{ as } t\downarrow 0\right)$$

$$\Rightarrow \overline{F}(t) = e^{-tq_{ii}} \Rightarrow F(t) = 1 - e^{-tq_{ii}}.$$

Note If $q_{ii} = 0$, then $P(T_i > t \mid X(0) = i) = e^0 = 1$ for all $t \geq 0$. Hence $P(T_i = \infty \mid X(0) = i) = 1$, i.e. the state i is absorbing.

6.6 Some Population Models

6.6.1 Non-homogeneous birth and death process

Here we make the following assumptions regarding birth and death processes.

A.1: During small interval of time $(t, t+\Delta t)$, the probability that a birth occurs is $\lambda_n(t)\Delta t + o(\Delta t)$. The probability of no birth in $(t, t+\Delta t)$ is $1 - \lambda_n(t)\,\Delta t + o(t)$ and probability of more than one birth occurring is $o(\Delta t)$. Births occurring in $(t, t+\Delta t)$ are independent of time since the last occurrence.

A.2: The probability that a death will occur in small interval of time $(t, t+\Delta t)$ is $\mu_n(t)\Delta t + o(\Delta t)$. The probability of no death is $1 - \mu_n(t)\,\Delta t + o(\Delta t)$ and the probability of more than one death is $o(\Delta t)$.

A.3: $n = 0$ is an absorbing state of the process.

A.4: For the same population size, the birth and death occur independently of each other.

Let $X(t)$ be the population size at time $t\,(0 < t < \infty)$.

Define $p_{nN}(t) = P[X(t) = n \mid X(0) = N]$. Then the Chapman-Kolmogorov's equation reduces to

$$\begin{aligned} p_{nN}(t+\Delta t) = {} & [1 - (\lambda_n(t) + \mu_n(t))\Delta t]p_{nN}(t) + [\lambda_{n-1}(t)\,\Delta t + o(\Delta t)] \\ & [1 - \mu_{n-1}(t)\Delta t + o(\Delta t)]p_{1,N}(t) + [\mu_{n+1}(t)\Delta t + o(\Delta t)] \\ & [1 - \lambda_{n-1}(t)\Delta t + o(\Delta t + o(\Delta t)]\, p_{n+1N}(t) \quad (n > 1) \end{aligned}$$

Neglecting the terms of $o(\Delta t)$

$$\begin{aligned} p_{nN}(t+\Delta t) = {} & [1 - (\lambda_n(t) + \mu_n(t))\Delta t]p_{nN}(t) + \lambda_{n-1}(t)\Delta t\, p_{n-1,N}(t) \\ & + \mu_{n+1}(t)\Delta t\, p_{n+1,N}(t) + o(\Delta t) \quad (n > 1). \end{aligned}$$

Since '0' is an absorbing state, $\mu_0(t) = 0$ and we get

$$p_{0N}(t+\Delta t) = (1 - \lambda_0(t)\Delta t)p_{0N}(t) + \mu_1(t)\Delta t\, p_{1N}(t)$$

Taking limits as $\Delta t \to 0$, we get

$$\frac{\delta p_{nN}(t)}{\delta t} = -[\lambda_n(t) + \mu_n(t)]p_{nN}(t) + \lambda_{n-1}(t)p_{n-1N}(t) + \mu_{n+1}(t)p_{n+1,N}(t) \quad (n \geq 1) \qquad (6.26)$$

$$\frac{\delta p_{0N}(t)}{\delta t} = -\lambda_0(t)p_{0N}(t) + \mu_1(t)p_{1N}(t) \qquad (6.27)$$

with initial condition $p_{nN}(0) = \delta_{nN}$, the Kroncker's delta.

Let $G(z, t) = \sum_{n=0}^{\infty} p_{nN}(t)z^n, |z| < 1$ be the generating function of $\{p_{nN}(t)\}$. Multiplying (6.26) by z^n and summing over n we get

$$(\delta/\delta t)\left(\sum_{n=0}^{\infty} p_{nN}(t)z^n\right) = -\left[\sum_{n=0}^{\infty} z^n\{p_{nN}(t)(\lambda_n(t) + \mu_n(t)\}\right]$$

$$+ \sum_{n=0}^{\infty} z^n p_{n-1N}(t)\lambda_{n-1}(t) + \sum_{n=0}^{\infty} z^n P_{n+1N}(t)\mu_{n+1}(t)$$

$$\delta G(z,t)/\delta t = -\sum_{n=1}^{\infty} z^n \lambda_n(t)p_{nN}(t) - \sum_{n=0}^{\infty} z^n \mu_n(t)p_{nN}(t)$$

$$+ z\sum_{n=0}^{\infty} z^n \lambda_n(t)p_{nN}(t) + z^{-1}\sum_{n=0}^{\infty} z^n \mu_n(t)p_{nN}(t)$$

$$= \sum_{n=0}^{\infty} z^n p_{nN}(t)\{(z-1)\lambda_n(t) + (z^{-1} - 1)\mu_n(t)\} \quad (6.28)$$

with initial condition $G(z, 0) = z^N$ if $P(X(0) = N) = 1$.

In the simple linear birth and death process (Generalized B-D processes : see Kendall-1948) $\lambda_n(t) = n\lambda(t)$ and $\mu_n(t) = n\mu(t)$ and (6.28) becomes

$$\delta G(z, t)/\delta t = \delta G(z, t)/\delta z\{(z-1)[z\lambda(t) - \mu(t)]\} \quad (6.29)$$

The auxiliary equation of (6.29) is

$$dz/dt + (z-1)\,[z\lambda(t) - \mu(t)] = 0 \quad (6.30)$$

which can be obtained from the characteristic equation

$$dt/1 = dz/\{(1-z)[z\lambda(t) - \mu(t)]\} = dG/0.$$

Making a change of variable $1/u = 1 - z$, (6.30) can be written as

$$du/dt - [\lambda(t) - \mu(t)]u + \lambda(t) = 0$$

which now involves a linear function of u. The integrating factor for the above equation is $e^{\Delta(t)}$, where $\Delta(t) = -\int_0^t [\lambda(\tau) - \mu(\tau)]d\tau$.

Then we have

$$e^{\Delta(t)}\frac{du}{dt} - [\lambda(t) - \mu(t)]ue^{\Delta(t)} + \lambda(t)e^{\Delta(t)} = 0$$

or
$$d/dt[ue^{\Delta(t)}] + \lambda(t)\,e^{\Delta(t)} = 0.$$

Integrating this and replacing u by $1/1-z$, we get

$$e^{\Delta(t)}/(1-z) + \int_0^t \lambda(\tau)e^{\Delta(t)}d\tau = \text{a constant } C_2.$$

Another equation we get from the characteristic equation is $G(z, t) = C_1$.

The general solution can be written as

$$G(z,t) = f\left(\frac{e^{\Delta(t)}}{1-z}\right) + \int_0^t \lambda(\tau)e^{\Delta(t)}d\tau$$

If $P(X(0) = N) = 1$, then $G(z, 0) = f(1/1 - z) = z^N, |z| < 1$.
Writing $\theta = 1/1 - z$ or $z = 1/1 - \theta, f(\theta) = (1 - 1/\theta)^N$ if $|1 - 1/\theta| < 1$.
The required G.F. is

$$G(z, t) = [1 - 1/\{e^{\Delta(t)}/(1 - z) + \int_0^t \lambda(\tau)e^{\Delta(t)}d\tau\}]^N$$

$$= \left[\frac{(e^{\Delta(t)} + \int_0^t \lambda(\tau)e^{\Delta(\tau)}d\tau - 1) + z\left(1 - \int_0^t \lambda(\tau)e^{\Delta(\tau)}d\tau\right)}{e^{\Delta(t)} + \int_0^t \lambda(\tau)e^{\Delta(\tau)}d\tau - z\int_0^t \lambda(\tau)e^{\Delta(\tau)}d\tau}\right]^N$$

Letting $\alpha(t) = 1 - \left[e^{\Delta(\tau)} + \int_0^t e^{\Delta(\tau)}\lambda(\tau)d\tau\right]^{-1}$ and $\beta(t) = 1 - e^{\Delta(\tau)}[-\alpha(t)]$, the G.F. can be written as

$$G(z, t) = \left[\frac{\alpha(t) + [1 - \alpha(t) - \beta(t)]z}{1 - \beta(t)z}\right]^N \tag{6.31}$$

Let us first calculate $p_n(t) = P[X(t) = n \mid X(0) = 1]$ in the simpler case when $\lambda_n(t) = n\lambda(t)$ and $\mu_n(t) = n\mu(t)$ and population start with one member ($N = 1$). Then the Kolmogorov's equation becomes

$$\frac{\partial p_n(t)}{\partial t} = (n + 1)\mu(t)p_{n+1}(t) + (n-1)\lambda(t)p_{n-1}(t) - n(\lambda(t) + \mu(t))p_n(t), \; n \geq 1$$

and $\frac{\partial p_0(t)}{\partial t} = \mu(t)p_1(t)$. Equation (6.29) becomes

$$\frac{\partial G(z, t)}{\partial t} = \frac{dG(z, t)}{\partial z}\{(z - 1)[z\lambda(t) - \mu(t)]\}$$

with initial conditions $G(z, 0) = z$. The auxiliary equation is

$$\frac{dz}{dt} = -\mu(t) + (\lambda(t) + \mu(t))z - \lambda(t)z^2.$$

The fundamental property of this Recatti equation is that the general solution is a homographic function of the constant of integration, so that $z = \frac{f_1 + cf_2}{f_3 + cf_4}$ and equivalently $c_2 = c = \frac{zf_3 - f_1}{f_2 - zf_4}$, where f_1, f_2, f_3 and f_4 are all functions of time t. Thus, the general solution is of the form $G(z, t) = \psi\left\{\frac{zf_3 - f_1}{f_2 - zf_4}\right\}$ and from the boundary condition $\psi\left\{\frac{zf_3(0) - f_1(0)}{f_2(0) - zf_4(0)}\right\} = z$ and puting $u = \frac{zf_3(0) - f_1(0)}{f_2(0) - zf_4(0)}$ we get $z = \frac{f_1(0) + f_2(0)u}{f_3(0) + f_4(0)u}$.

Hence

$$G(z,t) = \frac{f_1(0) + f_2(0)\dfrac{zf_3 - f_1}{f_2 - zf_4}}{f_3(0) + f_4(0)\dfrac{zf_3 - f_1}{f_2 - zf_4}} = \frac{g_1(t) + zg_2(t)}{g_3(t) + zg_4(t)}$$

where g_1, g_2, g_3 and g_4 are functions of t.

On expansion, we get

$$p_0(t) = \alpha(t) \text{ and } p_n(t) = \{1 - p_0(t)\}\,(1 - \beta(t))[\beta(t)]^{n-1} \quad (n \geq 1)$$

where $\alpha(t)$ and $\beta(t)$ are functions of t. Thus for generalized birth death process the population size at any time t is distributed in a geometric distribution with a modified zero term.

In the generalized birth and death process starting with N individuals the probability of extinction is given by

$$p_{0N}(t) = \left[1 - \frac{1}{e^{\Delta(t)} + \int \lambda(\tau)e^{\Delta(\tau)}d\tau}\right]^N$$

Now

$$e^{\Delta(t)} + \int_0^t \lambda(\tau)e^{\Delta(\tau)}d\tau = e^{\Delta(t)} + \int_0^t \mu(\tau)e^{\Delta(\tau)}d\tau - \int_0^t e^{\Delta(\tau)}[u(\tau) - \lambda(\tau)]d\tau$$

$$= e^{\Delta(t)} + \int_0^t \mu(\tau)e^{\Delta(\tau)}d\tau - [e^{\Delta(\tau)}]_0^t = 1 + \int_0^t \mu(\tau)e^{\Delta(\tau)}d\tau.$$

$$\text{Hence, } p_{0N}(t) = \left[\frac{\int_0^t \mu(\tau)e^{\Delta(\tau)}d\tau}{1 + \int_0^t \mu(\tau)e^{\Delta(\tau)}d\tau}\right]^N$$

We observe that chance of extinction tends to one as $t \to \infty$ iff

$$\lim_{t\to\infty} \int_0^t \mu(\tau)e^{\Delta(\tau)}d\tau = \infty.$$

The first two moments of $X(t)$ are given by

$$m(t) = Ne^{\Delta(t)}$$

$$\sigma^2(t) = Ne^{-2\Delta(t)} \int_0^t [\lambda(\tau) + \mu(\tau)]e^{\Delta(\tau)}d\tau.$$

In Arley process (1943) the birth rate $\lambda(t) = \lambda$ is a constant, death rate is linear $\mu(t) = \mu t$ and $\Delta(t)$ is given by

$$\Delta(t) = \int_0^t [\mu(\tau) - \lambda(\tau)]d\tau = 1/2\,\mu t^2 - \lambda t.$$

Also the mean and variance functions are given by

$$m(t) = \exp[\lambda t(-\mu t^2)/2],\ \sigma^2(t) = m(t)\,[1 - m(t) + 2\lambda m(t)]^2 \int_0^t d\tau / m(\tau).$$

Since $\lim_{t\to\infty} \int_0^t \mu\tau \exp\,[\mu t^2/2 - \lambda\tau]d\tau = \infty$, the extinction is certain in Arley process. There are four important special forms for birth and death rates. If $\lambda_n = \alpha$, a constant we have immigration at rate α. If $\lambda_n = n\lambda$, we have simple linear birth process with birth rate λ per individual. If $\mu_n = \beta$ and $\mu_0 = 0$, we have emigration at rate β. If $\mu_n = n\mu$ we have linear death process with rate μ per individual.

6.6.2 Linear birth and death process

If $\lambda_n(t) = n\lambda$ and $\mu_n(t) = n\mu$ in the generalized birth and death process, proceeding as before, we get from equation (6.31) the P.G.F.

$$\left.\begin{aligned} G(z, t) &= \left[\frac{\alpha(t) + (1 - \alpha(t) - \beta(t))z}{1 - \beta(t)z}\right]^N \text{ where} \\ \alpha(t) &= \frac{\mu(1 - e^{(\lambda-\mu)t})}{\mu - \lambda(e^{(\lambda-\mu)t})} \text{ and } \beta(t) = (\lambda/\mu)\alpha(t) \end{aligned}\right\} \tag{6.32}$$

The general solution $p_{nN}(t)$ is the coefficient of z^n in the expansion of $G(z, t)$.

Now $$\{\alpha(t)+[1-\alpha(t)-\beta(t)]z\}^N = \sum_{j=0}^{\infty} \binom{N}{j} [(\alpha(t))^{N-j}[1-\alpha(t)-\beta(t)]^j z^j] \tag{6.33}$$

Since $|\,\beta(t)z\,| < 1$ and $\binom{-N}{i} = (-1)^i \binom{N+i-1}{i}$

$$[1 - \beta(t)z]^{-N} = \sum_{i=0}^{\infty} \binom{-N}{i} (-1)^i [\beta(t)z]^i = \sum_{j=0}^{\infty} \binom{N+i-1}{i} [\beta(t)]^i z^i \tag{6.34}$$

Using equations (6.33) and (6.34), we get

$$p_{nN}(t) = \sum_{j=0}^{\min(N,n)} \binom{N}{j} \binom{N+n-j-1}{n-j} [\alpha(t)]^{N-j} [\beta(t)]^{n-1} [1 - \alpha(t) - \beta(t)]$$

$$(n \geq 1)$$

and $p_{0N}(t) = [\alpha(t)]^N$, the probability of extinction of the process. Now from (6.26) the asymptotic behaviour of the probability of extinction $p_{0N}(t)$ depend on the parameters μ and λ. Hence

$$\lim_{t\to\infty} p_{0N}(t) = \begin{cases} 1 \text{ if } \mu \geq \lambda \\ (\mu/\lambda)^N \text{ if } \mu < \lambda \end{cases} \tag{6.35}$$

From (6.32) $p_{0N}(t) = [\alpha(t)]^N = G(0, t)$. Hence for $|z| < 1$

$$\lim_{t\to\infty} G(0,t) = \begin{cases} 1 \text{ if } \mu \ge \lambda \\ (\mu/\lambda)^N \text{ if } \mu < \lambda \end{cases} \tag{6.36}$$

Equation (6.35) shows that it is certain that the population dies out when the death rate exceeds the birth rate, but when the birth rate is greater than the death rate, the probability of eventual extinction is equal to $(\mu/\lambda)^N$. From (6.36), since the limiting P.G.F. is constant, the probability that the population will increase without bound is $1 - (\mu/\lambda)^N$. The relative magnitude of λ and μ also influences the asymptotic value of the expectation and variance of $X(t)$.

The moments can be calculated for linear birth and death processes by taking derivative of $G(z, t)$ at $z = 1$. We get

$$m(t) = EX(t) = N\frac{1-\alpha(t)}{1-\beta(t)} = Ne^{(\lambda-\mu)t},$$

$$\sigma^2(X(t)) = \frac{N[1-\alpha(t)][\alpha(t)+\beta(t)]}{[1-\beta(t)^2]} = N\left(\frac{\lambda+\mu}{\lambda-\mu}\right)[e^{(\lambda-\mu)t}-1]e^{(\lambda-\mu)t}$$

Hence
$$\lim_{t\to\infty} m(t) = \begin{cases} 0 & \text{if } \mu > \lambda \\ N & \text{if } \lambda = \mu \\ \infty & \text{if } \lambda > \mu \end{cases}$$

and
$$\lim_{t\to\infty} \sigma^2(X(t)) = \begin{cases} 0 & \text{if } \mu > \lambda \\ \infty & \text{if } \mu \le \lambda \end{cases}$$

When $\lambda = \mu$ we have an interesting case in which though the probability of extinction tends to one, yet the expected population size tends to N. Most population will eventually become extinct while a few others will attain huge sizes, so that the average size is N.

6.6.3 Effect of imigration

The Kendall Process

If with a birth and death process we take into account immigration, we have to assume that during an interval t, there is a probability $\gamma\Delta t + o(\Delta t)$ of a new individual being added to the population from outside the country, we have a birth and death process for which $\lambda_n = n\lambda + \gamma$ and $\mu_n = n\mu$ $(n \ge 0)$, the forward equations being

$$p_n'(t) = -(n(\lambda+\mu)+\gamma)p_n(t) + [\lambda(n-1)+\gamma]p_{n-1}(t) + \mu(n+1)p_{n+1}(t) \ge 0)$$

with initial conditions $p_0(t) = 0$, $p_n(0) = 0$ for $n \ge 1$, i.e. $X(0) = 0$ a.s. Like Feller-Arley process the P.G.F. $G(z, t)$ satisfies the partial differential equation

$$\frac{\partial G}{\partial t} - (z-1)(\lambda z - \mu)\frac{\partial G}{\partial z} = \gamma(z-1)G$$

The auxiliary equation is

$$\frac{dt}{1} = \frac{dz}{-(z-1)(\lambda z - \mu)} = \frac{dG}{\gamma(z-1)G}$$

Now $\frac{dz}{dt} = -(z-1)(\lambda z - \mu)$ and hence, we get

$$(\mu - \lambda)t = \log\frac{z-1}{\lambda z - \mu} + \text{constant or } \frac{\lambda z - \mu}{z-1}e^{(\mu-\lambda)t} = C_2 \tag{6.37}$$

From

$$\frac{dz}{(\lambda z - \mu)(z-1)} = \frac{dG}{\gamma(z-1)G} \text{ we get } G(z, t)(\lambda z - \mu)^{\gamma/\lambda} = C_1 \tag{6.38}$$

or

$$G(z, t) = C_1(\lambda z - \mu)^{-\gamma/\lambda}$$

From (6.37) and (6.38) $G = \phi\left(\frac{\lambda z - \mu}{z-1}e^{(\mu-\lambda)t}\right)(\lambda z - \mu)^{-\gamma/\lambda}$ for some unknown function ϕ to be determined from initial conditions.

Initial conditions give $G(z, 0) = 1$.

Hence

$$\phi\left(\frac{\lambda z - \mu}{z-1}\right)(\lambda z - \mu)^{-\gamma/\lambda} = 1.$$

Put $\theta = \frac{\lambda z - \mu}{z-1}$ i.e. $z = \frac{\theta - \mu}{\theta - \lambda}$, then we have

$$\phi(\theta) = \left(\lambda\left(\frac{\theta-\mu}{\theta-\lambda}\right) - \mu\right)^{\gamma/\lambda} = \theta^{\gamma/\lambda}(\lambda-\mu)^{\gamma/\lambda}(\theta-\lambda)^{-\gamma/\lambda}$$

Now replacing θ by $\frac{\lambda z - \mu}{z-1}e^{(\mu-\lambda)t}$ we get the general solution

$$G(z,t) = \left[\frac{\lambda z - \mu}{z-1}e^{(\mu-\lambda)t}\right]^{\gamma/\lambda}(\lambda-\mu)^{\gamma/\lambda}\left[\lambda - \frac{\lambda z - \mu}{z-1}e^{(\mu-\lambda)t}\right]^{-\gamma/\lambda}(\lambda z-\mu)^{\gamma/\lambda}.$$

Simplifying the above expression we get the generating function $G(z, t)$ as

$$G(z, t) = \left(\frac{p}{1-qz}\right)^{\gamma/\lambda} \tag{6.39}$$

where

$$p = \begin{cases}(\lambda-\mu)/(\lambda e^{(\lambda-\mu)} - \mu) \text{ if } \lambda \neq \mu \\ (1+\lambda t)^{-1} \text{ if } \lambda = \mu\end{cases} \tag{6.40}$$

(applying L' Hospital's rule)

and $q = 1 - p$.

From (6.39) we see that $X(t)$ has a negative binomial distribution. As $t \to \infty$ (6.40) shows that $p \to 0$, $q \to 1$ if $\lambda \geq \mu$, while $p \to 1 - \lambda/\mu$, $q \to \lambda/\mu$ if $\lambda < \mu$.

Therefore, if $\lambda \geq \mu$, no limiting distribution exists for $X(t)$ as $t \to \infty$, i.e. $X(t) \to \infty$ with probability one. If $\lambda < \mu$ we find from (6.39) that

$$\lim_{t\to\infty} G(z,t) = \left(1 - \frac{\lambda}{\mu} z\right)^{-\gamma/\lambda} \left(1 - \frac{\lambda}{\mu}\right)^{\gamma/\lambda} \tag{6.41}$$

which gives in particular $\lim_{t\to\infty} p_0(t) = (1 - \lambda/\mu)^{\gamma/\lambda} < 1$ (constant term in the expansion (6.41) which shows that there is a positive probability that the population avoids ultimate extinction).

The mean number of individuals is given by

$$\frac{\partial G}{\partial z}\Big|_{z=1} = m(t) = \frac{\gamma}{\lambda - \mu}\left(e^{(\lambda-\mu)t} - 1\right) + e^{(\lambda-\mu)t} \text{ for } \lambda \neq u$$

$$= \gamma t + 1 \text{ for } \lambda = \mu \text{ (apply L'Hospital's rule)}$$

Therefore, $\lim_{t\to\infty} m(t) = \dfrac{\gamma}{\mu - \lambda}$ if $\lambda \neq \mu$.

These formulae relate to the stable distribution of population size which immigration can just maintain against the excess of μ over λ. It is clear from (6.39) that the distribution will still be negative binomial in form for every finite value of t when $\mu = 0$, the process being one of immigration and reproduction only. On the other hand, when $\lambda = 0$ so that the process is one of immigration and mortality (or emmigration) only the distribution assume a Poisson form.

Exercises and Complements

Exercise 6.1 Let $\{X(t), t \geq 0\}$ be a stationary Markov process with state space $S = [-1, 1\}$, an initial distribution $\pi(1) = P(X(0) = 1) = 1/2$ and a transition function $p_{ij}(t)$ satisfying $P_{1,1}(t) = p_{-1,-1}(t) = \alpha(t)$ where $\alpha(t)$ is a continuous function on $[0, \infty)$ such that $\alpha(0) = 1$. Find $P_\pi(X(t) = 1) = P(X(t) = 1 \mid X(0) = 1)$. Show that $\alpha(t) = (1 + e^{-\lambda t})/2$, $\lambda > 0$ for all $t \geq 0$.

Exercise 6.2 A stochastic process $X(t)$ $(t \geq 0)$ may take the value 0 and 1 with probabilities $p_0(t)$ and $p_1(t)$, respectively. Also

$$P[X(t+h) = 1 \mid X(t) = 0] = \alpha h + o(h)$$

and

$$P[X(t+h) = 0 \mid X(t) = 0] = \beta h + o(h) \text{ as } h \to 0, \ \alpha \text{ and } \beta$$

being positive constants. Show that

$$p_0(t) = \frac{\beta}{\alpha+\beta} + \left\{p_0(0) - \frac{\beta}{\alpha+\beta}\right\} e^{-(\alpha+\beta)t} = \frac{\beta}{\alpha+\beta} + \frac{\alpha}{\alpha+\beta} e^{-(\alpha+\beta)t}$$

if $p_0(0) = 1$.

Exercise 6.3 (Continuation of Exercise 6.2) Let $\mu_{ij}(i, j = 0, 1)$ be the expected length of time in $(0, t)$ that the Markov process $\{X(t), t \geq 0\}$ with two states $\{0, 1\}$ spends in state j, having started at $X(0) = i$ $(i = 1, 2)$ initially. Then show that $\mu_{ij}(i, j = 0, 1)$ are given by

$$\mu_{00}(t) = \beta t/(\alpha+\beta) + \frac{\alpha}{(\alpha+\beta)^2}\left[1 - e^{-(\alpha+\beta)t}\right]$$

$$\mu_{01}(t) = \frac{\alpha t}{(\alpha+\beta)} - \frac{\alpha}{(\alpha+\beta)^2}[1 - e^{-(\alpha+\beta)t}]$$

$$\mu_{10}(t) = \frac{\beta t}{(\alpha+\beta)} - \frac{\beta}{(\alpha+\beta)^2}[1 - e^{-(\alpha+\beta)t}]$$

$$\mu_{11}(t) = \frac{\alpha t}{(\alpha+\beta)} + \frac{\beta}{(\alpha+\beta)^2}[1 - e^{-(\alpha+\beta)t}]$$

Exercise 6.4 (Contagion or Accident Proneness) Suppose that the probability of an accident occuring in t to $t + \Delta t$ is $(\mu + \lambda_j)\,\Delta t + o(\Delta t)$ given that $j \geq 0$ accidents have occured previous to time t, for $v, \lambda_j > 0$. Show that this is a pure birth process on the set of integers $S = \{0, 1, 2, \ldots\}$ which satisfies the forward equations

$$p'_{0n}(t) = -(v + n\lambda)p_{0n}(t) + (v + (n-1)\lambda)p_{0,n-1}(t),\ n \geq 1 \tag{1}$$

$$p'_{00}(t) = -v\,p_{00}(t)$$

and the transition probabilities of $X(t)$ are given by

$$P_{jk}(t) = e^{-(v+j\lambda)t}\binom{-j - v/\lambda}{k-j}(e^{-\lambda t} - 1)^{k-j}(k \geq j) \tag{2}$$

In particular show that

$$P_{0n+1}(t) = \frac{v(v+\lambda)\ldots(v+n\lambda)}{(n+1)!\lambda^{n+1}}\,e^{-vt}(1 - e^{-\lambda t})^{n+1} \tag{3}$$

Exercise 6.5 (Branching Birth and Death Process) Consider a collection of particles which act independently in giving rise to succeeding generation of particles. Suppose that each particles, from the time it appears, waits a random length of time having an exponential distribution with parameter θ and the splits into two identical particles with probability p and disappears with probability $1 - p$. Let $X(t)$, $0 \leq t \leq \infty$ be the number of particles present at time t. This branching process is a birth and death process. Find the birth and death rates.

Exercise 6.6 In exercise 6.5 suppose that new particles immigrate into the system at random times from a Poisson process with parameter λ and then give rise to succeeding generations as described in the previous exercise.

Find the birth and death rates of this birth and death process.

Exercise 6.7 (Application of Birth and Death Process in a Telephone Traffic Problem) Consider a telephone exchange where the number of available lines is large that for all practicle purposes it can be considered infinite. Assume that $Y(t)$, the number of lines in use at time t follows a birth and death process with $P(Y(t+h) = i + 1 \mid Y(t) = i) = \lambda h + o(h)$, $t > 0$, $h > 0$ and $i \geq 1$.

Also $P(Y(t + h) = i - 1 \mid Y(t) = i) = \mu_i h + o(h)$ as $h \to 0$

(In the first case we have assumed that a call will occur in $(t, t + h)$ is independent of the number of busy lines at time t): Asume also that $\mu_i = i\mu$. Find the probability generating function (P.G.F.) of $Y(t)$, $P(Y(t) = k)$ and $E(Y(t))$.

Exercise 6.8 Consider the *Feller-Arley process (linear growth process)* which is a birth and death process with birth and death rates $\lambda_n = n\lambda$ and $\mu_n = n\mu$ respectively for all $n \geq 1$ and initial condition $P(X(0) = n) = \delta_{n1}$.

(a) Show that the generating function $G(z, t)$ of the size distribution at time t satisfies the partial differential equation

$$\frac{\partial G}{\partial t} = \{\mu - (\lambda + \mu)z + \lambda z^2\}\frac{\partial G}{\partial z}.$$

(b) Show that the size distribution is given by

$$p_n(t) = P\{X(t) = n \mid X(0) = 1\} = (1 - \xi)(1 - \eta)\eta^{n-1},\ n \geq 1$$

where $\xi = \xi(t)$ and $\eta = \eta(t)$ are functions of t, λ and μ.

(c) What is $p_n(t)$ if $\lambda = \mu$?

(d) Discuss the limits $p_n(t)$ as $t \to \infty$, for $\lambda > \mu$, $\lambda = \mu$ and $\lambda < \mu$.

(e) Compute the mean population size and its limit for the three cases $\mu < \lambda$, $\mu = \lambda$ and $\mu > \lambda$. Also do the same for variance.

(f) Classify the states of this process.

Exercise 6.9 (Feller) Let $\{X(t), t \geq 0\}$ be a pure birth process with distinct birth rates λ_i and $\lambda_0 \geq 0$. Then show that: (a) the transition function $p_{ij}(t)$ satisfies the recursion relation $p_{ii+1}(t) = \lambda_i \int_0^t \exp(-(t-s)\lambda_{i+1})e^{-s\lambda_i}\,ds$

$$= \begin{cases} \lambda_i[\exp(-t\lambda_i) - \exp(-t\lambda_{i+1})]/(\lambda_{i+1} - \lambda_i) \text{ if } \lambda_i \neq \lambda_{i+1} \\ \lambda_i \exp(-t\lambda_i) \text{ if } \lambda_i = \lambda_{i+1} \end{cases}$$

(b) Show that the forward equations have the unique solution given by

$$p_{ij}(t) = \begin{cases} \sum_{k=i}^{j} B(k,i,j)\exp(-t\lambda_k) \text{ if } j \geq i \\ 0 \quad \text{if } j < i \end{cases}$$

where $$B(k,i,j) = \frac{\lambda_i \lambda_{i+1} \ldots \lambda_{j-1}}{(\lambda_i - \lambda_k)\ldots(\lambda_{k-1} - \lambda_k)(\lambda_{k+1} - \lambda_k)\ldots(\lambda_j - \lambda_k)}$$

(c) Show that in the linear growth process ($\lambda_k = k\lambda$)

$$p_{ij}(t) = \exp(-i\lambda t)\binom{j-i}{j-1}(1 - e^{-t\lambda})^{j-i}$$

Exercise 6.10 (Feller and Lundberg) Let $\{X(t), t \geq 0\}$ be a pure birth process with distinct birth rates λ_i and $\lambda_0 \geq 0$. Then show that only finite number of transitions occur with probability one, i.e.

$$\sum_{j=k}^{\infty} p_{ij}(t) = 1 \text{ iff } \sum_{j=k}^{\infty} 1/\lambda_j = \infty \text{ for all finite } k. \tag{1}$$

Also prove that if (1) holds, the nth jump time

$$W_n \to \infty \text{ a.s. as } n \to \infty.$$

Exercise 6.11 Let $\{X(t), t \geq 0\}$ be a pure birth process with birth rates $\{\lambda_k\}_0^\infty$. (a) show that $X(t) \to \infty$ *a.s.* as $t \to \infty$ iff $\sum_{k=0}^{\infty} \frac{1}{\lambda_k} < \infty$. Hence show that a Yule process or a linear growth process (i.e. $\lambda_k = k\lambda$) does not move out to infinity.

Exercise 6.12 (a) Show that for a stationary distribution to exist in a birth and death process it is necessary and sufficient that the series $\sum_{k=1}^{\infty} \frac{\lambda_0 \lambda_1 \ldots \lambda_{k-1}}{\mu_1 \mu_2 \ldots \mu_k}$ $(\mu_k > 0,\ k \geq 1)$ converges (λ_k's and μ_k's have their usual meanings).

(b) (The servicing of lathes) Suppose that m lathes are serviced by a crew of a repairmen. When a lathe fails to function properly it is repaired immediately unless all of the repairmen are working on lathes that have already failed in which case the lathe must await repair. The following assumptions are made about functioning of a lathe.

(1) The probability of getting out of order during an interval of time $(t, t + \Delta t)$ is indpependent of t and is equal to $\lambda(\Delta t) = \lambda \Delta t + o(\Delta t)$, independently of the "history" of its operation. Similarly, if a lathe is being repaired, the probability of its being putback into operation during an interval of time $(t, t + \Delta t)$ is equal to $\mu(\Delta t) = \mu \Delta t + o(\Delta t)$ and is independent of the nature of its work and its length of service upto the instant t. Let E_k denote the event that at a given instant the number of lathes being repaired or awaiting repair is equal to k subsequent removal of a single lathe from service denotes transition to the state E_{k+1}, and completion of the repair of one of the lathes indicates transition into the state E_{k-1}. Show that this is a birth and death process with finitely many states E_0, $E_1, \ldots, E_m$. Find also its stationary distribution.

Exercise 6.13 Let M be a discrete time a periodic Markov chain with state space $S = \{0, 1, 2, \ldots\}$, transition matrix (p_{ij}) and initial distribution $\{\alpha_i, i \in S\}$. Let λ be a fixed positive real number and s any real number of $[0, \infty)$. Define for all $i, j \in S$

$$\tilde{p}_{ij}(s) = \sum_{n=0}^{\infty} e^{-\lambda s} (\lambda s)^n / n! p_{ij}^{(n)}.$$

1. Show that $(\tilde{p}_{ij}(s))$ is a transition matrix of a continuous time Markov chain $\tilde{M}$, with state space S, and initial distribution $\{\alpha_i, i \in S\}$. [Note : Only check the conditions for transition functions of a continuous parameter Markov process. Do not prove actual existence of $\tilde{M}$.]

2. Evaluate the infinitesimal generator $\tilde{A}$ of $\tilde{M}$ in terms of the original matrix P of M Is $\tilde{M}$ conservative?

3. Prove the following statements:

(a) $\tilde{M}$ is irreducible iff M is irreducible.
(b) State $j \in \tilde{S}$ is positive recurrent iff $j \in S$,
(c) State $j \in \tilde{S}$ is transient iff $j \in S$ is transient.

Exercise 6.14 Continuation of Exercise 6.13.4. Assume that M is irreducible and positive recurrent with $\{\pi_j, j \in S\}$ as stationary distribution and there exists a constant ρ, $0 \leq \rho < 1$ such that for all $n \geq 0$

$$| p_{ij}^{(n)} - \pi_j | \leq \rho^n \text{ for all } i, j \in S.$$

Prove that for all $s \geq 0$

$$| \tilde{p}_{ij}(s) - \pi_j | \leq \exp[-\lambda(1 - \rho)s].$$

Is there any anologous result for M transient ? Null recurrent ?

Remark In the proof of the next exercise, use the following theorem of Chung: If the infinitesimal generator $A = (a_{ij})$ satisfies $\operatorname{Sup} | a_{ij} | \leq K < \infty$, then A determines the transition matrix $P(t)$ uniquely.

Show that a Markov chain with $\sup_i |a_{ij}| \leq K < \infty$ is positive iff there exists a solution $\{X_i, i \in S\}$ for $\sum_{i=0}^{\infty} x_i a_{ij} = 0$ for which $0 < \sum_{i=0}^{\infty} |x_i| < \infty$.

** **Exercise 6.15** Let $P(t)$ be the transition matrix of a continuous time Marikov chain with infinitesimal generator A given by

$$A = \begin{pmatrix} -\alpha & \beta & \gamma \\ \gamma & -\alpha & \beta \\ \beta & \gamma & -\alpha \end{pmatrix}, \text{ where } \beta + \gamma = \alpha > 0.$$

By the remark to Exercise 6.14, we can determine $P(t)$ as follows:

1. Prove that the eigenvalues of A are $0, -\rho + i\sigma, -\rho - i\sigma$ where

$$\rho = \frac{3\alpha}{2}, \text{ and } \alpha = 1/2\,(\beta - \gamma)\sqrt{3}.$$

2. We put for all $t \geq 0$

$$P(t) = R + S\exp(-(\rho + i\sigma)t) + T\exp(-(\rho - i\sigma)t)$$

where R, S and T are 3×3 matrices with (possibly) complex entries.

Use the following facts to compute R, S and T.

(a) A is symmetric; (b) $P(t)$ is real for all $t \geq 0$;

(c) $P(0) = I$: (d) $P'(0) = A$; (e) Refer to Exercise 6.14. Show that all entries of the matrix R are 1/3.

3. Using the above results, indicate how it is possible for an infinite number of different continuous time Markov Chains, to have a common skeleton chain.
4. Use the above results to prove: If the one-step transition matrix of a descrete time markov chain is of the form

$$\begin{pmatrix} p & q & r \\ r & p & q \\ q & r & p \end{pmatrix}, \text{ where } p + q + r = 1,\ p \neq 1, \text{ then there exist infinitely many}$$

solutions to Elfving's problem.

****Exercise 6.16** Consider a generalised birth and death process with $\lambda_n(t) = n\lambda t$, $\lambda > 0$ and $\mu_n(t) = n\mu t$, $\mu > 0$ for all $n \geq 0$. Also $p_n(0) = \delta_{1n}$.

(a) Solve this birth and death process by exhibiting an explicit expression for $p_n(t) = P(X(t) = n \mid X(0) = 1)$.

(b) Compute the mean population size, and evaluate its limit as $t \to \infty$.

(c) If T denotes the time till absorption, then show that $P(T \leq t) = p_0(t)$. Compute $E(T)$ if $\lambda = 0$.

Exercise 6.17 Show that Poisson processes are the only renewal processes which are Markov Chains.

7

Poisson Process

7.1 Definitions and Their Equivalence

Definition 7.1 Let N_t be a non-negative integer-valued stochastic process, where $t \in T = [0, \infty]$, satisfying the conditions:

(1) $P[N_0 = 0] = 1$

(2) $\{N_t, t \in T\}$ has independent increments, i.e., for

$$t_0 < t_1 < \ldots < t_n \text{ in } T \; N_{t_n} - N_{t_{n-1}}, N_{t_{n-1}} - N_{t_{n-2}}, \ldots, N_{t_1} - N_{t_0}$$

are independent random variables.

(3) $\{N_t, t \in T\}$ has stationary increments, i.e. for $t > 0$, $N_t - N_s$ has same distribution as $N_{t+h} - N_{s+h}$ for all h such that $t + h$ and $s + h$ are in T.

(4) For $t > s$, $P[N_t - N_s = k] = e^{-\lambda(t-s)} \dfrac{[\lambda(t-s)]^k}{k!}$, $\lambda > 0, k = 0, 1, 2, \ldots$

λ is called the intensity of the Poisson process.

Let $\{X_t, t \in T\}$, be a stationary independent increment process with $P[X_0 = 0] = 1$.

Then $\varnothing_{X_{t_1+t_2}}(u) = E(e^{iuX_{t_1+t_2}})$

$$= E(e^{iu(X_{t_1+t_2} - X_{t_1})} e^{iuX_{t_1}})$$

$$= E(e^{iu(X_{t_1+t_2} - X_{t_1})} E(e^{inX_{t_1}})$$

(because of independence of increments)

$$= [e^{in(X_{t_2} - X_0)}] E(e^{iux_{t_1}}) \text{ (because of stationarity)}$$

$$= \varnothing_{X_{t_2}}(u) \cdot \varnothing_{X_{t_1}}(u)$$

Let $f(t) = \varnothing_{X_t}(u)$. Then $f(t)$ satisfies the following functional equation (known as Cauchy equation)

$$f(t_1 + t_2) = f(t_1) \cdot f(t_2) \quad \text{for all } t_1 \text{ and } t_2.$$

If f is continuous, the solution is $f(1) = [f(1)]'$ for all real t. Hence we get the following proposition.

Proposition 7.1 If $\{X_t, t \in T\}$ be a stationary independent increment stochastic process with $P[X_0 = 0] = 1$ and $\varnothing_{X_t}(u) = f(t)$ is continuous in t for every u, then $\varnothing_{X_t}(u) = [\varnothing_{X_1}(u)]^t$, where $(\varnothing_{X_t}(u)$ is called the characteristic functional of X_t.

Definition 7.1a A characteristic function $\varnothing(u)$ (random variable X with characteristic function $\varnothing(u)$) is called infinitely divisible if $[\varnothing(u)]^t$ is a characteristic function for all $t > 0$. Thus, the Poisson random variable is infinitely divisible. Similarly, the Poisson process has infinitely divisible characteristic functional, since $\varnothing_{X_t}(u) = e^{-\lambda t(1-e^{iu})}$ is infinitely divisible.

Poisson process can be characterized as pure birth process. Hence we have the following alternative definition of a Poisson process.

Definiton 7.2 It is a continuous time Markov chain (Birth process) with $\lambda_n = \lambda$ for all n and $\mu_n = 0$, i.e. $dp_n(t)/dt = -\lambda p_n(t) + \lambda p_{n-1}(t)$ is its Kolmogorov's differential equation, where $p_n(t) = P[N_t = n \mid N_0 = 0]$.

The G.F. $G(t, z) = \sum_{n=0}^{\infty} p_n(t) z^n, |z| < 1$ of Poisson process is given by

$$\frac{dG}{dt} = \sum_{n=0}^{\infty} [-\lambda p_n(t) + \lambda p_{n-1}(t)] z^n$$

$$= -\lambda G + \lambda z G = \lambda(z - 1)\, G$$

$$\Rightarrow \frac{dG}{G} = \lambda(z - 1)\, dt \Rightarrow G(t, z) = c.e^{\lambda(z-1)t}.$$

The boundary condition $1 = G(0, z)$ gives $c = 0$. Hence

$$G(t, z) = e^{-\lambda t} \sum_{n=0}^{\infty} \frac{(\lambda t)^n}{n!} z^n.$$

Therefore $P[N_t = n] = p_n(t) = \dfrac{e^{-\lambda t}(\lambda t)^n}{n!}$.

Axiomatic Definition 7.3 It is a continuous parameter stochastic process satisfying the conditions:

(1) $P[N_0 = 0] = 1$

(2) $\{N_t, t \in T\}$ has independent increments

(3) $\{N_t, t \in T\}$ has stationary increments

(4) $P[N_t > 0] > 0$ for all $t > 0$

(5) $\lim_{h \downarrow 0} \dfrac{P[N_h \geq 2]}{P[N_n = 1]} = 0$

or

(4′) $P[N_h = 1] = \lambda h + o(h),\ h \downarrow 0,\ \lambda > 0$

(5′) $P[N_h \geq 2] = o(h)$ as $h \downarrow 0$

Conditions (4) and (5) or (4′) and (5′) are known as regularity or oderly properties of Poisson process.

Definition 7.4 Poisson process is a Renewal process with interarrival times as exponential distributions (see Renewal theory).

Let N_t denote the number of occurrences upto time t.

Theorem 7.1 Under the axioms (postulates) (1) to (5) N_t has Poisson distribution with mean function λt, i.e. $p_n(t)$ is given by the Poisson law

$$e^{-\lambda t}(\lambda t)^n/n!\ n = 0, 1, 2, \ldots$$

Proof Let $p_n(t) = P(N_t = n|N_0 = 0) = P(N_t = n)$ (by axiom 1). Actually $P(N_{t+h} - N_t = y|N_t = x)$ is independent of x. Axioms (4′) and (5′) $\Rightarrow p_0(t) = 1 - \lambda t + o(t)$. Consider $p_n(t + h)$ for $n \geq 0$.

The n events by epoch $t + h$ can happen in the following mutually exclusive ways $A_1, A_2, A_3, \ldots, A_n$.

For $n \geq 1$

$A_1 = \{n$ events occur by epoch t and no event occurs between t and $t + h\}$

we have $$P(A_1) = P[N_t = n]P[N_{t+h} - N_t = 0|N_t = n]$$
$$= p_n(t)p_0(h) = p_n(t)(1 - \lambda h) + o(h),$$

$A_2 = \{(n - 1)$ events occur by epoch t and 1 event occurs between t and $t + h\}$

$$P(A_2) = P[N_t = n - 1]P[N_{t+h} - N_t = 1|N_t = n - 1]$$
$$= p_{n-1}(t)p_1(h) = p_{n-1}(t)(\lambda h) + o(h),$$

$A_3 = \{(n - 2)$ events occur by time t and 2 events occur between t and $t + h\}$

$$P(A_3) = p_{n-2}(t)\{o(h)\} = o(h)$$

and so on for $A_4, A_5, \ldots, A_n$ i.e. $P(A_k) = o(h)$ for $4 \leq k \leq n$.

Therefore $p_n(t + h) = p_n(t)(1 - \lambda h) + p_{n-1}(t)\lambda h + o(h)$ for $n \geq 1$

or $$\frac{p_n(t + h) - p_n(t)}{h} = -\lambda p_n(t) + \lambda p_{n-1}(t) + \frac{o(h)}{h}$$

Taking

$$\lim h \to 0 \text{ yields } p_n'(t) = -\lambda p_n(t) + \lambda p_{n-1}(t), n \geq 1 \ldots \qquad (7.1)$$

as the Kolmogorov's differential equation for Poisson process.

For $n = 0$, $p_0(t + h) = p_0(t)p_0(h) = p_0(t)(1 - \lambda h) + o(h)$.

Therefore $$\frac{p_0(t + h) - p_0(t)}{h} = -\lambda p(t) + \frac{o(h)}{h}$$

Taking limit $h \to 0$. we get

$$p_0'(t) = -\lambda p_0(t) \qquad (7.2)$$

Initial conditions: $N_0 = 0$, i.e. $p_0(0) = 1, p_n(0) = 0$ for $n \neq 0$ (7.3)

Equations (7.1), (7.2) and (7.3) give by generating function method (done after definition 7.2)

$$p_n(t) = \frac{e^{-\lambda t}(\lambda t)^n}{n!},\ n \geq 1. \qquad (7.4)$$

Alternatively by induction on n, for $n = 0$ (7.2) gives $p_0(t) = c.e^{-\lambda t}$, since $p_0(0) = 1$, $c = 1$.

Therefore $$p_0(t) = e^{-\lambda t} \tag{7.5}$$

Consider equation (7.1) for $n = 1$ and substituting the value of p_0 and solving the equation by using (7.3), we get

$$p_1(t) = \lambda t e^{-\lambda t}.$$

Assume (7.4) holds for $n = 1$. In a similar manner it can be shown that it holds in general for $n + 1$ if it holds for n.

7.2 Poisson Process and Renewal Theory

Theorem 7.2 If $\{N_t, t \in T\}$ is a Poisson process with intensity $\lambda > 0$. Then

(a) $\{N_i\}$ is a Renewal process with exponential interarrival times T_i i.e. $P(T_i \leq t) = F(t) = 1 - e^{-\lambda t}$.

(b) $W_n \left(= \sum_{i=1}^{\infty} T_i \right)$, the waiting times are gamma (Erlang), i.e.

$$P(W_n \leq t) = F^{(n)}(t) = \frac{\lambda}{\sqrt{n}} \int_0^t (\lambda x)^{n-1} e^{-\lambda x} dx.$$

(Waiting time W_n = Time from origin to the nth subsequent occurrence)

(c) $H(t) \left(= \sum_1^{\infty} F^{(n)}(t) \right)$, the renewal function in this case is given by λt.

Proof

(a) See Theorem 7.5. (b) This is trivial.

(c) Note that $P(N_t \leq n - 1) = P(W_n > t) = 1 - P(W_n \leq t)$

$$= 1 - \int_0^t \frac{\lambda^n x^{n-1}}{\Gamma n} e^{-\lambda x} dx = \frac{1}{\Gamma n} \int_{\lambda t}^{\infty} y^{n-1} e^{-y} dy$$

$$= \sum_{k=0}^{n-1} e^{-\lambda t} (\lambda t)^k / k! \text{ (integrating by parts)}$$

Therefore, $P(N_t = n) = e^{-\lambda t}(\lambda t)^n/n!$

Hence $H(t) = \sum_{n=1}^{\infty} F^{(n)}(t) = E(N_t) = \lambda t$.

Recall the definitions from Renewal theory that the *Excess life time* or *Residual life-time* or *Forward recurrence time* is defined by $Y(t) = W_{N_{t+1}} - t$. The *Backward recurrence time* is $Z(t) = t - W_{N_t}$. (Note that we have used the notation W_n instead of S_n used in the Renewal Theory Chapter 4).

Recall the Theorems 4.5 and 4.6 from Renewal theory.

Theorem 7.3 (i) $P[Y(t) \leq x] = \int_t^{t+x} [1 - F(t + x - u)] dH(u)$ for $t > 0$,

$$= F(t + x) - \int_0^t [1 - F(t + x - u)] dH(u) \text{ if } x \geq 0$$

(ii) $P[Z(t) \le x] = \int_{t-x}^{t} [1 - F(t-u)]dH(u)$

$$= \begin{cases} F(t) - \int_0^{t-x} [1 - F(t-u)]dH(u) & \text{if } x \le t \\ 1 \text{ if } x > 1 \end{cases}$$

Proof
Let $S_1 < S_2 < \ldots < S_n$ be positive renewal points with S_n be the largest such points. Then

$[Y(t) \le x]$ = [At least one renewal point fall in the interval $(t, t + x]$].

Hence $P(Y(t) \le x) = P(\bigcup_{n \ge 1} \{t < S_n \le t + x < S_{n+1}\})$

$$= \sum_{n=1}^{\infty} P(t < S_n \le t + x < S_{n+1}).$$

By conditioning on $S_n = u$ and taking expectation, we get

$$P(t < S_n \le t + x < S_{n+1}) = \int_t^{t+x} P[S_{n+1} - S_n > t + x - u]dF^{(n)}(u),$$

where $F^{(n)}$ is the n-fold convolution of F.

Therefore

$$P[Y(t) \le x] = \int_t^{t+x} P[S_{n+1} - S_n > t + x - u]dF(u),$$

$$= \int_t^{t+x} [1 - F(t + x - u)]dH(u), \text{ where } H(u) = \sum_{n=1}^{\infty} F^{(n)}(u).$$

The rest of the proof is given in Theorems 4.5 and 4.6.

Corollary 7.1 If $\{N_t, t \in T\}$ is a Poisson process, then

$$P[Z(t) \le x] = \begin{cases} 1 - e^{-\lambda x} & \text{if } x \le t \\ 1 & \text{if } x > t \end{cases}$$

Proof $P[Z(t) \le x] = 1 - e^{-\lambda t} - \lambda \int_0^{t-x} e^{-\lambda(t-s)} ds = 1 - e^{-\lambda x}$.

Note If $x > t$ $[Z(t) \le x]$ is a sure event.

Corollary 7.2 $P(Y(t) \le x) = 1 - e^{-\lambda x}$ if $x > 0$.
Thus the excess life processes have the same exponential distribution as every life, a manifestation of the memoryless property of the exponential distribution. The spent life has a truncated exponential distribution.

Note (1) This shows that the Poisson process is a stationary or equilibrium process.

(2) $Y(t)$ and Z(t) are independent iff $\{N(t), t \in T\}$ is a Poisson process (see Exercise 7.9).

7.3 Properties of Poisson Process

Elementary Properties of Poisson Process

(1) Sum of two independent Poisson processes is a Poisson process.

(2) The difference of two independent Poisson processes $N_1(t)$ and $N_2(t)$ $N_t = N_1(t) - N_2(t)$ has probability distribution given by

$$P[N_t = n] = e^{-(\lambda_1 \lambda_2)t} \left(\frac{\lambda_1}{\lambda_2}\right)^{n/2} I_{|n|}\left(2t\sqrt{\lambda_1 \lambda_2}\right) \tag{7.6}$$

$$n = 0, \pm 1, \pm 2, \ldots$$

where $$I_n(x) = \sum_{r=0}^{\infty} \frac{(x/2)^{2r+n}}{r!\,\overline{|n + r + 1}}$$

is the modified Bessel function of order $n (n \geq -1)$.

Proof The p.g.f. of N_t is $E(s^{N_t}) = E(s^{N_1(t) - N_2(t)})\, E(s^{N_t})$

$$= E(s^{N_1(t)})\, E\left(\frac{1}{s}\right)^{N_2(t)} = \exp\{\lambda_1 t(s-1)\} \exp\left\{\lambda_2 t\left(\frac{1}{s} - 1\right)\right\}$$

$$= \exp\{-(\lambda_1 + \lambda_2)t\} \exp\{\lambda_1 ts + \lambda_2 t/s\} \tag{7.7}$$

$P[N_t = n]$ is the coefficient of s^n in the series expansion of (7.7) in positive and negative powers of s

or directly, $$P[N_t = n] = \sum_{r=0}^{\infty} P[N_1(t) = n + r] P[N_2(t) = r]$$

$$= \sum_{r=0}^{\infty} \frac{e^{-\lambda_1 t}(\lambda_1 t)^{n+r}}{(n+r)!} \frac{e^{-\lambda_2 t}(\lambda_2 t)^r}{r!}$$

$$= e^{-(\lambda_1 + \lambda_2)t} \left(\frac{\lambda_1}{\lambda_2}\right)^{n/2} \sum_{r=0}^{\infty} \frac{\left(t\sqrt{\lambda_1 \lambda_2}\right)^{2r+n}}{r!(n+r)!}$$

Note (1) $I_{-n}(t) = I_n(t) = I_{|n|}(t)$.

(2) $EN_t = (\lambda_1 - \lambda_2)t$, $E(N_t)^2 = (\lambda_1 + \lambda_2)t + (\lambda_1 - \lambda_2)^2 t^2$ and Var $(N_r) = (\lambda_1 + \lambda_2)t$.

Example 7.1 If passengers arrive (singly) at a taxi-stand in accordance with a Poisson process with parameter λ_1, and taxi arrive in accordance with an independent Poisson process with parameter λ_2 then $N_t = N_1(t) - N_2(t)$ gives the excess of passengers over taxis in an interval of length t. The distribution of N_t $P[N_t = n]$, $n = 0, \pm 1, \pm 2, \ldots$, is given by (7.6). Then Mean of N_t is $(\lambda_1 - \lambda_2)t \gtreqless 0$ according as

$$\lambda_1 > = < \lambda_2 \text{ and Var } (N_t) = (\lambda_1 + \lambda_2)t.$$

The Poisson type of occurrences are called purely random events and the poisson process is called a purely (or completely) random process the occurrence is equally likely to happen anywhere in [0, T] given that only one occurrence has taken place in that interval. We state this by the following theorem.

Theorem 7.4 Given that only one occurrence of a Poisson process N_t has occurred by epoch T, the distribution of the time interval τ, in $[0, T)$ in which it occurred is uniform in $[0, T]$, i.e.

$$P[t < \tau < t + dt | N_T = 1] = \frac{dt}{T}, 0 < t < T$$

Proof Since τ is exponential, $P[t < \tau < t + dt] = \lambda e^{-\lambda t}\, dt$. Also

$$P[N_T = 1] = e^{-\lambda T}(\lambda T) \quad \text{and} \quad P[N_T = 1 | \tau = t] = e^{-\lambda(T-t)}$$

which is the probability of no occurrence in the intrval of length $(T - t)$.

$$P[t < \tau < t + dt | N_T = 1] = \frac{P[t < \tau < t + dt, N_T = 1]}{P[N_T = 1]}$$

$$= P[t < \tau < t + dt] P[N_T = 1 \mid \tau = t] / P[N_T = 1]$$

$$= \frac{(\lambda e^{-\lambda t}\, dt)(e^{-\lambda(T-t)})}{(e^{-\lambda T})(\lambda T)} = \frac{dt}{T}$$

Note It may be said that a Poisson process distributes points at random over the infinite interval $[0, \infty]$ in the same way as the uniform distribution distributes points at random over a finite interval $[a, b]$.

Interarrival time

With a Poisson process $\{N_t\}$, the number of occurrences of an event E by epoch t, there is associated a random variable—the interarrival time X between two successive occurrences of E.

Theorem 7.5 The interarrival time between two successive occurrences of a Poisson process $\{N_t\}$ having parameter λ has a negative exponential distribution with mean $1/\lambda$.

Proof Let X be the r.v. representing the interval between two successive occurrences of $\{N_t\}$ with $P[X \le x] = F(x)$ be its distribution function. Let E_i and E_{i+1} be the successive events and E_i occurred at time t_i. Then

$$P[X > x] = P[E_{i+1} \text{ did not occur in } (t_i, t_i = x) \text{ given that } E_i \text{ occurred at instant } t_i]$$

$$= P[E_{i+1} \text{ did not occur in } (t_i\ t_i + x) | N_{t_i} = i]$$

$$= P[\text{no occurrence takes place in an interval } (t_i, t_i + x) \text{ of length } x | N_{t_i} = i].$$

$$= P[N_{t_i + x} - N_{t_i} = 0 | N_{t_i} = i]$$

$$= P[N_x = 0] \text{ (since } N_t \text{ is stationary and has independent increments)}$$

$$= p_0(x) = e^{-\lambda x}, x > 0.$$

$$F(x) = P[X \le x] = 1 - P[X > x] = 1 - e^{-\lambda x}, x > 0.$$

Actually the property of the Poisson process stated in Theorem 7.5 is a characterizing property which will be discussed in next theorem.

7.4 Characterization of Poisson Process

Theorem 7.6 If T_i denotes the interval between E_i and E_{i+1}, $i = 1, 2, 3, \ldots$, in a renewal process then $T_1, T_2, T_3, \ldots$, are i.i.d. r.v.s with

(a) $P[T_i \le x] = 1 - e^{-\lambda x}$, $x \ge 0$ iff the renewal process is a Poisson process.

(b) If $\tau_n = T_i + T_{i+1} + \ldots + T_{n-1}$, then $\tau_n \to \infty$ a.s.

Proof (*J.L. Doob*) Consider only two interarrival times T_1 and T_2. By previous Theorem 7.5

$$P[T_1 \le t_1] = 1 - e^{-\lambda t_1} = P[N(t_1) - N(0) \ge 0],\ t_1 \ge 0.$$

Also

$$P[T_1 \le t_1, T_2 \le t_2] \le \sum_{j=0}^{n-1} P\left[N\left(\frac{jt_1}{n}\right) - N(0) = 0,\ N\left(\frac{j+1}{n}t_1\right) - N\left(\frac{jt_1}{n}\right)\right.$$

$$\left. = 1,\ N\left(\frac{j+1}{n}t_1 + t_2\right) - N\left(\frac{j+1}{n}t_1\right) \ge 1\right] + \sum_{j=0}^{n-1} P\left[N\left(\frac{jt_1}{n}\right) - N(0)\right.$$

$$\left. = 0,\ N\left(\frac{j+1}{n}t_1\right) - N\left(\frac{jt_1}{n}\right) \ge 2\right]$$

$$\text{Now the second sum} \le \sum_{j=0}^{n-1} e^{-\lambda jt_1/n}\left(1 - e^{-\frac{\lambda t_1}{n}} - e^{-\frac{\lambda t_2}{n}} \lambda \frac{t_1}{n}\right)$$

$$= \frac{1 - e^{-\lambda t_1}}{1 - e^{-\lambda t_1/n}} \cdot o(n^{-2}) = o\left(\frac{1}{n}\right).$$

First sum $\le \sum_{j=0}^{n-1} e^{(-\lambda jt_1/n - \lambda t_1/n)}\left(\frac{\lambda t_1}{n}\right)(1 - e^{-\lambda t_2})$ (by L'Hospital's rule and convergence of G.P. series) $\to (1 - e^{-\lambda t_1})(1 - e^{-\lambda t_2})$ as $n \to \infty$.

Hence $P[T_1 \le t_1, T_2 \le t_2] \le (1 - e^{-\lambda t_1})(1 - e^{-\lambda t_2})$

$$= P[T_1 \le t_1]P[T_2 \le t_2].$$

To prove the reverse inequality,

$$P[T_1 \le t_1, T_2 \le t_2] \ge \sum_{j=0}^{n-1} P\left[N\left(\frac{jt_1}{n}\right) - N(0)\right.$$

$$\left. = 0,\ N\left(\frac{j+1}{n}t_1\right) - N\left(\frac{jt_1}{n}\right) = 1,\ N\left(\frac{j}{n}t_1 + t_2\right) - N\left(\frac{j+1}{n}t_1\right) \ge 1\right]$$

$$= \sum_{j=0}^{n-1} e^{(-\lambda j t_1/n)}\left(\frac{t_1}{n}\right) e^{-\lambda t_1/n}\left(1 - e^{-\lambda\left(t_2 - \frac{t_1}{n}\right)}\right)$$

$$\to (1 - e^{-\lambda t_1})(1 - e^{-\lambda t_2}) \text{ as } n \to \infty.$$

Therefore $P[T_1 \le t_1, T_2 \le t_2] \ge P[T_1 \le t_1]P[T_2 \le t_2]$.
Hence by induction $T_1, T_2, T_3, \ldots$ are independent.

Note Implicitly the idea of Separability of stochastic process (see Appendix I) is used here.

(b) Define $\tau_n = T_j + T_{j+1} + \ldots + T_{n-1}, j = 1, \ldots, n-1$

The characteristic function of T_j is $\int_0^\infty \lambda e^{i\theta t}\lambda e^{-\lambda t} dt = \left(1 - \frac{i\theta}{\lambda}\right)^{-1} = \varnothing(\theta)$.

Since $T_j, T_{j+1}, \ldots$ are independent random variables the ch. function of τ_n is

$\varnothing_n(\theta) = \left(1 - \frac{i\theta}{\lambda}\right)^{j-n}$, $n > j$ and hence $|\varnothing_n(\theta)| = \left(1 - \frac{\theta^2}{\lambda^2}\right)^{\frac{j-n}{2}} \to 0$ as $n \to \infty$ for all $\theta > 0$. τ_n Now, being sums of non-negative independent r.v.s, $\tau_n \to \infty$ in distribution as $n \to \infty$. By Levy's theorem τ_n being sums of independent r.v.s, $\tau_n \to \infty$ a.s. as $n \to \infty$.

Note . . . $T_i = \inf [t \mid N(t) = i + 1]$ if $P[N(0) = i] = 1$, i.e. T_i is the time of occurrence of the first jump after the ith step.

Alternative proof of (b)

$$P(\tau_{n+1} - \tau_n \ge \alpha) \ge \exp(-\lambda\alpha),\ \alpha \ge 0.$$

It follows by Fatou's lemma that

$$P(\limsup_{n\to\infty} (\tau_{n+1} - \tau_n) \ge \alpha) \ge \limsup_{n\to\infty} P(\tau_{n+1} - \tau_n \ge \alpha) \ge \exp(-\lambda\alpha).$$

This implies $P(\lim_{n\to\infty} \tau_n = \infty) \ge \exp(-\lambda\alpha)$ for every $\alpha \ge 0$.
In particular it follows that $P(\lim \tau_n = \infty) = 1$.

Now we give **an alternative proof of Theorem 7.6(a) based on renewal argumant**.

Proof of (a) Let $0 < S_1 < S_2 < S_3 \ldots$ are the renewal points (arrival points) of the Poisson process. We prove the theorem by induction. For T_1, the theorem has already been proved in Theorem 7.5. Now calculate the joint distribution of T_1 and T_2. After conditioning on $T_1 = x$ and taking expectation, we get

$$P(T_1 \le u, T_2 > v) = \int_0^u P(S_2 > v + x \mid T_1 = x)\lambda e^{-\lambda x} dx$$

(since T_1 has exponential distribution).

Now given that $T_1 = S_1 = x$, $\{S_2 > v + x\} = \{\text{No renewals in } (x, v + x]\} = \{N(v) = 0\}$ and hence

$$P(S_2 > v + x \mid T_1 = x) = P(N(v) = 0) = e^{-\lambda v}.$$

Therefore $$P(T_1 \le u, T_2 > v) = e^{-\lambda v} \int_0^u \lambda e^{-\lambda v}\, dx = e^{-\lambda v}(1 - e^{-\lambda u}).$$

Hence the theorem is proved for T_1 and T_2.

Next, assume that the theorem has been proved for $T_1, T_2, \ldots T_k$. Then the joint distribution of T_j, $1 \le j \le k+1$, calculated by first conditioning on $T_j = x_j$, $1 \le j \le k$ and then taking expectation

$P(T_j \le u_j, 1 \le j \le k, T_{k+1} > x_{k+1})$

$$= \lambda^k \int_0^{u_1} \ldots \int_0^{u_k} P(S_{k+1} > \sum_{i=1}^{k+1} x_i | T_i = x_i, 1 \le i \le k) \exp\left(-\lambda \sum_1^k x_i\right) dx_1 \ldots dx_k.$$

Now $$\{T_i = x_i, 1 \le i \le k\} = \left\{S_1 = x_1, S_2 = x_1 + x_2, \ldots, S_k = \sum_1^k x_i\right\}$$

Hence $P(S_{k+1} > \sum_{j=1}^{k+1} x_j | T_i = x_i, 1 \le i \le k)$

$$= P\left\{N\left(\sum_1^{k+1} x_i\right) - N\left(\sum_1^k x_i\right) = 0\right\} = P\{N(x_{k+1}) = 0\}$$

(By stationarity and independence of increment of Poisson process $N(x)$) $= e^{-\lambda x_{k+1}}$. Evaluating the multiple integral, we get

$$P(T_j \le u_j, 1 \le j \le k, T_{k+1} > x_{k+1}) = e^{-\lambda x_{k+1}}(1 - e^{-\lambda u_1}) \ldots (1 - e^{-\lambda u_k})$$

Therefore $T_1, T_2, \ldots, T_{k+1}$ are i.i.d. exponential r.v.s.

7.5 Generalization of Poisson Process

Using the notation $N(t)$ instead of N_t it can be easily shown that a Poisson process $\{N(t), t \ge 0)$ satisfies the following axioms.

axiom 0: $P[N(0) = 0] = 1$
axiom 1: $\{N(t), t \ge 0\}$ has independent increments
axiom 2: $P[N(t) > 0] > 0$ for any $t > 0$
axiom 3: for any $t > 0$

$$\lim_{h \to 0} \frac{P[N(t+h) - N(t) \ge 2]}{P[N(t+h) - N(t) = 1]} = 0$$

axiom 4: $N(t)$ has stationary increments.

Following generalization of the Poisson process are very well known and useful:

I. Non-homogeneous Poisson Process or Time Dependent Poisson Process

Let $\{N(t), t \ge 0\}$ be an integer valued stochastic process satisfying axioms 0, 1, 2, 3, and

$$\text{axiom } 4' \lim_{h \to 0} \frac{1 - P[N(t+h) - N(t) = 0]}{h} = v(t),$$

is a nonrandom function of time. In this case Kolmogorov difference differential equations become $p_n'(t) = -v(t)p_n(t) + v(t)p_{n-1}(t), n \geq 1$.

Define $\psi(z, t)$ as the generating function of $P(N(t) = n)$ and $p_0'(t) = v(t)p_0(t)$ as the initial equation.

Since $\{N(t), t \geq 0\}$ has independent increments,

$$E[z^{N(t+h)}] = E[z^{N(t+h)-N(t)}]E[z^{N(t)}].$$

Now $$\psi[z, t] = \sum_{n=0}^{\infty} z^n P[N(t) = n]$$

Therefore, $$\psi(z, t+h) = \psi(z, t)E[z^{N(t+h)-N(t)}].$$

Hence

$$\frac{1}{h}[E\{z^{N(t+h)-N(t)}\}-1] = \frac{1}{h}\{P[N(t+h) - N(t) = 0] - 1\}$$

$$+z\frac{1}{h}P\{N(t+h) - N(t) = 1\} + \frac{1}{h}\sum_{n-2}^{\infty} z^n P[N(t+h) - N(t) = n]$$

Therefore

$$\lim_{h\to 0}\frac{1}{h}\{E[z^{N(t+h)-N(t)}] - 1\} = v(t)[z-1] \quad \text{by axioms 3 and 4}'.$$

Hence

$$\lim_{h\to 0}\frac{\psi(z, t+h) - \psi(z, t)}{h} = \lim_{h\to 0}\frac{\psi(z, t)}{h}\{E[z^{N(t+h)-N(t)}] - 1\}.$$

Thus for $|z| < 1, t \geq 0$

$$\frac{\partial}{\partial t}\psi(z, t) = \psi(z, t)v(t)[z-1].$$

Under initial condition $\psi(z, 0) = 1$ and therefore $\psi(z, t) = \exp[m(t)[z-1]]$,

where $m(t) = \int_0^t v(t')dt'$ is the mean function of $N(t)$, because $m(t) - EN(t) = d\psi/dz \mid z = 1$.

II. Generalised Poisson Process

Let $\{N(t), t > 0\}$ be an integer valued stochastic process satisfying axioms 0, 1, 2, 4 and axiom 3′: there exists sequence $\{p_k\}$ $k = 1, 2, \ldots, t \geq 0$ such that

$$\lim_{h\to 0} P[N(t+h) - N(t) = k | N(t+h) - N(t) \geq 1\} = p_k.$$

Again desired $\psi(z, t)$ has to be determined.

Since $\{N(t), t \geq 0\}$ has independent increments,

$$E[z^{N(t+h)}] = E[z^{N(t+h)-N(t)}]E[z^{N(t)}].$$

Due to stationary increments $\psi(z, t+h) = \psi(z, t)\ \psi(z, h)$

Therefore $\frac{1}{h}\{\psi(z, t+h) - \psi(z, t)\} = \psi(z, t)\frac{1}{h}\{\psi(z, h) - 1\}$

Taking limit as $h \to 0$

$$\frac{\partial}{\partial t}\psi(z,t) = \psi(z,t)\lim_{h\to 0}\frac{1}{h}\{\psi(z,h)-1\}$$

But $\frac{1}{h}\{\psi(z,h)-1\} = \frac{1}{h}\{P[N(h)=0]-1\}$

$$+\frac{P[N(h)\geq 1]}{h}\sum_{n=1}^{\infty} z^n P[N(h)=n|N(h)\geq 1].$$

Then we can show (as in the case of ordinary Poisson process) that there exists a constant $v > 0$ such that

$$\lim_{h\to 0}\frac{1}{h}\{1-P[N(h)=0]\} = \lim_{h\to 0}\frac{1}{h}P[N(h)\geq 1] = v.$$

Applying axiom 3′ and Lebesgue dominated convergence theorem

$$\lim_{h\to 0}\sum_{n=1}^{\infty} z^n[P[N(h)=n|N(h)\geq 1] = \sum_{n=1}^{\infty} z^n p_n$$

Hence, $\frac{\partial}{\partial t}\psi(z,t) = \psi(z,t)v\left[\sum_{n=0}^{\infty} z^n p_n - p_0\right]$ or $\psi(z,t) = e^{vt[\psi(z)-p_0]}$

Note $\psi(z)$ can be the characteristic function of a non-negative integer valued r.v. with probability distribution $\{p_k\}$.

III. Compound Poisson Process

A stochastic process $\{X(t), t \geq 0\}$ is said to be a compound Poisson process if for $t \geq 0$ it can be represented by $X(t) = \sum_{n=1}^{N(t)} Y_n$ = sum of random number of independent and identically distributed r.v.s, where $\{N(t), t \geq 0\}$ is a Poisson process,

$\{Y_n, n = 1, 2, \ldots\}$ are independently and identically distributed r.vs. and $\{N(t), t \geq 0\}$, $\{Y_n\}$ are independent of each other.

Let $\psi_{X(t)}(u)$ be the characteristic function of $X(t)$.

Since $\{N(t), t \geq 0\}$ has independent increments and $\{Y_n\}$ is a sequence of independent and identically distributed r.v.s, it is obvious that $X(t)$ had independent increments.

For $n = 0, 1, 2, \ldots$ and for any $t > s \geq 0$

$$E[e^{iu\{X(t)-X(s)\}}|N(t)-N(s)=n] = \{\varphi_Y(u)\}^n$$

where $\varphi_Y(u)$ is the ch.f. of Y.

Since given n events have occurred in $(s, t]$, $X(t) - X(s)$ is the sum of n independent r.v.s distributed as Y's.

Then

$$\psi_{X(t)-X(s)}(u) = \sum_{n=0}^{\infty} E[e^{iu\{X(t)-X(s)\}}|N(t)-N(s)=n]\cdot P[N(t)-N(s)=n]$$

$$= \sum_{n=0}^{\infty} \{\varphi_Y(u)\}^n e^{-v(t-s)} \frac{\{v[t-s]\}^n}{n!} = e^{-v(t-s)} \exp[v[t-s]\varphi_Y(u)]$$

Therefore
$$\psi_{X(t)}(u) = e^{vt\{\varphi_Y(u)-1\}}$$

where $\varphi_Y(u)$ is the common characteristic function of $\{Y_n\}$.

Thus we have shown that $\{X(t), t \geq 0\}$ has independent stationary increments with the above characteristic function for any $t \geq 0$.

Note If Y is integer-valued then the compound Poisson process is a *generalised Poisson process.*

Compound Poisson process arises in bulk queues where customers arrive or are served in groups. An important application arises in "collective risk theory". Suppose that claims against a company occur in accordance with a Poisson process with mean λt and the individual claims X_i's are i.i.d. with distribution $\{P_k\}$. Then $X(t)$ represents the total claims at time t. If A represents initial reserve and c the rate of increase of the reserves in the absence of claims, then the total reserve at time t is $A + ct - X(t)$ and negative reserve implies "ruin". The moments of $X(t)$ are given by Wald's identity as

$$E(X(t)) = E(N(t))E(X_1) = \lambda t(EX_1),$$

$$\text{Var}(X(t)) = E(N(t))\,\text{Var}(X_1) + E^2(X_1)\,\text{Var}(N(t)) = \lambda t\, EX_1^2 \quad (EX_1 = 0)$$

and
$$\text{Cov}(X(t), X(s)) = \min(s, t)\, \lambda EX_1^2$$

(since $\text{Cov}(N(t), N(s)) = \lambda \min(s, t)$).

IV. Poisson Increment Process

Define $Y_1 = N_{t+l} - N_t$ for all $l > 0$. $\{Y_t, t \in T\}$ is called the Poisson increment process.

Theorem 7.7

(i) $EY_1 = \lambda l$

(ii) $$\text{Cov}(Y_t, Y_s) = \begin{cases} \lambda(l - |t-s|) & \text{if } l > |t-s| \\ 0 & \text{if } l < |t-s| \end{cases}$$

(iii) $\{Y_t, t \in T\}$ is covariance stationary.

Proof (i) $EY_t = E(N_{t+l}) - E(N_t) = \lambda)t + l) - \lambda t = \lambda l$

(ii) $$\begin{aligned}\text{Cov}(Y_1, Y_s) &= \text{Cov}(N_{t+l} - N_t, N_{s+l} - N_s) \\ &= \text{Cov}(N_{t+l} - N_{s+l} + N_{s+l} - N_t, N_{s+l} - N_t + N_t - N_s) \\ &\quad + \text{Cov}(N_{t+l} - N_{s+l}, N_{s+l} - N_t) \\ &\quad + \text{Cov}(N_{s+l} - N_t\, N_{s+l} - N_t) \\ &= \text{Var}(N_{s+l} - N_t) \text{ (stationarity and independence of increment)} \\ &= \text{Var}(N_{s+l-t} - N_0) \\ &= \text{Var}(N_{s+l-1}) = \lambda(s + l - t) = \lambda(l - (t - s)).\end{aligned}$$

(iii) The proof is obvious and hence the proof is omitted.

7.6 Order Statistics and Poisson Process

There is a close connection between Poisson process and so-called order statistics.

Let $Y_1, Y_2, \ldots, Y_n$ be i.i.d. random variables with common d.f. $F(t)$. Assume, for simplicity that $F(t)$ is strictly increasing.

$Y_{(i)}$ = ith smallest among $Y_1, Y_2, \ldots, Y_n$ is called ith order statistics. Consider $X_i = F(Y_i)$, then X_i's are uniform on [0, 1] i.i.d. random variables.

Let $X_1^*, X_2^*, \ldots, X_n^*$ be the order statistics of $X_1, X_2, \ldots X_n$. Let us state a well-known lemma from statistics for our later use.

Lemma 7.1 The joint probability density function of the order statistics $(X_1^*, \ldots, X_n^*)$ is given by

$$f_{X_1^* \ldots X_n^*}(X_1, \ldots X_n) = \begin{cases} n!, \text{ if } 0 \le x_1 \le \ldots \le x_n \le 1 \\ 0, \text{ otherwise.} \end{cases}$$

Theorem 7.8 Let $\{N_t, t \in T\}$ be a Poisson process with intensity λ. Let t be a fixed real number. If $N_t = n$, then the arrival times $\tau_1, \tau_2, \ldots, \tau_n$ are distributed as the n order statistics of the n independent random variables uniformly distributed over $[0, t]$.

Proof We note that the interarrival times are independent and exponentially distributed with parameter λ. Hence

$$P[t_1 \le \tau_1 \le t_1 + h_1, \ldots, t_n \le \tau_n < t_n + h_n | N_t = n]$$

$$= \frac{1}{P[N_t = n]} P[t_1 \le \tau_1 < t_1 + h_1, \ldots t_n \le \tau_n < t_n + h_n, N_t = n]$$

$$= \frac{(e^{-\lambda t_1} \lambda h_1)(e^{-\lambda h_1})(e^{-\lambda(t_2 - t_1 - h_1)})(\lambda h_2)e^{-\lambda h_2} \ldots (\lambda h_n)e^{-\lambda h_n} e^{-\lambda(t - t_n - h_n)}}{e^{-\lambda t}(\lambda t)^n / n!}$$

$$= \frac{\lambda^n h_1 h_2 \ldots h_n e^{-\lambda t} n!}{\lambda^n t^n e^{-\lambda t}} = \frac{n!}{t^n} h_1 h_2 \ldots h_n.$$

Let $Z_{(1)} \le Z_{(2)} \le\le Z_{(n)}$ denote the order statistics in a sample of size n from the exponential distribution

$$P[Z \le t] = 1 - e^{-t}, 0 \le t < \infty \tag{7.8}$$

The joint p.d.f. of $Z_{(n)}$'s is

$$n! \exp\left(-\sum_{r=1}^{n} z_r\right) \quad \text{if} \quad 0 \le z_1 \le \ldots \le z_n < \infty$$

which may be written as

$$n! \exp\left[-\sum_{r=1}^{n} (n - r + 1)(z_r - z_{r-1})\right], z_0 = 0.$$

Let
$$Y_r = (n - r + 1)(Z_{(r)} - Z_{(r-1)}), r = 1, 2, \ldots, n \tag{7.9}$$

Note that range of each Y_r is $(0, \infty)$ and Y_r are independent random variables.

Now
$$Z_{(r)} = \sum_{i=1}^{r} (Z_{(i)} - Z_{(i-1)}) = \sum_{i=1}^{r} \frac{Y_i}{n - i + 1} \tag{7.10}$$

a linear function of independent exponential variates and hence $Z_{(1)}, Z_{(2)}, \ldots, Z_{(n)}$ is a Markov chain (cf. Renyi, 1953).

As a general case, let $X_{(1)} \le X_{(2)} \le \ldots \le X_{(n)}$ be the order statistics from a strictly increasing distribution function $F(t)$.

The transformation $U_k = F(X_k)$ converts $X_{(r)}$ into $U_{(r)}$, $r = 1, 2, \ldots n$, the order statistics from uniform $R[0, 1]$ distribution. Let $Z_k = -\log U_k$.

$$\begin{aligned} P[Z_k \le x] &= P[-\log U_k \le x] = P[U_k \ge e^{-x}] \\ &= P[F(X_k) \ge e^{-x}] = P[X_k \ge F^{-1}(e^{-x})] \\ &= 1 - F(F^{-1}(e^{-x})) = 1 - e^{-x} \end{aligned}$$

i.e. Z_k's are exponential random variables.

Hence, $Z_{(r)} = \log U_{(n-r+1)}$ are order statistics from exponential distribution.

Now $\quad X_{(n-r+1)} = F^{-1}(U_{(n-r+1)}) = F^{-1}(e^{-z_{(r)}})$

$$= F^{-1}\ [\exp - \left[\left(\frac{Y_1}{n} + \frac{Y_2}{n-1} + \ldots + \frac{Y_r}{n-r+r}\right)\right] \tag{7.11}$$

Thus, $\quad X_{(n-r)} = F^{-1}\ [\exp\ (-(Y_1/n + \ldots + Y_{r+1}/n - r))]$

$$= F^{-1}\left[\exp\left\{\log F(X_{(n-r+1)}) - \frac{Y_{r+1}}{n-r}\right\}\right] \tag{7.12}$$

Since Y_r's are independent, hence by (7.11) $X_{(n-r+1)}$ and Y_{r+1} are independent.

Therefore, by (7.12), $X_{(n)}, X_{(n-1)}, \ldots, X_{(1)}$ form a Markov chain and hence $\{X_{(1)}, X_{(2)}, \ldots, X_{(n)}\}$ is also a Markov chain.

From this the following theorem follows.

Theorem 7.9 For random sample of n from a continuous distribution function the conditional distribution of $X_{(s)}$ given $X_{(r)} = x_r (s\ .\ r)$ is just the distribution of $(s - r)$th order statistics in a sample of $(n - r)$ drawn from the parent distribution truncated on the left at $x = x_r$.

Proof

$$\frac{X_{(r)}}{X_{(r+1)}} = \frac{e^{-Z_{(n-r+1)}}}{e^{-Z_{(n-r)}}} = e^{(Z_{(n-r)} - Z_{(n-r+1)})}$$

$$= e^{(Y_{n-r+1})/r},\ Y_{n+1} = 1. \text{ are independent random variables.}$$

Hence, $\left[\dfrac{X_{(r)}}{X_{(r+1)}}\right]^r = e^{-Y_{n-r+1}}$ is a sequence of i.i.d. uniform [0, 1] random variables (Malmequist 1950).

Lemma 7.2 The density of $Z_{(k)}$ is given by

$$f_k(t) = \binom{n}{k} k e^{-nt} [e^t - 1]^{k-1}$$

Proof For $k = 1, f_1(t) = ne^{-nt}$.
Then prove by induction on k.

Proposition 7.2 Let $\{N_t, t \in T\}$ be a Poisson process with intensity λ. If $N_t = n$, then the density of W_r (waiting times of Poisson process $= \sum_{i=1}^{r} T_i \,;\, T_i = \tau_i - \tau_{i-1}$) for $r < n$ is given by

$$\frac{n!}{(n-r)!(r-1)!}\left(\frac{x}{t}\right)^{r-1}\left(1-\frac{x}{t}\right)^{n-r}\frac{1}{t}.$$

Proof The density of kth order statistic $X_{(k)}$ from uniform $[0, t]$ distribution is given by

$$\frac{n!}{(n-k)!(k-1)!}\left(\frac{x}{t}\right)^{k-1}\left(1-\frac{x}{t}\right)^{n-k}$$

Now from Theorem 7.8 given $N_t = n$ arrival times are distributed as order statistics from uniform $[0, t]$ distribution.

Therefore, W_r is distributed as rth order statistics from uniform $[0, t]$ distribution.

Theorem 7.10 Let $\{N_t, t \in T\}$ be a Poisson process with intensity λ. Let $s \in T$, $s < t \in T, k \le n$.

Then $P[N_s = k \mid N_t = n] = \binom{n}{k}\left(\frac{s}{t}\right)^{k}\left(1-\frac{s}{t}\right)^{n-k}$

Proof $P[N_s = k \mid N_t = n] = \dfrac{P[N_s = k, N_t = n]}{P[N_t = n]}.$

$$= \frac{P[N_t - N_s = n-k, N_s = k]}{P[N_t = n]}$$

$$= \frac{P[N_t - N_s = n-k]P[N_s = k]}{P[N_t = n]} \quad \text{(due to independent increment)}$$

$$= \frac{\dfrac{e^{-\lambda(t-s)}[\lambda(t-s)]^{n-k}}{(n-k)!}\,\dfrac{e^{-\lambda s}(\lambda s)^k}{k!}}{e^{-\lambda t}(\lambda t)^n/n!}$$

$$= \binom{n}{k}\frac{(t-s)^{n-k}s^k}{t^n}.$$

Exercises and Complements

Exercise 7.1 Assume a device fails when a cumulative effect of k shocks occur. If the shocks happen according to Poisson process with parameter λ, find the density function for the life T of the device.

Exercise 7.2 Let $\{X(t), t \ge 0\}$ be a Poisson process with intensity parameter λ. Suppose that each arrival is "registered" with probability p, independent of other arrivals. Let $\{Y(t), t \geq 0\}$ be the process of registered arrivals. Prove that $Y(t)$ is a Poisson process with parameter λp.

Exercise 7.3 Suppose $g(t)$ is the conditional failure rate of an article at time t, given that it has not failed upto time t, i.e. P (failure in time $(t, t+h)$/no failure upto time $t\} = g(t)h + o(h)$ as $h \downarrow 0$. Assume that $g(t)$ is positive and continuous on $(0, \infty)$. Find an expression for $F(t) = P$ (failure at some time τ, $\tau < t$) in terms of $g(t)$.

Exercise 7.4 Let $X(t)$ and $Y(t)$ be two independent Poisson processes with intensity parameters λ and μ respectively. Let W and W^* be the times of two successive occurrences of Poisson events of the processes $X(t)$, i.e. $W < W^*$, $X(t) = X(W)$ for $W \leq t < W^*$ and $X(W^*) = X(W) + 1$. Define $N = Y(W^*) - Y(W)$. Find the probability distribution of N.

Exercise 7.5 Let $Y(t) = \sum_{n=1}^{X(t)} \xi_n$ be a compound Poisson process. Then show that (a) the process $Y(t)$ has independent increments. (b) The characteristic functional $\phi_{Y(t)}$ of $Y_{(t)}$ is given by $\phi_{Y(t)}(u) = \exp[\lambda t(\phi_\xi(u) - 1)]$ where $\phi_\xi(u)$ is the common characteristic function of $\{\xi_n\}$ and λ is the paramcter of the poisson events. (c) If $E(\xi^2) < \infty$ then show that $E(Y(t)) = \lambda t E(\xi)$ and Var $(Y(t)) = \lambda t E\xi^2$.

Exercise 7.6 Consider a Poisson process with rate $\lambda\,(> 0)$. Let T_1 be the time of occurrance of the first event and let $N(T_1)$ denote the number of events in the next T_1 units of time. Show that $E(N(T_1)T_1) = 2/\lambda$ and find the variance of $N(T_1)T_1$.

Exercise 7.7 Let $X(t)$, $-\infty < t < \infty$, be a Poisson process with rate λ. Let $Y(t) = X(t+1) - X(t)$, $-\infty < t < \infty$. Find the mean and covariance function of $Y(t)$ process and show that it is a second order stationary process.

Exercise 7.8 Let $X(t)$ be a Poisson process with parameter λ and T_a be the first time $X(t)$ reaches $a > 0$. Calculate $E(T_a)$, Var (T_a) and also m.g.f. of T_a. Can you guess what type of random variable will be T_a?

Exercise 7.9 Prove that if the sum of two independent renewal processes is a Poisson process, then both renewal processes must be Poisson processes.

Exercise 7.10 A telephone exchange has m channels. Calls arrive in the pattern of a Poisson process with parameter λ; they are accepted if there is an empty channel, otherwise they are lost. The duration of each call is a r.v. where distribution function is exponential with parameter μ. The life times of separate calls are independent r.v.s. Find the stationary probabilities of the number of busy channels.

Exercise 7.11 Show that if the interarrival distribution function $F(x)$ of a renewal process $Y(t)$ is continuous and if the expected residual life time $E(Y(t))$ is a finite constant for $t > 0$, then show that the renewal process is a Poisson process.

Exercise 7.12 In a factory, accidents happen according to a Poisson process with intensity λ. The times that the victims stay in the hospital are independent random variables with common distribution function $F(t)$. Let $X(t)$ be the number of persons in the hospital at time $t \in (0, \infty)$.

(a) Show that $P(X(t) = k) = e^{-\lambda\alpha(t)}[\lambda\alpha(t)]^k/k!$, $k = 0, 1, 2, \ldots$

where $$\alpha(t) = \int_0^t [1 - F(x)]dx$$

(b) Derive the distribution of $\lim_{t\to\infty} X(t)$ if $F(t)$ has a finite first moment.

(c) Compute $P(X(t) = k)$ if all the victims stay in the hospital for a constant length of time L.

8

Continuous Time and Continuous State Markov Process

8.1 Diffusion Processes

Let us consider a continuous time and continuous state space stochastic process $\{X(t), t \in T\}$ with $T = [0, \infty)$, $S = (-\infty, +\infty)$.

More specifically let $\{X(t), t \in T\}$ be a Markov process on a probability space $(\Omega, \mathscr{B}, P)$ assuming values in a set S. Denote by $\mathscr{L}$ a σ-algebra of subsets of S. The measure space $(S, \mathscr{L})$ will be called the "state space" of the stochastic process $X(t)$. The process $X(t)$ is said to be a homogeneous Markov process if its transition probability has the property $P(X(t + s) \in B|X(s) = x) = P(x, t, B)$, $B \in \mathscr{L}$. One may imagine a continuous parameter Markov process $\{X(t), t \in T\}$ that is not a process with independent increments. Suppose that given $X(s) = x$, for small times t, the displacement $X(s + t)–X(s) = X(s + t)–x$ has mean and variance approximately $t\mu(x)$ and $t\sigma^2(x)$, respectively. Here $\mu(x)$ and $\sigma^2(x)$ are functions of the state of x, and not constants as in the case of Brownian motion $W(t)$. The distinction between $\{W(t)\}$ and $\{X(t)\}$ is analogous to that between simple random walk and a birth-death chain. More precisely, suppose

$$\left.\begin{aligned} &E(X(s+t) - X(s)|X(s) = x) = t\mu(x) + o(t) \\ &E((X(s+t) - X(s)^2\,|X(s) = x) = t\sigma^2(x) + o(t) \\ &E(|X(s+t) - X(s)|^3\,|X(s) = x) = o(t) \end{aligned}\right\} \tag{8.1}$$

hold, as $t \downarrow 0$ for every $x \in \mathscr{L}$.
Note that (8.1) holds for Brownian motion.

A more general formulation of the existence of infinitesimal mean and variance parameters, which does not require the existence of finite moments will now be discussed. Using Kolmogorov's consistency theorem (see A. Basu) it can be seen that a Markov Process can be defined by the following simpler alternative form.

Let
$$t_1 < t_2 < t_3 < \ldots < t_n < t \in T, \tag{8.2}$$

and $X, X_1, X_2, \ldots, X_n$ are values from state space S. Then $\{X(t), t \in T\}$ is said to be a Markov Process if it satisfies

$$P[X(t) \le x \mid X(t_n) = x_n, X(t_{n-1}) = x_{n-1} \dots, X(t_1) = x_1]$$
$$= P[X(t) \le x \mid X(t_n) = x_n] \text{ (Markov Property)}$$

for all $t_1, t_2, \dots, t_n$ satisfying (8.2) for all $n \ge 1$. We have seen in a continuous time discrete state space Markov Process the transition function $p_{ij}(t) = P[X(t + s) = j \mid X(s) = i]$ satisfies the following properties:

(i) $p_{ij}(t) \ge 0,\ i, j \in S$

(ii) $\sum_{j \in s} p_{ij}(t) = 1$ for all $t \in T, j \in S$

(iii) $p_{ij}(t+s) = \sum_{k \in s} p_{ik}(t)\, p_{kj}(s)$ (Chapman-Kolmogorov equation)

(iv) $P(0) = I$, i.e. $p_{ij}(0) = \delta_{ij}$, the Kronecker's delta, and I is the Identity Matrix.

We shall develop an analogous theory in the continuous state case. To avoid many measure theoretic difficulties let us define transition distribution function as

$$F(y, s; x, t) = P[X(t) \le x \mid X(s) = y],\ s < t,\ s, t \in T.$$

To avoid many pathological difficulties we shall assume that the transition distribution function of a continuous time Markov process satisfies the following axioms:

(i) $F(y, s; x, t) \ge 0$

(ii) $F(y, s; -\infty, t) = 0,\ F(y, s; +\infty, t) = 1.$

(iii) Chapman-Kolmogorov equations: For every τ, s and t satisfying $s < \tau < t$, we have

$$F(y, s; x, t) = \int_{-\infty}^{\infty} F(z, \tau; x, t)\, dF_z(y, s; z, \tau).$$

(iv) $F(y, t; x, t) = \begin{Bmatrix} 0 \text{ if } x < y \\ 1 \text{ if } x \ge y \end{Bmatrix} = \delta(y, x),$ Dirac delta function.

(v) For all $\delta > 0$ the limits

(a) $$\lim_{\Delta t \to 0} \frac{1}{\Delta t} \int_{\{|x-y| > \delta\}} (y - x)\, dF_y(x, t; y, t + \Delta t) = 0,$$

(b) $$\lim_{\Delta t \to 0} \frac{1}{\Delta t} \int_{\{|x-y| \le \delta\}} (y - x)\, dF_y(x, t; y, t + \Delta t) = a(x, t)$$
(drift coefficient)

(c) $$\lim_{\Delta t \to 0} \frac{1}{\Delta t} \int_{\{|x-y| \le \delta\}} (y - x)^2\, dF_y(x, t; y, t + \Delta t) = b(x, t) > 0$$
(diffusion coefficient)

exist, the convergence being uniform in x.

Condition (v (a)) implies that small changes of state occur continuously during small intervals of time, i.e.

$$P[|X(t) - X(s)| > \delta \mid X(s) = x] = o(t - s)\ (s < t).$$

Definition 8.1 A Markov process that satisfies axioms (i) to (v) is called a Diffusion process.

Special Case (When density function exists)

If $f(y, s; x, t) = \dfrac{\partial}{\partial x} F(y, s; x, t)$ exists then

$f(y, s; x, t)$ is called the *Transition density function*. Transition density of a M.P. satisfies.

(iii) $$f(y, s; x, t) = \int_{-\infty}^{\infty} f(z, \tau; x, t) f(y, s; z, \tau)\, dz$$

(Chapman-Kolmogorov equation)

(iv) $f(y, t; x, t) = \delta(x - y)$, Dirac delta function,

where $\delta(x) = 0$ for $x \neq 0$ but $\int_{-\infty}^{\infty} \delta(x)\, dx = 1$

($\delta(x)$ is also called Generalized Distribution in Mathematical Physics).

If $F(y, s; x, t) = F(y, 0; x, t - s)$ depends on $(t-s)$ only, then the Diffusion process is called *Homogeneous*. For homogeneous process we write $F(y; x, t) = P(X(t) \leq x \mid X(0) = y)$.

Example 8.1 A Diffusion process can be considered as the continuous limit of a simple (homogeneous) random walk.

Informal Proof

Consider the n-step transition probabilities in a simple random walk $P(S_{m+n} = j \mid S_m = i) = P(X_1 + \ldots + X_n = j - i \setminus S_0 = 0) = P(S_n = j - i)$, where X_i's are i.i.d. Bernoulli random variables taking values 1 and –1 with $P(X_i = 1) = p$ and $q = 1 - p$.

$$\text{Then}\quad p_{ij}^{(n)} = \begin{cases} 0 \text{ if } n + j - i \text{ is odd} \\ \dbinom{n}{(n + j - i)/2} p^{(n+j-i)/2} q^{\frac{n-j+i}{2}} & \text{if } n + j - i \text{ is even} \end{cases}$$

$$= \upsilon(j - i, n) \quad (-\infty < i, j < \infty),$$

where $\upsilon(x, n) = \dbinom{n}{(n + x)/2} p^{\frac{n+x}{2}} q^{\frac{n-x}{2}}$ if $x = j - i$.

Here x is the displacement of the particle (performing random walk) in n-steps.

Consider a (Brownian) particle performing a random walk such that X is the displacement in time t in n-steps.

Now we assume that each step has length Δx and time between two consecutive steps is Δt. Then $n = \dfrac{t}{\Delta t}$ i.e. during the interval $(0, t)$, the interval of length t is divided into n equal sub-intervals of length Δt. Note that $E(X) - n(p - q)$ and $V(X) = 4npq$.

If the total displacement during $(0, t)$ be $X(t) = \sum_1^n Z_i$ where Z_i are i.i.d. Bernoulli r.v.s. taking values $\pm\Delta x$, then

$$E[\text{Total displacement}] = n(p-q)\Delta x = \frac{t}{\Delta t}(p-q)(\Delta x)$$

and

$$\text{Var [Total displacement]} = n \cdot 4\,pq(\Delta x)^2 = \frac{t}{\Delta t}(4\,pq)(\Delta x)^2$$

In order to get a meaningful result we have to assume that

$$(\Delta x)^2/\Delta t \to a \text{ limit } 2D \text{ if } \Delta t \to 0 \text{ and } \Delta x \to 0$$

and $p-q = p-(1-p) = 2p-1 \sim \frac{C}{D}\Delta x$ (a multiple of Δx).

$$\left.\begin{aligned} \Rightarrow p &\sim \frac{C}{D}\frac{\Delta x}{2} + \frac{1}{2} = \frac{1}{2} + \frac{C}{2D}\Delta x \\ q &\sim \frac{1}{2} - \frac{C}{2D}\Delta x. \end{aligned}\right\} \text{ where } C \text{ and } D \text{ are constants.} \tag{8.3}$$

$$E\,[\text{Total displacement}] \sim t\frac{C}{D}\frac{(\Delta x)^2}{\Delta t} \to \frac{Ct}{D}2D$$

$$= 2Ct \sim a(x, t) = m(t), \text{ the mean function}$$

$$\text{Var [displacement]} \sim 4t\left(\frac{1}{4} - C^2\Delta t\right)\frac{(\Delta x)^2}{\Delta t}$$

$$\to 2Dt \sim b(x, t) = \sigma^2(t), \text{ the variance function.}$$

Thus the Binomial distribution $v(x, n)$ of displacement tends to $f(x, t)\,dx$, where

$$f(x,t) = \frac{1}{\sqrt{2\pi}\sigma(t)} \exp\left[-\frac{1}{2\sigma^2(t)}(x - m(t))^2\right] \quad \text{(by CLT)}$$

that is, the displacement during $(0, t)$ $X(t) - X(0)$ and the increments in the displacement during $(0, t)$ $X(t) - X(0)$ and the increments in the displacement during $0 < s < t$ that is, $N((t-s)\,2C, 2D(t-s))$. $p_{ij}^{(n)}$ satisfies recurrence relation $p_{ij}^{(n+1)} = p \cdot p_{i,j-1}^{(n)} + qp_{i,j+1}^{(n)}$ which can be written as $v\,(j, n+1) = pv\,(j-1, n) + qv(j+1, n)$ and in our new units

$$v(x, t+\Delta t) = pv(x-\Delta x, t) + qv(x+\Delta x, t) \tag{8.4}$$

By Taylor's expansion and substituting p, q from (8.3) in (8.4) and proceeding to the limit as $\Delta x \to 0$,

$$v(x, t+\Delta t) = v(x,t) + \Delta t\frac{\partial v}{\partial x} + o(\Delta t) \tag{8.5}$$

$$v(x \pm \Delta x, t) = v(x,t) \pm \Delta t\frac{\partial v}{\partial x} + \frac{1}{2}(\Delta x)^2\frac{\partial^2 v}{\partial x^2} + o(\Delta x)^2 \tag{8.6}$$

By (8.3), (8.4) and (8.5) and letting $\Delta t \to 0$, we get

$$\frac{\partial v(x,t)}{\partial t} = D\frac{\partial^2 v}{\partial^2 x} - 2C\frac{\partial v}{\partial x},$$

the forward Kolmogorov partial differential equation for Diffusion process as the continuous limit of the difference equations in two variables of homogeneous random walk.

8.1.1 Alternative method to show Brownian motion (most important example of a diffusion process) as an approximation of simple random walk

Suppose Brownian particle starts at the origin and jumps every Δt seconds, moving to the right a small distance Δx with probability 1/2 or to the left a distance Δx with probability 1/2. If $X_n(t)$ is the position of the particle at time $t = n\Delta t$ (n being large and Δt being small time interval), then $X_n(t)$ is the sum of independent r.v.s. $Y_1, Y_2, \ldots, Y_n$, where $P(Y_i = \pm \Delta x) = 1/2$ for all $i \geq 1$. Now Var $(X_n(t)) = n(\Delta x)^2$ and hence

$$X_n(t) = \frac{Y_1 + \ldots + Y_n}{\sqrt{n}\Delta x} \sqrt{n}\Delta x = Z_n \sqrt{n}\Delta x,$$

where $E(Z_n) = 0$, Var $(Z_n) = 1$.

Now $n = t/\Delta t$ and $\sqrt{n}\Delta x = \sqrt{t}\Delta x/\sqrt{\Delta t}$. If we assume that $(\Delta x)^2/\Delta t \to \sigma^2 > 0$ as $\Delta t \to 0$, the CLT for i.i.d. r.v.s implies $X_n(t) \xrightarrow{d} N(0, \sigma^2 t)$, that is $X_n(t) \xrightarrow{d} W(t)$. Now we show that all the finite dimensional distribution of $\{X_n(t)\}$ converges to that of $(W(t), t \geq 0\}$.

If $0 \leq t_1 < \ldots < t_n$, then

$$X_n(t_j) - X_n(t_{j-1}) \xrightarrow{d} W(t_j) - W(t_{j-1}) \text{ for each fixed } j \geq 1.$$

Since the random walk and the Brownian motion $\{W(t)\}$ have independent increments, the joint ch. f.s of $X_n(t_1, \ldots, X_n(t_k)$ is

$$\phi_n(u_1, \ldots, u_k) = E\left[\exp\left\{i \sum_{j=1}^{k} u_j X_n(t_j)\right\}\right]$$

$$= E\left[\exp\left\{i\left(\sum_{j=1}^{k} u_j\right) X_n(t_1)\right\}\right] E\left[\exp\left\{i \sum_{j=2}^{k} u_j (X_n(t_2) - X_n(t_1))\right\}\right] \ldots$$

$E[\exp\{iu_k(X_n(t_k) - X_n(t_{k-1}))\}]$

$\to \phi(u_1, \ldots, u_k)$, the joint ch. f. of $W(t_1), \ldots, W(t_k)$.

Hence by continuity theorem for ch.f. the random vectors $(X_n(t_1), \ldots, X_n(t_k)) \xrightarrow{d} (W(t_1), \ldots, W(t_k))$.

In principle, Kolmogorov's consistency theorem proves the result. But complete proof of convergence of the entire process to Brownian motion needs techniques (Invariance principle and tightness-see Billingsley (1968)) beyond the scope of this book.

8.2 Kolmogorov Backward and Forward Diffusion Equations

There are various methods for determining transition probability function or transition distribution of a Markov process, ranging from purely analytical to purely probabilistic. The method presented here was developed by Kolmogorov in 1931.

Theorem 8.1 Let $\{X(t),\ t \in T\}$ be a diffusion process. Assuming that $\frac{\partial}{\partial y} F(y, s; x, t)$ and $\frac{\partial^2}{\partial y^2} F(y, s; x, t)$ exist and are jointly continuous in all the variables, we get

$$\frac{\partial}{\partial s} F(y, s; x, t) = -a(y, s)\frac{\partial}{\partial y} F(y, s; x, t) - \frac{1}{2} b(y, s)\frac{\partial^2}{\partial y^2} F(y, s; x, t),$$

the Generalized Heat equation or the Backward Kolomogorov's Differential equation.

Assume that $f(y, s; x, t)=\frac{\partial}{\partial x} F(y, s; x, t)$ exists and that $f(x, t) = f(y, s; x, t)$. Also assume that $\frac{\partial f}{\partial t}, \frac{\partial}{\partial x}[a(x, t) f(x, t)]$ and $\frac{\partial^2}{\partial x^2}[b(x, t) f(x, t)]$ exist and are continuous in all the variables. Then we get

$$\frac{\partial f(x, t)}{\partial t} = -\frac{\partial}{\partial x}[a(x, t) f(x, t)] + \frac{1}{2}\frac{\partial^2}{\partial x^2}[b(x, t) f(x, t)],$$

the Kolmogorov's forward differential equation or the Fokker-Planck equation. Proof of Backward Equation

$$F(y, s - \Delta s; x, t) - F(y, s; x, t) = \int_{-\infty}^{\infty} F(z, s; x, t)\, dF_z(y, s - \Delta s; z, s) - F(y, s; x, t)\int_{-\infty}^{\infty} dF_z(y, s - \Delta s; z, s)$$

Let $$I = \int_{-\infty}^{\infty} [F(z, s; x, t) - F(y, s; x, t)]\, dF_z(y, s - \Delta s; z, s)$$

By assumptions for $|z - y| \le \delta$ and for some $\delta > 0$,

$$F(z, s; x, t) = F(y, s; x, t) + (z - y)\frac{\partial F(y, s; x, t)}{\partial y} + \frac{1}{2}(z - y)^2\frac{\partial^2 F(y, s; x, t)}{\partial y^2} + o((z - y)^2).$$

Then

$$I = \int_{|z-y|>\delta} [F(z,s;x,t) - F(y,s;x,t)]\, dF_z(y,s-\Delta s;z,s) + \int_{|z-y|\le\delta} [(z-y)\frac{\partial F}{\partial y} + \frac{1}{2}(z-y)^2\frac{\partial^2 F}{\partial y^2} F(y, s; x, t)\ o((z - y)^2)]\, dF_z(y, s - \Delta s; z, s)$$

$$= I_1 + I_2.$$

where $|I_1| \le 2 \left| \int_{|z-y|>\delta} dF_x(y, s - \Delta s; z, s) \right|$, and

$$I_2 = \frac{\partial F}{\partial y} \int_{|z-y|\le\delta} (z-y)\, dF_z(y, s - \Delta s; z, s)$$

$$+ \frac{1}{2} \partial^2 F/\partial y^2 \int_{|z-y|\le\delta} (z-y)^2\, dF_z(y, s - \Delta s; z, s) + o((z-y)^2).$$

Dividing by Δs and letting $\Delta s \to 0$ we get the left hand derivative

$$\lim_{\Delta s \to 0} (F(y, s - \Delta s; x, t) - F(y, s; x, t))/\Delta s = -\frac{\partial F}{\partial s}(y, s; x, t)$$

as $a(y, s) \frac{\partial}{\partial y} F(y, s; x, t) + \frac{1}{2} b(y, s) \frac{\partial^2}{\partial y^2} F(y, s; x, t)$

(first term I_1 tends to 0 due to $v(a)$).

Similarly, the right hand derivative exists and is given by the last expression. The proof of Kolmogorov's forward equation is difficult and is given at the end of Ornstein and Unlenbeck process in section 8.5.

8.3 Wiener Process (Brownian Motion)

Definition 8.2

Let $\{X(t), t \in T\}$ be a continuous space and continuous time stochastic process satisfying the following conditions:

(i) $P\{X(0) = 0\} = 1$

(ii) $\{X(t), t \in T\}$ has independent increments

(iii) $\{X(t), t \in T\}$ has stationary increments

(iv) $P[X(t) - X(s) \le x] = \frac{1}{(2\pi D(t-s))^{1/2}} \int_{-\infty}^{\infty} \exp[-u^2/2D(t-s)]\, du$

Then $\{X(t), t \in T\}$ is called Wiener Process (or Wiener-Levy process or Brownian Motion).

The erratic movement of small particles suspended in fluids, caused by impacts of the molecules of the fluids, is called Brownian motion (after Robert Brown, 1827, an English botanist). Statistical treatment of this was given by N. Wiener (1923). Before that Einstein and Smoluchowski (1905) also used it. We shall outline the theory as given by Uhlenbeck and Ornstein (1930).

Theorem 8.2 Wiener process as defined above is a Diffusion process.

Proof The transition distribution is

$$F(y, s; x, t) = P[X(t) \le x \mid X(s) = y]$$

$$= P[X(t) - X(s) \le x - y \mid X(s) - X(0) = y] \text{ (from (i))}$$

$$= P[X(t) = X(s) \le x - y] \quad \text{(by independent increments)}$$

$$= \frac{1}{(2\pi D(t-s))^{1/2}} \int_{-\infty}^{x-y} \exp[-u^2/2\,D(t-s)]\,du$$

$$= \frac{1}{\sqrt{2\pi}} \int_{-\infty}^{(x-y)/D(t-s)} \exp(-u^2/2)\,du \tag{8.7}$$

Obviously (i) $F(y, s; x, t) \geq 0$; (ii) $F(y, s; -\infty, t) = 0$ and $F(y, s; \infty, t) = 1$.

$$\text{(iii)} \quad F(y, s; x, t) = \begin{cases} 0 \text{ if } x < y \\ 0 \text{ if } x \geq y \end{cases}$$

From condition (iv) of Wiener process, the transition density exists and equals

$$f(y, s; x, t) = \frac{1}{(2\pi D(t-s))^{1/2}} e^{-(x-y)^2/[2D(t-s)]}$$

and satisfies Chapman-Kolmogorov equation

$$\text{(iv)} \quad f(y, s; x, t) = \int_{-\infty}^{\infty} f(z, \tau, x, t) f(y, s; z, \tau)\,dz$$

Check that by putting, for example, $y = s = 0$ we get

$$f(0, 0; x, t) = \frac{1}{(2\pi Dt)^{1/2}} e^{-x^2/2Dt}$$

and using the simple additive property of normal distribution, viz. that sum of two independent normal distributions with 0 means and variances σ_1^2, σ_2^2 respectively is again normal with mean zero and variance $\sigma_1^2 + \sigma_2^2$

$$\text{(v(a))} \lim_{\Delta t \to 0} \frac{1}{\Delta t} \int_{|y-x|>\delta} dF_y(x, t; y, t+\Delta t)\,dy,\ \delta > 0$$

$$= \lim_{\Delta t \to 0} \frac{1}{\Delta t} \int_{|y-x|>\delta} \frac{1}{\sqrt{2\pi D(\Delta t)}} e^{-(x-y)^2/[2D\Delta t)]}\,dy$$

$$= \lim_{\Delta t \to 0} \frac{1}{\Delta t} \frac{1}{\sqrt{2\pi}} \int_{|u\sqrt{D\Delta t}|>\delta} e^{u^2/2}\,du \quad \left(u = \frac{y-x}{\sqrt{D\Delta t}}\right)$$

$$= \lim_{\Delta t \to 0} \frac{1}{\Delta t} \sqrt{\frac{2}{\pi}} \int_{\delta/\sqrt{D\Delta t}}^{\infty} e^{-u^2/2}\,du \quad \text{(by symmetry)}$$

$$= \lim_{\Delta t \to 0} \frac{1}{\Delta t} \sqrt{\frac{2}{\pi}} \int_{\delta/\sqrt{D\Delta t}}^{\infty} \frac{u^2 D\Delta t}{\delta^2} e^{-u^2/2}\,du \left(\because \frac{\delta}{\sqrt{D\Delta t}} \leq u \Rightarrow 1 \leq \frac{u^2 D\Delta t}{\delta^2}\right)$$

$= 0$ (by absolute continuity of integral)

$$(vb) \lim_{\Delta t \to 0} \frac{1}{\Delta t} \int_{|y-x|\leq\delta} (y-x) \frac{1}{\sqrt{2\pi D\Delta t}} e^{-(y-x)^2/(2D\Delta t)}\,dy$$

$= 0$ (by symmetry)

$= a\,(x, t)$.

$$\text{(v(c))} \lim_{\Delta t \to 0} \frac{1}{\Delta t} \int_{|y-x| \le \delta} (y-x)^2 \frac{1}{\sqrt{2\pi D\Delta t}} e^{-(y-x)^2/(2D\Delta t)} dy$$

$$= \lim_{\Delta t \to 0} \frac{1}{\Delta t} \frac{1}{\sqrt{2\pi}} \int_{|u| \le \delta|} u^2 \frac{D\Delta t}{\sqrt{D\Delta t}} e^{-u^2/2} du \quad (\text{put } y - x = u\sqrt{D\Delta t})$$

$$= \frac{D}{\sqrt{2\pi}} \int_{-\infty}^{\infty} u^2 e^{u^2/2} du \quad \text{(by Monotone Convergence Theorem)}$$

$$= D = b(x\ t)$$

Note (a) If the Diffusion process is homogeneous, then $a(x, t)$ and $b(x, t)$ are independent of t.

(b) If the Diffusion process is additive (i.e. given $X(t_0) = x_0$ the increment $X(t) - X(t_0)$ depends on t and t_0 and not on x_0) then $a(x, t)$ and $b(x, t)$ are independent of x. For example Wiener process is additive.

8.4 Simple Properties of Wiener Process and First Passage Time Distribution

Some Simple Properties

(1) A Wiener process with drift coefficient $\mu = a(x, t)$ and diffusion coefficient $\sigma^2 = b(x, t) = D$ satisfies Kolmogorov's forward partial differential equation $\frac{\partial f}{\partial t} = -\mu \frac{\partial f}{\partial x} + \frac{1}{2}\sigma^2 \frac{\partial^2 f}{\partial x^2}$.

(2) It is not covariance stationary, since
Cov $(X(t), X(s)) = \sigma^2 \min(t, s)$ (compare with Poisson process)

(3) Inversion
$Y(t) = tX(1/t)$, $0 < 1 \le 1$ with $Y(0) = 0$ a.s. is again a Wiener process.
Since $P(Y(t) \le y] = P[tX(1/t) \le y]$

$$= P\left[\frac{X(1/t)}{\sigma\sqrt{1/t}} \le \frac{y/t}{\sigma\sqrt{1/t}}\right]$$

$$= \Phi(y/\sigma\sqrt{t}), \text{where } \Phi \text{ is the c.d.f. of } N(0, 1).$$

Hence, $E(Y(t)) = 0$ and $\text{Var}(Y(t)) = t^2\sigma^2/t = \sigma^2 t$. In fact $X(1/t)$ and $X(1/u) - X(1/t)$ are independent for $1/t < 1/u$. The independence of increments can be established by showing that they are uncorrelated.

$$E(uX(1/u)[tX(1/t) - uX(1/u)\}$$

$$= E(utX(1/u)\, X(1/t) - u^2X^2(1/u)) = ut(1/t) - u^2(1/u) = 0$$

(since $E(X(s)X(t)) = \min(s, t)$). Hence for $0 \le u < t$.

$$E(\exp[i\theta\{tX(1/t) - uX(1/u)\}])$$

$$= E\exp\{i\theta((t-u)\,X(1/t) - u[X(1/u) - X(1/t)])\}$$

$$= \exp\left\{-\theta^2/2(t-u)^2\,1/t - \left(\frac{\theta u}{2}\right)^2 (1/u - 1/t)\right\}$$

$$= \exp\{-\theta^2/2(t-u)\}.$$

Hence $Y(t) - Y(u)$ has Gaussian distribution.

(4) (a) $X(t)$, $t \le 0$ is a Wiener Process (follows by symmetry).

(b) (*Origin change and strong Markov property*). $\{X(t+\tau) - X(t), t \ge 0, \tau \ge 0$ may be a stopping time$\}$ is a Wiener process

(c) $\left\{\frac{1}{\sqrt{\alpha}} X(\alpha t), t \ge 0, \alpha \text{ (constant)} > 0\right\}$ is a Wiener process (scale change).

(d) $\{X(T) - X(T-t), 0 \le t \le T \text{ (fixed)}\}$ is a Wiener process.

Proof of (2) We shall show Cov $(X(t), X(t')) = \sigma^2 \min(t, t')$.
If $0 < t' < t$, Cov $(X(t), X(t'))$

$$= \text{Cov}(X(t'), X(t') + X(t) - X(t'))$$
$$= \text{Cov}(X(t'), X(t')) + \text{Cov}(X(t'), X(t) - X(t'))$$
$$= \text{Var}(X(t')) + 0 \quad \text{(due to independent increments)}$$
$$= \sigma^2 t'.$$

Similarly Cov $(X(t'), X(t)) = \sigma^2 t$ if $0 < t < t'$
Therefore Cov $(X(t), X(t')) = \sigma^2 \min(t, t')$.

Hence Corr $(X(t), X(t')) = \rho(t, t') = \dfrac{\text{Min}(t, t')}{\sqrt{t, t'}}$.

8.4.1 First passage time distribution for Wiener process

The possible realizations of a stochastic process are called sample path or Trajectories. For a Wiener process the sample paths are almost surely continuous functions (see Appendix 1). Let us state the following well known result in Probability Theory which will be useful for our next result.

D. Andre's (1887) Reflection Principle There is a 1:1 correspondance between all paths from $A(a_1, a_2)$ to $B(b_1, b_2)$ which touch or cross the x-axis and all paths from $A'(a_1, -a_2)$ to $B(b_1, b_2)$.

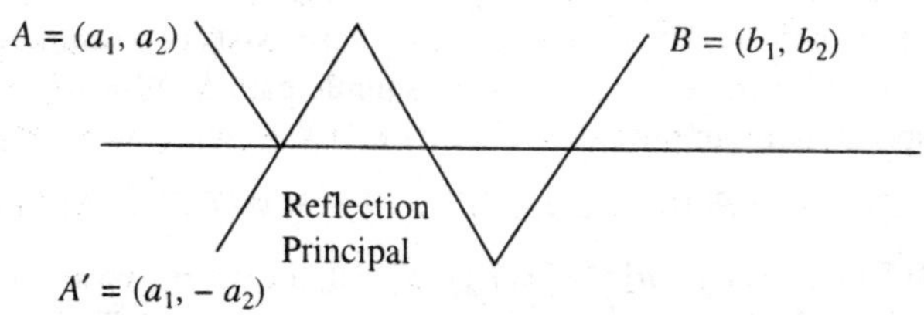

Let $T_a = \inf\{t: X(t) \ge a\}$ be a random variable and in fact a stoping time (continuous valued). Then $\{T_a, a \in R^+\}$ is a stochastic process.

Lemma 8.1 Let $\{X(t), 0 \le t \le T)$ be a Wiener Process with $X(0) = 0$ a.s. and $\mu = 0$. Then for any a > 0,

$$P[M(T) = \max_{0<t\le T} X(t) \ge a] = 2P[X(T) \ge a].$$

(This is due to Bachlier (1900)).

Proof Consider the collection of sample paths $\{X(t),\ 0 \le t \le T\}$ such that $X(T) \ge a$, $a > 0$. Now $X(0) = 0$ and $X(t)$ is continuous, there exists a random time τ at which $X(t)$ first attains the value a, denoted by T_a. The hitting time (or first passage time) T_a is a random variable.

For $t > T_a$, we reflect $X(t)$ about the line $x = a$ to define

$$X_a(t) = \begin{cases} X(t) \text{ if } t < T_a \\ a - [X(t) - a] = 2a - X(t) \text{ if } t > T_a \end{cases}$$

i.e. the new Stochastic process $X_a(t)$ gives for $t > T_a$ the reflection of $X(t)$ about the line $x = a$.

By symmetry, $X(t)$ and $X_a(t)$ have same probability distribution.

$$[M(T) \ge a] = [M(T) \ge a, X(T) > a]\ U\ [M(T) \ge a, X(T) < a]\ U\ [M(T) \ge a, X(T) = a]$$

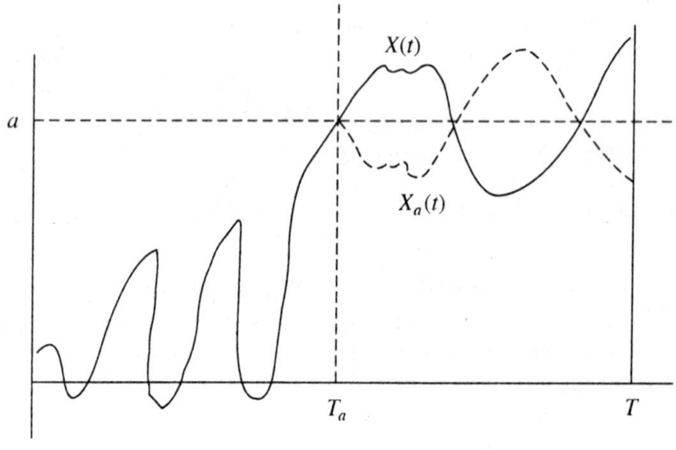

Note that $X_a(T) \le a$, since $X(T) \ge a$. Because the probability law of the path for $t \ge T_a$ given $X(T_a) = a$, is symmetrical with respect to the values $x > a$ and $x < a$ and is independent of the history prior to time T_a. From reflection principle to every sample path $X(t)$ for which $X(T) \ge a$, there exists two sample points such that $M(T) \ge a$ and vice versa i.e. to every sample path $X(t)$ for which $M(T) \ge a$ there corresponds two sample paths $X(T) \ge a$ and $X(T) < a$ with equal probability.

$$P[M(T) \ge a] = P[M(T) \ge a,\ X(T) > a] + P[M(T) \ge a,\ X(T) < a]$$

(Since $P[M(T) \ge a, X(T) = a] \le P[X(T) = a] = 0$, the other two are mapped onto one another by reflection about the line $x = a$ after the time T_a)

$$= 2P[M(T) \ge a,\ X(T) > a] = 2P[X(T) \ge a].$$

Note T_a, $a > 0$ is a.s. right continuous but may not be continuous. $P[M(t) \ge a] = 2P[X(t) \ge a]$ with $\mu = 0$, $X(0) = 0$ and $\sigma = 1$.

$$= \frac{2}{\sqrt{2\pi t}} \int_a^\infty \exp(-x^2/2t)\,dx$$

$$= \frac{2}{\sqrt{2\pi}} \int_{a/\sqrt{t}}^\infty \exp(-y^2/2)\,dy \left(\text{put } \frac{x}{\sqrt{t}} = y\right)$$

$$= 2\left[1 - \Phi\left(\frac{a}{\sqrt{t}}\right)\right]$$

Hence $P[T_a \le t] = P[M(t) \ge a] = 2\left[1 - \Phi\left(\frac{a}{\sqrt{t}}\right)\right]$

$$= \frac{a}{\sqrt{2\pi}} \int_0^t s^{-3/2} \exp\left(-\frac{a^2}{2s}\right) ds,\ s > 0 \left(s = \frac{a^2 t}{x^2} = \frac{a^2}{y^2}\right)$$

$$f_{T_a}(t) = F'_{T_a}(t) = \frac{a}{\sqrt{2\pi}}\, t^{-3/2} \exp\left(-\frac{a^2}{2t}\right),\ t > 0$$

is the density function for first passage time of the standard Brownian motion.

$$E(T_a) = \infty, \text{ sin ce } E(T_a) = \int_0^\infty P(T_a > x)\,dx$$

$$= \int_0^\infty \{1 - (2/\pi)^{1/2} \int_{a/\sqrt{x}}^\infty e^{-(1/2)z^2}\,dz\}\,dx$$

$$= (2/\pi)^{1/2} \int_0^\infty dx \int_0^{a/\sqrt{x}} e^{-(1/2)z^2}\,dz = \left(\frac{2}{\pi}\right)^{1/2} \int_0^\infty e^{-(1/2)z^2}\,dz \int_0^{a^2/z^2} dx$$

$$= a^2 (2/\pi)^{1/2} \int_0^\infty \frac{1}{z^2} e^{-(1/2)z^2}\,dz \ge a^2 (2/\pi)^{1/2} e^{-1/2} \int_0^1 \frac{dz}{z^2} = \infty.$$

In fact Laplace transform of T_a is $\varnothing(s) = \exp(-a\sqrt{2s})$ and hence no moment of T_a exists.

But if $\mu \ne 0$ and $\sigma > 0$, i.e. if $X(t)$ is an Wiener process with drift μ and diffusion coefficient σ then the first passage time r.v. T_a^* has density

$$f_{T_a^*}(t) = \frac{a}{\sigma\sqrt{2\pi}}\, t^{-3/2} \exp\left[-\frac{(a-\mu t)^2}{2\sigma^2 t}\right],\ t > 0.$$

Then, Laplace transform will be $\exp\left[\frac{a}{\sigma^2}\{\mu - \sqrt{\mu^2 + 2s\sigma^2}\}\right]$.

Hence $E(T_a^*) = a/\mu$ and $\text{Var}(T_a^*) = a\sigma^2/\mu^3$.

If $\mu > 0$, then $f_{T_a^*}(t)$ is called Inverse Gaussian density. For properties of this distribution, see Johnson and Kotz (1970) (Statistical Distribution, Vol. I) and for statistical applications, see Folks and Chhikara (1978) in J.A.S.A.

8.5 Ornstein and Uhlenbeck Process and Gaussian Processes

8.5.1 Ornstein and Uhlenbeck process (1930)

We have seen that for a Wiener process $X(t)$, the displacement Δx in a small interval of time Δt is small, being of the order $0(\sqrt{\Delta t})$, the velocity of which is $0(\sqrt{\Delta t}/\Delta t) = 0(1/\sqrt{\Delta t}) \to \infty$ as $\Delta t \to 0$.

Thus the Wiener process does not provide a satisfactory model for motion of small values of t although for moderate and large values of t it does so. An alternative model which holds for small t was proposed by Ornstein and Uhlenbeck in 1930.

The theory of Brownian motion developed by Einstein and Smoluchowski were not based on Newtonian mechanics. Langevian's approach, on the other hand, relies heavily on Newton's Second law of motion. In what follows we shall give a brief account of Langevian's model. Let $V(t)$ denote the random velocity of a large solute molecule immersed in a liquid at rest. For simplicity, let $V(t)$ denote the vertical component of the velocity. The solute molecule is subject to two forces: gravity and friction (resistence). It turns out that the gravitational force is negligible compared to the frictional force exerted on the solute molecule by the liquid. The frictional force is directed opposite to the direction of motion and is proportional to the velocity in magnitude. In the absence of statistical fluctuations, one would to the direcion of motion and is proportional to the velocity in magnitude. In the absence of statistical fluctuations, one would therefore have $mdv(t)\,dt = -\beta\,dv(t)$, where m is the mass of the solute molecule and β the constant of proportionality known as the coefficient of friction. However, this frictional force $-\beta V$ may be thought of as the mean of a large number of random molecular collisions. Assuming that the central limit theorem applies to the superposition of displacement due to a large number of such collisions, the change in momentum mdv (t) over an interval $(t, t + \Delta t)$ is approximately Gaussian, provided dt is such that the mean number of collisions during $(t, t + dt)$ is large.

In the Ornstein and Uhlenbeck process instead of the displacement $X(t)$, $V(t) = X'(t)$ is assumed to be a Brownian motion.

Let $v_0 = V(0)$ and $\{V(t), t \in T\}$ be a stochastic process representing velocity of unit mass molecule. The equation of motion of Brownian particle can be written as

$$\frac{dV(t)}{dt} = -\beta V(t) + f(t) \qquad \text{(Langevian equation)}$$

Since the derivative $\frac{dV(t)}{dt}$ may not exist we replace it by $\Delta V(t) = -\beta V(t)\Delta t$, where $f(t)dt$ is replaced by $\Delta F(t)$. Here $-\beta V(t)$ represent a symmetric part due to resistance of the medium and $dF(t)$ represents the random component.

Assume that F and V are independent and $F(t)$ is a Wiener Process with $\mu = 0$, Var $(F(t)) = \sigma^2 t = 2\,Dt$.

$$E(\Delta F(t)) = E[F(t + \Delta t) - F(t)] = 0$$

$$V(\Delta F(t)) = 2D\Delta t,\ 2D = \sigma^2,\ D > 0$$

$$E(\Delta F(t)\Delta F(s)) = 0 \text{ if } t < t + \Delta t \le s < s + \Delta s.$$

Then

$$a(v, t) = \lim_{\Delta t \to 0} \frac{1}{\Delta t} E[\Delta V(t) | V(t) = v]$$

$$= \lim_{\Delta t \to 0} \frac{1}{\Delta t} \{-\beta E[V(t)\Delta t | V(t) = v] + E[\Delta F(t) | V(t) = v]\}$$

$$= \lim_{\Delta t \to 0} \frac{1}{\Delta t} [-\beta E[(v\Delta t)] + E(\Delta F(t))] \quad \text{(Since } F(t) \text{ is independent of } V(t))$$

$$= -\beta v$$

$$b(v, t) = \lim_{\Delta t \to 0} \frac{1}{\Delta t} E[(\Delta V(t))^2 | V(t) = v]$$

$$= \lim_{\Delta t \to 0} \frac{(\Delta t)^2}{\Delta t} + \lim_{\Delta t \to 0} \frac{\text{Var}(\Delta F(t))}{\Delta t} = 2D.$$

Since these limits exist, the stochastic process $V(t)$ is a diffusion process and its transition density function $g(v_0, v, t)$ satisfies the forward Kolmogorov equation.

More specifically let $g(v, t)dv = P[v \le V(t) \le v + dv \mid V(0) = v_0]$.

Hence, the Kolmogorov's forward equation is

$$\frac{\partial g}{dt} = \beta \frac{\partial}{dv}(vg) + D \frac{\partial^2 g}{dv^2} \tag{8.8}$$

with initial conditions:

$g(v, 0) = \partial(v - v_0)$, Dirac delta function.

$$vg(v, t) \to 0 \text{ if } v \to \pm\infty \text{ and } \frac{\partial g(v, t)}{\partial v} \to 0 \text{ if } v \to \pm\infty.$$

To solve the differential equation for Ornstein-Uhlenbeck process one of the methods used is called separation of variable, i.e. we assume $g(v, t) = V(v)T(t)$, say and use the so-called Weber's function. But we shall use the method of Laplace Transform (ch.f.). Let $\varnothing(v, t)$ be the ch. function of g, i.e.

$$\int_{-\infty}^{\infty} e^{i\theta v} g(v, t)\, dv = \varnothing(v, t).$$

From equatin (8.8) we get

$$\int_{-\infty}^{\infty} e^{i\theta v} \frac{dg}{dt} dv = \beta \int_{-\infty}^{\infty} e^{i\theta v} \frac{\partial}{dv}(vg)\, dv + D \int_{-\infty}^{\infty} e^{i\theta v} \frac{d^2 g}{d^2 v} dv.$$

Integrating by parts and using the boundary conditions we get

$$\frac{\partial \varnothing(\theta, t)}{\partial t} = -\beta\theta \frac{\partial \varnothing}{\partial \theta} - D\theta^2 \varnothing, \text{ a linear p.d.e of Lagrange type.}$$

The characteristic equation is given by

$$\frac{dt}{1} = \frac{d\theta}{\beta\theta} = \frac{d\varnothing}{-D\theta^2 \varnothing}$$

Hence, $\beta dt = \dfrac{d\theta}{\theta} \Rightarrow \beta t = \log \theta / c_1 \Rightarrow e^{\beta t} = \theta / c_1.$

Therefore $c_1 = \theta e^{-\beta t}$. (8.9)

Also $$\frac{d\varnothing}{\varnothing} = -\frac{D}{\beta}\theta\, d\theta \Rightarrow \log \varnothing = -\frac{D\theta^2}{\beta 2} + \log c.$$

Therefore $$\varnothing(\theta, t) = c_2 e^{-D\theta^2/2\beta}. \tag{8.10}$$

Also $c_2 = \psi(c_1)$, say.

Eliminating one of the constants c_1 and c_2, we get,

$$e^{D\theta^2/2\beta}\varnothing(\theta, t) = \psi(\theta e^{-\beta t}) \tag{8.11}$$

Initial condition gives $g(v, 0) = \delta(v - v_0)$.

$$\int_{-\infty}^{\infty} e^{i\theta v}\, g(v, 0)\, dv = \int_{-\infty}^{\infty} e^{i\theta v}\, \delta(v - v_0)\, dv$$

$$= e^{i\theta v_0}$$

Therefore $$\varnothing(\theta, 0) = e^{i\theta v_0} = \exp(-D\theta^2/2\beta)\psi(\theta) \tag{8.12}$$

and $\psi(\theta) = e^{(D\theta^2/2\beta) + i\theta v_0}$ (δ is the Dirac delta function)

Hence $$\varnothing(\theta, t) = e^{-D\theta^2/2\beta} \exp\left[\frac{D}{2\beta}(\theta e^{-\beta t})^2 + i(\theta e^{-\beta t})v_0\right] \tag{8.13}$$

From (8.11) and (8.12) (8.13) can be written as

$$\varnothing(\theta, t) = \exp\,[i\theta m(t) - 1/2\ \theta^2 \sigma^2(t)],$$

where $$m(t) = v_0 e^{-\beta t} \quad \text{and} \quad \sigma^2(t) = \frac{D}{\beta}(1 - e^{-2\beta t}).$$

Note $Y(t) = e^{-\beta t} X(ae^{2\beta t})$ is the Ornstein-Uhlenbeck process if $X(t)$ is a Brownian Motion.

Therefore $Y(t) = \dfrac{1}{g(t)} X(ag(t))$, where $g(t) = e^{2\beta t}$, $\beta > 0$.

So, Ornstein-Uhlenbeck process is a time transformation of Brownian motion. Now we want to study what happens if $t \to \infty$.

$$m(t) \to 0, \text{ for } \beta > 0 \text{ and } \sigma^2(t) \to \frac{D}{\beta} = \sigma_0^2 \text{ (say) when } t \to \infty.$$

Then $g_0(v)\, dv = \lim\limits_{t\to\infty} g(v, t)\, dv$

$$= \frac{1}{\sqrt{2\pi}\,\sigma_0} e^{-\frac{1}{2}(v/\theta_0)^2} dv, \text{ i.e. the distribution of velocity}$$

is normal with mean zero and variance D/β. In Maxwell Boltzman distribution (kinetic theory of gases), $\sigma_0^2 = \dfrac{KT}{m}$, where K is the Boltzman constant, T the temperature and m the mass of the particle.

They assume the initial distribution is MB (Maxwell-Boltzman), i.e.

$$P[v_0 \le V_0 \le v_0 + dv_0] = \frac{1}{\sqrt{2\pi}} e^{-1/2(v_0/\sigma)^2} dv_0 .$$

The unconditional density of velocity is given by

$$g(v, t) = P[v \le V(t) \le v + dv]$$

$$= \int_{-\infty}^{\infty} P[v \le V(t) \le v + dv | V(0) = v_0] \, P[v_0 \le V(t) \le v_0 + dv_0] dv_0$$

$$= \int_{-\infty}^{\infty} \frac{1}{\sqrt{(2\pi\sigma^2(t))}} e^{-\frac{1}{2}\left(\frac{v-m(t)}{\sigma(t)}\right)^2} \frac{1}{\sqrt{(2\pi\sigma_0^1)}} e^{-\frac{1}{2}\left(\frac{v_0}{\sigma_0}\right)^2} dv_0$$

$$= \frac{e^{-\frac{1}{2}\frac{v^2}{\sigma_0^2}}}{\sqrt{(2\pi\sigma^2(t))}\sqrt{(2\pi\sigma_0^2)}} \int_{-\infty}^{\infty} e^{-\frac{1}{2}\left(\frac{v_0 - ve^{-\beta t}}{\sigma(t)}\right)} dv_0$$

$$= \frac{1}{\sqrt{2\pi}\sigma_0} e^{-\frac{1}{2}(v/\sigma_0)^2} dv_0 = g_0(v) dv_0$$

which is known as the Maxwell distribution of velocity of gases ($\sigma_0^2 = D/\beta$).

Hence, if v_0 is Maxwell Boltzman then $E(V(t)) = 0$, $\text{Var}(V(t)) = \sigma_0^2$ and $\text{Cov}(V(t), V(s)) = \sigma_0^2 e^{-\beta|t-s|}$.

Thus, $V(t)$ is covariance stationary.

But $V(t) = \frac{dx}{dt}$ where X is the displacement. Hence

$$E(X(t)) = v_0, \ \text{Var}(X(t)) = \frac{2\sigma_0^2}{\beta^2}(e^{-\beta t} - 1 + \beta t).$$

****Proof of Kolmogorov's forward differential equations for diffusion process**

Let $Q(X)$ be an arbitrary continuous function with $Q'(x)$, $Q''(x)$ and $Q(x) = 0$ outside $[a, b]$. Then from continuity $Q(x)$, $Q'(x)$ and $Q''(x)$ vanish at $x = a$ and $x = b$.

Consider

$$I = \int_{-\infty}^{\infty} Q(y) \frac{\partial}{\partial y} f(y, t) dy = \frac{\partial}{\partial t} \int_a^b Q(y) f(x_0, \tau; y, t) dy$$

(since $Q(y)f(y, t)$ is continuous in y and t and $\partial f/\partial t$ is continuous)

$$= \lim_{\Delta t \to 0} \int_a^b Q(y) \frac{f(x_0, \tau; y, t + \Delta t) - f(x_0, \tau; y, t)}{\Delta t} dy$$

$$= \lim_{\Delta t \to 0} \frac{1}{\Delta t} \left\{ \int_a^b dy \, Q(y) \int_{-\infty}^{\infty} f(x_0, \tau; y, t) f(x, t, y, t + \Delta t) \, dx \right.$$

$$\left. - \int_a^b Q(x) f(x_0, \tau; y, t) \, dy \right\} \text{ (by Chapman-Kolmogorov)}$$

$$= \lim_{\Delta t \to 0} \frac{1}{\Delta t} \int_{-\infty}^{\infty} f(x_0 \tau; y, t) \left\{ \int_{-\infty}^{\infty} f(x, t; y, t + \Delta t) Q(y) dy - Q(x) \right\} dx$$

[For interchanging the order of integration:

$$g(x, y) = Q(y) f(x, t; y, t + \Delta t)\, f(x, \tau; y, t)$$

is continuous in

$$(x, y) \in (-\infty, \infty) \times [a, b], \mid \varnothing(y) \mid = \left| \int_{-\infty} g(x, y) dx \right|$$

$$\leq \mid Q(y) \mid \int_{-\infty}^{\infty} f(x, t; y, t + \Delta t) dx$$

is uniformly continuous]

$Q(x)$ being continuous on $[a, b]$, it is bounded by diffusion condition $v(a)$ and hence we get

$$\int_{|y-x|>\delta} f(x, t; y, t + \Delta t) dy = o(\Delta t), \text{ as } \Delta t \to 0$$

Thus,

$$I = \lim_{\Delta t \to \infty} \frac{1}{\Delta t} \int_{-\infty}^{\infty} dx\, f(x_0, \tau; x, t) \left\{ \int_{|y-x|\leq\delta} f(x, t, y; t + \Delta t) Q(y) dy - Q(x) \right\} dy$$

Expanding $Q(y)$ in the neighbourhood of x, we get

$$Q(y) = Q(x) + (y - x) Q'(x) + \frac{1}{2} (y - x)^2 Q''(x) + o(y - x)^2 \quad (8.14)$$

Using (8.14) and diffusion conditions (a), (b) and (c), we get

$$\int_{|y-x|\leq\delta} f(x, t; y, t + \Delta t) Q(y) dy - Q(x)$$

$$= \int_{|y-x|\leq\delta} f(x, t; y, t + \Delta t)[Q(x) + (y - x) Q'(x) + \frac{1}{2} (y - x)^2 Q''(x)$$

$$+ o(y - x)^2]\, dy - Q(x)$$

$$= o(\Delta t) + Q'(x) \int_{|y-x|\leq\partial} (y - x) f(x, t; y, t + \Delta t) dy$$

$$+ \frac{1}{2} Q''(x) \int_{|y-x|\leq\partial} [(y - x)^2 + o(y - x)^2] f(x, t; y, t + \Delta t) dy$$

(since by diffusion condition (v(a))

$$\int_{|y-x|\leq\partial} f(x, t; y, t + \Delta t) dy = 1 - \int_{|y-x|\leq\partial} f(x, t; t, + \Delta t) dy = 1 - o\,\Delta t))$$

$$\text{Therefore } I = \int_{-\infty}^{\infty} f(x_0, \tau; x, t)[Q'(x) a(x, t) + 1/2\, Q''(x) b(x, t)] \quad (8.15)$$

[by diffusion conditions (v(b)) and (v(c))]

Integrating by parts

$$\int_{-\infty}^{\infty} (af)Q'(x)\,dx = [afQ]_a^b - \int_a^b Q(x)\frac{\partial}{\partial x}(af)\,dx$$

$$= -\int_a^b Q(x)\frac{\partial}{\partial x}(af)\,dx \tag{8.16}$$

(since $Q(a) = Q(b) = 0$)

and

$$\int_{-\infty}^{\infty} (bf)Q''(x)\,dx = [bfQ']_a^b - \int_a^b Q'(x)\frac{\partial}{\partial x}(bf)\,dx$$

$$= -\int_a^b Q'(x)\frac{\partial}{\partial x}(bf)\,dx \;(\because Q'(a) = Q'(b) = 0)$$

$$= -\int_a^b Q(x)\frac{\partial}{\partial x}(bf)\,dx$$

$$= -\frac{\partial}{\partial x}(bf)Q]_a^b + \int_a^b Q(x)\frac{\partial^2}{\partial x^2}(bf)\,dx$$

$$= \int_a^b Q(x)\frac{\partial^2}{\partial x^2}(bf)\,dx \tag{8.17}$$

By (8.15), (8.16) and (8.17), we get

$$I = \int_{-\infty}^{\infty} \left\{\frac{\partial}{\partial x}(-af) + 1/2\,\frac{\partial^2}{\partial x^2}(bf)\right\}Q(x)\,dx$$

Since $I = \int_{-\infty}^{\infty} \frac{\partial f}{\partial t}Q(x)\,dx,$

$$\int_{-\infty}^{\infty} \left\{\frac{\partial f}{\partial t} + \frac{\partial}{\partial x}(af) - 1/2\,\frac{\partial^2}{\partial x^2}(bf)\right\}Q(x)\,dx = 0.$$

Since the last expression in $\{\cdot\}$ is a continuous function and $Q(x)$ is arbitrary,

$$\frac{\partial f}{\partial t} + \frac{\partial}{\partial x}(af) - 1/2\,\frac{\partial^2}{\partial x^2}(bf) = 0.$$

8.5.2 Gaussian process

Definition 8.2 The stochastic process $\{X(t), t \in T\}$ is said to be a "Gaussian process" if for each $\{t_1, \ldots, t_n\} \subset T$, $n \geq 1$, $(X(t_1), \ldots, X(t_n))$ is multivariate normal.

The covariance function $\Gamma(s, t) = E(X(s) - \mu(s))(X(t) - \mu(t))$ possess the following properties:

(i) $\Gamma(s, t) = \Gamma(t, s)$;

(ii) For any $\{t_1, \ldots, t_n\} \subset T, n \geq 1$, and any real numbers $u_1 \ldots, u_n$, $\sum_{i=1}^{n} \sum_{i=1}^{n} u_i u_j \Gamma(t_i, t_j) \geq 0$ i.e. the covariance function is non-negative definite. This follows from the fact that the covariance matrix of the multivariate normal random vector is positive definite.

Theorem 8.3 The Gaussian Stochastic process $\{X(t), t \in T\}$ is stationary iff the covariance function $\Gamma(m, n) = \Gamma(m - n, 0)$ for all m, n.

Proof If $\{X(t)\}$ is stationary then $E(X(m)X(n)) = E(X_{m-n}X_{n-m}) = \Gamma(m - n, 0)$. Conversely, if true, then the characteristic function of $X(1), X(2), \ldots, X(n)$ is

$$\exp\left(1/2 \sum_{k=1}^{n} \sum_{j=1}^{n} u_j u_k \Gamma(k-j, 0)\right).$$

But this is exactly the characteristic function of $X_{1+m}, \ldots, X_{n+m}$.
From uniqueness theorem of characteristic function the result follows.

Theorem 8.4 If a Gaussian process $\{X(t), t \in T\}$ have mean function $EX(t) = \mu(t) = 0$, and $\Gamma(s, t) = EX(s)X(t)$, then it is a Markov process iff

$$\Gamma(t_1, t_3) = \frac{\Gamma(t_1, t_2)\Gamma(t_2, t_3)}{\Gamma(t_2, t_2)} \text{ for any } t_1, t_2, t_3 \in T \text{ with } t_1 < t_2 < t_3 \tag{8.5.1}$$

Proof
Let us first prove that if $X = (X_1, \ldots, X_n)$ is multivariate normal $N(0, \Sigma)$, $\Sigma = (\sigma_{ij})^{-1}$ then

$$E(X_n | X_1 = x_1, \ldots, X_{n-1} x_{n-1}) = -\sum_{i=1}^{n-1} \left(\frac{\sigma_{in}}{\sigma_{nn}}\right) x_i \tag{8.5.2}$$

Now
$$\sum_{i=1}^{n} \sum_{j=1}^{n} \sigma_{ij} x_i x_j = \sum_{i=1}^{n-1} \sum_{j=1}^{n-1} \sigma_{ij} x_i x_j + x_n^2 \sigma_{nn} \hat{+} 2x_n \sum_{i=1}^{n-1} \sigma_{in} x_i \tag{8.5.3}$$

Then

$$f(x_n | x_1, \ldots, x_{n-1}) = \frac{\exp(-1/2\sigma_{nn} x_n^2 - x_n \sum_{i=1}^{n-1} \sigma_{in} x_i)}{\int_{-\infty}^{\infty} \exp(-1/2\sigma_{nn} x_n^2 - x_n \sum_{i=1}^{n-1} \sigma_{in} x_i) dx_n}$$

$$= \frac{\exp\left\{-1/2\sigma_{nn}\left[x_n + \sum_{i=1}^{n-1} (\sigma_{in}/\sigma_{nn}) x_i\right]^2\right\}}{\int_{-\infty}^{\infty} \exp\left\{-1/2\sigma_{nn}\left[x_n + \sum_{i=1}^{n-1} (\sigma_{in}/\sigma_{nn}) x_i\right]^2\right\} dx_n}$$

$$\times C \exp\left\{-1/2\sigma_{nn}\left[x_n + \sum_{i=1}^{n-1} (\sigma_{in}/\sigma_{nn}) x_i\right]^2\right\} \tag{8.5.4}$$

where C is a constant independent of $x_1, \ldots, x_{n-1}$. Equation (8.5.4) is clearly a normal p.d.f. with mean $-\sum_{i=1}^{n-1} (\sigma_{in}/\sigma_{nn}) x_i$.

Now if $X(t)$ is a Markov process, it follows from (8.5.2) that for any $s < t$ from T,

$$E(X(t)|X(s) = x) = -\sigma_{12}/\sigma_{22}x \text{ (with } n = 2).$$

Now σ_{ij} are elements of $\Sigma^{-1} = \begin{bmatrix} \Gamma(s,s) & -\Gamma(s,t) \\ -\Gamma(t,s) & \Gamma(t,t) \end{bmatrix}^{-1}$

$$= \frac{1}{\Gamma(s,s)\Gamma(t,t)-\Gamma^2(s,t)} \begin{bmatrix} \Gamma(t,t) & -\Gamma(s,t) \\ -\Gamma(s,t) & \Gamma(s,s) \end{bmatrix}.$$

Therefore
$$E(X(t)|X(s) = x) = \frac{\Gamma(s,t)}{\Gamma(s,s)} x \tag{8.5.5}$$

Since given the present, in a Markov process the past is independent of the future,

$$\Gamma(t_1, t_3) = E(X(t_1)X(t_3)) = E\{E(X(t_1)X(t_3) \mid X(t_2))\}$$
$$= E\{E(X(t_1) \mid X(t_2))E(X(t_3 \mid X(t_2))\},\ t_1 < t_2 < t_3$$

Hence by (8.5.5) we have

$$\Gamma(t_1, t_3) = E\left\{\left(\frac{\Gamma(t_1,t_2)}{\Gamma(t_2,t_2)}\right) X(t_2) \left(\frac{\Gamma(t_2,t_3)}{\Gamma(t_2,t_2)}\right) X(t_2)\right\}$$

$$= \left(\frac{\Gamma(t_1,t_2)\Gamma(t_2,t_3)}{(\Gamma(t_2,t_2))^2}\right) E(X(t_2))^2 = \left(\frac{\Gamma(t_1,t_2)\Gamma(t_2,t_3)}{\Gamma(t_2,t_2)^2}\right) \tag{8.5.1}$$

for $t_1 < t_2 < t_3$.

To prove the converse, i.e. a zero mean Gaussian process whose covariance function satisfies (8.5.1) must have the Markov property, we first prove that a zero mean Gaussian process is Markov iff, for any

$$t_1 < t_2 < \ldots < t_n,\ n \geq 2,$$

$$E\{X(t_n) \mid X(t_1) = x_1, \ldots, X(t_{n-1}) = x_{n-1}\} = E(X(t_n) \mid X(t_{n-1}) = x_{n-1}\} \tag{8.5.6}$$

The necessity is obvious. To prove its sufficiency note that from (8.5.2) and (8.5.6), it follows that

$$\sigma_{1n} = \sigma_{2n} = \ldots = \sigma_{n-2,\,n} = 0$$

so that by (8.5.4) $f(x_n | x_1, x_2, \ldots, x_{n-1}) = C \exp\left\{-\frac{1}{2}\sigma_{nn}\left[x_n + \frac{\sigma_{n-1n}}{\sigma_{nn}} x_{n-1}\right]^2\right\}.$

Obviously $f(x_n | x_{n-1}) = C.\exp\left\{-\frac{1}{2}\sigma_{nn}\left[x_n + \frac{\sigma_{n-1n}}{\sigma_{nn}} x_{n-1}\right]^2\right\}.$

Therefore $f(x_n \mid x_1, \ldots, x_{n-1}) = f(x_n|x_{n-1})$ i.e. $X(t)$ is a Markov process.

Let us now show that (8.5.1) implies (8.5.6). Equation (8.5.1) implies that for $1 \leq k \leq n-1$.

$$\Gamma(t_k, t_n) = \frac{\Gamma(t_k, t_{n-1})\Gamma(t_{n-1}, t_n)}{\Gamma(t_{n-1}, t_{n-1})} \quad \text{(since } t_k < t_{n-1} < t_n)$$

and that implies

$$E\left\{\left[X(t_n) - \left(\frac{\Gamma(t_{n-1}, t_n)}{\Gamma(t_{n-1}, t_{n-1})}\right) X(t_{n-1})\right] X(t_k)\right\} = 0$$

for all $k = 1, 2, \ldots, n-1$, $y_n = X(t_n) - \dfrac{\Gamma(t_{n-1}, t_n)}{\Gamma(t_{n-1}, t_{n-1})} X(t_{n-1})$ are normally distributed r.v.s, are independent of $X(t_1), \ldots, X(t_{n-1})$. Therefore,

$$E\left\{X(t_n) - \frac{\Gamma(t_{n-1}, t_n)}{\Gamma(t_{n-1}, t_{n-1})} X(t_{n-1}) \middle| X(t_1) = x_1, \ldots, X(t_{n-1}) = x_{n-1}\right\}$$

$$= E\left\{X(t_n) - \frac{\Gamma(t_{n-1}, t_n)}{\Gamma(t_{n-1}, t_{n-1})} X(t_{n-1})\right\} = 0$$

or $\quad E\{X(t_n) \mid X(t_1) = x_1, \ldots, X(t_{n-1}) = x_{n-1}\}$

$$= \frac{\Gamma(t_{n-1}, t_n)}{\Gamma(t_{n-1}, t_{n-1})} X_{n-1} = E\{X(t_n) | X(t_{n-1}) = x_{n-1}\}$$

This proves (8.5.6) and hence the desired result.

8.5.3 *k*-Dimensional Brownian motion

Let $\{X_1(t); t \geq 0\}, \ldots, \{X_k(t), t \geq 0\}$, be k independent Brownian motions, i.e. for any finite set of points

$$t_{11}, t_{12}, \ldots, t_{1n_1}, t_{21}, t_{22}, \ldots, t_{2n_2}$$

and $\quad t_{k1}, t_{2k}, \ldots, t_{k,n_k},$

the k vectors

$$X_1(t_1) = (X_1(t_{11}), \ldots, X_1(t_{1n_1})),$$

$$X_2(t_2) = (X_2(t_{21}), \ldots, X_2(t_{2n_2})),$$

$$\ldots\ldots\ldots\ldots\ldots\ldots\ldots\ldots\ldots$$

$$X_k(t_k) = (X_k(t_{k1}), \ldots, X_k(t_{kn_k})),$$

are independent. The vector-valued process defined by

$$X(t) = (X_1(t), \ldots, X_k(t)), t \geq 0,$$

is called k-dimensional Brownian motion.

The movement of a particle undergoing Brownian motion in the plane and in space are described by two- and three-dimensional Brownian motions, respectively.

8.5.4 Radial Brownian motion (Bessel process)

Let $\{X(t); t \geq 0\}$ be a k-dimensional Brownian motion process. The stochastic process defined by

$$R(t) = [X_1^2(t) + \ldots + X_k^2(t)]^{1/2}, t \geq 0,$$

is called the Radial Brownian motion or the Bessel process with parameter (1/2) k–1. It is a Markov process having continuous sample paths in $(0, \infty)$. The transition probability density function from x to y is

$p(t; x, y) = t^{-1} \exp\{-(x^2 + y^2)/(2t)\}(xy)^{1-k/2} I_{[k/2]-1}(xy/t)y^{k-1}, t > 0, x, y > 0$

where $I_j(z)$ is the modified Bassel function

$$I_j(z) = \sum_{k=0}^{\infty} (z/2)^{2k+j} / [k! \overline{|k + j + 1}].$$

We shall prove the case $k = 2$.

Use the polar coordinates by defining

$$R(t) = \sqrt{X_1^2(t) + X_2^2(t)},\ \theta(t) = \tan^{-1}(X_2(t)/X_1(t)).$$

Since Brownian motion is Markov,

$$P[R(t_n + t) \le y \mid X(t_0) = \underset{\sim}{x}_0, \ldots, X(t_{n-1}) = \underset{\sim}{x}_{n-1}, X(t_n) = \underset{\sim}{x}]$$
$$= P[R(t_n + t) \le y \mid X(t_n) = \underset{\sim}{x}] = P[R(t) \le y \mid X(0) = \underset{\sim}{x}],$$

where $0 < t_0 < \ldots < t_n, t > 0$, and $\underset{\sim}{x}_0, \ldots, \underset{\sim}{x}_{n-1}$ are arbitrary points in the plane, $x = (x_1, x_2)$. Then

$P[R(t) \le y \mid X(0) = x]$

$$= \iint_{u_1^2 + u_2^2 \le y^2} \frac{1}{2\pi t} \exp\left\{\frac{(u_1 - x_1)^2 + (u_2 - x_2)^2}{2t}\right\} du_1 du_2$$

$$= \int_0^y \left[\int_0^{2\pi} \frac{1}{2\pi t} \exp\left\{\frac{(r \sin\theta - x_1)^2 + (r\cos\theta - x_2)^2}{2t}\right\} d\theta\right] r dr$$

where transformations $u_1 = r\cos\theta$ and $u_2 = r\sin\theta$ gives $du_1\, du_2 = r dr d\theta$. Since

$$(r\sin\theta - x_1)^2 + (r\cos\theta - x_2)^2 = r^2 + 2r(x_1 \sin\theta + x_2\cos\theta) + \| x \|^2$$

where $\|x\| = x_1^2 + x_2^2$,

$$P(R(t) \le y \mid X(0) = x) = \int_0^y \frac{r}{2\pi t} \exp\left\{\frac{r^2 + \|x\|^2}{2t}\right\} I(r, x)\, dr,$$

where $$I(r, x) = \int_0^{2r} \exp\left\{\frac{r}{t}(x_1 \sin\theta + x_2 \cos\theta)\right\} d\theta.$$

Defining an angle ψ by writing

$\sin\psi = x_1/\| x \|$, $\cos\psi = x_2/\| x \|$, we get

$$I(r, x) = \int_0^{2\pi} \exp\left\{\frac{r\|x\|}{2t}\cos(\psi + \theta)\right\} d\theta$$

$$= \int_0^{2\pi} \exp\left\{\frac{r\|x\|}{2t}\cos\theta\right\} d\theta,$$

Since $\cos(\psi + \theta)$ is periodic over the interval $[0, 2\pi]$.

$$\int_0^{2\pi} \exp(a\cos\theta)\, d\theta = \int_0^{2\pi} \sum_{k=0}^{\infty} \frac{(a\cos\theta)^k}{k!}\, d\theta = \sum_{k=0}^{\infty} \frac{a^k}{k!} \int_0^{2\pi} \cos^k\theta\, d\theta$$

But $$\int_0^{2\pi} \cos^k\theta\, d\theta = \begin{cases} 0 & \text{if } k = 1, 3, \ldots \\ \dfrac{k!\,2\pi}{2^k[(k/2)!]^2} & \text{if } k = 0, 2, 4, \ldots \end{cases}$$

Therefore $$\int_0^{2\pi} \exp\{a\cos\theta\}\, d\theta = 2\pi \sum_{k=0,2,\ldots} a^k/k! \left\{\frac{a^k}{2^k[(k/2)!]^2}\right\}$$

$$= 2\pi \sum_{k=0} \frac{a^{2k}}{2^{2k}(k!)^2} = 2\pi I_0(a)$$

Hence $$P(R(t) \le y \mid X(0) = x) = \int_0^y \frac{r}{2\pi t} \exp\left\{\frac{r^2 + \|x\|^2}{2t}\right\} 2\pi I_0 \left\{\frac{r\|x\|}{t}\right\} dr$$

$$= \int_0^y f_t(\|x\|, r)\, dr \quad \text{(when } k = 2)$$

Therefore

$$P(R(t_n + t) \le y \mid X(t_0) = x_0, \ldots, x(t_n) = x) = \int_0^y f_t(\|x\|, r)\, dr$$

Let $r_i^2 = (x_{1i}^2 + x_{2i}^2)$, $i = 0, \ldots, n$, where $\underset{\sim}{x}_i = (x_{1i}, x_{2i})$ and let $\theta_j = \tan^{-1}(x_{2j}/x_{1j})$. The conditions $X(t_0) = \underset{\sim}{x}_0, \ldots, X(t_n) = \underset{\sim}{x}_n$, are equivalent to the conditions

$$R(t_0) = r_0,\ \theta(t_0) = \theta_0, \ldots, R(t_n) = r_n,\ \theta(t_n) = \theta_n.$$

Denoting the joint probability density of $\theta(t_0), \ldots, \theta(t_n)$ by $f(\theta_0, \ldots, \theta_n)$, we have

$P(R(t_n + t) \le y \mid R(t_0) = r_0, \ldots, R(t_n) = r_n)$

$$= \int_0^{2\pi} \cdots \int_0^{2\pi} P(R(t_n + t) \le y \mid R(t_0) = r_0,\ \theta(t_0) = \theta_0, \ldots, R(t_n) = r_n$$

$$\theta(t_n) = \theta_n\} f(\theta_0, \ldots, \theta_n)\, d\theta_0, \ldots, d\theta_n$$

$$= \int_0^{2\pi} \cdots \int_0^{2\pi} \left\{\int_0^y f_1(r_n, r)\right\} f(\theta_0, \ldots, \theta_n)\, d\theta_0 \ldots d\theta_n$$

$$= \int_0^y f_t(r_n, r)\, dr$$

This verifies Markov property of $R(t)$ and its transition density when $k = 2$. Similarly, the general case can be proved by induction.

Exercises and Complements

Exercise 8.1 If $X(t)$ is a standard Brownian motion process, show that the conditional density of $X(t)$ given $X(0) = X(1) = 0$ a.s. is given by

$$(2\pi/t(1-t))^{-1/2}\exp\left[-\frac{1}{2}x^2/(t(1-t))\right], -\infty < x < \infty, 0 < t < 1.$$

Complements The conditional process of $X(t)$ given $X(0) = X(1) = 0$ is known as *Brownian Bridge* or *Tidedown Brownian* motion and is denote by $B(t)$. This can also be written as $B(t) = X(t) - tX(1)$, $0 \le t \le 1$. It is easy to see that $B(1) = B(0) = 0$ a.s. and Cov $(B(t), B(s)) = s(1-t)$ if $0 < s < t \le 1$ and Var $(B(t)) = t(1-t)$, $0 \le t \le 1$.

Exercise 8.2 (*Interpolation formula for Brownian motion*) Show that the conditional density of $X(t)$ for $t_1 < t < t_2$ given $X(t_1) = A$ and $X(t_2) = B$ is a normal density with

$$\text{Mean} = A + \frac{B-A}{t_2 - t_1}(t - t_1) \quad \text{and} \quad \text{Var} = (t_2 - t)(t - t_1)/(t_2 - t_1).$$

Exercise 8.3 Let $\{X(t), t \ge 0\}$ be a Brown motion process with $X(0) = 0$ a.s. Then show that

(a) $p = P[X(t)$ has at least one zero in an interval $(a, b)]$
$= 2/\pi \cos^{-1}(a/b)^{1/2}$.

(b) $P[T < t] = 2/\pi \sin^{-1}(a/t)^{1/2}$ where T is the largest zero of $X(t)$ with $T \le t$.

Exercise 8.4 Show that a stochastically continuous stationary Gaussian process $\{X(t),\ t \ge 0\}$ with $EX(t) = 0$ $E(X(t))^2 = 1$ is Markovian iff the Correlation function $\rho(t) = e^{-c|t|}$, $c > 0$.

Exercise 8.5 Let Z and θ be independent r.v.s such that Z has a density function f_z given by

$$f_z(Z) = \begin{cases} 0 & \text{if } z < 0 \\ ze^{-(1/2)z^2} & \text{if } z \ge 0 \end{cases}$$

and θ is uniformly distributed in the interval $[0, 2\pi]$.

Define $X(t) = Z\cos(2\pi t + \theta)$.

Show that $\{X(t), -\infty < t < \infty\}$ is a Gaussian process.

Exercise 8.6 Suppose that $\{X(t), -\infty < t < \infty\}$ is given by the form

$$X(t, \omega) = A(\omega)\cos[2\pi t + \theta(\omega)]$$

where A and θ are independent. A is nonnegative and θ is uniformly distributed on $[0, 2\pi)$.

(a) Show that $\{X(t), -\infty < t < \infty\}$ is a stationary process.

(b) Show that $E(X(t)) = 0$ for all t provided that $E(A) < \infty$. Does

$$M(t) = 1/2T\int_{-T}^{T} X(t)\,dt \text{ converges a.s. to } E(X(t)) \text{ as } T \to \infty.$$

(c) Show that $\{X(t), -\infty < t < \infty\}$ is a Gaussian process iff A has a Rayleigh distribution, that is, A has a density f_A given by

$$f_A(a) = \frac{a}{\sigma^2}\exp(-1/(2\sigma^2 a^2)),\ a \ge 0.$$

Is $\{X(t),\ -\infty < t < \infty\}$ a Markov process?

Exercise 8.7 Determine the Covariance function of $U(t) = e^{-t}\,W(e^{2t})$, $t \ge 0$ where $W(t)$ is the standard Brownian motion.

Exercise 8.8 Let $X(t) = \int_0^t W(s)\,ds$. Calculate $EX(t)$ and Var $(X(t))$. Also find the conditional distribution of $X(t)$ given that $W(t) = w$.

****Exercise 8.9** Prove Kolmogorov's inequality for a separable Brownian motion

$$P[\sup_{0\le u\le t} |W(u)| > C] \le t/\varepsilon^2, \varepsilon > 0.$$

Exercise 8.10 Prove that $\lim_{t\to\infty} \frac{1}{t} W(t) = 0$ a.s.

Exercise 8.11 Let $W_1(t)$ and $W_2(t)$ are two independent Brownian motions and T_x be the first time t at which $W_1(t) = x$. Find the distribution of $Y(x) = W_2(T_x)$. Show that the stochastic process $\{T_x: x \ge 0\}$ and $\{Y(x); x \ge 0\}$ have stationary independent increments.

Complements In general, given any Markov process $\{X(t), t \ge 0\}$ and a process $\{T(x), x \ge 0\}$ having stationary independent increments and increasing sample paths, with $T(0) = 0$, it is possible to derive a new process $Y(x) = X(T(x)), x \ge 0$.

The process of forming Y from X is called *subordination*, and the process $\{T(x)), x \ge 0\}$ is called *subordinator*. Under the above, conditions $\{Y(x)), x \ge 0\}$ will be a Markov process. If, in addition, $\{X(t), t \ge 0\}$ has stationary independent increments, then so will $\{Y(x), x \ge 0\}$.

Exercise 8.12 (*Geometric Brownian motion*) Let $\{X(t), t \ge 0\}$ be a Wiener process with drift μ and diffusion coefficient σ^2. Find the $E[Y(t)| Y(0) = y]$ and Var $[Y(t) \mid Y(0) = y]$ of the Geometric Brownian motion $Y(t) = e^{X(t)}, t \ge 0$.

Also calculate the diffusion coefficients

$$\lim_{h\downarrow 0} \frac{E[\{Y(t+h) - Y(t)\}| Y(t) = y]}{h} = a(y), 0 < y < \infty$$

and

$$\lim_{h\downarrow 0} \frac{E[\{Y(t+h) - Y(t)\}^2 | Y(t) = y]}{h} = a(y), 0 < y < \infty$$

Exercise 8.13 Let $M(t) = \max_{0\le s\le t} W(s)$ and $Y(t) = M(t) - W(t)$

Find the joint p.d.f. of $M(t)$ and $T(t)$.

Exercise 8.14 Let $\{W(t)t \ge 0\}$ be a standard Brownian motion. Let $a < 0 < b$ and T_{ab} be the first time $W(t)$ reach a or b. Find $E(T_{ab})$. If $a < x < b$, then find $P[X(T_{ab}) = b \mid W(0) = x]$ and $E[T_{ab} \mid W(0) = x]$.

Exercise 8.15 Let $\{W(t)\ t \ge 0\}$ be a Wiener process with drift $\mu \ne 0$ and diffusion coefficient σ^2 and suppose $W(0) = x$. Find the probability that $W(t)$ reaches level $b > x$ before hitting $a < x$ i.e. $P[W(T_{ab}) = b \mid W(0) = x]$. If $\mu < 0$ and $M = \max_{0\le s\le t} W(s) - W(0)$, show that M has exponential distribution.

Exercise 8.16 Let T be the first time the Brownian motion process crosses the line $b(t) = a + bt (a > 0, b > 0)$. Find the Laplace transform of T.

9

Time Series Analysis

9.1 Introduction and Stationary Time Series

A typical time series can be considered to be the result of four distinct components.

(1) Trend

A smooth long-term movement covering a number of years reflecting the general tendency of the series. Growth curves are good examples of time series exhibiting trend. Naturally, characteristics of the process depend very much on the time of observation. One method of eliminating the trend factor from a time series is the method of moving averages which is well known and we shall not discuss it in some detail in this book.

(2) Cyclical Component (Seasonal Component)

The result of a periodic movement in definite time periods and can be represented by a strictly periodic function of time i.e. $X_t = X_{t+\theta} = X_{t+2\theta} = \ldots$ (the period is θ). Fourier analysis techniques may be employed in the investigation of a known periodic movement like a seasonal variation. Periodogram analysis is a method based on the harmonic analysis of the Fourier representation. We shall consider only sine and cosine functions later on in our studies.

(3) Oscillatory Component

It is the Irregular periodic movement of a time series. Economists and many others go for this type of non seasonal cyclic patterns. The sunspot cycle and outfall of many rivers are good examples.

(4) Random Component

Random disturbance term is the basis for the stochastic Process representation of the time series.

Difference Between (2) and (3)

In cyclical series, apart from the disturbances due to superimposition of the random element, the maximum and minimum occur at equal intervals. Moreover, the amplitude also remains the same.

In an oscillatory series the distance from peak-to-peak and trough-to-trough varies. The oscillation also does not remain the same around some constant value. (See the following figures).

We would, however, consider only those oscillatory series which have a constant amplitude. When the trend and cyclical component are eliminated from a time series we are left with a series which apart from the disturbance term fluctuates round some constant value.

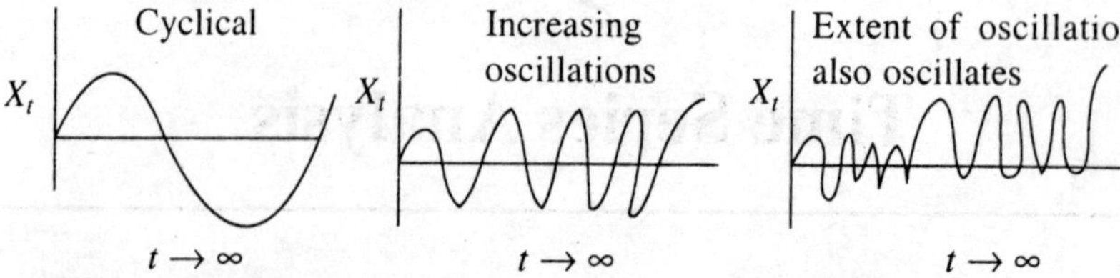

The question would arise whether it is possible to represent such a series by some function of t and in fact three systemmatic procedures have been developed by which an oscillatory series can be represented appropriately. These procedures are called the *Schemes*.

Consider, three schemes which can account for a typical oscillatory series.

(i) *Moving Average of a purely random component* may generate an oscillatory series.

(ii) *Sum of cyclical components*, e.g.

$$X_t = a_1 \sin \frac{2\pi t}{\omega_1} + a_2 \sin \frac{2\pi t}{\omega_2}$$

where $\omega_1 \neq c\omega_2$ (c being, a constant multiplier).

Thus the sum of a number of cyclical components may generate an oscillatory series (since the amplitudes change due to superimposition).

(iii) *Autoregressive scheme* Let us consider the representation of a time series as

$$X_t = g(X_{t-1}, X_{t-2}, \ldots, X_{t-k}) + \varepsilon_t$$

where g is some mathematical function and ε_t is the random component. Then the series X_t is an oscillatory series.

9.1.2 Stationary time series

Denote by $(\Omega, \mathscr{F}, P)$ the probability space and let T be an index set. Then a real valued time series (stochastic process) is a real-valued function $X(t, \omega)$ belonging to the product space $TX\ \Omega$.

For fixed $t \in T$, $X(t, \omega)$ is a random variable defined over Ω and is denoted by X_t. The time series $\{X_t, t \in T\}$ is therefore a collection of all such random variables.

The series $\{X_t\}$ is said to be strictly stationary if

1. $F_{X_{t_1} \ldots X_{t_n}}(x_{t_1}, \ldots, x_{t_n}) = F_{X_{t_1+h} \ldots X_{t_n+h}}(x_{t_1}, \ldots, x_{t_n})$ for $(t_1, \ldots, t_n)$ and $(t_1 + h, \ldots, t_n + h) \subset T$ and $(X_{t_1}, \ldots, X_{t_n})$ lies in the range of the r.v. X_t.

Note

(i) The distribution of any point in the index set remains the same,

(ii) From (i) it is clear that the joint distribution of a finite number of points in the index set do not involve the points themselves, instead it is a function of the distance between the points,

(iii) $t_1, \ldots, t_n$ are not necessarily consecutive.

(iv) If the second order moment $E(X_t^2) < \infty$, then

$$\left.\begin{aligned} E(X_t) &= \text{constant for all } t \\ V(X_t) &= \text{constant for all } t \end{aligned}\right\} \text{for such a stationary time series.}$$

Recall that $\{X_t, t \in T\}$ is *covariance stationary* if

$$E(X_t) = \text{constant}$$

and Cov $(X_t, X_{t+h}) = \gamma(h)$ for all $t \in T$.

9.1.3 Stationary vector stochastic process

In many areas of statistical application such as economics, business, engineering, earth sciences, environmentals sciences, data are collected over time, typically quarterly, daily, hourly, in second etc. Often such data is available on several related variables. An example is the daily measurement of dissolved oxygen concentration and biochemical oxygen demand in a river or ocean coast as indicators of water quality. Another example may concern the study of the relation between immigration and emigration from a region of a country as they relate to economic variables such as capital growth.

There are two basic reasons for analyzing the data jointly as opposed to singly:

1. To understand the dynamic relations among the series, the effect of change in our series may show in another series simultaneously or over a period of times and with varying amplitude effect. Some series may lead others and there may be feedback connections between them.

2. To provide more accurate forecasts, the joint modelling allows the use of joint information; intuitively this enables forecast errors to be reduced.

We assume the availability of observations on a set of variables, say $\underset{\sim}{X}_t = (X_{1t}, \ldots, X_{kt})$. In order to model the data generation process, we associate the random vector $\underset{\sim}{X}_t$ with each integer $t = 0, \pm 1, \pm 2, \ldots$. This section is mainly concerned with the first two moments of the random vector $\underset{\sim}{X}_t$ We call them stationary (second order) if these moments are finite and constant through time t. Formally $\{\underset{\sim}{X}_t\}$ is stationary if:

1. $E(\underset{\sim}{X}_t) = \underset{\sim}{\mu} < \infty$ for all t
2. $E[(\underset{\sim}{X}_t - \underset{\sim}{\mu})(\underset{\sim}{X}_t - \underset{\sim}{\mu})'] = \Sigma < \infty$ for all t
3. $E[(\underset{\sim}{X}_t - \underset{\sim}{\mu})(\underset{\sim}{X}_{t+h} - \underset{\sim}{\mu})'] = \Gamma(h)$ for all t and h.

We define strict stationarity in the same way as in the univariate case (see Chapter 1, Section 1.2).

Note Strictly stationary may not imply covariance stationary. Since the variance restriction may not be fulfilled.

A covariance stationary series is also known as *weakly* or *second order stationary series* (generally referred as a stationary time series). The following notation will be used in this chapter.

$$\gamma(h) = \text{autocovariance function and}$$

$$\rho(h) = \frac{\gamma(h)}{\gamma(0)}, \text{ autocorrelation function.}$$

9.1.4 Some important results for a Weakly stationary time series involving autocovariance (autocorrelation) function

A function $f(x)$, $x \in S$ is said to be *positive semidefinite* (p.s.d.) if for any set of real number $(a_1, \ldots, a_n)$

$$\sum_{i=1}^{n} \sum_{j=1}^{n} a_i a_j f(t_i - t_j) \geq 0 \text{ for } (t_i - t_j) \in S.$$

Result (1) Autocovariance function is p.s.d.

Proof Let $\{X_t, t \in T\}$ be a time series.
Consider a finite number of points $(t_1, \ldots, t_n) \in T$ and real numbers $(a_1, \ldots, a_n)$ i.e. observe $X_{t_1}, \ldots, X_{t_n}$. Then to show $\gamma(h)$ is p.s.d., it is enough to show that $\sum_{i=1}^{n} \sum_{j=1}^{n} a_i a_j \gamma(t_i - t_j) \geq 0$.

$$\text{Now } 0 \leq V\left[\sum_{i=1}^{n} a_i X_{t_i}\right] = \sum_i \sum_j a_i a_j \operatorname{cov}(X_{t_i}, X_{t_j}) = \sum_i \sum_j a_i a_j \gamma(t_i - t_j).$$

Hence the result.

Result (2) Autocovariance function is uniformly continuous if it is continuous at $h = 0$, i.e.

$$|\gamma(\delta) - \gamma(0)| < \eta \Rightarrow |\gamma(h + \delta) - \gamma(h)| < \varepsilon \text{ if } \delta < \delta_0.$$

Proof Let $\gamma(h) = \text{Cov}(X_{t+h}, X_t)$. Since this is a weakly stationary time series, without loss of generality assume $E(X_t) = 0$.

Then $\gamma(h) = E(X_{t+h} X_t)$ and $\gamma(h + \delta) = E(X_{t+h+\delta} X_t)$.

$$\{\gamma(h + \delta) - \gamma(h)\}^2 = [E\{X_t(X_{t+h+\delta} - X_{t+h})\}]^2$$

$$\leq E(X_t^2) E(X_{t+h+\delta} - X_{t+h})^2$$

(by Cauchy-Schwartz inequality).

Therefore $\quad E(X_{t+h+\delta} - X_{t+h})^2 = E(X_{t+h+\delta}^2 + X_{t+h}^2 - 2 X_{t+h+\delta} X_{t+h})$

$$= E(X_{t+h+\delta}^2) + E(X_{t+h}^2) - 2E(X_{t+h+\delta} X_{t+h})$$

$$= 2[\gamma(0) - \gamma(\delta)] \leq 2\eta.$$

Therefore $\quad \{\gamma(h + \delta) - \gamma(h)\}^2 \leq \{\gamma(0)\} \{2/\gamma(0) - \gamma(\delta))\}$

$$< \varepsilon^2 \text{ by taking } \eta \le \frac{\varepsilon^2}{2\gamma(0)}.$$

Hence $|\gamma(h + \delta) - \gamma(h)| < \varepsilon$ if $\delta < \delta_0$.

Example 9.1 Non-stationary time series

Consider a time series $X_t = \hat{\beta}_1 + \hat{\beta}_2 t$, $t \in [0, 1]$, where

$$(\hat{\beta}_1, \hat{\beta}_2) \sim N_2(\binom{\beta_1}{\beta_2}, \Sigma), \Sigma = (\sigma_{ij}).$$

Then $\left.\begin{array}{r} E(X_t) = \text{a function of } t = f_1(t) \\ \text{Cov}(X_t, X_{t+h}) = f_2(t) \end{array}\right\}$ i.e. they are not constant.

Any realization of this process is a continuous function in the interval [0, 1].

Such a process can be considered as the result of an experiment to estimate the linear response due to a variable t measured on the interval [0, 1]. Each experiment is conducted to obtain a set of the regression coefficients $\hat{\beta}_1, \hat{\beta}_2$. Since $(\hat{\beta}_1, \hat{\beta}_2) \sim N_2(\binom{\beta_1}{\beta_2}, \Sigma)$, $E(X_t) = \beta_1 + \beta_2 t$ is not a constant for all t, and

$$\begin{aligned} \text{Cov}(X_t, X_{t+h}) &= E((X_t - E(X_t))(X_{t+h} - E(X_{t+h}))) \\ &= E(((\hat{\beta}_1 - \beta_1) + (\hat{\beta}_2 - \beta_2)t)((\hat{\beta}_1 - \beta_1) + (\hat{\beta}_2 - \beta_2)(t + h))) \\ &= \sigma_{11} + \sigma_{12} t + \sigma_{12}(t + h) + \sigma_{22} t(t + h) \end{aligned}$$

is not a constant for all t.
Hence it is not covariance stationary.

Result (3) Autocovariance function is an even function of h, i.e.

$$\gamma(h) = \gamma(-h).$$

Proof $\gamma(h) = \text{Cov}(X_t, X_{t+h}) = \text{Cov}(X_{t'-h}, X_{t'}) = \gamma(-h)$ where $t = t' - h$.

9.1.5 Useful stationary Processes representing different time series models

1. *Purely Random Process* A time series $\{X_t, t \in T\}$ will be called a purely random process if the r.v.s X_t's are a collection of i.i.d. random variables with $E(X_t) = \mu$, $V(X_t) = \sigma^2$ and $\text{Cov}(X_t, X_{t+s}) = \gamma(s) = 0$.

A purely random process $\{X_t, t \in T)$ is also called *white noise* (W.N.) when the process is a discrete time process.

For example, in autoregressive process $Z_t = \alpha_1 Z_{t-1} + \ldots + \alpha_p Z_{t-p} + X_t$ where X_t is a White noise. For a continuous parameter process we have a 'continuous white noise'. Purely random processes are used as a constituent of more complicated stationary process viz. moving average process, Autoregressive process, etc.

2. *Moving Average Process MA (q)* A time series $\{X_t, t \in T\}$ is called a moving average process of order q if it can be expressed in the form

$$X_t = \beta_0 \varepsilon_t + \beta_1 \varepsilon_{t-1} + \ldots + \beta_q \varepsilon_{t-q},$$

β's are constant, and where $\{\varepsilon_t\}$ is a white noise with $E(\varepsilon_t) = 0$ and $V(\varepsilon_t) = \sigma^2$.

This is simply a moving average of a random series generating a stationary process with $E(X_t) = 0$ for all t and $V(X_t) = \sigma^2 \sum_{i=0}^{q} \beta_i^2$ for all t.

$$\begin{aligned}\gamma(k) &= \text{Cov}(X_t, X_{t+k})\\ &= \text{Cov}(\beta_0\varepsilon_t + \ldots + \beta_q\varepsilon_{t-q},\ \beta_0\,\varepsilon_{t+k} + \ldots + \beta_q\varepsilon_{t+k-q})\\ &= \begin{cases} 0 & \text{if } k > q \\ \sigma^2 \sum_{i=0}^{q-k} \beta_i\beta_{i+k} & \text{if } k \le q \text{ for all } t \\ \gamma(-k) & \text{if } k < 0 \end{cases}\end{aligned}$$

Hence this process $\{X_t\}$ is covariance stationary.

Also
$$\rho(k) = \begin{cases} 0 \text{ if } k > q \\ 1 \text{ if } k = 0 \\ \dfrac{\sum_{i=0}^{q-k} \beta_i\beta_{i+k}}{\sum_{i=0}^{q} \beta_i^2} \text{ if } k = 1, 2, \ldots, q. \end{cases}$$

and $\rho(k) = \rho(-k)$ if $k < 0$.

3. *Autoregressive Process AR (p)* A time series $\{X_t\}$ is called an autoregressive process of order p if

$$X_t = \alpha_1 X_{t-1} + \ldots + \alpha_p X_{t-p} + \varepsilon_t,$$

where ε_t is a white noise with

$$E(\varepsilon_t) = 0,\ \ \text{Var}(\varepsilon_t) = \sigma^2, \text{Cov}(\varepsilon_t, \varepsilon_{t'}) = 0, t \ne t'.$$

Here the process arises from an oscillatory series giving rise to an autoregressive scheme.

(i) When $p = 1$, this is a Markov process

(ii) When $p = 2$, this is a Yule process.

For
$$\begin{aligned} p = 1, X_t &= \alpha X_{t-1} + \varepsilon_t \\ &= \alpha(\alpha X_{t-2} + \varepsilon_{t-1}) + \varepsilon_t \\ &= \alpha^2 X_{t-2} + \alpha\varepsilon_{t-1} + \varepsilon_t \\ &= \varepsilon_t + \alpha\varepsilon_{t-1} + \alpha^2\varepsilon_{t-2} + \ldots \text{ if } |\alpha| < 1. \\ &= \beta_0\varepsilon_t + \beta_1\varepsilon_{t-1} + \beta_2\varepsilon_{t-2} + \ldots \text{ (where } \beta_i = \alpha^i, i \ge 0) \end{aligned}$$

is a MA (∞) process.

In general an AR(p) is an MA (∞) (a duality property between AR(p) and MA (∞) exists) if all the roots of the equation $z^p - \alpha_1 z^{p-1} - \alpha_2 z^{p-2} - \ldots - \alpha_p = 0$ are less than one in absolute value. Back to AR(1) we have now

and
$$\left.\begin{aligned} E(X_t) &= \mu \sum_{i=0}^{\infty} \beta_i = 0 \\ \text{Var}(X_t) &= \sigma^2 \sum_{i=0}^{\infty} \beta_i^2 \end{aligned}\right\} \text{ if } \sum_{i=0}^{\infty} \beta_i^2 < \infty \text{ or } \sum_{i=0}^{\infty} |\beta_i| < \infty$$
(these conditions are satisfied if $-1 < \alpha < 1$).

$$\gamma(k) = \text{Cov}(X_t, X_{t+k}) = \sigma^2 \sum_{i=0}^{\infty} \beta_i \beta_{i+k} < \infty \text{ if } |\alpha| < 1$$

(by Cauchy-Schwartz inequality).

Therefore $\{X_t\}$ is a covariance stationary process.

To show AR(p) is an MA (∞) process, let us write

$$X_t = \alpha_1 X_{t-1} + \ldots + \alpha_p X_{t-p} + \varepsilon_t$$
$$= \alpha_1 BX_t + \ldots + \alpha_p B^p X_t + \varepsilon_t$$

where B is the Back-shift operator such that $X_{t-p} = B^p X_t$.

Therefore white noise

$$\varepsilon_t = (1 - \alpha_1 B - \ldots - \alpha_p B^p) X_t$$
$$= f(B) X_t, \text{ say.}$$

Therefore if $f(B)$ is invertible we have $X_t = f^{-1}(B)\, \varepsilon_t$

$$= (1 - \alpha_1 B - \ldots - \alpha_p B^p)^{-1}\, \varepsilon_t = \varepsilon_t + \beta_1 \varepsilon_{t-1} + \beta_2 \varepsilon_{t-2} + \ldots \beta_0 = 1$$

which gives $E(X_t) = 0$ for all $t = 0, \pm 1, \pm 2, \ldots$,

$$V(X_t) = \sigma^2 \sum_{i=0}^{\infty} \beta_i^2 \text{ for all } t = 0, \pm 1, \pm 2, \ldots, \text{ and } \gamma(k) = \sigma^2 \sum_{i=0}^{\infty} \beta_i \beta_{i+k}.$$

This again shows that $\{X_t\}$ generates a second order stationary process.

4. *Autoregressive Moving Average ARMA (p, q)* Consider the stochastic process

$$X_t = \alpha_1 X_{t-1} + \ldots + \alpha_p X_{t-p} + \beta_0 \varepsilon_t + \beta_1 \varepsilon_{t-1} + \ldots + \beta_q \varepsilon_{t-q}.$$

This is also second order stationary by means of results (2) and (3) To describe a autoregressive series or a moving average process with a large order (which is difficult to handle) we take recourse to the ARMA process.

(5) *Random Walk* $X_t = X_{t-1} + \varepsilon_t$ with

$$\begin{cases} E(\varepsilon_t) = \mu, \\ V(\varepsilon_t) = \sigma^2 \text{ and} \\ \text{Cov}(\varepsilon_t, \varepsilon_s) = 0 \text{ if } t \neq s \end{cases}$$

Take $X_0 = 0$. Now

$$X_1 = \varepsilon_1$$
$$\vdots$$
$$X_t = \sum_{i=1}^{t} \varepsilon_i.$$

Then $E(X_t) = t\mu$ and $V(X_t) = t\sigma^2$.

Hence the process is not second order stationary.

However, the first order difference of $\{X_t\}$ is a stationary process, since $\varepsilon_t = X_t - X_{t-1}$, with expectation μ and variance $= \sigma^2$ (independent of t). For example, share price on tth day = Share price on $(t-1)$th day + error.

9.1.6 Stationarity and ergodicity

Consider a stationary process $\{X_t\}$ with $E(X_t) = \mu$ for all t and $\text{Var}(X_t) = \sigma^2$ for all t.

The expectation and variance of the averages of realization averages are $E\left[\frac{1}{m}\sum_{j=1}^{m}\left(\frac{1}{n_j}\sum_{t=1}^{n_j} X_t\right)\right] = E\left(\frac{1}{m}\sum_{j=1}^{m}\overline{X}_{(j)}\right) = \mu$, the ensemble average and $\text{Var}\left(\frac{1}{m}\sum_{j=1}^{m}\overline{X}_{(j)}\right) = \frac{\sigma^2}{m}$.

If the number of realizations is large enough then $\frac{1}{m}\sum_{j=1}^{m}\overline{X}_{(j)} \xrightarrow{P} \mu$ as $m \to \infty$,

since $E\left(\frac{1}{m}\sum_{j=1}^{m}\overline{X}_{(j)}\right) = \mu$ and $\text{Var}\left(\frac{1}{m}\sum_{j=1}^{m}\overline{X}_{(j)}\right) \to 0$ as $m \to \infty$.

If we have a number of realizations then the average of all the realization averages would converage in 'quadratic mean' (q.m.) to μ as the number of realizations get large (cf. Theorem 9.1).

Question If instead of a large number of realizations we have only a single realization, then will it be possible to estimate μ on the basis of it? The answer to this question gives rise to Ergodicity.

Ergodicity Property of a Time Series

A stationary time series $\{X_t\}$ is said to be ergodic if a single realization of the series can make all possible conclusions about the probability law that generates the process. Specifically, the sample moments obtained from a single realization converge to the population moments with probability 1 or in mean square.

Ergodicity of μ

Consider a discrete stationary time series $\{X_t\}$ with $E(X_t) = \mu$ for all t. Then the series is mean ergodic if the sequence of sample means $\overline{X}_n = \frac{1}{m}\sum_{t=1}^{n} X_t$, $n = 1, 2, \ldots$ converges in q.m. to μ.

Theorem 9.1 A stationary time series $\{X_t\}$ whose covariance function as $\gamma(h) \to 0$ as $h \to \infty$ is ergodic for μ. Hence we can suggest that $\overline{X}_n = \frac{1}{n}\sum_{t=1}^{n} X_t$ is an optimal estimate of μ.

Proof To show that $\overline{X}_n \xrightarrow{\text{q.m.}} \mu$ we are to show that $\lim_{n\to\infty} E(\overline{X}_n - \mu)^2 = 0$.

Now
$$\text{Var}(\overline{X}_n) = \text{Var}\left(\frac{1}{n}\sum_{t=1}^{n} X_t\right) = \frac{1}{n^2}\sum_{t=1}^{n}\sum_{t'=1}^{n}\text{Cov}(X_t, X_{t'})$$
$$= \frac{1}{n^2}\sum_{t=1}^{n}\sum_{t'=1}^{t}\text{Cov}(X_t, X_{t'}) - \frac{1}{n^2}\sum_{t=1}^{n}\text{Var}(X_t).$$

Since Var $(X_t) = \sigma^2$, second term of right hand side $\to 0$ as $n \to \infty$.

Now $$\left| \frac{2}{n^2} \sum_{t=1}^{n} \sum_{t'=1}^{t} \text{Cov}(X_t, X_{t'}) \right| \le \frac{2}{n^2} \sum_{t=1}^{n} t \left| \frac{1}{t} \sum_{t'=1}^{t} \gamma(t-t') \right|$$

$$\le \frac{2}{n} \sum_{t=1}^{n} \left| \frac{1}{t} \sum_{h=0}^{t-1} \gamma(h) \right| \left(\text{since } \gamma(h) = \gamma(-h) \text{ and } \frac{t}{n} \le 1 \right)$$

$$\le \frac{2}{n} \sum_{t=1}^{n} \left| \frac{1}{t} \sum_{h=0}^{t-1} \gamma(h) \right| + \frac{2}{n} \sum_{t=N+1}^{n} \left| \frac{1}{t} \sum_{h=0}^{t-1} \gamma(h) \right|, (n > N).$$

The second term on right hand side can be made

$$< \frac{\varepsilon}{2} \frac{n-N}{n} \text{ for } t > N \left(\text{since } \left| \frac{1}{t} \sum_{h=0}^{t-1} \gamma(h) \right| < \frac{\varepsilon}{4} \right).$$

Now take $n > N$, so that the second term $< \frac{\varepsilon}{2}$. Also for $n > N$, first term $< \frac{\varepsilon}{2}$. Hence $\text{Var}(\overline{X}_n) = E(\overline{X}_n - \mu)^2 \to 0 \Rightarrow X_n \xrightarrow{\text{q.m.}} \mu$.

Corollary 9.1 A stationary time series with absolutely summable covariance function is ergodic for mean. Also

$$\lim_{n \to \infty} n \, \text{Var}(\overline{X}_n) = \sum_{h=-\infty}^{\infty} \gamma(h).$$

Proof $\sum_h | \gamma(h) | < \infty \Rightarrow \gamma(h) \to 0$ as $h \to \infty$. Ergodicity follows from the theorem.

Also we have

$$n \, \text{Var}(\overline{X}_n) = \frac{1}{n} \sum_{j=1}^{n} \sum_{t=1}^{n} \gamma(t-j) = \frac{1}{n} \sum_{h=-(n-1)}^{n-1} (n - |h|) \gamma(h)$$

$$= \sum_{h=-(n-1)}^{n-1} (1 - |h|/n) \gamma(h) \to \sum_{h=-\infty}^{\infty} \gamma(h)$$

(by Kronecker's lemma $\lim_{n\to\infty} \sum_h \frac{|h|}{n} |\gamma(h)| = 0$.)

An alternative expression of $\lim_{n\to\infty} n\text{Var}(\overline{X}_n)$ given by the following corollary should be studied after section 9.2.4.

****Corollary 9.2** If the spectral density of a stationary time series X_t is continuous, then $\lim_{n\to\infty} n\text{Var}(\overline{X}_n) = 2\pi f(0)$, where $f(\lambda)$ is the spectral density of X_t (to be defined in section 9.2.4).

Proof Since $f(\lambda)$ is a continuous periodic function on $[-\pi, \pi]$ the Cesaro sum $C_n(\lambda) = \frac{1}{n} \sum_{r=2}^{n} \sum_{k=1}^{r-1} (a_k \cos k\lambda + b_k \sin k\lambda) + a_0/2$ of the Fourier series of $f(\lambda) \to f(\lambda)$ uniformly in $\lambda \in [-\pi, \pi]$.

Therefore

$$2\pi f(0) = \lim_{n\to\infty} 2\pi\left[a_0/2 + \frac{1}{n} \sum_{r=2}^{n} \sum_{k=1}^{r-1} a_k\right]$$

$$= \lim_{n\to\infty}\left[\gamma(0) + \frac{2}{n} \sum_{r=2}^{n} \sum_{k=1}^{r-1} \gamma(k)\right]\left(\text{note } a_k = \frac{1}{\pi}\int_{-\pi}^{\pi} \cos k\lambda\, f(\lambda)\, d\lambda = \gamma(k)\right)$$

$$= \lim_{n\to\infty}\left[\gamma(0) + \frac{2}{n} \sum_{r=1}^{n} (n-r)\gamma(r)\right]$$

$$= \lim_{n\to\infty} n\text{Var}(\overline{X}_n)$$

(from the proof of corollary 9.1).

9.2 Different Approaches of Time Series

Analysis of time series is done by two distinct approaches.

Analysis of Time Series

A: {Analysis in time domain | Analysis in frequency domain} : B

A : For fixed $\omega \in \Omega$, the observations used are X_t in

(i) determining the internal structure of a given time series $\{X_t\}$.

(ii) to estimate the parameters of the given stochastic model of $\{X_t\}$ and is done using the ergodicity property.

In (i) the basic tool used in drawing any inference about the given time series $\{X_t\}$ is *autocovariance (autocorrelation) function.*

The analysis here is known as *correlogram analysis.*

B : For fixed t, observations are functions of ω, X_t. $\omega \in \Omega$.

The basic tool used in drawing any inference from frequency domain approach is the *spectral density function (spectrum).* The analysis here is known as *periodogram analysis.*

In order to understand analysis of time series in frequency domain, let us introduce some of the tools and terminologies pertinent to time series analysis through system engineering vocabularies. Let X_t denote an electric current at an instant t andto know the energy dissipated by this current per unit of resistance, we have the following natural definitions. The quantity $\int_{t_1}^{t_2} X_t^2\, dt$ is called the *energy* of the stochastic process $\{X_t\}$ in the course of the interval

$[t_1, t_2]$. The integral $\int_{-\infty}^{\infty} X_t^2\, dt$, if it converges, is called the *total energy of the series* $\{X_t, -\infty < t < \infty\}$. The limit $\lim_{T\to\infty} \frac{1}{2T}\int_{-T}^{T} X_t^2 dt$ is known as the *average power of the random process*. If the stochastic process is complex valued, we write $|X_t|^2$ in place of X_t^2.

In many situations time series are simulated by the sum of harmonics with given frequencies and random amplitudes and phases. In other words, one considers processes of the form

$$X_t = \sum_{k=1}^{n} a_k \cos(u_k t + \theta_k) \tag{9.2.1}$$

where u_k are given numbers, amplitudes a_k's and phases θ_k's are random quantities. The probabilistic structure of this process is completely determined by the joint distribution of the r.v.s a_k and θ_k for $k = 1, 2, \ldots, n$. If a second order process X_t is the sum of n harmonic oscillations with amplitudes $|a_k|$ and frequencies u_k, then all the frequencies $\{u_k\}$, for $k = 1, 2, \ldots, n$, considered as a set of points on the real line, is called the *spectrum* of the process $\{X_t\}$. In many cases it is convenient to consider complex-valued processes of oscillatory nature

$$Z_t = \sum_{k=1}^{n} \gamma_k \theta^{iu_k t} \tag{9.2.2}$$

with complex amplitudes $\gamma_k = a_k + ib_k$ for $k = 1, 2, \ldots, n$. The process Z_t can be split into real and imaginary parts

$$Z_t = X_t + iY_t \text{ where}$$

$$\left.\begin{aligned} X_t &= \sum_{k=1}^{n} (a_k \cos u_k t - b_k \sin u_k t) \\ Y_t &= \sum_{k=1}^{n} (a_k \sin u_k t + b_k \cos u_k t) \end{aligned}\right\} \tag{9.2.3}$$

If we set $a_k = |\gamma_k| \cos\theta_k$ and $b_k = |\gamma_k| \sin\theta_k$, we obtain an expression for the function X_t and Y_t in the form

$$X_t = \sum_{k=1}^{n} |\gamma_k| \cos(u_k t + \theta_k) \text{ and } Y_t = \sum_{k=1}^{n} |\gamma_k| \sin(u_k t + \theta_k).$$

The mean value of the process Z_t

$$\frac{1}{2T}\int_{-T}^{T} Z_t\, dt = \sum_{k=1}^{n} \gamma_k \frac{\sin Tu_k}{Tu_k} \tag{9.2.4}$$

where $\frac{\sin t}{t} = 1$ at $t = 0$ and $\gamma_k = |\gamma_k| e^{i\theta_k}$ and its average power is given by

$$\frac{1}{2T}\int_{-T}^{T} |Z_t|^2\, dt = \frac{1}{2T}\int_{-T}^{T} \sum_{k=1}^{n}\sum_{r=1}^{n} \gamma_k \bar{\gamma}_r e^{it(u_k - u_r)}\, dt$$

$$= \sum_{k=1}^{n} |\gamma_k|^2 + \sum_{k,r=1}^{n} \gamma_k \bar{\gamma}_r \frac{\sin T(u_k - u_r)}{T(u_k - u_r)}, k \neq r$$

Therefore $$\lim_{T\to\infty} \frac{1}{T} \int_{-T}^{T} |Z_t|^2 \, dt = \sum_{k=1}^{n} \gamma_k^2 \text{ a.s.} \tag{9.2.5}$$

Thus the average power of the oscillatory stochastic process Z_t is equal to the sum of the average power of the harmonic components of the process.

From (9.2.4) we see that if 0 is a point of the spectrum, then

$$\lim_{T\to\infty} \frac{1}{T} \int_{-T}^{T} Z_t dt = \gamma_0,$$

where γ_0 is the amplitude corresponding to the frequency at $t = 0$.

Suppose that complex amplitudes γ_k have $E(\gamma_k) = 0$, $E(\gamma_k \bar{\gamma}_r) = 0$, $k \neq r$. Then $E(Z_t) = 0$. Note that if 0 is not a point of the spectrum of a stochastic process then $E(Z_t)$ coincides with the mean value over the infinite interval $(-\infty, \infty)$. On the other hand, if 0 is a point of the spectrum of the process, then the mean value of Z_t over time $\lim_{T\to\infty} \frac{1}{T} \int_{-T}^{T} Z_t \, dt$ is a random variable.

Now Z_t has covariance function

$$R(t_1, t_2) = E(Z_{t_1} \bar{Z}_{t_2}) = E\left(\sum_{k=1}^{n} \sum_{r=1}^{n} \gamma_k \bar{\gamma}_r e^{i(u_k t_1 - u_r t_2)} \right) = \sum_{k=1}^{n} C_k^2 e^{iu_k (t_1 - t_2)},$$

where $C_k^2 = E|\gamma_k|^2$. Thus covariance function $R(t_1, t_2)$ depends only on the difference, i.e.

$$R(t_1, t_2) = R(t_1 - t_2) \tag{9.2.6}$$

where $$R(t) = \sum_{k=1}^{n} C_k^2 e^{iu_k t} \tag{9.2.7}$$

Thus, if the amplitudes are uncorrelated with zero mean, then the process Z_t is a stationary process in the wide sense and the covariance function is given by the formula (9.2.7) called the *spectral distribution* of the covariance function. It determines the spectrum of the random process, i.e. the set of frequencies $\{u_k\}_1^n$, of the harmonic oscillation constituting the process $\{Z_t\}$ and $C_k^2 = E|\gamma_k|^2$ of the process in a unit time. The quantity C_k^2 can be called simply the *mean value of the power of the harmonic components* of the second order process with frequency u_k. It is obtained by first averaging the power over time and then averaging it in the probabilistic sense, that is, by taking expectation. In connection with these energy representation we introduce the important characteristic of a second order stochastic process, the *spectral function* $F(u)$ of the process (9.2.2) defined by

$$F(u) = \sum_{u_k \leq u} C_k^2.$$

This shows that $F(u)$ is equal to the average power of the harmonic components of the process Z_t (whose frequencies are less than or equal to u) in a unit time. It completely characterizes both the average power of each harmonic component of the process Z_t and the overall average power of the harmonic components of the process, whose frequencies lie in a given interval. Thus

$$C_k^2 = F(u_k) - F(u_k - o), \quad \sum_{u_1 < u_k \le u_2} C_k^2 = F(u_2) - F(u_1).$$

By means of spectal function, the covariance function of the process Z_t can be written in the form

$$R(t) = \int_{-\infty}^{\infty} e^{iut}\, dF(u) \tag{9.2.8}$$

Therefore, spectral function is a non-negative nondecreasing right continuous function that is piecewise constant with finitely many jumps of magnitude C_k^2. This concept of spectral function will be introduced for arbitrary weakly stationary stochastic processes by passing to the limit of the processes $Z_t = \sum_{k=1}^{n} \gamma_k e^{iu_k t}$ and decreasing the complex amplitudes γ_k whereas the spectrum of the process (i.e. the set of all frequencies u_k) fills the whole real line $(-\infty, \infty)$ over more densely. For arbitrary second order process in passing to the limit we obtain a process that has a continuous spectrum for which analogue of (9.2.8) is valid. This is given by the celebrated *(Wiener) Khinchine's theorem* for *spectral representation of covariance function.*

9.3 Correlogram Analysis

Consider a covariance stationary time series $\{X_t\}$. Let ρ_k be the autocorrelation of lag k. Different covariance stationary time series possesses different forms of correlograms ρ_k when plotted against k is called the *correlogram*. By analysing the shape of the correlogram we can infer about the internal structure of $\{X_t\}$. For different standard stochastic models we present the correlograms :

1. MA (q)

$$X_t = \beta_0 \varepsilon_t + \beta_1 \varepsilon_{t-1} + \ldots + \beta_q \varepsilon_{t-q}$$

where ε_t is a white noise with

$$E(\varepsilon_t) = 0, \ \mathrm{Var}(\varepsilon_t) = \sigma^2 \text{ and}$$

$$\rho_k = \begin{cases} 1 & \text{for } k = 0 \\ \dfrac{\sum_{i=0}^{q-k} \beta_i \beta_{i+k}}{\sum_{i=0}^{q} \beta_i^2} & \text{for } k = 1, 2, \ldots, q \\ 0 & k > q \end{cases}$$

In particular, $\beta_i = \dfrac{1}{q+1}$ for all i, indicates simple MA process.

Then $$\rho_k = \begin{cases} 1 - \dfrac{k}{q+1} & \text{if } k = 0, 1, \ldots, q \\ 0 & \text{if } k > q \end{cases}$$

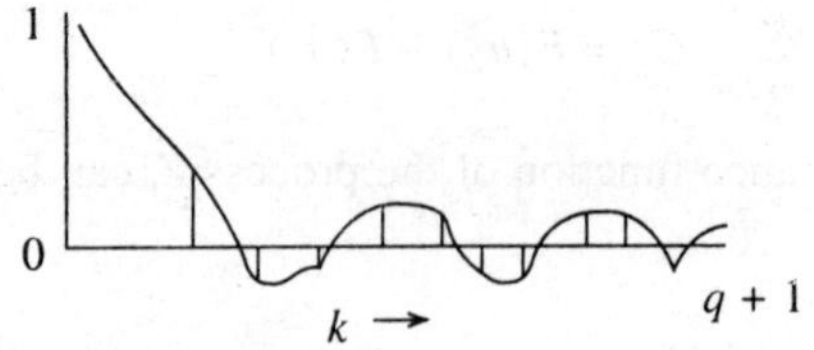

The autocorrelation oscillates between the points (0, 1) and $(q + 1, 0)$ and then coincides with k-axis.

If it is a simple MA process then autocorrelation will decline smoothly. Otherwise it may have some oscillatory movement depending on the coefficients β_i, and it will ultimately coincide with k-axis for $k = q + 1$.

2. First order autoregressive (or Markov) process

$X_t = \alpha X_{t-1} + \varepsilon_t$, $|\alpha| < 1$ (this assumption is needed to make the time series stationary) where ε_t's are white noise with $E(X_t) = 0$ and $\mathrm{Var}(X_t) = \sigma^2$.

$$\begin{aligned} X_t &= \alpha(\alpha X_{t-2} + \varepsilon_{t-1}) + \varepsilon_t \\ &= \alpha^n X_{t-n} + \alpha^{n-1}\varepsilon_{t-n+1} + \ldots + \alpha\varepsilon_{t-1} + \varepsilon_t \\ &= \alpha^n X_{t-n} + \sum_{i=0}^{n-1} \alpha^i \varepsilon_{t-i} \end{aligned}$$

will be stationary and convergent, if $|\alpha| < 1$.

$$X_t \sim \sum_{i=0}^{\infty} \alpha^i \varepsilon_{t-i} \text{ is a MA } (\infty) \text{ process if } |\alpha| < 1.$$

Now $$E(X_t) = 0, \mathrm{Var}(X_t) = \sigma^2 \sum_{i=0}^{\infty} \alpha^{2i} = \frac{\sigma^2}{1-\alpha^2} \quad (|\alpha| < 1)$$

and $\gamma_k = E(X_t X_{t+k})$

$$= E\left(\sum_{i=0}^{\infty} \alpha^i \varepsilon_{t-i}\right)\left(\sum_{j=0}^{\infty} \alpha^j \varepsilon_{t-k-j}\right)$$

$$= E\left[\sum_{l=-\infty}^{t} \alpha^{t-l}\varepsilon_l\right]\left[\sum_{m=-\infty}^{t+k} \alpha^{t+k-m}\varepsilon_m\right] \text{ (where } l = t - i,\ m = t + k - j)$$

$$= \sigma^2 \alpha^k \sum_{l=-\infty}^{t} \alpha^{2(t-l)} \text{ (only those terms which satisfy } l = m \text{ contribute)}$$

$$= \sigma^2 \alpha^k \sum_{i=0}^{\infty} \alpha^{2i}$$

$$= \sigma^2 \alpha^k / (1 - \alpha^2).$$

Therefore, $\rho_k = \alpha^k$.

To get an even function defined for all integer k, $\rho_k = \alpha^{|k|}$, $k = 0, \pm 1, \pm 2, \ldots$. As $k \to \infty$, $\rho_k \to 0$ if $|\alpha| < 1$.

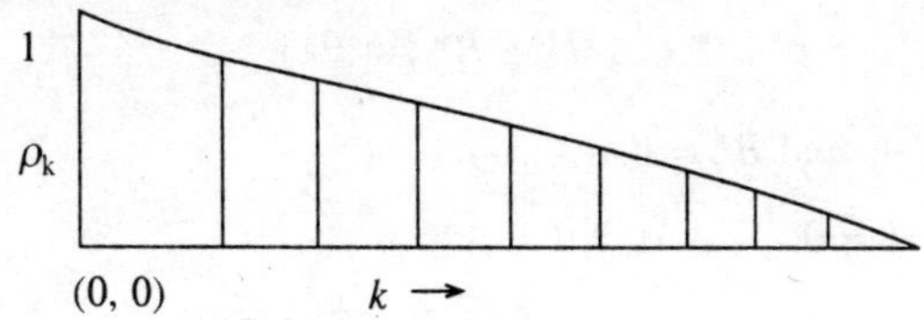

3. Correlogram of AR(2) (Yule process)
Consider

$$X_t = aX_{t-1} + bX_{t-2} + \varepsilon_t,$$

where ε_t is a white noise with $E\varepsilon_t = 0$, $\text{Var}(\varepsilon_t) = \sigma^2$.

Multiplying both sides by X_{t-k}, taking expectation and dividing by $\text{Var}(X_t)$ we get

$$\rho_k = a\rho_{k-1} + b\rho_{k-2} \text{ (Yule-Walker equation).}$$

The general solution is given by

$$\rho_k = A_1 r_1^k + A_2 r_2^k, \; |r_1| < 1 \text{ and } |r_2| < 1. \tag{9.3.1}$$

where A_1 and A_2 has to be found from initial conditions and r_1, r_2 are the roots of $r^2 - ar - b = 0$ (characteristic equation of the process).

Case (i) r_1 and r_2 are real, i.e. $\dfrac{a \pm \sqrt{(a^2 + 4b)}}{2}$ are real, i.e. if $a^2 + 4b > 0$.

A_1 and A_2 are found as follows :
Since $\rho_0 = 1 = A_1 + A_2$. From first Yule-Walker equation, we have

$$\rho_1 = a\rho_0 + b\rho_{-1} = a + b\rho_{-1} = a + b\rho_1$$

Therefore $\rho_1 = \dfrac{a}{1-b}$.

Also $\rho_1 = A_1 r_1 + A_2 r_2$ (from (9.3.1))

$$= A_1 r_1 + (1 - A_1) r_2 = A_1 (r_1 - r_2) + r_2$$

Hence $$A_1 = \left(\frac{a}{1-b} - r_2\right) \Big/ (r_1 - r_2) \text{ and } A_2 = 1 - A_1$$

For real equal roots of the characteristic equation $r_1 = r_2 = r$ and

$$\rho_k = \left\{1 + k\frac{(1 - r^2)}{1 + r^2}\right\} r^k \to 0 \text{ as } k \to \infty.$$

The correlogram is oscillatory and decaying in character.

Case (ii) r_1, r_2 are complex conjugates if $a^2 + 4b < 0$.
Write $r_1 = p(\cos\theta + i\sin\theta)$ and $r_2 = p(\cos\theta - i\sin\theta)$.

Therefore $\rho_k = p^k[A_1 (\cos\theta + i\sin\theta)^k + A_2 (\cos\theta - i\sin\theta)^k]$

$$= p^k[A_1 (\cos\theta k + i\sin\theta k) + A_2 (\cos\theta k - i\sin\theta k)]$$

$$= p^k [A^* \cos\theta k + B^* \sin\theta k]$$

where $A^* = A_1 + A_2$ and $B^* = i(A_1 - A_2)$.

If $k = 0, \quad \rho_0 = 1 = A^*$

$k = 1, \quad \rho_1 = p [\cos\theta + B^* \sin\theta]$

$k = -1, \quad \rho_{-1} = \frac{1}{p}[\cos\theta - B^* \sin\theta]$

since $\rho_1 = \rho_{-1}$, equating the above two equations

$$R^* = \frac{1-p^2}{1+p^2}\frac{\cos\theta}{\sin\theta} = \frac{1-p^2}{1+p^2}\cot\theta = \cot\psi \text{ (say)}$$

Now $\rho_k = p^k [\cos\theta k + \cot\psi \sin\theta k]$

$$= p^k \frac{\sin(\theta k + \psi)}{\sin\psi} \to 0 \left(\text{since} \left|\frac{\sin(\theta k + \psi)}{\sin\psi}\right| \leq c\right)$$

as $k \to \infty$ and ρ_k is a damped sine curve.

Now $p^2 = r_1 r_2$ and also from the theory of quadratic equations

$$r_1 r_2 = -b \text{ since } a^2 + 4b < 0 \Rightarrow b < 0$$

Therefore, $p = \sqrt{b} < 1$ if $|r_1| < 1, |r_2| < 1$.

4. Autoregressive model of order p, AR(p)

$X_t = \alpha_t X_{t-1} + \ldots + \alpha_p X_{t-p} + \varepsilon_t$ with white noise ε_t,

$$E(\varepsilon_t) = 0,\ \mathrm{Var}(\varepsilon_t) = \sigma^2.$$

Multiplying by X_{t-k}, taking expectation, dividing by σ_x^2 and using the fact $\rho_k = \rho_{-k}$, we get the Yule-Walker equation

$$\rho_k = \alpha_1\rho_{k-1} + \alpha_2\rho_{k-2} + \ldots + \alpha_p\rho_{k-p} (k > 0).$$

The general solution is $\rho_k = \sum_{i=1}^{P} A_i r_i^k$, $|r_i| < 1$ for all $i = 1, 2, \ldots, p$ where r_i's are the roots of the equation $r^P = \alpha_1 r^{p-1} + \alpha_2 r^{p-2} + \ldots + \alpha_p$.

Here also $|r_i| < 1, i = 1, \ldots, p$ are necessary and sufficient conditions for $\{X_t\}$ to be a stationary process.

In case of real roots $\rho_k \to 0$ as $k \to \infty$.

9.3.1 Interpreting the sample correlogram

Sample autocovariance is defined as

$$c_k = \frac{1}{n}\sum_{t=1}^{n-k}(X_t - \overline{X})(X_{t+k} - \overline{X}) \text{ or } \frac{1}{n-k}\sum_{t=1}^{n-k}(X_t - \overline{X})(X_{t+k} - \overline{X})$$

and sample autocorrelation is similarly defined as

$$r_k = \sum_{t=1}^{n-k} (X_t - \overline{X})(X_{t+k} - \overline{X}) \Big/ \sum_{t=1}^{n-k} (X_t - \overline{X})^2.$$

A useful aid in interpreting a set of auto-correlation coefficients is a graph called a *sample correlogram* in which r_k is plotted against lag k.

(a) A completely random series

In this case $r_k \sim 0$ for large n and for all non-zero values of k. This follows from sampling distribution of $r_k \cdot r_k \sim N(0, 1/n)$ for large n (Quennoule, 1957).

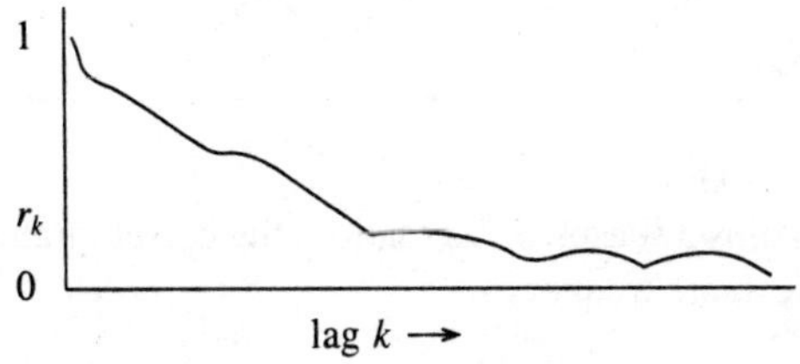

(b) Short term series

Stationary series often exhibit short term correlation characterised by a fairly large value (also from theory of equation) of r_1 followed by 2 or 3 more coefficients which, while significantly greater than zero, tend to get successively smaller values of r_k for larger lags tends to be approximately 0. A time series which gives rise to such a correlogram, is one for which an observation above the mean tends to be followed by one or more further observations above the mean and similarly for observations below the mean. This is an indication of autoregressive model.

(c) Non-stationary series

If a time series contains a trend, then the values of r_k will not come down to zero except for very large values of lag k. This is because an observation on one side of the overall mean tends to be followed by a large number of further observations on the same side of the mean because of trend. Little can be inferred from a correlogram of this type, as the trend dominates all other features. Simple autocorrelation should be calculated only for the stationary series and so any trend should be removed before calculating $\{r_k\}$.

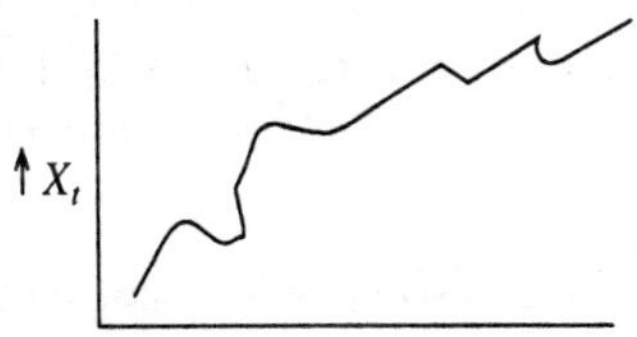

Upward trend

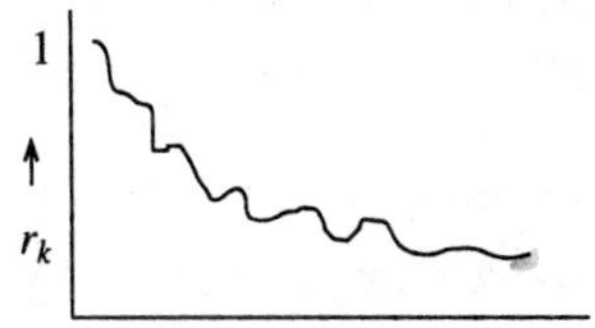

Downward trend

(d) Alternating Series

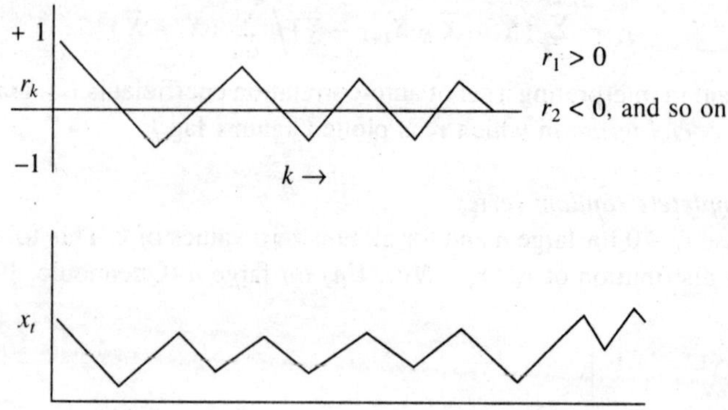

(e) Seasonal fluctuations

If a time series contains a seasonal fluctuation, the correlogram will also exhibit an oscillation at the same frequency.

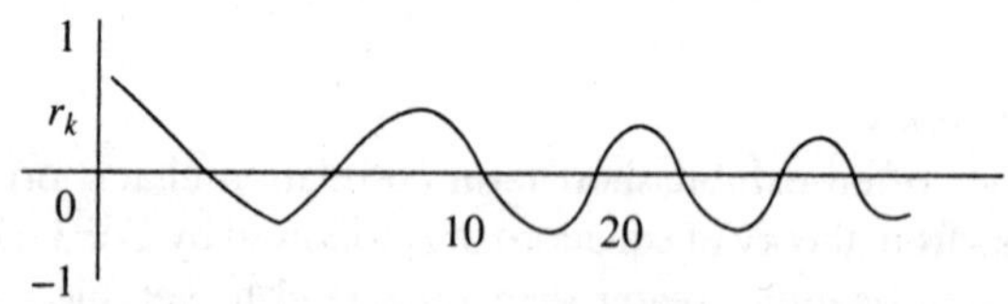

If X_t follows a sinusoidal pattern then so does r_k, e.g.

$$X_t = a \cos t\omega, \text{ frequency } \omega,\ 0 < \omega < n$$

then $\overline{X}_n \rightarrow 0$ and $r_k \sim \cos k\omega$ for large n.

(f) ARMA

The correlogram of a MA(q) process is easy to recognise as it "cuts off" at large q whereas AR(p) process is a mixture of damped exponential and sinusoids and dies out slowly (or attenuates). The correlogram of a mixed ARMA model also generally attenuates rather than cuts off.

Example Show that the infinite MA (∞) process $\{X_t\}$ defined by

$$X_t = \varepsilon_t + c(\varepsilon_{t-1} + \varepsilon_{t-2} + \ldots)$$

where ε_t's are white noise with $E\varepsilon_t = 0$, Var $\varepsilon_t = \sigma^2$ and c is a constant, is a non-stationary process. Also show that the series of 1st differences $\{Y_t\}$ defined by $Y_t = X_t - X_{t-1}$ is a first order MA(1) process and also 2nd order stationary. Find the autocovariance function of $\{Y_t\}$.

Solution Var $(X_t) = \sigma^2[1 + c^2 \Sigma\, 1]$ is infinite, obviously X_t is not stationary.

$$\begin{aligned} Y_t &= \varepsilon_t + c(\varepsilon_{t-1} + \ldots) - \varepsilon_{t-1} - c(\varepsilon_{t-2} + \ldots) \\ &= \varepsilon_t + (c - 1)\, \varepsilon_{t-1} \end{aligned}$$

is MA (1) process with

$$E(Y_t) = 0 + (c-1)\,0 = 0,\ \text{Var}(Y_t) = \sigma^2 + (c-1)^2\,\sigma^2 = \sigma^2(1 + (c-1)^2)$$

do not depend on t and hence $\{Y_t\}$ is second order stationary.

$$\begin{aligned}\gamma(k) = E(Y_tY_{t+k}) &= (c-1)\,\sigma^2 \text{ if } k = \pm 1\\ &= \sigma^2(1+(c-1)^2) \text{ if } k = 0\\ &= 0 \quad \text{otherwise}\end{aligned}$$

$$\rho_Y(k) = \begin{cases} 1 \text{ if } k = 0 \\ (c-1)/[1+(c-1)^2] \text{ if } k = \pm 1 \\ 0 \quad \text{otherwise} \end{cases}$$

9.3.2 Integrated models (ARIMA)

If the observed time series is non-stationary in the mean then we can difference the series and this approach is widely used in econometrics. If X_t is replaced by $\nabla^d X_t$ in the ARMA (p, q) model we have a model capable of describing certain types of non-stationary series. Such a model is called an "integrated" model because the stationary model which is fitted to the differenced data has to be summed or "integrated" to provide a model for the non-stationary model. Writing $W_t = \nabla^d X_t$, the general autoregressive integrated moving average process (ARIMA) is of the form

$$W_t = \alpha_1 W_{t-1} + \ldots + \alpha_p W_{t-p} + \varepsilon_t + \beta_1 \varepsilon_{t-1} + \ldots + \beta_q \varepsilon_{t-q}$$

and is denoted by ARIMA (p, d, q). Usually $d = 1$. Also random walk is a ARIMA (0, 1, 0) process.

ARMA (p, q) model can be generalised as

$$\sum_{j=0}^{p} \alpha_j X_{t-j} = \sum_{i=0}^{q} \beta_i \varepsilon_{t-1} \tag{9.3.2}$$

with $\alpha_0 = \beta_0 = 1$ and $\{\varepsilon_t\}$'s are uncorrelated $(0, \sigma^2)$ r.v.s. We assume that the characteristic equation

$$z^p + \alpha_1 z^{p-1} + \ldots + \alpha_p = 0$$

has all the roots less than one in absolute value.

(i) Then X_t has the MA (∞) representation

$$X_t = \sum_{j=0}^{\infty} \upsilon_j \varepsilon_{t-j} \text{ where } \upsilon_0 = \alpha_0 = 1,\ \upsilon_1 = \beta_1 - \alpha_1,$$

$$\upsilon_2 = \beta_2 - \alpha_2 - \alpha_2 \upsilon_1, \ldots$$

$$\upsilon_j = \beta_j - \sum_{i=1}^{\min(j,p)} \alpha_j \upsilon_{j-i} \quad \text{if } j \le q$$

$$\upsilon_j = -\sum_{i=1}^{\min(j,p)} \alpha_j \upsilon_{j-i} \quad \text{if } j > q$$

(ii) If ε_t's are white noise

$$\rho_k = \sum_{j=1}^{p} \alpha_j z_j^k \text{ if } k > q.$$

Here z_j are the roots of the characteristic equation $\alpha(z) = 0$.

Proof Let $u_t = \sum_{j=0}^{p} \alpha_j X_{t-j}$ be an AR (p).

The $$X_t = \sum_{j=1}^{\infty} w_j u_{t-j} \text{ is a MA. } (\infty). \tag{9.3.3}$$

The case $p = 1$ has been done before.

In case of general p by partial fraction method we can prove

$$X_t = \sum_{j=1}^{\infty} w_j \left(\sum_{i=0}^{q} \beta_l \varepsilon_{t-j-i} \right) = \sum_{r=0}^{\infty} \upsilon_r \varepsilon_{t-r} \tag{9.3.4}$$

Now $$\sum_{i=0}^{q} \beta_i \varepsilon_{t-i} = \sum_{j=0}^{p} \alpha_j X_{t-j} = \sum_{j=0}^{p} \alpha_j \left(\sum_{r=0}^{\infty} \upsilon_r \varepsilon_{t-j-r} \right)$$

$$= \sum_{r=0}^{\infty} \upsilon_r \sum_{j=0}^{p} \alpha_j \varepsilon_{t-j-r}$$

and obtain υ_j's by equating coefficients of ε_{t-j}, $j = 0, 1, 2, \ldots$.

(ii) Equation (9.3.2) can be written as

$$\phi_P(B) X_t = \psi_q(B) \varepsilon_t, \tag{9.3.5}$$

where $\phi_P(B)$ and $\psi_q(B)$ are the polynomials in Backshift operator B. Since $E(\varepsilon_t X_{t-k}) = 0$, multiplying (9.3.5) by $X_{t-k} (k > 0)$ and taking expectation on both sides

$$\phi_p(B) \rho_k = 0 (k > q)$$

and so, $\sum_{i=0} \alpha_i \rho_{k-i} = 0$. Now $\alpha_0 = 1$ and the last equation is a homogeneous difference equation of order p. Hence $\rho_k = \sum_{r=1}^{p} \alpha_r z_r^k \, (k > q)$.

9.4 Wold's Decomposition and Wiener-Khintchine and Bochner's Theorem for Stationary Time Series

9.4.1 Deterministic and indeterministic processes

Deterministic process A time series $\{X_t\}$ will be called deterministic if there exists a function $g(t)$ of past and present values of X_t

$$g(t) = g(X_{t-j}, j = 0, 1, 2, \ldots)$$

such that $$E(X_{t+1} - g(t))^2 = 0 \tag{9.4.1}$$

If a function $g(t)$ is a linear function of X_{t-j}, $j \geq 0$, then $\{X_t\}$ will be called linear deterministic; e.g. consider the series $X_t = ae^{bt}$ so that $X_{t+1} = e^b X_t$. If b is not known, it can be perfectly estimated from the past of the series, but the estimate will be a nonlinear function of the (sequence) series; for instance $b = 1/2\ [\log X_t^2 - \log X_{t-1}^2]$ allowing for the fact that a may be negative. Other linear deterministic functions are the periodic functions $X_t = a \cos(\omega t + \theta)$ provided a is known and there exists an integer k such that $2\pi k/\omega$ is an integer. For example, if $X_t = a \cos(2\pi t/12 + \theta)$, then $X_{t+1} = X_{t-11}$ so that X_t is clearly linearly deterministic. If ω has to be estimated from past data, the series becomes nonlinear deterministic. Let $X_t = \sum_{k=0}^{n} d_k t^k$ so that the sequence is a polynomial in t. Then $\Delta^n X_t = n! d_n$ and $\Delta^{n+1} X_t = 0$, i.e. X_t series obeys a homogeneous linear difference equation and hence is linearly deterministic. To use this, procedure one has to know the value of n or $n' > n$.

A typical time series consists of a deterministic part and a random part, e.g.

$$X_t = A\sin(\theta t) + \varepsilon_t$$

$$\downarrow \qquad\qquad \downarrow$$

Deterministic white noise

Indeterministic Process Consider a discrete time stationary time series $\{X_t\}$, $E(X_t) = 0$ and $\text{Var}(X_t) = \sigma^2$. Suppose the residual variance obtained by regressing X_t on $(X_{t-q}, X_{t-q-1}, \ldots)$ is $\sigma_q^2 (\leq \sigma^2)$. Certainly σ_q^2 is a non-decreasing bounded sequence, i.e.

$$\lim_{q\to\infty} \sigma_q^2 = \sigma_0^2 = \begin{cases} \sigma^2 \\ 0 \end{cases} \text{(considering two extreme situations)}$$

(i) If residual variance $\sigma^2 > 0$ then it is useless to consider the regression of X_t on $\{X_{t-q}, X_{t-q-1}, \ldots\}$. Such a process $\{X_t\}$ is then called indeterministic, since the process cannot be used for forecasting purposes (MA, AR, ARMA, satisfy this property). Then linear regression on the remote past is useless for prediction purpose.

(ii) If residual variance is 0 then it can be used for forecasting purposes and hence is deterministic.

Theorem 9.2 Wold Decomposition Theorem (1930)
Any discrete stationary time series $\{X_t\}$ can be expressed as the sum of two uncorrelated processes, one purely deterministic V_t and the other purely indeterministic process U_t. Further, U_t may be represented as a infinite moving average MA (∞), i.e.

$$U_t = \varepsilon_t + \theta_1 \varepsilon_{t-1} + \theta_2 \varepsilon_{t-2} + \ldots$$

where ε_t's are uncorrelated. The proof is given in Appendix II(B), p.284.

9.4.2 Analysis in frequency domain

Inference based on the conceptual tool known as spectral density function (spectrum) is called analysis in frequency domain. The spectral density is the natural tool for considering the frequency properties of a time series.

Theorem 9.3 Wiener-Khintchine's Theorem Suppose $\{X_t\}$ is a covariance stationary time series with autocovariance function $\gamma(k)$. Then there is a function $F(\lambda) \uparrow$ in λ such that

$$\gamma(k) = \int_0^{\pi} \cos(\lambda k)\, dF(\lambda) \tag{9.4.2}$$

This is called the *spectral representation of autocovariance function*, where λ = frequency of the series $y_t = \cos(\lambda t)$ and λ can be written as $\lambda = \frac{2\pi}{\theta}$ where θ is the period of oscillation. Note that $\lambda \in (0, \pi)$.

Following is the direct physical interpretation of spectral representation (9.4.2):

If $k = 0$, $\gamma(0) = \int_0^{\pi} dF(\lambda) = F(\pi) = \sigma^2$ (from equation (9.4.2))

As $F(\lambda) \uparrow$ in λ, $\max\limits_{0 \le \lambda \le \pi} F(\lambda) = F(\pi) = \sigma^2 = \text{Var}(X_t)$.

Therefore $F(\lambda)$ is the contribution to the variance of the series which is accounted for frequency $\lambda \in (0, \pi)$. $F(\lambda)$ is an absolutely continuous function for any discrete stationary time series $\{X_t\}$ satisfying $\Sigma \mid \gamma(k) \mid < \infty$. Therefore $f(\lambda)$ = spectral density = $\frac{dF(\lambda)}{d\lambda}$ exists in this case and (9.4.2) becomes $\gamma(k) = \int_0^{\pi} \cos(\lambda k) f(\lambda)\, d\lambda$.

The autocovariance function and the spectral density function are equivalent in representing underlying time series. In some situations, the autocovariance function is seen to be useful in determining the underlying structure of the time series while in some situations the spectral density function $f(\lambda)$ is seen to be more useful. Fourier Analysis of spectral density $f(\lambda)$ plays an important role.

9.4.3 Fourier Analysis

Suppose that a function $f(t)$ is defined on $(-\pi, \pi)$ and satisfies the Dirichlet conditions (It is absolutely integrable over the domain $(-\pi, \pi)$, has a finite number of discontinuities and a finite number of maxima & minima). Then $f(t)$ may be approximated by the Fourier series

$$\frac{a_0}{2} + \sum_{r=1}^{\infty} (a_r \cos rt + b_r \sin rt),$$

where $$a_0 = \frac{1}{\pi}\int_{-\pi}^{\pi} f(t)\, dt,\ a_r = \frac{1}{\pi}\int_{-\pi}^{\pi} f(t) \cos(rt)\, dt$$

and $$b_r = \frac{1}{\pi}\int_{-\pi}^{\pi} f(t) \sin(rt)\, dt \quad r = 1, 2, 3, \ldots$$

It can be shown that the finite Fourier series $s_k = \frac{a}{2} + \sum_{r=1}^{k} (a_r \cos rt + b_r \sin rt)$ converges to $f(t)$ as $k \to \infty$ except at points of discontinuities where it converges to

$$\frac{1}{2}[f(t+) + f(t-)].$$

Theorem 9.4 (*From Fourier Analysis*) If $\sum_{k=1}^{\infty} (\mid a_k \mid + \mid b_k \mid) < \infty$, then the associated trigonometric series $a_0/2 + \sum_{k=1}^{\infty} (a_k \cos k\lambda + b_k \sin k\lambda)$ converges absolutely and uniformly to a continuous function of period 2π of which it is the

Fourier seires. Let us recall the definition " A stochastic process $\{X_t\}$ is said to be *mean-square continuous* if $\lim_{h\to 0} \text{Var}(X_{t+h} - X_t) = 0$."

Theorem 9.5 (*Bochner-Khintchine*) If the function ρ_h is the correlation function of a mean-square continuous stationary time series with index set $T = (-\infty, \infty)$, then it is representable in the form

$$(1)\ \rho_h = \int_{-\infty}^{\infty} \theta^{i\omega h}\, dF(\omega) \text{ where } F(\omega) \text{ is a d.f.}$$

$$= \int_{-\infty}^{\infty} e^{i\omega h}\ \mu(d\omega),\ \mu \text{ is a finite measure on } (R, \mathscr{B}).$$

If the index set T is discrete, i.e. $T = \{0, \pm 1, \pm 2, \ldots\}$ then we have a theorem due to Herglotz.

Let the correlation function ρ_k of a stationary time series be absolute summable (i.e. $\sum_k |\rho_k| < \infty$). Then it is representable in the form

$$(2)\ \rho_k = \int_{-\pi}^{\pi} e^{i\lambda k}\, dF(\lambda)$$

$$= \int_{-\pi}^{\pi} e^{i\lambda k}\ \mu(d\lambda) \text{ where } \mu \text{ is a finite measure on } (E = [-\pi, \pi], \mathscr{B}).$$

In fact more is true.

There exists a continuous function $f(\lambda)$ such that

(i) $\rho_k = \int_{-\pi}^{\pi} f(\lambda) \cos \lambda k\, d\lambda$,

(ii) $f(\lambda) \neq 0$,

(iii) $\int_{-\pi}^{\pi} f(\lambda)\, d\lambda = 1$

(iv) $f(\lambda)$ is an even function of λ.

Proof From Fourier Analysis

$$g(\lambda) = \frac{1}{2} + \sum_{k=1}^{\infty} \rho_k \cos k\lambda$$

is a well defined continuous (since $\sum_k |\rho_k| < \infty$) function.

By positive semi-definite property of correlation functions ρ_k

$$\sum_{m=1}^{n} \sum_{q=1}^{n} \rho_{m-q} \cos(m\lambda) \cos(q\lambda) \geq 0$$

and

$$\sum_{m=1}^{n} \sum_{q=1}^{n} \rho_{m-q} \sin(m\lambda) \sin(q\lambda) \geq 0.$$

Hence $$\sum_{m=1}^{n}\sum_{q=1}^{n} \rho_{m-q} [\cos(m\lambda)\cos(q\lambda) + \sin(m\lambda)\sin(q\lambda)]$$

$$= \sum_{m=1}^{n}\sum_{q=1}^{n} \rho_{m-q} \cos(m-q)\lambda \geq 0$$

Note $$\begin{bmatrix} m-q = 0 \rightarrow n \text{ times} \\ = 1 \rightarrow n-1 \text{ times} \\ \dots \\ = n-1 \rightarrow 1 \text{ times} \end{bmatrix}$$

Letting $m - q = k$, we have

$$\sum_{k=-(n-1)}^{(n-1)} \left(\frac{n-|k|}{n}\right) \rho_k \cos(k\lambda) \geq 0$$

Now $\{\rho_k \cos k\lambda\}$ is absolutely summable ($\sum_k |\rho_k \cos(k\lambda)| < \infty$) and hence by Kronecker's lemma

$$\lim_{n\to\infty} \sum_{k=-(m-1)}^{(n-1)} \frac{|k|}{n} \rho_k \cos(k\lambda) = 0.$$

Thus, $$\lim_{n\to\infty} \sum_{k=-(n-1)}^{(n-1)} \frac{n-|k|}{n} \rho_k \cos(k\lambda) = \sum_{k=-\infty}^{\infty} \rho_k \cos(k\lambda) = 2g(\lambda) \geq 0$$

Now ρ_k are the Fourier coefficients of g shows that

$$\rho_k = \int_{-\pi}^{\pi} g(\lambda)\cos(\lambda k)\, d\lambda$$

We have only to multiply $g(\lambda)$ by a constant to get (iii). Since

$$\frac{1}{\pi}\int_{-\pi}^{\pi} g(\lambda)\, d\lambda = 1,$$

the appropriate constant is $\frac{1}{\pi}$ and we define

$$f(\lambda) \text{ by } f(\lambda) = \frac{1}{\pi}\left[\frac{1}{2} + \sum_{k-1}^{\infty} \rho_k \cos(k\lambda)\right]$$

Note that $\frac{1}{2\pi}\sum_{k=-\infty}^{\infty} \rho_k \cos k\lambda = \frac{\sigma^2}{\pi}\left[1 + 2\sum_{k=1}^{\infty} \rho_k \cos(k\lambda)\right]$ if $\sigma^2 = 1$

(iv) The function $f(\lambda)$ is an even function since it is the uniform limit of a sum of even functions (cosines).

The stationary process $\{X_t\}$ itself has a spectral representation (Cramer 1942),

$$X_t = \int_{-\infty}^{\infty} e^{it\lambda} dZ(\lambda) = \int_{-\infty}^{\infty} e^{it\lambda} \mu(d\lambda) \tag{9.4.3}$$

where $Z(\lambda)$ $(-\infty < \lambda < \infty)$ is a stochastic process with finite second moments has orthogonal increments such that $E \mid Z(\lambda_1) - Z(\lambda_2) \mid^2 < \infty$ and

$$E[(Z(\mu_2) - Z(\mu_1)) \overline{(Z(\lambda_2) - Z(\lambda_1))}] = 0, \mu_1 < \mu_2 \le \lambda_1 < \lambda_2;$$

also
$$Z(t) = \begin{cases} \mu(-\infty, t) \text{ if } T = R^1 \\ \mu[-\pi, t] \text{ if } T = [-\pi, \pi]. \end{cases}$$

The representation (9.4.3) is called the spectral representation of the stationary process $\{X_t\}$ and the function $F(\lambda) = E \mid Z(\lambda) \mid^2$ is the spectral distribution function. The measure μ is uniquely determined by the spectual function via the equality $\mu[a, b) = F(b) - F(a)$.

9.5 Spectral Densities of some Standard Processes and Autocovariance Generating Function

9.5.1 Determination of $f(\lambda)$ in terms of $\gamma(k)$ and ρ_k

From the spectral representation theorem

$$f(\lambda) = \frac{\sigma^2}{\pi}\left[1 + 2\sum_{k=1}^{\infty} \rho_k \cos(k\lambda)\right]$$

$$= \frac{1}{\pi}\left[\sigma^2 + 2\sum_{k=1}^{\infty} \gamma_k \cos(k\lambda)\right] \qquad (\gamma_0 = \sigma^2)$$

Exercise 9.1 MA(1)

$$X_t = \varepsilon_t + \beta\varepsilon_{t-1}, \quad \rho_k = \begin{cases} 1 \text{ if } k = 0 \\ \dfrac{\beta}{1+\beta^2} \text{ if } k = \pm 1 \\ 0 \text{ otherwise} \end{cases}$$

$$f(\lambda) = \frac{\sigma^2}{\pi}\left[1 + \frac{2\beta}{1+\beta^2}\cos\lambda\right] \text{ for } 0 < \lambda < \pi$$

$$= \frac{1}{\pi}\frac{\sigma^2}{1+\beta^2}[1 + 2\beta\cos\lambda + \beta^2]$$

and $f(0) = \dfrac{\sigma^2}{\pi}\left[1 + \dfrac{2\beta}{1+\beta^2}\right], f(\pi) = \dfrac{\sigma^2}{\pi}\left[1 - \dfrac{2\beta}{1+\beta^2}\right].$

Here $\dfrac{\sigma^2}{1+\beta^2}$ is the variance of ε_t process and not of X_t.

Special Case $\beta = 1$ and $\begin{cases} f(0) = \dfrac{2\sigma^2}{\pi} \\ f(\pi) = 0 \end{cases}$

If $\beta = -1$, then $\begin{cases} f(0) = 0 \\ f(\pi) = 2\sigma^2/\pi. \end{cases}$

The shape of the spectrum depends on the value of β. When $\beta > 0$ the power is concentrated at low frequencies giving what is called a low frequency spectrum,

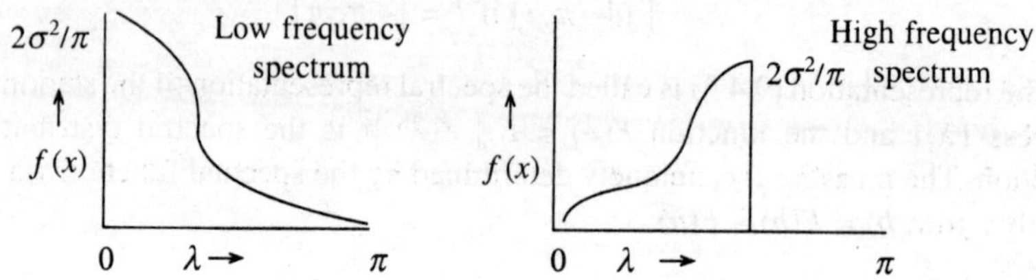

Exercise 9.2 AR(1) $X_t = \alpha X_{t-1} + \varepsilon_t$, $|\alpha| < 1$

Here we have $\rho_k = \alpha^{|k|}$.

$$f(\lambda) = \frac{\sigma^2}{2\pi}\left[1 + 2\sum_{k=1}^{\infty} \alpha_k \cos(\lambda k)\right]$$

$$= \frac{\sigma^2}{2\pi}\left[1 + \sum_{k=1}^{\infty} \alpha^k (e^{i\lambda k} + e^{-i\lambda k})\right]$$

$$= \frac{\sigma^2}{2\pi}\left[1 + \sum_{k=1}^{\infty} (\alpha^k e^{i\lambda})^k + \sum_{k=1}^{\infty} (\alpha e^{-i\lambda})^k\right]$$

$$= \frac{\sigma^2}{2\pi}\left[1 + \frac{\alpha e^{i\lambda}}{1 - \alpha e^{i\lambda}} + \frac{\alpha e^{-i\lambda}}{1 - \alpha e^{-i\lambda}}\right]$$

$$= \frac{\sigma^2}{2\pi}\left[\frac{1 + \alpha^2 - \alpha(e^{i\lambda} + e^{-i\lambda}) + \alpha e^{i\lambda} - \alpha^2 + \alpha e^{-i\lambda} - \alpha^2}{1 + \alpha^2 - \alpha(e^{i\lambda} + e^{-i\lambda})}\right]$$

$$= \frac{\sigma^2}{2\pi}\left[\frac{1 - \alpha^2}{1 + \alpha^2 - 2\alpha\cos\alpha}\right]$$

$$= \frac{\sigma^2(1 - \alpha^2)}{2\pi}\left[\frac{1}{1 + \alpha^2 - 2\alpha\cos\lambda}\right]$$

$$f(0) = \frac{\sigma^2}{2\pi}\frac{1 - \alpha^2}{(1 - \alpha)^2} = \frac{\sigma^2}{2\pi}\frac{1 + \alpha}{1 - \alpha}\left(> \frac{\sigma^2}{\pi}\right) \text{ and } f(\pi) = \frac{\sigma^2}{2\pi}\frac{1 - \alpha}{1 + \alpha}\left(> \frac{\sigma^2}{\pi}\right)$$

Note here X_t has variance $\sigma^2(1 - \alpha^2)$ and σ^2 is the variance of ε_t.

The shape of the spectrum depends on the value of α. For $\alpha > 0$, the power is low frequency spectrum and for $\alpha < 0$ a high frequency spectrum.

If $\alpha < 0$, the values of X_t will tend to oscillate: a positive value of X_t will be followed by a negative value and vise versa. The rapid oscillations correspond to high frequency variation.

Exercise 9.3 Let $\{X_t\}$ be a white noise process with $E(X_t) = 0$ for all t and

$$E(X_s X_t) = \begin{cases} \sigma^2 & \text{if } s = 1 \\ 0 & \text{if } s \neq t \end{cases}$$

Now covariance function is given by $\gamma_k = \int_{-\pi}^{\pi} e^{-ik\lambda}\, dF(\lambda)$.

In the discrete case, by inverse Fourier transform, the spectral density is given by

$$f(\pi) = \frac{1}{2\pi} \sum_{k=-\infty}^{\infty} \gamma_k e^{-ik\lambda}.$$

Here

$$\gamma_k = 0 \text{ if } k \neq 0$$
$$= \sigma^2 \text{ if } k = 0.$$

So, $f(\lambda)$ becomes, $\sigma^2/2\pi,\ -\pi \leq \lambda \leq \pi$.

Autocovariance generating functions are very useful for finding spectral densities of many time series models.

9.5.2 Autocovariance and autocorrelation generating function (AGF)

Define $C(z) = \sum_{k=-\infty}^{\infty} \gamma_k z^k$, where z is such that $C(z)$ is convergent and $C(z)$ is called autocovariance generating function.

Replacing γ_k by $\rho_k \left(= \frac{\gamma_k}{\rho_0} \right)$, we get autocorrelation generating function

$$S(z) = \sum_{k=-\infty}^{\infty} \rho_k z^k = \frac{C(z)}{\text{Var}(X_t)}.$$

The spectral density can be expressed in terms of AGF as follows:

$$f(\lambda) = \frac{1}{2\pi} \sum_{k=-\infty}^{\infty} \gamma_k e^{-ik\lambda} = \frac{1}{2\pi} C(e^{-i\lambda})$$
$$= \frac{\text{Var}(X_t)}{2\pi} S(e^{-i\lambda}).$$

Theorem 9.6 Let Y_t be a linear process, i.e.

$$Y_t = \sum_{j=0}^{\infty} a_j X_{t-j} \text{ with } \sum_{j=0}^{\infty} |a_j| < \infty$$

is a stationary process where X_t is a white noise.

If $C_y(z)$, $C_x(z)$ are the AGF's and $f_y(\lambda)$ and $f_x(\lambda)$ are spectral density functions of $\{Y_t\}$ and $\{X_t\}$ respectively, then

$$C_y(z) = h(z)h(z^{-1})\,C_x(z)$$

and $$f_y(\lambda = h(e^{i\lambda})h(e^{-i\lambda})\,f_x(\lambda)$$

where $h(z) = \sum_{j=0}^{\infty} a_j z^j$.

Proof Assume $E(X_t) = 0$ for all t, then $E(Y_t) = 0$ for all t. Denote

$$\gamma(k) = E(X_t X_{t+k}),\ R(k) = E(Y_t Y_{t+k}).$$

then $$R(k) = E\left\{\left(\sum_{j=0}^{\infty} a_j X_{t-j}\right)\left(\sum_{l=0}^{\infty} a_l X_{t+k-1}\right)\right\}$$

$$= \sum_{j=0}^{\infty}\sum_{l=0}^{\infty} a_j a_l \gamma(k+j-l)$$

Hence $C_y(z) = \sum_{k=-\infty}^{\infty} R(K) z^k$

$$= \sum_{k=-\infty}^{\infty} z^k \left\{\sum_{j=0}^{\infty}\sum_{l=0}^{\infty} a_j a_l \gamma(k+j-l)\right\}$$

$$= \sum_{j=0}^{\infty}\sum_{l=0}^{\infty}\sum_{k=-\infty}^{\infty} a_j a_l z^{-j+l}\gamma(k+j-l)\,z^{k+j-l}$$

$$= \left(\sum_{l=0}^{\infty} a_l z^l\right)\left(\sum_{j=0}^{\infty} a_j z^{-j}\right)\left(\sum_{k=-\infty}^{\infty} \gamma(s)\,z^s\right), \text{ setting } s = k+j-1$$

[Note: $\sum_j |a_j| < \infty \Rightarrow \sum_j a_j^2 < \infty$. Hence changing the order of summation is valid.]

$$= h(z)h(z^{-1})C_x(z).$$

Again, $$f_y(\lambda) = \frac{1}{2\pi} C_y(e^{-i\lambda}) = \frac{1}{2\pi} h(e^{-i\lambda})h(e^{i\lambda})C_x(e^{-i\lambda})$$

$$= h(e^{i\lambda})h(e^{-i\lambda})f_x(\lambda)$$

Exercise 9.4 MA(q) $X_t = \sum_{j=0}^{q} a_j \varepsilon_{t-j}$, $\text{Var}(\varepsilon_t) = \sigma^2$, where ε_t is a white noise. Then

$$f_x(\lambda) = \frac{\sigma^2}{2\pi} h(e^{i\lambda})h(e^{-i\lambda})$$

$$= \frac{\sigma^2}{2\pi}\left(\sum_{l=0}^{q} a_l e^{i\lambda l}\right)\left(\sum_{r=0}^{q} a_r e^{-ir\lambda}\right)$$

$$= \frac{\sigma^2}{2\pi}\sum_{k=-q}^{q}\sum_{r=0}^{q-k} a_r a_{r+k} \cos(k\lambda)$$

$[l - r = k \text{ or } l - k = r]$.

Since spectral density is a real function of λ only cosine terms appear.

Exercise 9.5 AR (*P*)

$$\sum_{j=0}^{p} \alpha_j X_{t-j} = \varepsilon_t, \alpha_0 = 1$$

Assume that roots of the equation $z^p + \alpha_1 z^{p-1} + \ldots + \alpha_p = 0$ all lie within the unit circle. Applying Theorem 9.6 to $\varepsilon_t = \sum_{j=0}^{p} \alpha_j X_{t-j}$, spectral density of $\{\varepsilon_t\}$

$$f_\varepsilon(\lambda) = \left(\sum_{j=0}^{p} \alpha_j e^{ij\lambda}\right)\left(\sum_{k=0}^{p} \alpha_k e^{-ik\lambda}\right) f_x(\lambda),$$

where $f_x(\lambda)$ is spectral density of X_t.

Hence
$$f_x(\lambda) = \frac{\sigma^2}{2\pi}\left[\left(\sum_{j=0}^{p} \alpha_j e^{ij\lambda}\right)\left(\sum_{k=0}^{p} \alpha_k e^{-ik\lambda}\right)\right]^{-1}$$

Particular case p = 2

$$f_x(\lambda) = \frac{\sigma^2}{2\pi}[(1 + \alpha_1 e^{i\lambda} + \alpha_2 e^{2i\lambda})(1 + \alpha_1 e^{-i\lambda} + \alpha_2 e^{-2i\lambda})]^{-1}$$

$$= \frac{\sigma^2}{2\pi}\frac{1}{1 + \alpha_1^2 + \alpha_2^2 + 2\alpha_1(1 + \alpha_2)\cos\lambda + 2\alpha_2\cos 2\lambda}$$

Exercise 9.6 ARMA (*p*, *q*)

$$\sum_{r=0}^{p} \alpha_r X_{t-r} = \sum_{r=0}^{q} \beta_r \varepsilon_{t-r}, \alpha_0 = 1, \{\varepsilon_t\} \text{ is a white noise.}$$

Assume roots of $Z^p + \alpha_1 Z^{p-1} + \ldots + \alpha_p = 0$ all lie within the unit circle. Putting $Z_t = \sum_{r=0}^{q} \beta_r \varepsilon_{t-r}$ and applying Theorem 9.6 we get

$$f_z(\lambda) = \frac{\sigma^2}{2\pi}\left(\sum_{r=0}^{q} \beta_r e^{ir\lambda}\right)\left(\sum_{s=0}^{q} \beta_s e^{-is\lambda}\right)$$

where $f_z(\lambda)$ is the spectral density of $\{Z_t\}$.

Again from $Z_t = \sum_{r=0}^{q} \alpha_r X_{t-r}$ we get

$$f_z(\lambda) = \left(\sum_{r=0}^{p} \alpha_r e^{ir\lambda}\right)\left(\sum_{s=0}^{p} \alpha_r e^{-is\lambda}\right) f_x(\lambda)$$

Equating the two expressions we get

$$f_x(\lambda) = \frac{\sigma^2}{2\pi}\frac{\left(\sum_{r=0}^{q} \beta_r e^{ir\lambda}\right)\left(\sum_{s=0}^{q} \beta_s e^{-is\lambda}\right)}{\left(\sum_{r=0}^{p} \alpha_r e^{ir\lambda}\right)\left(\sum_{s=0}^{p} \alpha_s e^{-is\lambda}\right)}$$

Exercise 9.7 *Ornstein-Uhlenbeck Process* Unlike other time series process, this is a continuous parameter process.

It s correlation function is given by

$$\rho(t) = ae^{-b|t|},\ a,\ b > 0$$

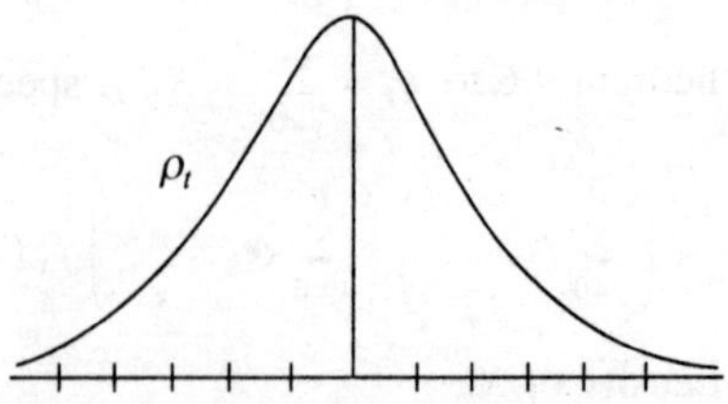

Falls off rapidly

Figure 1

The spectral density

$$f(\lambda) = \frac{1}{2\pi}\int_{-\infty}^{\infty} e^{-it\lambda}\rho_t\, dt$$

$$= \frac{1}{2\pi}\int_{-\infty}^{\infty} e^{-it\lambda}\, ae^{-b|t|}$$

$$= \frac{1}{2\pi}\left[\int_{-\infty}^{\infty} e^{(b-i\lambda)t}\, dt + \int_{0}^{\infty} e^{-|b+i\lambda|t}\, dt\right]$$

$$= \frac{a}{2\pi}\left(\frac{1}{b-i\lambda} + \frac{1}{b+i\lambda}\right) = \frac{ab}{\pi(b^2+\lambda^2)}$$

is a Cauchy type density.

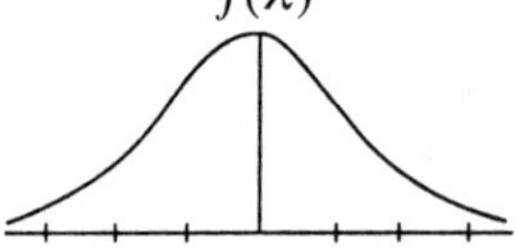

Figure 2

The $f(\lambda)$ has maximum at zero, remains constant when λ is small compared to b and then falls off slowly. In Figure 1, ρ_t in fact falls off exponentially with $|t|$ and can already be considered virtually zero when the distance from the origin is only a few multiples of b^{-1}. The b^{-1} characterizes the time needed for any correlation between $X(s)$ and $X(s + t)$ to die out.

9.6 Estimation of Spectrum

For a discrete parameter stationary time series the spectral density is given by

$$f(\lambda) = \frac{1}{2\pi}\sum_{k=-\infty}^{\infty} e^{-ik\lambda}\gamma_k.$$

We are interested in suggesting some estimate of the spectral density $f(\lambda)$ on the basis of n observations from $\{X_t\}$. Given any finite realization, a covariance stationary time series based on n observations can be represented as

$$X_t = \frac{a_0}{2} + \sum_{k=1}^{m} (a_k \cos \omega_k t + b_k \sin \omega_k t) \text{ assuming } n \text{ odd}, \ m = \frac{n-1}{2}$$

where $$\omega_k = \frac{2\pi k}{n}, k = 1, 2 \ldots, m$$

$$a_0 = \overline{X}, a_k = \frac{2}{n} \sum_{t=1}^{n} X_t \cos \omega_k t,$$

$$b_k = \frac{2}{n} \sum_{t=1}^{n} X_t \sin \omega_k t, \ k = 1, 2, \ldots, m.$$

Any function $f(t)$ defined on n integers can be expressed in a finite Fourier series as

$$f(t) = \sum_{k=0}^{L(n)} (a_k \cos \omega_k t + b_k \sin \omega_k t), t = 1, \ldots, n$$

where $\omega_k = \frac{2\pi k}{n}$, a_k, b_k are suitably chosen and $L(n)$ is the largest integer not exceeding $\frac{n}{2}$.

$$\begin{pmatrix} f(1) \\ f(2) \\ \vdots \\ f(n) \end{pmatrix} = \sum_{k=0}^{L(n)} \left\{ \begin{pmatrix} a_k \cos \omega_k 1 \\ \ldots \\ a_k \cos \omega_k n \end{pmatrix} + \begin{pmatrix} b_k \sin \omega_k 1 \\ \ldots \\ b_k \sin \omega_k n \end{pmatrix} \right\} \tag{9.6.1}$$

In each row, corresponding to each $f(i)$ there are thus $\frac{n-1}{2} + \frac{n-1}{2} + 1 = n$ terms. With these n-elements and n columns, we can thus form a $n \times n$ matrix of a_k, b_k, i.e.

$$\begin{pmatrix} f(1) \\ \vdots \\ f(n) \end{pmatrix} = \beta_1 \underset{\sim}{\alpha}_1 + \ldots + \beta_n \underset{\sim}{\alpha}_n,$$

where $\underset{\sim}{\alpha}_1 = \begin{pmatrix} \cos \omega_0 1 \\ \vdots \\ \cos \omega_0 n \end{pmatrix}, \ldots \underset{\sim}{\alpha}_{2k+1} = \begin{pmatrix} \cos \omega_k 1 \\ \vdots \\ \cos \omega_k n \end{pmatrix}, \underset{\sim}{\alpha}_{2k} = \begin{pmatrix} \sin \omega_k 1 \\ \vdots \\ \sin \omega_k n \end{pmatrix}$

Hence we can always represent the $f(t)$ as a linear combination of the orthogonal vectors $\underset{\sim}{\alpha}_1, \ldots, \underset{\sim}{\alpha}_n$. Since by the trigonometric results we have

1. $\sum_{t=1}^{n} \cos(\omega_k t) = \sum_{t=1}^{n} \sin(\omega_k t) = 0$

2. $\sum_{t=1}^{n} \cos(\omega_k t) \cos(\omega_{k'} t) = \begin{cases} 0 & \text{if } k \neq k' \\ n/2 & \text{if } k = k' \neq \frac{n}{2} \\ n & \text{if } k = k' = \frac{n}{2} \end{cases}$

3. $\sum_{t=1}^{n} \sin(\omega_k t)\sin(\omega_{k'} t) = \begin{cases} 0 & \text{if } k \neq k' \\ n/2 & \text{if } k = k' \neq \frac{n}{2} \\ n & \text{if } k = k' = \frac{n}{2} \end{cases}$

4. $\sum_{t=1}^{n} \cos(\omega_k t)\sin(\omega_{k'} t) = 0$ for all $\omega_k, \omega_{k'}$.

we can easily check that the above sine, cosine functions appearing in (9.6.1) are orthogonal functions. Thus a set of n functions defined on n integers can be obtained such that they form an orthogonal basis in n-dimensional space. Thus for suitable choices of a_k, b_k we can always represent any function $f(t)$ defined on n integers in the form (9.6.1).

To obtain a_k, b_k we apply the usual regression technique (least square), i.e. minimize

$$\sum_{t=1}^{n} \left\{ f(t) - \sum_{k=0}^{L(n)} (a_k \cos \omega_k t + b_k \sin \omega_k t) \right\}^2 \text{ with respect to } a_k \text{ and } b_k\text{'s.}$$

We get

$$\left.\begin{aligned} a_k &= \frac{2}{n} \sum_{t=1}^{n} f(t) \cos \omega_k t,\ k = 0, 1, 2, \ldots, L(n) \\ b_k &= \frac{2}{n} \sum_{t=1}^{n} f(t) \sin \omega_k t,\ k = 0, 1, \ldots, L(n) \end{aligned}\right\} \quad (9.6.2)$$

If n is even,

$$L(n) = \frac{n}{2}$$

$$a_{L(n)} = \frac{2}{n} \sum_{t=1}^{n} f(t) \cos(\pi t)$$

$$b_{L(n)} = \frac{2}{n} \sum_{t=1}^{n} f(t) \sin(\pi t) = 0$$

i.e. although $a_{L(n)}$ contributes something $b_{L(n)}$ does not. Hence we take n odd.

We have here represented n observations by a function having n parameters. As in usual regression analysis, we can partition the total sum of square (SS) into $(m + 1)$ components

$$\sum_{t=1}^{n} X_t^2 - n\bar{X}^2 = \sum_{k=1}^{m} \sum_{t=1}^{n} (a_k \cos \omega_k t + b_k \sin \omega_k t)^2,$$

where a_k and b_k are given by (9.6.2).

$$\begin{aligned} \text{Sum of squares} &= \sum_{t=1}^{n} (a_1 \cos \omega_1 t + b_1 \sin \omega_1 t)^2 + \sum_{t=1}^{n} (a_2 \cos \omega_2 t + b_2 \sin \omega_2 t)^2 \\ &\quad + \ldots + \sum_{t=1}^{n} (a_m \cos \omega_m t + b_m \sin \omega_m t)^2 \\ &= \frac{n}{2}(a_1^2 + b_1^2) + \ldots + \frac{n}{2}(a_m^2 + b_m^2) \end{aligned}$$

ANOVA Table

Source	d.f.	SS
Due to mean	1	$na_0^2/2$
frequency at ω_1	2	$n/2\,(a_1^2 + b_1^2)$
...	:	...
		...
...	.	...
... ω_m	2	$n/2\,(a_m^2 + b_m^2)$
Total	n	Total SS

The residual sum of squares expressed by the periodic component at frequency ω_k is given by the SS $n/2\,(a_k^2 + b_k^2)$ which is called the "periodogram".

Let $$I_n(\omega_k) = \frac{n}{2}(a_k^2 + b_k^2)$$

$$= \frac{2}{n}\left[\left(\sum_{t=1}^{n} X_t \cos \omega_k t\right)^2 + \left(\sum_{t=1}^{n} X_t \sin \omega_k t\right)^2\right]$$

$$= \frac{2}{n}\left|\sum_{t=1}^{n} X_t e^{-it\omega_k}\right|^2, \qquad k = 0, 1, \ldots, m$$

= Normalized SS at frequency ω_k.

The plot of $I_n(\omega_k)$ versus ω_k is usually called the *periodogram* even though $I_n(\omega_k)$ is a function of frequency rather than period. Anderson (1970) described the graph of $(a_k^2 + b_k^2)$ versus the period n/k as the periodogram and suggested the term *spectogram* to describe the graph of $I_n(\omega_k)$ against the frequency ω_k. Hannan (1970) defined the periodogram in terms of complex number as

$$\frac{1}{2\pi n}\left|\sum_{t=1}^{n} X_t e^{-it\omega_k}\right|^2.$$

The periodogram appears to be a natural way of estimating the spectral density function. But we shall see that for a process with a continuous spectrum, it provides a poor estimate and need to be modified.

Under normal autocovariance stationary time series $N(0, \sigma^2)$ we have

$$a_k = \frac{2}{n}\sum_{t=1}^{n} X_t \cos \omega_k t \sim N\left(0, \frac{2\sigma^2}{n}\right)$$

and $$b_k = \frac{2}{n}\sum_{t=1}^{n} X_t \sin \omega_k t \sim N\left(0, \frac{2\sigma^2}{n}\right) \text{ for } k \neq \frac{n}{2}$$

$$\text{cov}(a_k, b_k) = 4\,\text{cov}\left\{\sum_{t=1}^{n} X_t \cos \omega_k t, \sum_{t=1}^{n} X_t \sin \omega_k t\right\}\Big/ n^2$$

$$= \frac{4\sigma^2}{n^2} \sum_{t=1}^{n} \{\cos \omega_k t \sin \omega_k t\}$$

$$= \frac{2\sigma^2}{n^2} \sum_{t=1}^{n} \sin 2\omega_k t = 0 \text{ [since } X_i\text{'s are uncorrelated].}$$

Therefore, a_k, b_k's are independent (since being linear combinations of normal r.v.s a_k, b_k are normal).

Hence
$$\left.\begin{array}{l} \frac{n}{2} a_k^2 \sim \sigma^2 \chi_1^2 \\ \frac{n}{2} b_k^2 \sim \sigma^2 \chi_1^2 \end{array}\right\} \text{are independent}$$

and
$$I_n(\omega_k) = \frac{n}{2}(a_k^2 + b_k^2) \sim \sigma^2 \chi_2^2, k = 1, 2, \ldots, m$$

Thus
$$\text{Var}\,[I_n(\omega_k)] = 4\sigma^4, \; k = 1, 2, \ldots, m$$

Therefore periodogram ordinates $I_n(\omega_k)$ are independently and exponentially distributed with means $\sigma^2 = 2\pi f(\omega_k)$, where $f(.)$ is the spectral density of $\{X_t\}$.

Theorem 9.7 For any covariance stationary time series with absolutely summable autocorrelation function, $I_n^*(\omega) = \frac{1}{4\pi} I_n(\omega)$ is an asymptotically unbiased but inconsistent estimate of the spectral density

$$f(\omega) = \frac{1}{2\pi} \sum_{k=-\infty}^{\infty} \rho_k e^{-i\omega k}.$$

Proof
$$I_n(\omega) = \frac{2}{n} \left| \sum_{t=1}^{n} X_t e^{-it\omega} \right|^2$$

$$= \frac{2}{n} \left[\sum_{t=1}^{n} X_t e^{-it\omega} \sum_{t'=1}^{n} X_{t'} e^{it'\omega} \right]$$

$$= \frac{2}{n} \left[\sum_{t=1}^{n} \sum_{t'=1}^{n} X_t X_{t'} e^{-i(t-t')\omega} \right]$$

Now $EX_t = 0$. Therefore

$$E(I_n(\omega)) = \frac{2}{n} \left[\sum_{t=1}^{n} \sum_{t=1}^{n} E(X_t X_{t'})\, e^{-i(t-t')\omega} \right]$$

$$= \frac{2}{n} \sum_{t=1}^{n} \sum_{t'=1}^{n} \gamma_{t-t'}\, e^{-1(t-t')\omega}$$

$$= \frac{2}{n} \sum_{h=-(n-1)}^{n-1} \gamma_h e^{-ih\omega} (n - |h|) \quad [\gamma_{t-t'} \text{ is an even function}]$$

$$= 2 \sum_{h=-(n-1)}^{n-1} \gamma_h e^{-ih\omega} - 2 \sum_{h=-(n-1)}^{n-1} \frac{|h|}{n} \gamma_h e^{-ih\omega}$$

$t - t' = 0 \rightarrow n$ times,
$= 1 \rightarrow n - 1$ times,
$= \rightarrow n - 2$ times and
$= n - 1 \rightarrow 1$ times]

We have to show that $\left| \sum_{h=-(n-1)}^{n-1} \frac{|h|}{n} \gamma_h e^{-ih\omega} \right| < \varepsilon$ for $n > N$.

$$\left| \sum_{h=-(n-1)}^{n-1} \frac{|h|}{n} \gamma_h e^{-ih\omega} \right| \leq \sum_{h=-(n-1)}^{n-1} \frac{|h|}{n} \gamma_h |e^{-ih\omega}|$$

$$\leq 2 \sum_{h=0}^{n-1} \frac{h}{n} \gamma_h.$$

By applying the Kronecker's lemma (since autocovariances are absolutely summable), we get

$$\lim_{n\to\infty} \sum_{h=-(n-1)}^{n-1} \frac{|h|}{n} \gamma_h e^{-ih\omega} = 0.$$

Thus $\lim_{n\to\infty} E(I_n(\omega)) = 2 \sum_{h=-\infty}^{\infty} \gamma_h e^{-h\omega} = 4\pi f(\omega)$ (assuming $\gamma_0 = \sigma^2 = 1$).

Therefore, $\lim_{n\to\infty} E(I_n^*(\omega)) = f(\omega)$.

Hence $I_n^*(\omega)$ is an asymptotically unbiased estimate of the spectrum $f(\omega)$.

In case of normal error $N(0, \sigma^2)$, we have shown

$$\text{Var}(I_n^*(\omega)) = \frac{\sigma^4}{4\pi^2} \neq 0.$$

Bartlett (1966) even showed that

$$\lim_{n\to\infty} \text{Var}(I_n^*(\omega)) = \frac{\sigma^4}{4\pi^2} \neq 0$$

for stationary process with continuous spectrum. Thus $I_n^*(\omega)$ is not a consistent estimate of the spectrum $f(\omega)$.

9.6.1 Use of the periodogram

It is needed to search for "cycles" or "hidden periodicities" in the data. For example, let us assume that a time series is represented by

$$X_t = \mu + A \cos(\omega t) + B \sin(\omega t) + e_t \tag{9.6.3}$$

here e_t are normal independent $N(0, \sigma^2)$ random variables, A and B are fixed, and ω (known) is of the form, $\omega = \frac{2\pi k}{n}$, where k is an integer.

To test $H_0 : A = B = 0$ against

$$H_1 : A \neq 0 \text{ or } B \neq 0 \tag{9.6.4}$$

We can use

$$F_{2,2(m-1)} = \frac{(\text{Total SS} - \text{SS due to } \mu) - (\text{Total SS} - \text{SS due to } A, B - \text{SS due to } \mu)}{\text{Total SS } - \text{ SS due to } \mu - \text{ SS due to } A, B}$$

$$= \frac{(a_k^2 + b_k^2)/2}{\sum\limits_{j \neq k}^{m} (a_j^2 + b_j^2)/2(m-1)}, \text{ d.f. } 2(m-1) = n-3$$

If ω cannot be expressed as $\frac{2\pi k}{n}$, then the regression associated with (9.6.3) can be computed and the test be constructed by usual regression technique.

Fisher's test for hidden periodicities

To search for hidden periodicities we have to make the test mentioned in (9.6.4). When ω is unknown, we have to search for the largest periodogram ordinate and ask whether this ordinate can reasonably be considered the largest in a random sample of size m selected from a distribution function, i.e. a multiple of a χ_2^2. A statistic used to test the hypothesis is

$$\xi = \left[\frac{1}{m} \sum_{k=1}^{m} I_n(\omega_k) \right]^{-1} I_n(L),$$

where $I_n(L)$ is the largest periodogram ordinate in a sample of m periodogram ordinates each with 2 d.f. Fisher (1929, Proceedings of Royal Society, Series A) demonstrated that for $g > 0$,

$$P\left[\frac{1}{m} \xi_m > g\right] = \sum_{j=i}^{k} (-1)^{j-1} \binom{m}{j} (1 - gj)^{m-1}, \; \xi_m = m \max_{i \le j \le m} \frac{I(\omega_j)}{\sum\limits_{j=i}^{m} I(\omega_j)},$$

where ξ_m is constructed from the periodogram of a sequence of independent $N(\mu, \sigma^2)$ r.v.s and k the largest integer less than g^{-1} (Davis 1941, Analysis of Economic Time Series).

9.6.2 Some Consistent Estimation Procedures

We shall discuss some methods each of which will provide a consistent estimate of the spectrum. Although the periodogram is itself an inconsistent estimate the procedure considered here are essentially based on periodogram (by using some sort of smoothing technique).

We assume that the available time series data is free from trend or seasonal variation since otherwise the results of the spectral analysis are likely to be dominated by these effects, making any other effects different or impossible to see. Trend will produce a peak at zero frequency while seasonal variation produces peaks at the corresponding seasonal frequency and at integer multiples of the seasonal frequency.

The methods are: (i) transforming the truncated autocovariance function (ii) hanning, (iii) hamming and (iv) the fast Fourier transform.

Method (i) consists of taking a Fourier transform of truncated sample autocovariance function using a weighting procedure.

Using trigonometric result

$$\sum_{t=1}^{n} \cos \omega_p t = \sum_{t=1}^{n} \sin \omega_p t = 0$$

where $\omega_p = 2\pi p/n$,

$$I_n(\omega_p) = 2[(\sum_t X_t \cos \omega_p t)^2 + (\sum_t X_t \sin \omega_p t)^2]/n \quad \left(\text{also applies when } p = \frac{n}{p}\right)$$

$$= 2[(\sum_t (X_t - \overline{X})\cos \omega_p t)^2 + (\sum_t (X_t - \overline{X}) \sin \omega_p t)^2]/n$$

$$= 2 \sum_{t=1}^{n} \sum_{s=1}^{n} (X_t - \overline{X})(\cos \omega_p t \cos \omega_p s + \sin \omega_p t \sin \omega_p s)/n$$

$$= \frac{2}{n} \sum_{t=1}^{n} \sum_{s=1}^{n} (X_t - \overline{X}) \cos \omega_p (t - s)$$

$$= 2 \sum_{k=-(n-1)}^{(n-1)} C_k \cos \omega_p k, \text{ where } k = t - s \text{ and } C_k = \sum_{t=1}^{n-k} (X_t - \overline{X})(X_{t+k} - \overline{X})/n$$

$$I_n^*(\omega) = \frac{1}{4\pi} I_n(\omega_p) = \frac{1}{2\pi}\left[C_0 + 2 \sum_{k=1}^{n-1} C_k \cos \omega_p k\right] \tag{9.6.5}$$

which is the obvious estimate of

$$f(\omega) = \frac{1}{2\pi}\left[\gamma(0) + 2 \sum_{k=1}^{\infty} \gamma(k) \cos \omega k\right].$$

Equation (9.6.5) is obtained by simply replacing $\gamma(k)$ by its estimate C_k for $k = 0, 1, 2, \ldots, n - 1$ and $\gamma_k = 0$ for $k > n$. Hence $I_n^*(\omega)$ is the discrete Fourier transform for the complete sample autocovariance function.

The precision of the C_k decreases as k increases so that it would be intuitively reasonable to give less weight to the values C_k as k increases. An estimator is then constructed as

$$\tilde{f}(\omega) = \frac{1}{2\pi}\left\{b_0 C_0 + 2 \sum_{k=1}^{M(n)} b_k C_k \cos(\omega k)\right\}.$$

where $\{b_k\}$ are a set of weights called *the Lag Window* and M(n) is called the *truncation point.* Here we note that the values of C_k for $M < k < n$ are no longer used, while that of C_k for $k \le M$ are weighted by b_k.

In order to use the above estimator we shall have to choose a suitable lag window and a suitable truncation point. Two lag windows generally used in practice are (*a*) *Tukey-Window*

$$b_k = \frac{1}{2}(1 + \cos \pi k/M), k = 0, 1, 2, \ldots, M.$$

This window is also called the Tukey-Hanning or Blackman window.

(*b*) *Parzen-Window*

$$b_k = 1 - 6(k/M)^2 + \left(\frac{k}{M}\right)^3 \text{ if } 0 \le k \le \frac{M}{2}$$

$$= 2\left(1 - \frac{k}{M}\right)^3 \text{ if } \frac{M}{2} < k \le M$$

This has an advantage of not giving negative estimates for f.

9.7 Forecasting

Suppose we have an observed time series $X_1, X_2, \ldots, X_n$. The forecasting problem is to estimate X_{N+1} or more generally X_{N+h} (h is a positive integer). The prediction of X_{N+h} made at time N of the value h steps ahead will be denoted by $X(N, h)$. This integer h is called the *lead time.* There are many forecasting procedures which can be classified into three broad categories.

(*a*) *Subjective:* Forecasts can be subjective using judgement, intuition, commercial knowledge and any other relevant information procedures of this type (e.g. Delphi method) but they will not be discussed here.

(*b*) *Univariate:* These are forecasts entirely based on past observations of a given series, by fitting a model to the data and extrapolating. We shall discuss only two types.

(*c*) *Multivariate:* Such forecasts can be made by taking observations of other variables into account. For example, sales may depend on stocks. Regression models are of this type, as are econometric models. The use of a "leading indicator" also comes into this category. Multivariate models are sometimes called causal or prediction models. We shall not discuss this but only two of several forecasting procedures available for time series data.

9.7.1 Exponential smoothing (Holt-1958, Brown 1963)

Given a stationary, non-seasonal time series $X_1, X_2, \ldots, X_n$, it is natural to take as an estimate of X_{N+1} the weighted sum of past observations

$$\hat{X}(N, 1) = C_0 X_N + C_1 X_{N-1} + \ldots \tag{9.7.1}$$

where $\{C_j\}$ are weights. Geometric weights $C_j = \alpha(1 - \alpha)^j, j = 0, 1, \ldots$ where $0 < \alpha < 1$ are chosen to give more weights to recent observations and less weights to the past observations.

Since in reality we have only finite number of observations, let us modify observations and then (9.7.1) can be written as

$$\hat{X}(N, 1) = \alpha X_N + (1 - \alpha)[\alpha X_{N-1} + \alpha(1 - \alpha)X_{N-2} + \ldots]$$

$$= \alpha X_N + (1 - \alpha)\hat{X}(N - 1, 1) \tag{9.7.2}$$

If we set $\hat{X}(1, 1) = X_1$ then equation (9.7.2) can be used recursively to compute forecasts. Also it reduces the amount of arithmetic involved, since forecasts can easily be updated using only the latest observation and previous forecast. The procedure defined by (9.7.2) is called *exponential smoothing*.

The adjective *exponential* arises from the fact that the geometric weights lie on an exponential curve.

Equation (9.7.2) can be written as

$$\begin{aligned}\hat{X}(N, 1) &= \alpha[X_N - \hat{X}(N-1, 1)] + \hat{X}(N-1, 1)\\ &= \alpha e_N + \hat{X}(N-1, 1), \text{ where } e_N = X_N - \hat{X}(N-1, 1)\\ &= \text{prediction error at time } N.\end{aligned}$$

It can be shown that exponential smoothing is optimal if the underlying model for the time series is given by

$$X_t = \mu + \alpha \sum_{j<t} \varepsilon_j + \varepsilon_t.$$

This infinite MA(∞) is non-stationary, but the first differences $W_t = X_t - X_{t-1}$ form a MA(1) process so that X_t is ARIMA (0, 1, 1) process. The value of the smoothing constant α depends on the properties of the given time series. Values between 0.1 and 0.3 are often used. The value of α may be estimated from past data by a similar procedure as used in estimating the parameter of a MA process. The SS of prediction errors is computed for different values of α and the valuc is chosen which minimizes the S.S. i.e. $\hat{X}(1, 1) = X_1$, $\theta_2 = X_2 - \hat{X}(1, 1)$. $\theta_3 = X_3 - \hat{X}(2, 1)$, and so on. In general $\theta_N = X_N - \hat{X}(N-1, 1)$, $N = 4, 5, \ldots, N$.

Compute $\sum_{i=2}^{N} \theta_i^2$. Repeat for different values of α between 0 and 1 in steps of .01. Usually the SS surface is quite flat near the minimum and the choice of α is not critical.

9.7.2 The Box-Jenkins seasonal model (SARIMA)

In practice, many time series contain a seasonal periodic component which repeats after every s observations. For simplicity we consider monthly observations where $s = 12$. Box and Jenkins (1970) have generalized the ARIMA model to deal with seasonality and define a general multiplicative seasonal model in the form (using their notation).

$$\Phi_p(R)Ip(B^{12})W_t = \theta_q(B)P_Q(B^{12})a_t, \tag{9.7.3}$$

where B denotes the Backshift operator, Φ_p, I_p, θ_q, P_Q are polynomials of order p, P, q, Q respectively, and $\{a_t\}$ rather than $\{\varepsilon_t\}$ is the Box-Jenkins notation for a purely random process, with $Ea_t = 0$, $\text{Var}(a_t) = \sigma^2$ and $W_t = \nabla^d \nabla_{12}^D X_t$ which removes both trend and seasonality, e.g. $d = D = 1$,

$$\begin{aligned}W_t &= \nabla\nabla_{12}X_t = \nabla_{12}X_t - \nabla_{12}X_{t-1}\\ &= (X_t - X_{t-12}) - (X_{t-1} - X_{t-13})\end{aligned}$$

The model (9.7.3) is called Seasonal Integrated Autoregressive Moving Average (SARIMA $(p, d, q) X (P, D, Q)$).

Usually d and D do not exceed 1 and $B^{12}W_t$. $= W_{t-12}$

Suppose $p = Q = 1$ and $P = q = 0$ in (9.7.3). Then (9.7.3) becomes

$$(1 - \alpha B)W_t = (1 + \theta B^{12})a_t$$

so that

$$W_t = \alpha W_{t-1} + a_t + \theta a_{t-12} \tag{9.7.4}$$

If, for example $W_t = \nabla_{12} X_t$ then (9.7.4) becomes

$$X_t = X_{t-12} + \alpha(X_{t-1} - X_{t-13}) + a_t + \theta a_{t-12}.$$

9.7.3 Box-Jenkins forecasting procedure

This procedure is based on fitting an autoregressive integrated moving average (ARIMA) model to a given set of data and then taking conditional expectations. Following are the main stages in setting up a Box-Jenkins forecasting model:

I. Model identification

Examine data to see which member of the class of ARIMA Process appears to be the most appropriate.

II. Estimation

Estimate the parameters of the chosen model by L.S. (Least Square) or Maximum likelihood.

III. Diagnostic checking

Examine the residuals from the fitted model to see if it is adequate.

IV. Consider alternative models if necessary

If the first model appears to be inadequate for some reason, then other ARIMA models may be tried until a satisfactory model is found.

V. Forecasting

The first step in the Box-Jenkins procedure is to difference the data until it is stationary. This is achieved by examining the correlograms of various differenced series until one is found which comes down to zero "fairly quickly" and from which any seasonal effect has been largely removed. For non-seasonal data, first order differencing is usually sufficient. For seasonal data of period 12 month the operator $\nabla\nabla_{12}$ is often used if the seasonal effect is additive, while the operator ∇_{12}^2 may be used if the seasonal effect is multiplicative (e.g. $X_t = L_t \cdot S_t + \varepsilon_t$, where L_t is the linear trend, S_t is the seasonal component). Sometimes the operator ∇_{12} be itself will be sufficient. Over-differencing should be avoided. For quarterly data, the operator ∇_4 mey be used, and so on.

The differenced series will be denoted by $\{W_t, t = 1, \ldots, n-c\}$ where c terms are "lost" by differencing, e.g. if the operator ∇_{12} is used then $c = 13$.

If the data is non-seasonal, an ARIMA model can now be fitted to W_t. If the data is seasonal, then the general seasonal ARMA model may be fitted (done before). "Reasonable" values of p, P, q, Q are selected by examining the correlogram of the differenced series $\{W_t\}$. Values of p and q are selected by examining the first few value of r_k. Values of P, Q are selected by examining the values of r_k at $k = 12, 24 \ldots$ (where seasonal period is 12), e.g. r_{12} is large but r_{24} is small suggests one seasonal moving average term. So take $p = 0$, $Q = 1$.

After finding reasonable seasonal ARMA model, L.S. estimators of the parameters are obtained by minimizing residual S.S. in a similar way to that proposed for the ordinary ARMA model. In case of seasonal model, it is advisable to estimate initial values of a_t and W_t by back-forecasting (or back casting) rather than set them equal to zero (see Box-Jenkins section 9.2.4). If the model contains a seasonal moving average parameter which is close to one, several cycles of forward and backward iteration may be needed. The adequacy of the fitted model should be checked by what Box and Jenkins call "diagnostic-checking". This consists of examining the residuals from the fitted model to see if there is any evidence of non-randomness. The correlogram of the residuals is calculated and we can then see how many coefficients are significantly different from zero and whether any further terms are indicated in the ARIMA model. If the fitted model appears to be inadequate, then alternative ARIMA models may be tried until satisfactory one is found.

When a satisfactory model is found, forecasts may readily be computed. Given data upto time N these forecasts will involve the observations and fitted residuals upto and including time N. They rely on the fact that the best forecast of all future residuals is simple zero (i.e. $a_{N+1}, a_{N+2}, \ldots = 0$), e.g. consider the seasonal ARIMA model example

$$X_t = X_{t-12} + \alpha(X_{t-1} - X_{t-13}) + a_t + \theta a_{t-12}.$$

The forecast of X_{N+1} is obviously given by

$$\hat{X}(N, 1) = X_{N-11} + \hat{\alpha}(X_N - X_{N-12}) + \hat{\theta}\hat{a}_{N-11},$$

where $\hat{\alpha}, \hat{\theta}$ are estimators of α, θ and $\hat{a}_{N-11}$ is the estimated residual at time N-11.

The forecast of X_{N+2} will involve the (yet unknown) value of X_{N+1} and hence insert the forecast value instead, i.e.

$$\hat{X}(N, 2) = X_{N-10} + \hat{\alpha}^2(X_N - X_{N-12}) + \hat{\theta}\hat{\alpha}\hat{a}_{N-11} + \hat{\theta}\hat{a}_{N-10}$$

9.8 Estimation of Parameter in the First Order Autoregressive Model AR(1)

Let α be a real number and $\{\varepsilon_n, n \geq 1\}$ be an i.i.d. sequence of r.v.s defined on a probability space $(\Omega, \mathcal{F}, P)$ Given X_0 an arbitrary initial r.v. independent of $\{\varepsilon_n\}$, define recursively the sequence of r.v.s $\{X_n, n \geq 0\}$ as follows:

$$X_0, X_1 = \alpha X_0 + \varepsilon_1, \ldots, X_{n+1} = \alpha X_n + \varepsilon_{n+1} (n \geq 0). \tag{9.8.1}$$

Then $\{X_n; n \geq 0\}$ is a Markov process on the state space $S = R^1$, having the transition probability of going from x to A (a Borel set of R^1) in one step

$$p(x, A) = P(X_{n+1} \in A \mid X_0, X_1, \ldots, X_n)$$

$$= P(\alpha x + \varepsilon_{n+1} \in A \mid X_n = x)$$
$$= P(\varepsilon_{n+1} \in A - \alpha x \mid X_n = x) = F(A - \alpha X_n)$$

where F is the common d.f. of ε_n and the initial distribution given by the distribution of X_0. Without loss of generality we take $X_0 = 0$ a.s. Now X_n has the stationary representation

$$X_n = \sum_{j=0}^{\infty} \alpha^j \varepsilon_{n-j}, \mid \alpha \mid < 1. \tag{9.8.2}$$

Then the least square estimate of α is given by

$$\hat{\alpha}_n = \left(\sum_{t=1}^{n} X_{t-1} X_t\right) \Big/ \left(\sum_{1}^{n} X_{t-1}^2\right) = \alpha + \left(\sum_{t=1}^{n} \varepsilon_t X_{t-1}\right) \Big/ \left(\sum_{1}^{n} X_{t-1}^2\right)$$

or $\hat{\alpha}_n - \alpha = \left(\sum_{t=1}^{n} \varepsilon_t X_{t-1}\right) \Big/ \left(\sum_{t=1}^{n} X_{t-1}^2\right).$

We shall study the asymptotic optimal properties of $\hat{\alpha}$.

Theorem 9.8 If $\{\varepsilon_n, n \geq 0\}$ is an i.i.d. sequence of r.v.s with $P(\varepsilon_1 = 0) \neq 0$, $E\varepsilon_1 = 0$, $E\varepsilon_1^2 = \sigma^2 > 0$. Then $\hat{\alpha} \to \alpha$ a.s. as $n \to \infty$, i.e. $\hat{\alpha}$ is a strongly consistent estimator of α.

Without loss of generality we shall take $\sigma^2 = 1$.

From equation (9.8.1) we get

$$X_n = \alpha(\alpha X_{n-2} + \varepsilon_{n-1}) + \varepsilon_n = \ldots$$
$$= \alpha^{n-1}\varepsilon_1 + \alpha^{n-2}\varepsilon_2 + \ldots + \alpha\varepsilon_{n-1} + \varepsilon_n \tag{9.8.3}$$

To determine the convergence properties, one needs to consider three cases (1) $\mid \alpha \mid < 1$, (2) $\mid \alpha \mid = 1$ and (3) $\mid \alpha \mid > 1$. The corresponding processes are termed stable, unstable and èxplosive, respectively. Since $EX_n = 0$ and from (9.8.3) $\operatorname{Var}(X_n) = \sum_{k=1}^{n} \alpha^{2(k-1)}$, it follows that $\lim_{n\to\infty} \operatorname{Var}(X_n) < \infty$ iff $\mid \alpha \mid < 1$ since

$$\operatorname{Var}(X_n) = \begin{cases} (1 - \alpha^{2n})/(1 - \alpha^2) \text{ if } \alpha \neq 1 \\ n \text{ if } \alpha = 1 \end{cases}$$

Let

$$g(n, \alpha) = \begin{cases} n^{1/2}(1 - \alpha^2)^{-1/2} \text{ if } \mid \alpha \mid < 1 \\ n/\sqrt{2} \text{ if } \mid \alpha \mid = 1 \\ \mid \alpha \mid^n/(1 - \alpha^2)^{1/2} \text{ if } \mid \alpha \mid > 1 \end{cases} \tag{9.8.4}$$

Note that $g(n, \alpha)$ is called the Fisher information in statistics. To prove the theorem let us state a lemma.

Lemma 9.3 Let $V_n = \sum_{t=1}^{n} X_{t-1}^2$. Then $g(n, \alpha)^{-2} V_n = W_n \xrightarrow{d} V$ as $n \to \infty$ and $P(V = 0) = 0$.

Proof We shall prove the lemma only in the stable case $|\alpha| < 1$. The proofs of the other cases are quite involved and hence are omitted. It is enough to prove that $W_n \xrightarrow{P} \sigma^2 > 0$. By equation (9.8.3) we have

$$\frac{1}{n} V_n = \frac{1}{n} \sum_{t=1}^{n} X_{t-1}^2 = \frac{1}{n} \left[\varepsilon_1^2 + (\varepsilon_2 + \alpha\varepsilon_1)^2 + \ldots + \left(\sum_{k=1}^{n-1} (\alpha^{n-1-k} \varepsilon_k) \right)^2 \right]$$

$$= \frac{1}{n} \left[\varepsilon_1^2 \sum_{k=1}^{n-1} \alpha^{2(k-1)} + \varepsilon_2^2 \sum_{k=1}^{n-1} \alpha^{2(k-1)} + \ldots + \varepsilon_{n-1}^2 \right]$$

$$+ \frac{2}{n} [\alpha(\varepsilon_1\varepsilon_2 + \varepsilon_2\varepsilon_3 + \ldots + \varepsilon_{n-2}\varepsilon_{n-1}) + \ldots + \alpha^{n-2}\varepsilon_1\varepsilon_{n-1}]$$

The second expression has mean zero and variance

$$\frac{\sigma^4}{n^2} [\alpha^2(n-2) + \alpha^4(n-3) + \ldots + \alpha^{2(n-2)}] < \frac{\sigma^4}{n} \sum_{k=0}^{\infty} \alpha^{2k}$$

$$= \frac{\sigma^4}{n(1-\alpha^2)} \to 0 \text{ as } n \to \infty.$$

Hence the second term $\xrightarrow{P} 0$. The first term is equal to

$$\frac{1}{n} \left[\varepsilon_1^2 \frac{1-\alpha^{2(n-1)}}{1-\alpha^2} + \varepsilon_2^2 \frac{1-\alpha^{2(n-2)}}{1-\alpha^2} + \ldots + \frac{1-\alpha^2}{1-\alpha^2} \varepsilon_{n-1}^2 \right]$$

$$= \frac{1}{n(1-\alpha^2)} \sum_{i=1}^{n-1} \varepsilon_i^2 - \frac{1}{n(1-\alpha^2)} [\varepsilon_1^2 \alpha^{2(n-1)} + \ldots + \alpha^2 \varepsilon_{n-1}^2]$$

The second term has mean zero and variance $\to 0$ as $n \to \infty$, whereas the first term by the Kolmogorov's strong law for i.i.d. r.v.s tends to $E(\varepsilon_1^2)/(1-\alpha^2) = \dfrac{\sigma^2}{1-\alpha^2}$ a.s. Therefore, by Slutsky's Theorem as $[(1-\alpha^2)/n]V_n \xrightarrow{P} \sigma^2 > 0$ as $n \to \infty$.

****Proof of Theorem 9.8**

Since $V_n \geq 0$ a.s. for each $x > 0$,

$$P(V_n > x) = P(W_n > rg^{-2}(n, \alpha)) = 1 - F_n(xg^{-2}(n, \alpha)) \quad \text{(say)}$$

where F_n is the d.f. of W_n. By Lemma 9.3 $F_n \to F(x), x \in c(F)$ where F is the d.f. Hence given $\varepsilon > 0$, there exists $\eta > 0$ such that $F(\delta) < \eta$. Choose $n_0(\delta_\eta, x)$ such that $n \geq n_0 \Rightarrow xg^{-2}(n, \alpha) < \delta$.

Since $V_n \uparrow$, $\lim_{n\to\infty} P[V_n > x] = P[\lim V_n > x] \geq 1 - F(\delta) \geq 1 - \eta$.

Since $\eta > 0$ and $x > 0$, $P[\lim_n V_n = \infty] = 1$.

Hence $\lim_{n\to\infty} V_n^{-1} = 0$ a.s. and $V_n \geq k_0 > 0$ a.s. for large n.

Let $\mathscr{B}_n = \sigma(\varepsilon_1, \ldots, \varepsilon_n)$. Consider $Z_n = \sum_{k=1}^{n} V_{k-1}^{-1} X_{k-1} \varepsilon_k$

Since $E(\varepsilon_k) = 0$ and $E(|X_k| V_k^{-1}) \le [E(X_k^2/V_k)E(V_k^{-1})]^{1/2}$ $\le (EV_k^{-1})^{1/2} < \infty$ for $k \ge k_0$ and $\{Z_n, \mathscr{B}_n, n \ge n_0\}$ is a martingale.

$$EZ_n^2 = E\left(\sum_{k=1}^{n} V_{k-1}^{-2} X_{k-1}^2 \varepsilon_k^2)\right) + 2 \sum_{1 \le k < k' \le n} E(V_{k-1}^{-1} V_{k'-1}^{-1} X_{k-1} X_{k'-1} \varepsilon_k \varepsilon_{k'})$$

$$= \sum_{k=1}^{n} E(V_{k-1}^{-2} X_{k-1}^2 E(\varepsilon_k^2 | \mathscr{B}_{k-1}))$$

$$+2 \sum_{1 \le k < k' \le n}^{n} E(V_{k-1}^{-1} V_{k'-1}^{-1} X_{k-1} X_{k'-1} \varepsilon_k E(\varepsilon_{k'} | \mathscr{B}_{k'-1})$$

$$= \sigma^2 \sum_{k=1}^{n} E(X_{k-1}^2 V_{k-1}^{-2}) \text{ (since } E(\varepsilon_k^2 | \mathscr{B}_{k-1}) = E\varepsilon_k^2 = \sigma^2,$$

$$E(\varepsilon_k | \mathscr{B}_{k-1}) = E(\varepsilon_k) = 0))$$

$$= \sigma^2 E\left(\sum_{k=1}^{n} X_{k-1}^2 \; V_{k-1}^{-2}\right). \tag{9.8.5}$$

$$\sum_{k=1}^{n} X_k^2 \; V_k^{-2} = \sum_{k=2}^{n} \frac{V_k - V_{k-1}}{V_k^2} \le \sum_{k=2}^{n} \int_{v_{k-1}}^{v_k} \frac{dt}{t^2} \le \int_0^{\infty} \frac{dt}{t^2} < \infty \tag{9.8.6}$$

where $c = V_1^2 \ge k_0 > 0$ (taking $n_0 = 1$ for convenience).

For large n, $V_n \ge 1$ a.s. and from (9.8.5) and (9.8.6), we get $\sup_n EZ_n^2 < \infty$. Hence $\{Z_n, \mathscr{B}_n\}$ is L^2-bounded martingale and hence uniformly integrable. Thus by martingale convergence theorem $Z_n \to Z$ a.s. and in L_2 (mean-square). So by Kronecker's lemma $\frac{1}{V_n} \sum_{k=1}^{n-1} X_{k-1}\varepsilon_k \to 0$ a.s. as $n \to \infty$. Therefore, $\hat{\alpha}_n \to$ a.s. as $n \to \infty$.

Theorem 9.9 *(CLT for L.S. estimate $\hat{\alpha}_n$)*
Let $\sigma^2 = 1$.

(i) If $|\alpha| < 1$, then $(\hat{\alpha}_n - \alpha)g(n, \alpha) \xrightarrow{d} N(0, 1)$

and $$V_n^{\frac{1}{2}}(\hat{\alpha}_n - \alpha) \xrightarrow{d} N(0, 1).$$

(ii) If moreover ε_n's are $N(0, 1)$ and $|\alpha| > 1$, then

$$(\hat{\alpha}_n - \alpha)g(n, \alpha) \xrightarrow{d} \text{Cauchy}$$

and $$V_n^{\frac{1}{2}}(\hat{\alpha}_n - \alpha) \xrightarrow{d} N(0, 1).$$

Proof Now $\hat{\alpha}_n - \alpha = \sum_{t=1}^{n} X_{t-1}\varepsilon_t \Big/ \left(\sum_1^n X_{t-1}^2\right)$. If we assume $|\alpha| < 1$

then $$g(n, \alpha)(\hat{\alpha}_n - \alpha) = \sqrt{n} \sum_{t=1}^{n} \varepsilon_t X_{t-1} \Big/ \left(\sum_1^n X_{t-1}^2 (1 - \alpha^2)^{\frac{1}{2}}\right)$$

$$= (A_n/\sqrt{n}) \left(\frac{V_n}{n}\right)^{-1} (1 - \alpha^2)^{1/2} \tag{9.8.6}$$

where $A_n = \sum_{t=1}^{n} \varepsilon_t X_{t-1} = \sum_{t=2}^{n} \varepsilon_t \varepsilon_{t-1} + \alpha \sum_{t=3}^{n} \varepsilon_t \varepsilon_{t-2} + \ldots + \alpha^{n-2} \varepsilon_n \varepsilon_1$

Now by Lemma 9.3, $\sum_{t=1}^{n} X_{t-1}^2 / n \xrightarrow{P} \gamma(0) = \sigma^2/(1 - \alpha^2)$ and hence $g(n, \alpha)$ $(\hat{\alpha}_n - \alpha)$ has same asymptotic distribution as

$$n^{-1/2} A_n / (\sigma^2 (1 - \alpha^2)^{-1/2}).$$

Now $\{\varepsilon_n X_{n-1}, n \geq 1\}$ is a martingale difference sequence, since $E\varepsilon_n = 0$ and ε_n and X_{n-1} are independent.

Also $\text{Var}(X_t) = \dfrac{1 - \alpha^{2t}}{1 - \alpha^2} \sigma^2$ and $\text{Var}(A_n / \sqrt{n}) \sim \dfrac{\sigma^4}{1 - \alpha^2}$ for large n. Let us state the following Martingale CLT (due to Billingsley 1968).

Let $S_n = \sum_{t-1}^{n} Y_t$ be a martingale and Y_t's be stationary with finite (common) variance.

Then $S_n / (\text{Var}(S_n))^{1/2} = S_n / \sqrt{n} (\text{Var}(Y_1))^{1/2} \xrightarrow{d} N(0, 1)$ as $n \to \infty$. Take $Y_t = X_{t-1} \varepsilon_t$. Then $\text{Var}(X_{t-1} \varepsilon_t) \sim \dfrac{\sigma^4}{1 - \alpha^2}$. By Slutsky's theorem $(\hat{\alpha}_n - \alpha)\, g(n, \alpha) \xrightarrow{d} N(0, 1)$.

Again $V_n^{1/2} (\hat{\alpha}_n - \alpha) = \sqrt{n} (A_n)(V_n / n)^{-1/2}$ has the same asymptotic distribution as $\sqrt{n} A_n / \sigma (1 - \alpha^2)^{-1/2}$. This shows that

$$V_n^{1/2} (\hat{\alpha}_n - \alpha) \xrightarrow{d} N(0, 1).$$

Now ε_t's are independent $N(0, 1)$ and $\varepsilon_k = X_k - \alpha X_{k-1}$. The transformation from the ε_t to the X_t is one-to-one with jacobian unity, and hence the joint density of $X_1, \ldots, X_n$ is

$$f(x_1, \ldots, x_n) = (2\pi)^{-n/2} \exp\left\{ -\frac{1}{2} \sum_{k=1}^{n} (x_k - \alpha x_{k-1})^2 \right\} \tag{9.8.7}$$

$$= (2\pi)^{-n/2} \exp\left\{ -\frac{1}{2} (x' A x) \right\},$$

where A^{nxn} is a symmetric positive definite matrix given by $A = (a_{ij})$ with

$$a_{ii} = 1 + \alpha^2, \; i = 1, 2, \ldots, n - 1$$

$$a_{nn} = 1$$

$$a_{ii+1} = -\alpha, \; i = 1, 2, \ldots, n - 1$$

$a_{ij} = 0$ otherwise and $x' = (x_1, \ldots, x_n)$.

Also $\hat{\alpha}_n - \alpha$ can be written as

$$\hat{\alpha}_n - \alpha = \sum_{t=1}^{n} X_{t-1} \varepsilon_t \Big/ \sum_{t=1}^{n} X_{t-1}^2 = \sum_{t=1}^{n} (X_t X_{t-1} - \alpha X_{t-1}^2) \Big/ \sum_{t=1}^{n} X_{t-1}^2$$

$$= (X'BX)/(X'CX) \tag{9.8.8}$$

where $X' = (X_1, X_2, \ldots, X_n)$ and $B^{n\times n}$ is a symmetric matrix with 2α on the diagonal except for the nth element, which is zero, -1 for the first line parallel to the diagonal and zeros elsewhere, and

$$C^{n\times n} = \begin{pmatrix} 1 & 0 & 0 & \ldots & & & & 0 \\ 0 & 1 & 0 & & & & \ldots\ldots & 0 \\ 0 & 0 & 1 & 0 & & & \ldots & 0 \\ . & & . & & . & & .. & . \\ 0 & 0 & 0 & 0 & 0 & .. & 1 & 0 \\ 0 & 0 & 0 & 0 & 0 & & 0 & 0 \end{pmatrix}$$

The joint m.g.f. of the two r.v.s in (9.8.8) is given by

$$m(u, v) = E(\exp\{u(X'BX) + v(X'CX)\})$$

$$= (2\pi)^{-n/2} \iint \exp\left\{-\left(\frac{1}{2}\right)(x'Ax) + u(x'Bx) + v(x'Cx)\right\} dx_1 \ldots dx_n$$

$$= (2\pi)^{-n/2} \int \int \exp\left\{-\left(\frac{1}{2}\right)(x'Hx)\right\} dx_1 \ldots dx_n \qquad (9.8.9)$$

where $H = A - 2uB - 2vC$, which is positive definite if u, v are sufficiently small so that the integral in (9.8.9) exists. Since H is symmetric, there exists an orthogonal matrix P such that $H = P'DP$, where D is diagonal with eigen values $\lambda_j \geq 0$ of H. Setting $y = P'x$ and noting that the absolute value of the jacobian is equal to 1, we obtain from (9.8.9)

$$m(u, v) = (2\pi)^{m/2} \int \ldots \int \exp\left\{-\left(\frac{1}{2}\right)(y'Dy)\right\} dy_1 \ldots dy_n$$

$$= \left(\overset{n}{\underset{j=1}{\pi}} \lambda_j\right)^{-m/2} = |H|^{-1/2} \qquad (9.8.10)$$

Setting $p = 1 + \alpha^2 - 2v + 2\alpha u$, $q = -(\alpha + u)$, then writting H_n for the H which depends on n, we obtain by expansion (of $H_n = A_n - 2uB_n - 2vC_n$) the difference equation

$$H_n = pH_{n-1} - q^2 H_{n-2},\ n > 2 \qquad (9.8.11)$$

with initial values $H_1 = 1$ and $H_2 = p - q^2$. Let z_1 and z_2 be the roots the characteristic equation of the difference equation (9.8.11) $z^2 - pz + 1 = 0$ so that

$$z_1, z_2 = 1/2\, p \pm 1/2\, (p^2 - 4q^2)^{1/2} \qquad (9.8.12)$$

Hence
$$H_n = \frac{1 - z_2}{z_1 - z_2} z_1^n + \frac{1 - z_1}{z_2 - z_1} z_2^n \qquad (9.8.13)$$

Substituting (9.8.13) in (9.8.10) we get the m.g.f. of the r.v.s for each n.

Now
$$g(n, \alpha)(\hat{\alpha}_n - \alpha) = \left(\frac{X'BX}{g(n, \alpha)}\right) \Big/ \left(\frac{X'CX}{g^2(n, \alpha)}\right)$$

and we can obtain the joint m.g.f. from $m(u, v)$ by setting

$$M_n(\theta_1, \theta_2) = m\left(\frac{\theta_1}{g(n, \alpha)}, \frac{\theta_2}{g^2(n, \alpha)}\right) = |\tilde{H}_n|^{-1/2}, \tag{9.8.14}$$

where $\tilde{H}_n$ is H with u, v, replaced by $\theta_1/g(n, \alpha)$ and $\theta_2/g^2(n, \alpha)$, respectively. Recalling the values of $g(n, \alpha)$ and expanding the radical in z_1, z_2 of (9.8.12), we get the following expressions.

Case 1. $|\alpha| < 1$

$$z_1 = 1 - \frac{\theta_1^2 + 2\theta_2}{n} + 0\,(n^{-1/2})$$

$$z_2 = \alpha^2 + 2\alpha\,\theta_1[(1 - \alpha^2)/n]^{1/2} + \frac{\theta_1^2 + 2\alpha^2\theta_2}{n} + 0(n^{-3/2})$$

Case 2. $|\alpha| > 1$

$$z_1 = 1 + \frac{(\theta_1^2 + 2\theta_2)(\alpha^2 - 1)}{\alpha^{2n}} + 0\,(|\alpha|^{-3n})$$

$$z_2 = \alpha^2 + \frac{2\alpha\,\theta_1\,(\alpha^2 - 1)}{|\alpha|^n} - \frac{(\theta_1^2 + 2\alpha^2\theta_2)(\alpha^2 - 1)}{\alpha^{2n}} + 0(|\alpha|^{-3n})$$

Substituting these values in (9.8.14) and simplifying, we get

$$\lim_{n\to\infty} M_n(\theta_1, \theta_2) = \begin{cases} \exp\,(\theta_2 + \theta_1^2/2) & \text{if } |\alpha| < 1 \\ (1 - 2\theta_2 - \theta_1^2)^{-1/2} & \text{if } |\alpha| > 1 \end{cases}$$

Replacing θ_1, θ_2 by it_1 and it_2, we obtain the limiting joint ch.f.

$$\varnothing(t_1, t_2) = \lim_{n\to\infty} M_n(it_1, it_2) \tag{9.8.15}$$

$$= \begin{cases} \exp\,(it_2 - t_1^2/2 & \text{if } |\alpha| < 1 \\ (1 - 2\,it_2 + t_1^2)^{-1/2} & \text{if } |\alpha| > 1 \end{cases} \tag{9.8.16}$$

To get the limit d.f. of $g(n, \alpha)\,(\hat{\alpha}_n - \alpha)$, we see that

$$F(x) = \lim_{n\to\infty} P\left(\frac{X'B_nX}{g(n, \alpha)} \le x\,\frac{X'C_nX}{g^2(n, \alpha)}\right)$$

$$= \lim_{n\to\infty} P\left(\frac{X'B_nX}{g(n, \alpha)} - x\,\frac{X'C_nX}{g^2(n, \alpha)} \le 0\right)$$

$$= P(U - xV \le 0)\; P\left(\frac{U}{V} \le x\right)$$

where
$$\left(\frac{X'B_nX}{g(n, \alpha)}, \frac{X'C_nX}{g^2(n, \alpha)}\right) \xrightarrow{P} (U, V)$$

(from Lemma 9.3 and Theorem 9.8).

The ch.f. of (U,V) is given by (9.8.16) and hence by Cramer (1937) Theorem for ratio of two r.v.s and the joint ch.f., we get the density of U/V as follows.

Case 1. $|\alpha| < 1$

$$F'(x) = f(x) = \frac{1}{2\pi 1}\int_{-\infty}^{\infty} \frac{\partial \phi}{\partial u}(v, -xv)\,dv$$

$$= \frac{1}{2\pi i}\int_{-\infty}^{\infty} i\exp\{-(ivx + v^2/2)\}\,dv \quad \text{(by (9.8.15) and (9.8.16))}$$

$$= \frac{1}{\sqrt{2\pi}} e^{-x^2/2}$$

This was proved without the assumption of normality of ε_t.

Case 2. $|\alpha| > 1$

By Lemma 9.3 $P(V > 0) = 1$. Now $\phi(t_1, t_2)$ being the limit of ch.f.s is a ch.f. Again by Cramer's Theorem the density of U/V is given by

$$f(x) = \frac{1}{2\pi i}\int_{-\infty}^{\infty} \frac{\partial \phi}{\partial u}(t, -tx)\,dt$$

$$= \frac{1}{2\pi i}\int_{-\infty}^{\infty} \frac{i dt}{(1 + t^2 + 2itx)^{3/2}}$$

$$= \frac{1}{2\pi}\int_{-\infty}^{\infty} d\left(\frac{t + ix}{(1 + x^2)\sqrt{1 + t^2 + 2itx}}\right)$$

$$= \frac{1}{2\pi}\lim_{T\to\infty}\left[\frac{1 + (ix/T)}{(1 + x^2)(1 + T^2 + 2iTx)/T^2)^{1/2}}\right.$$

$$\left. - \frac{1 - (ix/T)}{(1 + x^2)(-1)([1 + T^2 - 2iT]/T^2)^{1/2}}\right]$$

$$= \frac{1}{2\pi}\frac{2}{1 + x^2} = \frac{1}{\pi}\frac{1}{1 + x^2}$$

which is the cauchy p.d.f.

Case 3. When ε_n's are normal and $|\alpha| > 1$ then $V_n^{1/2}(\hat{\alpha}_n - \alpha) \xrightarrow{d} N(0, 1)$. The proof is rather involved and hence is omitted.

Exercises and Complements

Exercise 9.1 Prove that for the ARMA (1, 1) process $X_t = \alpha X_{t-1} + \varepsilon_t + \beta\varepsilon_{t-1}$, ε_t is a $WN(0, \sigma^2)$, the autocovariances of X_t are given by

$$\gamma(0) = \frac{1 + 2\alpha\beta + \beta^2}{1 - \alpha^2}\sigma^2, \quad \gamma(1) = \frac{(1 + \alpha\beta)(\alpha + \beta)}{1 - \alpha^2}\sigma^2$$

and the autocorrelation $\rho(k) = \alpha\rho(k-1)$, $k \geq 2$. Hence show that autocorrelations of X_t obey $\rho(k) = \alpha^{k-1}\rho(1)$, $k \geq 2$ and $\rho(1) = \dfrac{(1+\alpha\beta)(\alpha+\beta)}{1+2\alpha\beta+\beta^2}$.

Exercise 9.2 If X_t and Y_t are two independent Moving average processes $MA(q_1)$ and $MA(q_2)$ respectively. Then show that $Z_t = X_t + Y_t$ is a $MA(q)$, where $q \leq \max(q_1, q_2)$.

Exercise 9.3 If X_t and Y_t are two independent ARMA (p_1, q_1) and ARMA (p_2, q_2) processes respectively. Then show that $Z_t = X_t + Y_t$ is also a ARMA (p, q) process, where $p \leq p_1 + p_2$ and $q \leq \max(p_1 + q_2, p_2 + q_1)$.

Exercise 9.4 (a) If X_t is an ARMA (1, 1) process given by $X_t = \phi X_{t-1} + \theta\varepsilon_{t-1} + \varepsilon_t$, $|\phi| < 1$ and ε_t is a WN $(0, \sigma^2)$ then express X_t as an infinite linear combination of $\{\varepsilon_t, \varepsilon_{t-1}, \varepsilon_{t-2}, \ldots\}$.

(b) If $|\theta| < 1$ then express ε_t as an infinite linear combination of $\{X_t, X_{t-1}, \ldots X_{t-k}, \ldots\}$.

Exercise 9.5 (a) If $\{X_t\}$ is zero mean second order stationary nondeterministic time series, then using Wold's representation theorem show that k-step optimal mean square predictor is given by $X_{t+k|t} = \sum_{i=0}^{\infty} \psi_{i+k}\varepsilon_{t-i}$, where $\sum_{0}^{\infty} \psi_t^2 < \infty$ and ε_t's are i.i.d. random variables. Also show that the minimum value of the mean square error is given by $\sigma^2 \sum_{i=0}^{k-1} \psi_i^2$. (b) In case of AR(1) process $X_t = \alpha X_{t-1} + \varepsilon_t$, $|\alpha| < 1$, show that the optimal mean square predictor is given by $X_{t+k|t} = \alpha^k X_t$. In case of MA(1) process $X_t = \varepsilon_t + \beta\varepsilon_{t-1}$, $|\beta| < 1$, show that the optimum one-step ahead predictor is given by

$$X_{t+k|t} = \beta X_t - \beta^2 X_{t-1} + \beta^3 X_{t\ 2} + \beta^4 X_{t-3} + \ldots$$

Exercise 9.6 Show that the real valued function $K(h)$ defined on the set of integers Z,

$$K(h) = \begin{cases} 1 \text{ if } h = 0 \\ \rho \text{ if } h = \pm 1 \\ 0 \text{ otherwise} \end{cases}$$

is an autocoveriance function iff $|\rho| \leq 1/2$.

Exercise 9.7 Let $\{S_t, t - 0, 1, 2, \ldots\}$ be the random walk with constant drift μ, defined by $S_0 = 0$ and

$$S_t = \mu + S_{t-1} + X_t,\ t = 1, 2, \ldots$$

where $X_1, X_2, \ldots$ are i.i.d. r.v.s with mean 0 and variance σ^2. Compute the mean of S_t and the autocovariance function of the process $\{S_t\}$. Show that $\{\Delta S_t = (1 - B)S_t\}$ is stationary and compute its mean and autocovariance function.

Exercise 9.8 Show that in order for an AR(2) process $X_t = \phi_{1X_{t-1}} + \phi_2 X_{t-2} + \varepsilon_t$, $\varepsilon_t \sim WN(0, \sigma^2)$ to have an infinite moving average representation, the parameters (ϕ_1, ϕ_2) may not lie in the triangular region determined by the intersection of the three regions $\phi_2 + \phi_1 < 1$, $\phi_2 - \phi_1 < 1$ and $|\phi_2| < 1$. Show that $\phi_1 = \dfrac{\rho(1)(1-\rho(2))}{1-\rho^2(1)}$ $\phi_2 = \dfrac{\rho(2)\rho^2(1)}{1-\rho^2(1)}$. Also prove that $\rho^2(1) < 1/2(\rho(2) + 1)$.

Exercise 9.9 Let $X_t = \theta X_{t-1} + \varepsilon_t$, $t = \ldots, -1, 0, 1, \ldots$ be an AR(1) process where ε_t are i.i.d. $N(0, \sigma^2)$ and $|\theta| < 1$. Show that autocovariance function is given by $\gamma(k) = \dfrac{\sigma^2|\theta|^k}{1-\theta^2}$, the spectral density is given by $f(\lambda) = \dfrac{\sigma^2}{2\pi} \mid 1 - \theta e^{-i\lambda} \mid^2$. Also show that the log-likelihood (n-dimensional p.d.f.)

$$L_n = \log p(x_1 <\ldots, x_n) \text{ of } X_1, \ldots, X_n \text{ is } - 1/2 \{n \log (2\pi\sigma^2) - \log (1 - \theta^2) + \sigma^{-2} [(1 - \theta^2) X_1^2 + \sum_{j=2}^{n} (X_j - \theta X_{j-1})^2]\}.$$

Exercise 9.10 Let $\{X_t\}$ be an ARMA (p, q) process given by $\sum_{r=0}^{p} \alpha_r X_{t-r} = \sum_{s=0}^{q} \beta_s \varepsilon_{t-s}$, $\alpha_0 \neq 0$, $\beta_0 \neq 0$ and $\{\varepsilon_t\}$ is a $WN(0, \sigma^2)$. Assume that the roots of the AR and MA polynomials are within the unit disc. Show that the variance of the prediction error is $E \mid X_t - \hat{X}_t \mid^2 = |^{\beta}{}_0/\alpha_0|^2 \sigma^2$ and $\hat{X}_t = E(X_t \mid X_s, s \leq t - 1)$. Also show that $\sigma^2 = \left|\dfrac{\alpha_0}{\beta_0}\right|^2 2\pi \exp\left\{1/(2\pi) \int_{-\pi}^{\pi} \log f(\lambda)\, d\lambda\right\}$ where $f(\lambda)$ is the spectral density of $\{X_t\}$.

****Exercise 9.11** Let $\{X_t\}$ be a MA(1) process $X_t = \theta_1\varepsilon_t - \varepsilon_{t-1}$, $\mid \theta_1 \mid \leq 1$ and ε_t is as in exercise 9.9. Show that the log-likelihood ratio $L_n = - 1/2$

$$\left\{n \log (2\pi\sigma^2) + \log \frac{1 - \mid \theta_1 \mid^{2(n+1)}}{1 - \mid \theta_1 \mid} + \sigma^{-2} \left[\sum_{j=1}^{n} \mid Y_j^2 \mid - \frac{1 - \mid \theta_1 \mid^2}{1 - \mid \theta_1 \mid^{2(n+1)}} \left| \sum_{j=1}^{n} \theta_1 X_j \right|^2 \right] \right\}$$

if $\mid \theta_1 \mid < 1$ and $Y_k = \sum_{j=1}^{k} \theta_1^{k-j} X_j$, $k = 1, 2, \ldots, n$ and $L_n = - 1/2$

$$\left\{n \log (2\pi\sigma^2) + \log (n + 1) + \sigma^{-2} \left[\sum_{j=1}^{n} Y_j^2 - \frac{1}{n + 1} \left(\sum_{j=1}^{n} Y_j \right)^2 \right] \right\}, \text{ if } \mid \theta_1 \mid = 1$$

where $Y_j = X_1 + X_2 + \ldots + X_j$, $j = 1, 2, \ldots, n$ and the spectral density is given by

$$f(\lambda) = \frac{\sigma^2}{2\pi} \mid e^{-i\lambda} - \theta_1 \mid^2.$$

**** Exercise 9.12** Let X_t be a MA(2) process given by $X_t = \varepsilon_t - 2\varepsilon_{t-1} + \varepsilon_{t-2}$ where ε_t's are as in Exercise 9.11. Show that the spectral density

$$f(\lambda) = \frac{\sigma^2}{2\pi} \mid 1 - e^{-i\lambda} \mid^4$$

and the log-likelihood

$$L_n = - 1/2 \left\{ n \log (2\pi\sigma^2) + \log d_n + \sigma^{-2} \sum_{j=1}^{n} Y_j^2 / d_n \left(\left[\sum_{j=1}^{n} Y_j \right]^2 \left[1 + \frac{n(n + 1)(2n + 1)}{6} \right] + (2 + n) \left(\sum_{j=1}^{n} jY_j \right)^2 + (2 - n - n^2) \left(\sum_{j=1}^{n} Y_j \sum_{j=1}^{n} j Y_j \right) \right) \right\}$$

where $Y_j = \sum_{1}^{n} k X_k$, $j = 1, 2, \ldots, n$ and $d_n = \dfrac{1}{12} (n + 1)(n + 1)(n^2 + 5n + 6)$.

**** Exercise 9.13** Let $\{X_t\}$ be an ARMA (1, 1) process given by

$$X_t = \theta_1 X_{t-1} + \theta_2 \varepsilon_t - \varepsilon_{t-1},$$

where $|\theta_1| \le 1, |\theta_2| \le 1$. Show that the spectral density is given by

$$f(\lambda) = (\sigma^2/2\pi)\,|\theta_1 - e^{i\lambda}|^2\,|\theta_2 - e^{i\lambda}|^2$$

and the log-likelihood

$$L_n = -1/2\left\{n\log(2\pi\sigma^2) + \log\left[1 + \frac{|\theta_1 - \theta_2|^2}{1-\theta_1^2}N_n(\theta_1)\right]\right.$$

$$+\sigma^{-2}\left[\sum_{j=1}^{n}(Y_j - \theta_1 Y_{j-1})^2 - |\theta_1 - \theta_2|^2 \sum_{j=1}^{n}\theta_2^{-(j-1)}(Y_j - \theta_1 Y_{j-1})^2\right.$$

$$\left.\left.(1-\theta_1^2) + |\theta_1 - \theta_2|^2 (N_n(\theta_2))^{-1}\right]\right\}, \text{ where } |\theta_2| < 1 \text{ and}$$

$$Y_j = \sum_{j=1}^{n} X_k \theta_2^{j-k}, j = 1, 2, \ldots n, Y_0 = 0$$

$$L_n = -1/2\left\{n\log(2\pi\sigma^2) + \log\left(1 + n\frac{1-\theta_1}{1+\theta_1}\right) + \sigma^{-2}\right.$$

$$\left.\left[\sum_{j=1}^{n} Y_{j,\theta_1} - \left(n + \frac{1+\theta_1}{1-\theta_1}\right)^{-1}\left(\sum_{j=1}^{n} Y_{j,\theta_1}\right)^2\right]\right\}, \text{ if } \theta_2 = 1$$

where $\quad Y_{1,\theta_1} = X_1,\ Y_{j,\theta_1} = X_j + (1-\theta_1)\sum_{k=1}^{j-1} X_k, j = 2, \ldots, n$

Exercise 9.14 (*Explosive AR(1) process*) Consider the Stochastic process generated by $X_t = \alpha X_{t-1} + \varepsilon_t$ where $|\alpha| > 1$ and ε_t is $WN(0, \sigma^2)$. Show that this may be written as $\alpha B(\alpha^{-1}B^{-1} - 1)X_t = \varepsilon_t$ and hence that $X_t = \alpha^{-1}B^{-1}(\alpha^{-1}B^{-1} - 1)^{-1}\varepsilon_t$ where B is the backshift operator and B^{-1} is the forward shift operator. Deduce that the original process can be modelled in terms of future ε_t's rather than past ones.

Exercise 9.15 Suppose $\varepsilon_1, \varepsilon_2, \ldots$ are independent r.vs. all having the same mean μ and variance σ^2. Define $X_1 = \varepsilon_1$,

$$X_n = \alpha X_{n-1} + \varepsilon_n \text{ for } n \ge 2,$$

where $-1 \le \alpha < 1$. Show that $n^{-1}\sum_{i=1}^{n} X_i \xrightarrow{L_2} \mu/(1-\alpha)$.

Exercise 9.16 Let $X_0 = 0$.

Define X_n as in exercise 9.15 and $\bar{X}_n = n^{-1}\sum_{i=1}^{n} X_i$. Prove that

$$\sqrt{n}(\bar{X}_n - \mu/(1-\alpha)) \xrightarrow{d} N(0, \sigma^2/(1-\alpha)^2), \text{ if } -1 < \alpha < 1,$$

$$\sqrt{n}(\bar{X}_n - \mu/2) \xrightarrow{d} N(0, \sigma^2/2), \text{ if } \alpha = -1.$$

What happens when $\alpha = 1$?

****Exercise 9.17** Let $X_1, X_2, \ldots,$ be i.i.d. r.v.s with mean and variance σ^2.

(a) Find the joint asymptotic distribution of $\bar{X}_n = n^{-1}\sum_{j=1}^{n} X_j$ and sample autocovariance

$\hat{\gamma}(1) = n^{-1} \sum_{i=1}^{n} (X_i - \overline{X}_n)(X_{i+1} - \overline{X}_n)$ [Hint : Find the asymptotic distribution of $a\overline{X}_n + b\overline{Z}_n$ for all real a and b, where $Z_n = n^{-1} \sum_{i=1}^{n} X_i X_{i+1}$. Then apply Cramer-Wold's device and express $\hat{\gamma}(1) = \overline{Z}_n - \overline{X}_n^2$.

(b) Now assume also $\mu = 0$ and $EX_1^4 < \infty$. Define the sample autocorrelation of lag 1 based on first $(n + 1)$ observations as

$$\hat{\rho}(1) = \left[n^{-1} \sum_{i=1}^{n} (X_i - X_{i+1}) - \overline{X}_n^2\right] \Big/ \left(n^{-1} \sum_{1}^{n} X_i^2 - \overline{X}_n^2\right)$$

(Note that the limiting distribution of $\hat{\rho}(1)$ does not depend on the assumption $\mu = 0$)

(i) Let $Z_i = (X_i, X_i^2, X_i X_{i+1})'$. Show that $\sqrt{n}(\overline{Z}_n - \theta) \xrightarrow{d} N_3(\theta, \Sigma)$,

where
$$\theta = \begin{pmatrix} 0 \\ \sigma^2 \\ 0 \end{pmatrix} \text{ and } Z = \begin{pmatrix} \sigma^2 & \mu_3 & 0 \\ \mu_3 & \mu_4 - \sigma^4 & 0 \\ 0 & 0 & \sigma^4 \end{pmatrix}$$

(ii) Show that $\sqrt{n}\,\hat{\rho}(1) \xrightarrow{d} N(0, 1)$.

Exercise 9.18 Let $\{X_n\}$ be a stationary time series with zero Means and autocovarience $\gamma(m)$. Show that

$$\text{Var}(\overline{X}n) = \gamma(0) \int_{-\pi}^{\pi} \left(\frac{\sin(n\lambda/2)}{n \sin(\lambda/2)}\right)^2 dF(\lambda)$$

where $F(\lambda)$ is the spectral d.f. Deduce that $\overline{X}n$ is mean square consistent (m.s) iff $F(0) - F(0-) = 0$ and show that $\gamma(0)\{F(0) - F(0-)\} = \lim_{n \to \infty} \frac{1}{n} \sum_{k=0}^{n-1} \gamma(k)$.

10

Queueing Theory

10.1 Introduction

Congestion is a natural phenomenon in real systems. A service facility gets congested if there are more people than the server can possibly handle. A footpath (side-walks) gets congested when there are too many pedestrians using the footpath at the same time. Queueing for milk, kerosin oil etc are examples of queueing system. In all these situations the uncertainties related to the system characteristics, such as *arrival of customers*, *time needed for service*, the processes occurring in the analysis of these systems are better represented by Stochastic Processes. "A system consisting of a servicing facility, a process of arrival of customers who wish to be served by the facility and the process of service, is called a queueing system". The four common characteristics of such systems are:

1. *Input process* Some factors needed for the complete specification of an input process are the source of arrivals, the type of arrivals and the interarrival times. The best can be done in this situation is to represent the input process in terms of r.v.s $Z_1, Z_2, \ldots, Z_n, \ldots$ Let the successive customers arrive at time t_0, $t_1, t_2, \ldots$ The interarrival times are then $Z_i = t_i - t_{i-1}$ with d.f.F and $\{Z_i\}$ is a renewal process.

2. *Service mechanism* The uncertainties involved in the Service Mechanism are the number of services, the number of customers getting served at any time, and the duration of service. Random variable representations of these characteristics seem to be essential.

3. *Queue discipline* All other factors regarding the rules of conduct of the queue can be pooled under this heading. In this context, rules as *first come first served* (*FCFS*) or *first come first out* (*FIFO*), *Last come first served* and *Random selection for service* are important. In many situation "Priority" disciplines need to be introduced so as to make the system more realistic.

3. *Number of queues* When there is only one server in the system, specifications of the above three factors give a complete description. However, in problems of net-work flows and job-shop scheduling, one has to deal with more than one server in series and/or in parallel.

This book will mainly deal with (i) negative exponential service times distribution, (ii) basic queue discipline will always be assumed to be first come first served and (iii) one server.

It is convenient to simplify the description of a queueing system by a notational representation (designed by Kendall in 1951). This can be written as :

Input distribution/service time distribution/number of servers in that order.

Some standard notations used are G (or GI if input) for an arbitrary distribution. M for Poisson (if arrivals) or negative exponential distributions (also called Markovian) (for interarrival or service times); D for constant length of time (for interarrival or service times), and E_k for the Erlang distribution (Gamma with one of the parameter, k an integer). For instance, suppose the arrivals are Poisson, service times are negative exponential, and there are s-servers, then the system is denoted by $M/M/s$. In this notation, $M/M/\infty$, $M/D/1$, are $M/G/1$, $M/G/1/R$ etc. same as $M/G/1$ with fourth descriptor R denoting that the system has a limited holding capacity R.

Queueing Process

The following r.v.s which arise in the study of Queueing theory provide important measures of performances and effectiveness of a stochastic Queueing system.

I. The number of customers $Q(t)$ at time t waiting in the queue including the one being served, if any, is also called Queue length.

II. Busy period $T(T_i)$ (initiated by i customers) which means the duration of the interval from the moment the service commences with arrival of an unit at an empty counter to the moment the server becomes free for the first time.

III. Waiting time $W(t)$ in the Queue i.e. the duration of time a unit has to spend in the Queue. Also the waiting time W_n of the nth arrival.

IV. The vartual waiting time $W'(t)$ i.e. the interval of time a unit would have to wait in the Queue, were it to arrive at the instant t. $Q(t)$, the number in the system at time t and its probability distribution, given by

$$\pi_n(t) = P(Q_x(t) = n/Q(0) = i)$$

are both time dependent. For a complete description of the queueing process we need to consider transient or time-dependent solutions. It is often difficult to obtain such solutions. In many practical situations one need to know the behaviour in steady state i.e. when the system reaches an equilibrium state after in operation for a pretty long time. It is easier to find $\pi_n = \lim_{t \to \infty} P_n(t)$ provided the limit exist. It is necessary to know the condition for the existence of the limit in the first place when the limit exists, it is said that the system has reached *equilibrium* or *steady state* and the problem then boils down to finding the steady state solutions.

There are certain useful statements and relationship in Queueing theory which hold under fairly general conditions. Though rigorous mathematical proofs of such relations are somewhat complicated, intutive and heuristic proofs are simple enough and have been known for a long time (see Morse, 1947). It has been argued in (Krakowski- 1973) also by Stidham (1974) that conservation methods could very well applied to supply proofs of some of these relations. Conservation principles have played a fundamental role in physical and engineering sciences as well as in economics etc. Similar principles may perhaps be applied in obtaining relations for queueing systems in steady state. Some such relations are given here. The most important one is $L = \lambda W$, where λ is the arrival rate, L the expected number of units in the system and W the expected waiting time in the system in steady state. A rigorous proof of the relation has been given by Little (1961) and so the relations is known as *Little's formula*. This result, of great generality, is independent of the form of interarrival and service time distributions, and holds under some very general conditions.

Denote the expected number in the queue and the expected waiting time in the queue in steady state by L_Q and W_Q, respectively.

These are related by the formula

$$L_Q = \lambda W_Q$$

A relation which holds for $G/G/1$ queue in steady state is $\lambda = (1 - \pi_0)\mu$ or $\pi_0 = 1 - p$, where $\rho = \lambda/\mu$, the traffic intensity.

Under some restrictive conditions for a system with Poisson input the *Pollaczek-Khinchine formula* holds. We shall prove this formula letter, while some of the other relations will be proved in some special cases.

10.1.1 A generalized Queueing model

Under certain assumptions on the arrival and service processes, a large class of queueing system can be modelled as a generalized birth and death process. Arrivals can be considered as 'birth' and service completion as 'death'.

1. *Arrivals* Let the number of customers in the system be n at time t, during the following infinitesimal interval $(t, t+\Delta t)$ no customer arrives be $1-\lambda_n\Delta t + o(\Delta t)$, and more than one customer arrives be $o(\Delta t)$. Also assume that the customer arrivals during $(t, t + \Delta t]$ are independent of the time since the last arrival.

2. *Service* Probability that there is one service completion in the interval $(t, t + \Delta t)$ be $\mu_n\Delta t + o(\Delta t)$, the probability that there is no more than one service completion is $o(\Delta t)$. Assume that the service completion epoch is independent of the time it started. Let

$$p_n(t) = P[Q(t) = n \mid Q(0) = 0].$$

Then

$$\begin{aligned}
p_{n,n-1}(t, t + \Delta t) &= [\mu_n\Delta t + o(\Delta t)][1 - \lambda_n\Delta t + o(\Delta t)] \\
&= \mu_n\Delta t \rightarrow o(\Delta t), \\
p_{nn}(t, t + \Delta t) &= [1 - \lambda_n\Delta t + o(\Delta t)][1 - \mu_n\Delta t + o(\Delta t)] \\
&= 1 - \lambda_n\Delta t - \mu_n\Delta t + o(\Delta t), \\
p_{n,n+1}(t, t + \Delta t) &= [\lambda_n\Delta t + o(\Delta t)][1 - \mu_n\Delta t + o(\Delta t)] \\
&= \lambda_n\Delta t + o(\Delta t), \text{ and}
\end{aligned}$$

$$\sum_{j\neq n-1,n,n+1} p_{nn-j}(t, t + \Delta t) = o(\Delta t).$$

For transitions occurring in nonoverlapping intervals $(0, t)$ and $(t, t + \Delta t)]$, by Chapman-Kolmogorov equations

$$p_n(t + \Delta t) = [1 - \lambda_n \Delta t - \mu_n \Delta t + o(\Delta t)]p_n(t)$$
$$+ [\lambda_{n-1}\Delta t + o(\Delta t)]p_{n-1}(t) + [\mu_{n+1}\Delta t + o(\Delta t)]p_{n+1}(t) + o(\Delta t)$$

which on rearranging and letting $\Delta t \to 0$, give the difference differential equation

$$p_n'(t) = -(\lambda_n + \mu_n)p_n(t) + \lambda_{n-1}p_{n-1}(t) + \mu_{n+1}p_{n+1}(t)$$

and
$$p_0'(t) = -\lambda_0 p_0(t) + \mu_1 p_1(t)$$

with initial conditions
$$p_n(0) = \begin{cases} 1 & \text{if } n = 0 \\ 0 & \text{if } n \neq 0 \end{cases}$$

Since for irreducible (i.e. when all states communicate) and positive recurrent (starting in any state, the mean time to return to that state is finite) Markov process $\lim_{t \to \infty} p_n(t)$ exists, it is independent of the time parameter and $p_n'(t) \to 0$ as $t \to \infty$.

Writing $\lim_{t \to \infty} p_n(t) = \pi_n$ and setting $\lim_{t \to \infty} p_n'(t) = 0$,

$$0 = -(\lambda_n + \mu_n)\pi_n + \lambda_{n-1}p_{n-1} + \mu_{n+1}\pi_{n+1} \text{ and}$$

$$0 = -\lambda_0 \pi_0 + \mu_1 \pi_1 \Rightarrow \pi_1 = \frac{\lambda_0}{\mu_1}\pi_0.$$

By induction
$$\pi_n = \frac{\lambda_0 \lambda_1 \ldots \lambda_{n-1}}{\mu_1 \mu_2 \ldots \mu_n}\pi_0.$$

Also
$$\sum_{j=0}^{\infty} \pi_j = 1 \Rightarrow \pi_0 = \left[1 + \sum_{n+1}^{\infty} \frac{\lambda_0 \lambda_1 \ldots \lambda_{n-1}}{\mu_1 \mu_2 \ldots \mu_n}\right]^{-1}$$

It should be noted that p_n is a nonzero solution only if

$1 + \sum_{n+1}^{\infty} \frac{\lambda_0 \lambda_1 \ldots \lambda_{n-1}}{\mu_1 \mu_2 \ldots \mu_n} < \infty$ i.e. limiting distribution of $Q(t)$ as $t \to \infty$ exists only if

$$1 + \sum_{n=1}^{\infty} \frac{\lambda_0 \lambda_1 \ldots \lambda_{n-1}}{\mu_0 \mu_1 \ldots \mu_n} < \infty \tag{10.1.1}$$

By this method we can analyze the $Q(t)$ processes of some simple queueing systems and solve realistic examples based on them. But there are other complex realistic models which cannot be solved by this method.

10.2 Standard Queueing Models

10.2.1 The Queue *M/M/*1 (Erlang's model)

Consider a queueing system in which arrivals occur one at a time in a Poisson process with parameter λ. These customers get served at a single counter and their service times are independent and identically distributed r.vs. Let the common distribution of these r.vs. be negative exponential with mean μ^{-1}. This implies that as long as there are enough customers for service, the service completions occur in a Poisson process with parameter μ. For waiting times of customers, we shall assume that the customers are served in the order of their arrival (first come first served). We assume that the service do not stop as long as there are customers to be served.

Hence $\lambda_n = \lambda$, $\mu_n = \mu$ for all n in the last generalized queueing model.

The difference differential equations for $p_n(t)$ take the form

$$p_0'(t) = -\lambda p_0(t) + \mu p_1(t)$$

$$p_n'(t) = -(\lambda + \mu) p_n(t) + \lambda p_{n-1}(t) + \mu p_{n+1}(t), n > 0.$$

Let $\rho = \dfrac{\lambda}{\mu} = \dfrac{\text{Mean Service time}}{\text{Mean interarrival time}}$ = Traffic intensity of the system (also called *loadfactor* or *utilization factor*). The limiting distribution of $Q(t)$ (queue length) exists if $1 + \rho + \rho^2 + \ldots < \infty$ i.e. if $\rho < 1$ (from 10.1.1).

Let $\pi_n = \lim_{t \to \infty} p_n(t)$, then $p_0 = [1 + \rho + \rho^2 + \ldots]^{-1} = 1 - \rho$ and $\pi_n = (1 - \rho)\rho^n$ for $n \geq 1$ (by 10.1.1).

Therefore the steady state solution (assuming the existence of a steady state of $M/M/1$ Queueing model) is given by a Geometric distribution. Writing $Q(\infty) = Q$,

$$E(Q) = \rho/1 - \rho \text{ and } V(Q) = \frac{\rho}{(1 - \rho)^2}.$$

Let $W(t)$ be the *virtual waiting time process* and it is the sum of the service times of those customers who are actually waiting and the remaining service time of the customers who are already in service. Here the service time distribution is negative exponential.

$$\left.\begin{array}{lll} \text{Hence} & W(t) = 0 \text{ if } Q(t) = 0 \\ \text{and} & W(t) = Z_1 + T_2 + \ldots + T_{Q(t)} \text{ if } Q(t) > 0 \end{array}\right\} \quad (10.2.1)$$

where Z_1 is the residual service time of the customers being served and $T_2, \ldots, T_{Q(t)}$ the service time of those waiting at time t and the distribution of the remaining service time Z_1 (by Markovian property) is the same as that of the original r.v. which is negative exponential. Therefore, if there are n customers in the system at time t, the waiting time $W(t)$ is given by the sum of n random variables which are i.i.d. with the negative exponential probability density (note that interarrival times are i.i.d. exponential in a Poisson process) $\mu e^{-\mu x}$, $x > 0$ and the probability distribution of the sum is the gamma density $g(x) = e^{-\mu x}\mu^n x^{n-1}/(n-1)!$, $x > 0$. (The total waiting time is gamma for Poisson process.)
Note $W(t)$ is a Markov process.

Writing $\quad W(\infty) = W$, let $F(x, t) = P[W(t) \leq x \mid W(0) = 0]$

and $\quad F(x) = P(W \leq x)$

(we get from (10.2.1))

$$F(0, t) = p_{00}(t) \quad (10.2.2)$$

Now let $Q(t) = j \geq 1$. Transition probabilities are

$$p_{ij}(t) = P(Q(t) = j \mid Q(0) = i).$$

The distribution of $W(t)$ is given by

$$d_xF(x, t) = P(x < W(t) < x + dx \mid W(0) = 0)$$

$$= \sum_{j=1}^{\infty} p_{0j}(t) \frac{\exp(-\mu x)\mu^j x^{j-1}}{(j-1)!} dx \, (0 < x < \infty) \qquad (10.2.3)$$

where $p_{0j}(t) \to \pi_j = (1 - \rho)\rho^j$ as $t \to \infty$.

The distribution of the waiting time thus has a discontinuity at the origin given by (10.2.2) and is continuous in the range $0 < x < \infty$. To obtain the limiting distribution of $W(t)$ as $t \to \infty$ taking limits inside the summation in (10.2.3) which is justified by dominated convergence theorem, we get

$F(x, t) \to F(x)$, where $P(Q = 0) = F(0) = 1 - \rho$ and

$$dF(x) = \sum_{n=1}^{\infty} (1 - \rho)\rho^n e^{-\mu x} \frac{(\mu x)^{n-1}}{(n-1)!} \mu dx$$

$$= \lambda(1 - \rho)e^{-(\mu-\lambda)x} dx \quad (0 < x < \infty) \qquad (10.2.4)$$

is valid for $\rho < 1$, $F(x, t) \to 0$ as $t \to \infty$ if $\rho \geq 1$.

These results imply, as in the case of queue length that if $\rho \geq 1$, the waiting time increases indefinitely as $t \to \infty$, whereas if $\rho < 1$, it tends to be distributed according to the negative exponential law in the range $0 < x < \infty$, the probability of not having to wait being $1 - \rho$. From equation (10.2.4) we get the moments of W as

$$E(W) = \rho/\mu(1 - \rho) \text{ and } V(W) = \frac{\rho(2 - \rho)}{\mu^2(1 - \rho)^2}.$$

In practical problems the distribution function of the waiting time W furnishes useful information regarding the system

$$F(x) = P(W \leq x) = P(W = 0) + P(0 < W \leq x)$$

$$= (1 - \rho) + \int_0^x \lambda(1 - \rho)e^{-(\mu-\lambda)y} dy$$

$$= (1 - \rho) + \frac{\lambda(1 - \rho)}{\mu - \lambda} \int_0^x (\mu - \lambda)^{-(\mu-\lambda)y} dy$$

$$= (1 - \rho) + \rho[1 - e^{-(\mu-\lambda)x}]$$

$$= 1 - \rho e^{-(\mu-\lambda)x}$$

It is the probability that the waiting time of the customer in the system is no more than x. If one wants to consider the total delay of the customer in the system (time spent in the system of *Sojurn time*), then the length (*Queueing time*) of service has to be added to the waiting time.

Let $G(x)$ be the probability that the total time spent by the customer in the system is no nore than x.

Then we have $G(x) = P[W + V \leq x]$, where V is the service time which is negative exponentially distributed with density $\mu e^{-\mu v}$. Hence

$$
\begin{aligned}
G(x) &= \int_0^x F(x - v)\mu e^{-\mu v} dv \\
&= \int_0^x [1 - \rho e^{-\mu(1-\rho e^{-\mu(1-\rho)(x-v)}}]\mu e^{-\mu v} dv \\
&= \int_0^x \mu e^{-\mu v} dv - \lambda \int_0^x e^{-\mu(1-\rho)x} e^{-\mu(1-\rho)v} e^{-\mu v} dv \\
&= 1 - e^{-\mu x} + e^{-\mu(1-\rho)x} \int_0^x (-\lambda e^{-\lambda v}) dv \\
&= 1 - e^{-\mu x} + e^{-\mu(1-\rho)x}[e^{-\lambda x} - 1] \\
&= 1 - e^{-\mu(1-\rho)x}
\end{aligned}
$$

The expected total stay of the customer in the system is

$$E(W + V) = \frac{1}{\mu(1 - \rho)}, \lambda E(W + V) = \frac{\lambda}{\mu(1 - \rho)} = \frac{\rho}{1 - \rho}$$

but $$EQ = \frac{\rho}{(1 - \rho)}. \text{ So } EQ = \lambda E(W + V).$$

Alternately, consider the customers $C_1, C_2, C_3, \ldots, C_n$ getting service in the order of their arrival. Let V_n be the service time of the customer C_n. Clearly, the number of customers in the system at the nth departure epoch (at which time C_n leaves the system) is the number of customers who have arrived during $W_n + V_n$, where W_n is the actual waiting time of C_n. If Q_n is the number of customer left by the nth departing customers, λ being the arrival rate, we can write

$$E(Q_n) = \lambda E(W_n + V_n),$$

when arrival process is Poisson W_n and $W(t)$ have the same form.

Therefore $L_Q = \lambda W_Q$ and $E(Q) = \lambda E(W + V)$ i.e. Little's (1961) formula holds, where $W = W(\infty)$.

10.2.2 Transition solution

Here $Q(t)$ is a birth and death process.

Let $p_{ij}(t) = P(Q(t) = j/Q(0) = i)$ for $t > 0$ and $p_{ij}(0) = \delta_{ij}$.

At $t = 0$, continuity and differentiability both hold, but from the right, and hence

$$
\left.
\begin{aligned}
p'_{i0}(t) &= -\lambda p_{c0}(t) + \mu p_{i1}(t) \\
p'_{ij}(t) &= -(\lambda + \mu)p_{ij}(t) + \lambda p_{ij-1}(t) + \mu p_{ij+1}(t)
\end{aligned}
\right\} \quad (10.2.5)
$$

is the forward Kolmogorov differential ($j > 0$) difference equations of the birth and death processes arise in $M/M/1$ Queue.

Let $$G(t, z) = \sum_{j=0}^{\infty} p_{ij}(t)z^j \ (|z| \le 1)$$

and $G(0, z) = z^i \ (i > 0)$ then $\frac{\partial G}{\partial t} = \sum_{j=0}^{\infty} p'_{ij}(t)z^j$.

Multiplying (10.2.5) successively by 1, z, z^2, . . . and adding, we obtain

$$\sum_{j=0}^{\infty} p'_{ij}(t)z^j = -(\lambda + \mu) \sum_{j=0}^{\infty} p_{ij}(t)z^j + \lambda \sum_{j=1}^{\infty} p_{ij-1}(t)z^j$$

$$+ \mu \sum_{j=0}^{\infty} p_{ij+1}(t) z^j + \mu p_{i0}(t).$$

Hence $$z\frac{\partial G}{\partial t} = [\lambda z^2 - (\lambda + \mu)z + \mu]G - \mu(1 - z)p_{i0}(t) \tag{10.2.6}$$

Laplace transforms of $G(t, z)$ and p_{ij} (t) and are

$$G^*(\theta, z) = \int_0^{\infty} e^{-\theta t} G(t, z)dt \tag{10.2.7}$$

and $$p^*_{ij}(\theta) = \int_0^{\infty} e^{-\theta t} p_{ij}(t)dt \ (\text{Re}(\theta) > 0) \tag{10.2.8}$$

respectively. Let us state a theorem from complex analysis which will be used soon.

Rouche's theorem If $f(z)$ and $g(z)$ are functions analytic inside and on a closed contour C, and if $|g(z)| < |f(z)|$ on C, the $f(z)$ and $f(z) + g(z)$ have the same number of zero(s) inside C.

Integrating (2.2.7) by parts yields the relation

$$\int_0^{\infty} e^{-\theta t}\delta G/\delta t = e^{-\theta t}G(t, z)]_0^{\infty} \ e^{-\theta t}G(t, z)dt$$

$$= - z^i + \theta G^*(\theta, z) \tag{10.2.9}$$

Multiplying both sides of (10.2.6) by $e^{-\theta t}$ and integrating over $(0 \le t < \infty)$, we obtain

$$\theta z G^*(\theta, z) - z^{i+1} = [\lambda z^2 - (\lambda + \mu) z + \mu] \ G^*(\theta, z)$$

$$-\mu(1 - z)p^*_{10}(\theta),$$

or $$G^*(\theta, z) = \frac{z^{i+1} - \mu(1 - z)p^*_{i0}(\theta)}{(\lambda + \mu + \theta)z - \mu - \lambda z^2} \tag{10.2.10}$$

The denominator on the right hand side of (10.2.10) has two zeros, namely

$$\xi = \xi(\theta) = \frac{\lambda + \mu + \theta - \sqrt{(\lambda + \mu + \theta)^2 - 4\lambda\mu}}{2\lambda} \tag{10.2.11}$$

$$\eta = \eta(\theta) = \frac{\lambda + \mu + \theta + \sqrt{(\lambda + \mu + \theta)^2 - 4\lambda\mu}}{2\lambda} \tag{10.2.12}$$

where the square root is taken so that its real part is positive. Clearly

$$|\xi| < |\eta|, \text{ and moreover } \xi + \eta = \frac{\lambda + \mu + \theta}{\lambda}, \xi\eta = \mu/\lambda \tag{10.2.13}$$

For $|z| = 1$ and Re $(\theta) > 0$, we get

$$|(\lambda + \mu + \theta)\, z| = |\lambda + \mu + \theta| > \lambda + \mu \geq |\mu + \lambda z^2|,$$

so that by Rouche's theorem $(\lambda + \mu + \theta)z - \mu - \lambda z^2$ has only one zero in the unit circle; this is clearly $z = \xi$. Now since the Laplace transform $G^*(\theta, z)$ converges in the region $|z| = 1$, Re $(\theta) > 0$, the zeros of the numerator and denominator on the right hand side of (10.2.10) must coincide and therefore

$$p_{i0}^*(\theta) = \frac{\xi^{i+1}}{\mu(1 - \xi)} \tag{10.2.14}$$

Substituting (10.2.14) in (10.2.10) we obtain the result

$$G^*(\theta, z) = \frac{z^{i+1}(1 - z)\xi^{i+1}/(1 - \xi)}{\lambda(z - \xi)(\eta - z)} \tag{10.2.15}$$

for the transform of the transition probabilities of the process $Q(t)$. This is due to Bailey (1954). The probabilities $p_{ij}(t)$ can be obtained by inverting transform (10.2.15). Alternately $p_{ij}(t)$'s can be obtained explicity by combinatorial method (due to Champernowne (1956)).

10.2.3 The basic process X(t)

Let $A(t)$ and $D(t)$ be the number of arrivals and departures, respectively, during $(0, t]$; their probability distributions being given by negative exponential law. Consider the process $N(t) = X(t) = A(t) - D(t)$, where $A(t)$ and $D(t)$ are independent. Let

$$\begin{aligned} k_j(t) &= P(X(t) = j), j = \ldots, -1, 0, 1, 2, \ldots \\ &= e^{-(\lambda+\mu)t}\rho^{j/2} I_j(2\sqrt{\lambda\mu}\, t) \end{aligned} \tag{10.2.16}$$

(from elementary properties of Poisson process)

where $I_j(x) = \sum_{n=0}^{\infty} (x/2)^{2n+j}/[n!(n+j)!]$ is the modified Bessel function of index j.

$$\begin{aligned} k_{-j}(t) &= e^{-(\lambda+\mu)t}\rho^{-j/2} I_j(2\sqrt{\lambda\mu}\, t) \\ &= \rho^{-j} k_j(t). \end{aligned}$$

When $\lambda = \mu$, this gives $k_{-j}(t) = k_j(t)$ $(\rho = 1)$ and $X(t)$ then is a symmetric stochastic process.

The P.G.F. of $X(t)$ is given by

$$\sum_{j=-\infty}^{\infty} k_j(t) z^j = \exp\{[\lambda z - (\lambda + \mu) + \mu/z]t\} \tag{10.2.16}$$

Letting $z \to 1$, $$P(X(t) < \infty) = \sum_{j=-\infty}^{\infty} k_j(t) = 1 \qquad (10.2.17)$$

which shows that $X(t)$ assumes only finite values. The d.f. of $X(t)$ is given by

$$K_j(t) = \sum_{\upsilon=-\infty}^{j} k_\upsilon(t) = \sum_{\ell \geq 0, m \geq 0} \sum_{\ell - m \leq j} e^{-(\lambda+\mu)t} \frac{\lambda^\ell \mu^m}{\ell!\, m!} (t)^{\ell+m} \qquad (10.2.18)$$

$$= P\{A(t) - D(t) \leq j\}.$$

For $j \geq 0$,

$$\left.\begin{aligned} \int_0^\infty e^{-\theta t} k_j(t)dt &= \eta^{-j} / (\lambda(\eta - \xi)) \\ \int_0^\infty e^{-\eta t} k_{-j}(t)dt &= \frac{\xi^j}{\lambda(\eta - \xi)} \end{aligned}\right\} \qquad (10.2.19)$$

10.2.4 Combinatorial methods

Note $X(t)$ increases or decreases by unity at random points of time and that the intervals between successive points of increase and the intervals between successive points of decrease are r.v.s distributed according to negative exponential laws with means λ^{-1} and μ^{-1}, respectively, all intervals being mutually independent. A typical realization (or sample path) of $X(t)$.

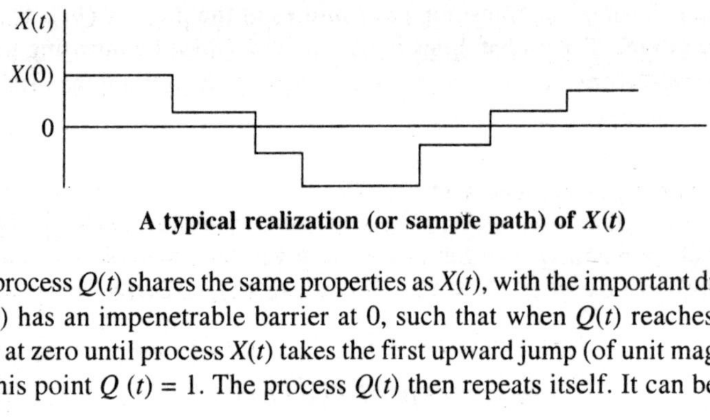

A typical realization (or sampte path) of $X(t)$

The process $Q(t)$ shares the same properties as $X(t)$, with the important difference that $Q(t)$ has an impenetrable barrier at 0, such that when $Q(t)$ reaches zero, it remains at zero until process $X(t)$ takes the first upward jump (of unit magnitude), and at this point $Q(t) = 1$. The process $Q(t)$ then repeats itself. It can be proved that

$$Q(t) = \max\{\sup_{0 \leq \tau \leq t} [X(t) - X(\tau)], Q(0) + X(t)\} \qquad (10.2.20)$$

This representation is typical of many of the processes which will be encountered henceforth and its regorous proof depends on *storage process* (due to Pyke and Gani (1960)) and additive process (separable infinitely divisible process)

$$Q(t) = \max\,[X(t) - \sigma(t), Q(0) + X(t)] \qquad (10.2.21)$$

where $$\sigma(t) = \inf_{0 \leq \tau \leq t} X(\tau) = \inf_{0 \leq \tau \leq t} [A(\tau) - D(\tau)].$$

Since $A(0) = D(0) = 0$, $\sigma(t) \leq 0$ a.s.

From (2.2.20) and some combinatorial arguments, reflection principle and the value of $K_j(t)$ gives

$$P(Q(t)\ j \mid Q(0) = i) = \sum_{\ell}\sum_{m} P\{A(t) = \ell, D(t) = m\} P(Q(t) \le j \mid Q(0) = i,$$
$$A(t) = \ell, D(t) = m) = K_{j-i}(t) - \rho^{j+1}K_{-j-i-2}(t).$$

Hence $P_{ij}(t) = P(Q(t) = j \mid Q(0) = i)$

$$= \{Q(t) \le j \mid Q(0) = i\} - P\{Q(t) \le j-1 \mid Q(0) = i\}$$
$$= \{K_{j-i}(t) - \rho^{j+1}K_{-j-i-2}(t)\} - \{K_{-j-1-i}(t) - \rho^{j}K_{-j-i-1}(t)\}\}$$
$$= k_{j-i}(t) + \rho^{j}k_{-j-i-1}(t) + (1-\rho)\rho^{j}K_{-j-i-2}(t) \qquad (10.2.22)$$

Using $k_{-j}(t) = \rho^{-j}k_j(t)$ we have

$$p_{ij}(t) = k_{j-i}(t) + \rho^{-i-1}k_{j+i+1}(t) + (1-\rho)\rho^{j}K_{-j-i-2}(t). \qquad (10.2.23)$$

When $\rho = 1$, this becomes

$$p_{ij}(t) = k_{j-i}(t) + k_{j+i+1}(t), \qquad (10.2.24)$$

where $k_j(t) = e^{-2\lambda t}I_j(2\lambda t)$.

By (10.2.22), we get

$$\sum_{j=0}^{\infty} p_{ij}(t) = \sum_{j=1}^{\infty} k_{j-i}(t) + \sum_{j=0}^{\infty} \rho^{j}k_{-j-i-1}(t) + (1-\rho)\sum_{j=0}^{\infty}\rho^{j}\sum_{\upsilon=-\infty}^{-j-i-2}k_{\upsilon}$$

$$= \sum_{\upsilon=-i}^{\infty} k_{\upsilon}(t) + \sum_{\upsilon=-\infty}^{-i-1}\rho^{-\upsilon-i-1}k_{\upsilon} + \sum_{\upsilon=-\infty}^{-i-1}k_{\upsilon}(1-\rho^{-\upsilon-i-1})$$

$$= \sum_{\upsilon=-\infty}^{\infty} k_{\upsilon}(t) = 1 \quad \text{(by (10.2.17))} \qquad (10.2.25)$$

Thus in finite time there is no possibility of an unduly long queue. Now an alternative derivation of limiting distribution $Q(t)$ as $t \to \infty$ can be given

$$k_j(t) \sim e^{-(\lambda+\mu)t}\rho^{j/2}\frac{\exp[2\sqrt{\lambda\mu}\,t]}{2(\pi t)^{1/2}(\lambda\mu)^{1/4}} \qquad \left(I_j(x) \sim \frac{e^{x}}{\sqrt{2\pi x}} \text{ as } x \to \infty\right)$$

$$= \frac{e^{-(\sqrt{\lambda}-\sqrt{\mu})^2 t}\rho^{j/2}}{2(\pi t)^{1/2}(\lambda\mu)^{1/4}} \text{ as } t \to \infty.$$

If $\rho > 1$, $K_j(t) = \sum_{-\infty}^{j} k_{\upsilon}(t) \sim \dfrac{\exp\{[\sqrt{\lambda}-\sqrt{\mu}]^2 t\}}{2(\pi t)^{1/2}(\lambda\mu)^{1/4}}\sum_{\upsilon=-\infty}^{i}\rho^{\upsilon/2}$ (10.2.27)

If $\rho < 1$, $1 - K_j(t) = \sum_{\upsilon=j+1}^{\infty} k_{\upsilon}(t) \sim \dfrac{\exp((\sqrt{\lambda}-\sqrt{\mu})^2 t\}}{2(\pi t)^{1/2}(\lambda\mu)^{1/4}}\sum_{j+1}^{\infty}\rho^{\upsilon/2}$ (10.2.28)

Therefore $K_j(t) \to \begin{cases} 0 & \text{if } \rho > 1 \\ 1 & \text{if } \rho < 1 \end{cases}$ as $t \to \infty$ for all j (10.2.29)

By (10.2.26), (10.2.29), (10.2.23) and (10.2.24), we get

$$\lim_{t\to\infty} p_{ij}(t) = \begin{cases} 0 \quad \text{if} \quad \rho \geq 1 \\ (1-\rho)\rho^j \ \text{if}\ \rho < 1 \end{cases} \tag{10.2.30}$$

This implies that independently of its initial length i, the queue grows unduly long at the end of an indefinitely long period of time if $\rho \geq 1$, whereas if $\rho < 1$, the Queue-length attains statistical equilibrium and tends to have the Geometric distribution. The limiting distribution is stationary in the sense that if the initial Queue-length $Q(0)$ has this distribution, so does this Queue-length $Q(t)$ for $t > 0$. For we have

$$P[Q(t) \leq j] = \sum_{i=0}^{\infty} P[Q(t) \leq j \mid Q(0) = i]\, P[Q(0) = i]$$

$$= \sum_{i=0}^{\infty} [K_{j-i}(t) - \rho^{j+1} K_{-j-i-a}(t)](1-\rho)\rho^i$$

$$= 1 - \rho^{j+1} \qquad \text{(after simplification).}$$

10.2.5 The busy period; zero-avoiding transitions

Suppose that initially the system contains $i\,(\geq 1)$ customer, and let T_i be the next subsequent epoch of time at which the server is free; T_i is called the *busy period initiated by i customers*.

$$\text{Then}\ \ T_i = \inf\{t \mid Q(t) = 0,\ Q(0) = i\}. \tag{10.2.31}$$

Suppose that T_i has d.f. $G_i(t) = P(T_i \leq t)$ $(0 \leq t < \infty)$.
Define the *zero-avoiding transition probabilities* of the Markov process

$$Q(t) \text{ as } {}^0p_{ij}(t) = P(Q(t) = j;\ T_i > t \mid Q(0) = i)\ (i, j \geq 1).$$

It is the probability that there will be j customers busy throughout the interval $[0, t]$.

By (10.2.20), (10.2.31) can be written as $T_i = \inf\{t \mid i + X(t) \leq 0\}$ and hence

$${}^0p_{ij}(t) = P[i + X(\tau) > 0 \quad (0 \leq \tau \leq t);\ i + X(t) = j] \tag{10.2.32}$$

$$= P\{i + \sigma(t) > 0,\ i + X(t) = j],$$

By (10.2.20) and (10.2.32), we get

$$P(Q(t) < j \mid Q(0) = i) = P(X(t) - \sigma(t) \ < j;\ i + X(t) = j - \upsilon)$$

$$= \sum_{\upsilon=i+1}^{\infty} P(\upsilon + \sigma(t) > 0;\ \upsilon + X(t) = \sum_{i+1}^{\infty} {}^0p_{kj}(t) \tag{10.2.33}$$

which is a relation connecting the zero-avoiding transition probability with the ordinary ones. In particular (putting $j = 1$ in (10.2.33)), we get

$$p_{i0}(t) = \sum_{i+1}^{\infty} {}^0p_{k1}(t) \tag{10.2.34}$$

From (10.2.33) and (10.2.22)

$$
\begin{aligned}
{}^0p_{ij}(t) &= P[Q(t) \le j-1 \mid Q(0) = i-1] - P[Q(t) \le j-1 \mid Q(0) = i] \\
&= K_{j-i}(t) - \rho^j K_{-j-i}(t)] - [K_{j-i-1}(t) - \rho^j K_{-j-i-1}] \\
&= k_{j-i}(t) - \rho^j k_{-j-i}(t) = k_{j-i}(t) - \rho^{-i} k_{j+i}(t)(i, j \ge 1) \qquad \text{(Bailey 1954)}
\end{aligned}
$$

The Laplace transform of ${}^0p_{ij}(t)$ is given by

$$
{}^0p^*_{ij}(\theta) = \int_0^\infty e^{-\theta t}\, {}^0p_{ij}(t)dt = \int_0^\infty e^{-\theta t} k_{j-i}(t)dt - \rho^{-i} \int_0^\infty e^{-\theta t} k_{j+i}(t)dt
$$

$$
\text{By (10.2.19) } {}^0p^*_{ij}(\theta) = \begin{cases} \dfrac{\eta^{-j}(\eta^i - \xi^i)}{\lambda(\eta - \xi)} \text{ if } j \ge i \\ \dfrac{\xi^i(\xi^{-j} - \eta^{-j})}{\lambda(\eta - \xi)} \text{ if } j \ge i \end{cases} \tag{10.2.35}
$$

where ξ, η are given by (10.2.11) and (10.2.12).

The distribution of busy period T_i is given by

$$
\begin{aligned}
dG_i(t) &= {}^0p_{i1}(t)\, \mu dt \qquad\qquad (10.2.36) \\
&= i/tk_{-i}(t)\, dt \text{ (after simplification)} \\
&= g_i(t)\, dt \text{ which is the density function of } T_i (i \ge 1, t > 0).
\end{aligned}
$$

($i = 1$ is due to Kendall (1951) and $i > 1$ due to Good (1948))

The Laplace transform of this p.d.f. is given by

$$
g_i^*(\theta) = \int_0^\infty e^{-\theta t} g_i(t)dt = \int_0^\infty e^{-\theta t}\, {}^0p_{i1}(t)\mu dt \tag{10.2.37}
$$

$$
- \frac{\xi^i(\xi^{-1} - \eta^{-1})}{\lambda(\eta - \xi)} \mu = \xi^i \text{ (from 10.2.35)}
$$

$$
P(T_i < \infty) = \lim_{\theta \to o} g_i^*(\theta) = \begin{cases} 1 \text{ if } \rho \le 1 \\ \rho^{-1} \text{ if } \rho > 1 \end{cases}
$$

Since $\xi(\theta) \to 1$ or ρ^{-1} according as $\rho \le 1$ or $\rho > 1$ (follows from 10.2.11). So the busy period will terminate eventually if $\rho \le 1$, whereas if $\rho > 1$ there is a positive probability that it will continue indefinitely. When $i = 1$ we have the busy period T initiated by a single customer, the Laplace transform of its distribution is $g_1^*(\theta) = \xi$, and from (10.2.37) we see that $g_i(t)$ is the i-fold convolution of $g_1(t)$. So T_i is the sum of ii.d. r.vs. each having same distribution as T_i.

10.3 The Model *M/M/c*

Here the input process and the service time distribution are the same as M/M/1 model but there are c servers where $1 \le c < \infty$ customers are served on a 'first come first served' basis (FCFS). Here again the number of customers $Q(t)$ present in the system at time t is a birth and death process, the transition densities are computed as follows.

Here for each of the c channels use i.i.d exponential service time distribution with mean rate μ. If n ($\le c$) channels, are busy, the number of services completed in the whole system is a Poisson process with mean $n\mu$ and thus the time between two successive service completions is exponential with mean $1/(n\mu)$ whereas if n ($> c$) channels are busy, the time between two successive service completions is exponential with mean $1/c\mu$.

$$P_{nn+1}(dt) = P(Q(t + dt) = n + 1 \mid Q(t) = n) = \lambda dt + o(dt)$$

$$p_{nn-1}(dt) = \mu_n dt + o(dt)$$

$$p_{nn}(dt) = 1 - (\lambda + \mu_n)dt + o(dt)$$

$$p_{nm}(dt) = P(Q(t + dt) = m \mid Q(t) = n) = o(dt) \text{ if } m \ne n - 1, n, n + 1, \tag{10.3.1}$$

where
$$\mu_n = \begin{cases} n\mu & \text{if } 0 \le n \le c \\ c\mu & \text{if } n \ge c \end{cases} \tag{10.3.2}$$

Thus $Q(t)$ is a B.D. process with constant arrival (birth) rate $\lambda_m = \lambda$ and state dependent service (death) rates as given above.

Substituting (10.3.1) in Chapman-Kolmogorov equation and proceeding as the model $M/M/1$ we find the forward Kolmogorov equations of the process to be

$$p'_{nm}(t) = -(\lambda + \mu_m)p_{nm}(t) + \lambda p_{n,m-1}(t) + \mu_{n+1}p_{n,m+1}(t)(n, m \ge 0).$$

Using (10.3.2) and writing $p_n(t) = P[Q(t) = n \mid Q(0) = 0]$, i.e. forward Kolmogorov equation of birth and death process becomes

$$p'_0(t) = -\lambda p_0(t) + \mu p_1(t)$$

$$p'_n(t) = -(\lambda + n\mu)p_n(t) + \lambda p_{n-1}(t) + (n + 1)\mu p_{n+1}(t)(0 < n < c)$$

$$p'_n(t) = -(\lambda + c\mu)p_n(t) + \lambda p_{n-1}(t) + c\mu p_{n+1}(t)(n \ge c).$$

These equations have been solved by Saaty (1960) by an extension of Bailey's method of generating functions. The transient solution for $c > 1$ is too complicated. We give only the steady state solution, putting the values of λ_n and μ_n in generalized Queueing model as $t \to \infty$, the limiting distribution π_n of the Queue length Q ($= Q(\infty)$) is obtained.

$$\pi_n = \frac{\lambda.\lambda \ldots \lambda}{\mu(2\mu) \ldots \ldots (n\mu)}\pi_0 = \frac{(\lambda/\mu)^n}{n!}\pi_0, n = 1, 2, \ldots, c$$

$$= \frac{\rho^n}{n!}\pi_0 \tag{10.3.3}$$

and
$$\pi_n = \pi_0 \frac{\lambda\lambda \ldots \lambda}{\mu(2\mu) \ldots c\mu\{(c\mu) \ldots (c\mu)\}} = \frac{\lambda^n}{c!\mu^c c^{n-c}\mu^{n-c}}\pi_0$$

$$= \frac{(\lambda/\mu)^n}{c!c^{n-c}}\pi_0 = \rho^n/(c!c^{n-c})\pi_0 (n \ge c). \tag{10.3.4}$$

The condition $\sum_{n=1}^{\infty} \pi_n = 1$ gives

$$\pi_0^{-1} = 1 + \sum_{n=1}^{c-1} (\lambda/\mu)^n/n! + \sum_{c}^{\infty} \frac{(\lambda/\mu)^n}{c!c^{n-c}} = \sum_{n=1}^{c-1} \frac{\rho^n}{n!} + \frac{c^c}{c!} \sum_{n=0}^{\infty} (\rho/c)^n.$$

For existence of steady state solutions, the series $\sum_{n=0}^{\infty} (\rho/c)^n$ must converge and for this to happen the relative traffic intensity ρ/c must be less than 1.

Then $$\pi_0^{-1} = \sum_{n=0}^{c-1} \frac{\rho^n}{n!} + \frac{\rho^c}{c!(1-\rho/c)}.$$

Also $\pi_n = (\rho/c)^{n-c}\pi_c$ for $n \geq c$.

Note

(1) π_n satisfy the following recurrence relations

$$\pi_n = \begin{cases} 1/n\rho\pi_{n-1}, n = 1, 2, \ldots, c-1 \\ 1/c\,\rho\pi_{n-1}, n = c, c+1, \ldots \end{cases}$$

(2) The probability that an arriving unit has to wait is given by

$$P(Q \geq c) = \sum_{n=c}^{\infty} \pi_n = \rho^c/[c!(1-\rho)]\pi_0 = \pi_c/(1-\rho).$$

The Mean Queue length is given by

$$E(Q) = \sum_{n=0}^{\infty} n\pi_n = \sum_{n=0}^{c} \pi_0 n/n!\rho^n + \sum_{n=c+1}^{\infty} \frac{n\pi_0\rho^n}{c!c^{n-c}}$$

$$= \rho\pi_0 \sum_{n=0}^{c-1} 1/n!\rho^n + \pi_0/c!(\rho)^c \sum_{1}^{\infty} (n+c)(\rho/c)^n$$

$$= \rho\pi_0 \sum_{0}^{c-1} 1/n!\rho^n + \pi_0/c!\rho^c \left[\frac{\rho/c}{(1-\rho/c)^2} + \frac{\rho}{(1-\rho/c)}\right]$$

$$= \rho + \frac{c\rho\pi_0}{(c-\rho)^2}.$$

Let Q' be the number of customers actually waiting in the long run. Then $Q' = \text{Max}(0, Q-c)$.

Therefore $$P(Q' = 0) = P(Q \leq c) = \pi_0 \sum_{0}^{c} 1/j!\rho^j$$

and $$P(Q' = j) = P(Q = j + c) = \frac{\pi_0}{c^j c!} \rho^{j+c} (j \geq 1).$$

$$E(Q') = \sum_{n=c+1}^{\infty} (n-c)\pi_n = \pi_0/c!\rho^c \sum_{1}^{\infty} j(\rho/c)^j = \frac{\rho^{c+1}}{c!(1-\rho/c)}\pi_0$$

$$= \frac{p/c}{(1-p/c)^2}\pi_0.$$

10.3.1 Distribution of W, the long run waiting time of a customer

If a customer on entering the system finds j customers ahead of him, then he will be served immediately if $j < c$, whereas j if $\geq c$, he will have to wait till the $(j - c + 1)$th customer completes his service. Therefore the probability that he does not have to wait is given by

$$F(0) = P(W = 0) = P(Q < c) = \pi_0 \sum_{n=0}^{c-1} \rho^n / n!$$

If all the servers are busy, then the intervals of time between successive departures are independent r.vs. having a negative exponential distribution with mean $1/c\mu$ and the total time until the $(j - c + 1)$th departure is the $(j - c + 1)$-fold convolution of this distribution and hence

$$dF(x) = P(x < W < x + dx) = \sum_{n=0}^{\infty} \pi_n \frac{e^{-c\mu x}(c\mu x)^{n-c}}{(n-c)!} c\mu dx$$

$$= \sum_{n=c}^{\infty} \rho^{n-c} \pi_0 e^{-c\mu x} \frac{(\mu x)^{n-c}}{(n-c)!} c\mu dx$$

$$= c\mu\, \pi_0 e^{-(c\mu - \lambda)x} dx \ (0 < x < \infty),\ \lambda = \mu\rho.$$

Hence, W has a discrete probability at the origin and a continuous frequency density in $0 < x < \infty$. The expected waiting time is given by

$$E(W) = \int_0^x x c\mu \pi_0 e^{-x(c\mu - \lambda)} dx$$

$$= \frac{c\mu\pi_c}{(c\mu - \lambda)^2} = \frac{\pi_0}{c\mu(1 - \rho^*)^2} \quad \text{where } \rho^* = \rho/c.$$

In a single server queue looking at the limiting probabilities ($\pi_0 = 1 - \rho$, $\pi_n = (1 - \rho)\rho^n$, $n \geq 0$) it is clear that the traffic intensity ρ is the probability that the server is busy at any time (or the fraction of time the server is busy). Is this true for the multichannel system $M/M/c$?

Consider the probability that a particular channel is busy at sometime in the queue $M/M/c$. It should be noted that when the number of customers in the queue is $\geq c$ all the channels are busy, and when there are $k < c$ customers in the queue the probability that the channel under consideration will be busy is k/c. Thus, the probability that all the channels are busy (by (10.3.3) and (10.3.4)) will be

$$\left\{ \sum_{n=c}^{\infty} \frac{\rho^n}{c^{n-c} c!} + \frac{c-1}{c} \frac{\rho^{c-1}}{(c-1)!} + \frac{c-2}{c} \frac{\rho^{c-2}}{(c-2)!} + \ldots + 1/c\rho \right\} \pi_0$$

$$= 1/c\, \pi_0 \left[\rho + \rho^2 + \ldots \frac{\rho^{c-2}}{(c-2)!} \rho + \frac{\rho^{c-1}}{(c-1)!} \rho + \sum_{n=c+1}^{\infty} \frac{\rho^{n-1}}{c^{n-1-c} c!} \rho \right]$$

$$= \rho / c\pi_0 \left[\sum_{n=0}^{c-1} \frac{\rho^n}{n!} + \sum_{n=c}^{\infty} \frac{\rho^n}{c^{n-c} c!} \right] = \frac{\rho}{c} \pi_0 \pi_0^{-1} = \rho^*.$$

Again the probability that a particular channel is busy at any time (in the long run) is the traffic intensity ρ.

10.4 The Concept of Imbeded Markov Chain in Queueing Models

In $M/M/1$, $M/M/c$, $M/M/\infty$, Queueing systems the Queue-length $Q(t)$ is a Markov process because the specification of the Queue-length at time γ is adequate to predict its length at time $\gamma + t$ on account of the Markovian property of the Poisson distribution. In the systems $M/E_k/1$ and $E_k/M/1$, $Q(t)$ is not Markovian, but its properties could be obtained from those of a second (another) process $Q_1(t)$, which is Markovian. When the input and service time distributions assume general forms, the process $Q(t)$ is no longer Markovian. The correct specification of the Queueing system is then given by the vector process $[Q(t), u(t), v(t)]$, where $u(t)$ is the time which is elapsed since the last arrival, and $v(t)$ is the expended service time of the customer being served at time t. Though rather complicated, with the help of this vector process it is possible to do something about the process $Q(t)$ by making use of its '*regeneration points*'

A set S of time-instants is said to be a set of regeneration points for a stochastic process $X(t)$ iff, for all $t > t_0$ we have

[Distribution $X(t) \mid X(t_0)$] = [Distribution $X(t) \mid X(\gamma)$ for all $\gamma \le t_0$] whenever $t_0 \in S$. This implies that the development of the process during $t > t_0$ is independent of the history of the process during $(0, t_0)$. For a Markov process the whole parameter space (range of t values) is a set of regeneration points. If a process has a set of regeneration points, then it is called **regenerative**. Suppose that there exists a denumerable S-set of regeneration points $\{t_n, n = 0, 1, 2, \ldots\}$ such that $t_0 < t_1 < t_2 < \ldots$ and put $X_n = X(t_n)$; then X_n is a M.C. This chain is said to be imbedded in the given process $X(t)$.

Note

1. In a single server queueing system, the set of points at which the counter is free is clearly an S-set.

2. Let S_x consists (for a given x) of those instants of time at which the customer at the counter has already been served for a period x.

3. In $M/G/1$ system, the instants at which the customers leave the system are points of regeneration. In the general case if there exists an increasing sequence $\{t_n\}$ of regeneration points such that the transition probabilities associated with $\{Q_{n+1} \mid Q_n\}$, where $Q_n = Q(t_n)$ can be calculated in a simple manner, then the analysis of the imbedded $M.C.$ $\{Q_n\}$ will yield valuable information regarding the process $Q(t)$.

10.4.1 The system with service in phases: The Queue model $M/E_k/1$

In this system customers arrive in a Poisson process with mean λt, and the service time V has the Erlangian distribution

$$dV(v) = \frac{1}{\Gamma k} e^{-\mu k v} (k\mu)^k v^{k-1} dv \quad (0 < v < \infty) \tag{10.4.1}$$

with $E(V) = \mu^{-1}$ and traffic intensity = $\rho = \lambda/\mu$.

Since the distribution (10.4.1) does not have the Markovian property, Queue-length process $Q(t)$ in this system is not Markovian. Note that the service here can be interpreted as being offered in k consecutive phases, the times required for these phases being mutually independent r.vs. with negative exponential distribution $k\mu e^{-k\mu v} dv (0 < v < \infty)$ and the service of a customer being completed at the end of the kth phase. The arrival of each customer introduces k phases into the system, and the phases are completed successively in a Poisson process at a rate k. If $Q_1(t)$ is the number of phases in the system at time t, it is clear that $Q_1(t)$ is a denumerable Markov process.

The transitions probabilities in an infinitesimal interval $(t, t + dt)$ are given by

$$p_{00}(dt) = 1 - \lambda dt + o(dt)$$

$$p_{ii}(dt) = 1 - (\lambda + k\mu)dt + o(dt) \quad (i \geq 1)$$

$$p_{i,i-1}(dt) = k\mu\, dt + o(dt) \quad (i \geq 1) \text{ and } p_{i,i+k}(dt) = \lambda dt + o(dt) \quad (i \geq 0),$$

$$p_{i,j}(dt) = o(dt) \quad (i \geq 0, j \neq 0, i - 1, k).$$

This stochastic process is a time homogenous birth and death process, slightly more general than Feller's process. The forward Kolmogorov equations are

$$\begin{cases} p'_{io}(t) = -\lambda p_{io}(t) + k\mu p_{i1}(t) \\ p'_{ij}(t) = -(\lambda + k\mu) p_{ij}(t) + k\mu p_{ij+1}(t)\ (1 \leq j < k) \\ p'_{ij}(t) = -(\lambda + k\mu) p_{ij}(t) + k\mu p_{ij+1}(t) + \lambda p_{ij-k}(t)\ (j \geq k) \end{cases}$$

Let $G(z, t) = \sum_j p_{ij}(t) z^j\ (|z| \leq 1)$ be the G.F. of $\{p_{ij}(t)\}$.

Then $\frac{\partial G}{\partial t} = \sum_j p'_{ij}(t) z^j$. Multiplying the last Kolmogorov's equation successively by 1, z, z^2, . . . and adding, we get, after some simplifications, the differential equation

$$z \frac{\delta G}{\delta t} = [k\mu - (\lambda + k\mu) z + \lambda^{k+1}] G - k\mu(1 - z) p_{io}(t).$$

Taking Laplace transforms with respect to time on both sides, we then get

$$G^*(z, \theta) = \frac{z^{i+1} - k\mu(1 - z) p^*_{io}(\theta)}{(\theta + \lambda + k\mu) z - k\mu - \lambda z^{k+1}} \quad (|\, z \,1 < 1, \text{Re}(\theta) > 0) \tag{10.4.2}$$

In the region $|z| = 1$, Re$(\theta) > 0$, we have

$$|(\theta + \lambda + k\mu) z| = |\theta + \lambda + k\mu| > \lambda + k\mu > |\lambda z^{k+1} + k\mu|,$$

so that, by Rouche's theorem, the denominator on the right hand side of (10.4.2), has only one zero in the unit circle. Let $\xi = \xi(\theta)$ denote this zero. The arguments used in case of $M/M/1$ Queue then give

$$p^*_{io}(\theta) = \frac{\xi^{i+1}}{k\mu(1-\xi)} \text{ and } G^*(z,\theta) = \frac{z^{i+1} - (1-z)\xi^{i+1}/(1-\xi)}{(\theta+\lambda+k\mu)z - k\mu - \lambda z^{k+1}}.$$

Since $G^*(z,\theta) \to 1/\theta$ as $z \to 1$, we get $\sum_{j=0}^{\infty} p_{ij}(t) = 1$.

Therefore $Q_1(t)$ remains finite in finite time.

10.4.2 Limiting distributions

Let us assume that the limiting probability distribution $\{\pi_j\}$ of $Q_1(t)$ exists, where $\lim_{t\to\infty} p_{ij}(t) = \pi_j \geq 0\ (j \geq 0)$ and $\pi_0 = \lim_{t\to\infty} p_{io}(t) = \lim_{\theta\to 0} \theta p^*_{io}(\theta)$From (10.4.2) we have

$$\lim_{t\to\infty} G(z,t) = \lim_{\theta\to 0} \theta G^*(z,\theta)$$

$$= \frac{k\mu(1-z)\pi_0}{k\mu - (\lambda+k\mu)z + \lambda z^{k+1}}\ (|z| \leq 1) \qquad (10.4.3)$$

$= p(z)$, say.

We invert the z transform to find the distribution of the number of phases in the system.

In case $k = 1, P(z) = \dfrac{\mu\pi_0(1-z)}{\mu + \lambda z^2 - (\lambda+\mu)z} = \dfrac{\pi_0(1-z)}{1+\rho z^2 - (1+\rho)z}$ (since $\rho = \lambda/\mu) = \dfrac{\pi_0(1-z)}{(1-z)(1-\rho z)} = \dfrac{\pi_0}{1-\rho z}$ which is the G.F. of geometric distribution $p_k = (1-\rho)\rho^k, k \geq 0$. ($P(1) = 1$ (10.4.3) and L'Hospital rule gives

$$P(1) = 1 = \frac{k\mu\pi_0}{k\mu - \lambda k}, \pi_0 = 1 - \lambda/\mu = 1 - \rho \text{ if } \rho < 1.$$

Expand $P(z)$ by partial fraction expansion and then invert each term by inspection. Denominator of (10.4.3) is equal to $(1-z)[k\mu - \lambda(z + z^2 + \ldots + z^k)]$. If $z_1, z_2, \ldots z_k$ are the remaining zeros of the denominator assuming distinct when ($\rho < 1$), the denominator $= k\mu(1-z)(1-/z_1)\ldots(1-z/z_k)$.

Then $P(z) = \dfrac{1-\rho}{(1-z/z_1)\ldots(1-z/z_k)}$. By Partial fraction expansion

$$P(z) = (1-\rho)\sum_{i=1}^{k}\frac{A_i}{(1-z/z_i)}, \text{ where } A_i = \prod_{\substack{n=1\\n\neq i}}^{k}\frac{1}{(1-z_i/z)}. \qquad (10.4.4)$$

Inverting (10.4.4) by inspection, the final solution for the distribution (the limiting distribution of $Q_1(t)$) of the number of phases in the system, we obtain

$$p_j = (1-\rho)\sum_{i=1}^{n} A_i Z_i^{-j}, j = 1, 2, \ldots, k$$

which is a weighted sum of geometric distribution [expanding (10.4.4) in power series in z, convergent in the region $|z| < z_i, |z_i| \geq 1$. z_i are the roots of the equation $z^k + z^{k-1} + \ldots + z = \rho/k$ and $\lambda z_1 . z_2 \ldots z_k = k\mu(-1)^{k+1}$

$$P(z) = (1-\rho)\sum_{i=1}^{k} A_i \sum_{j=0}^{\infty} (z/z_i)^j, |z| < \min_{1\le i\le k} |1/m_i|$$

$$= (1-\rho)/\pi_1^k (1 - zm_i) \text{ where } m_i = 1/z_i, |m_i| < 1].$$

Let W be the waiting time of a customer who joins the system after an indefinitely long period of time. If the queue is empty, $W = 0$; but if the system contains $j \ge 1$ phases then W is the time to complete these j phases.

Therefore $P(W = 0) = 1 - \rho$, $P(x < W < x + dx)$

$$= \sum_{j=1}^{\infty} (1-\rho) \sum_{i=1}^{k} A_i z_i^{-j} e^{-k\mu x} (k\mu)^j / (j-1)! \, x^{j-1} dx$$

$$= (1-\rho) \sum_{i=1}^{k} A_i z_i^{-1} k\mu \, e^{-k\mu(1-1/z_i)x} dx, 0 < x < \infty$$

is a continuous p.d.f. which is the weighted sum of k negative exponential terms. This concept of phase type service is due to Erlang (1920). Also note that the Queue-length $Q(t)$ and the number of phases $Q_1(t)$ are related by

$$Q(t) = [(Q_1(t) + k - 1)/k],$$

where $[x]$ is the greatest integer contained in x.

10.5 The Queueing Models *M/G/*1 and *GI/M/*1

We have discussed queueing processes which are either birth and death process or non-birth and death process. They are in either case Markovian and the theory of Markov chains and process could be applied in their studies. We now shall consider models where the distributions of the interarrival time or the service time do not posses the memory loss property i.e. are not exponential. Here the process $Q(t)$ will no longer be Markovian; however, the analysis of the process can be based on an associated process which is Markovian. Two techniques are generally being used for this purpose. D.G. Kendall (1951) used the concept of regeneration point (due to plam 1943) by suitable choice of regeneration points and extracts, from the process $Q(t)$, Markov chains in discrete time at those points. This is known as the *technique of imbedded Markov chains.* The second important technique (due to Keilson and Kooharian 1960) known as *supplementary variable technique,* involves inclusion of such variables. Let $R(t)$ be the remaining service time of the customer currently being served. The vector $(Q(t), R(t))$ is a vector Markov process. For a complete investigation into the behaviour of $Q(t)$ the vector process has to be studied by some means.

10.5.1 Queues with Poisson input: Model *M/G/*1

Assume that the input process is Poisson with intensity λ and that the service times are i.i.d. r.v.s $V_1, V_2, \ldots V_n$ with an arbitrary distribution with mean $1/\mu$. Denote the service time by V, its d.f. $G(v)$, p.d.f., if exists, by $g(v) = G'(v)$ and its Laplace transform by

$$G^*(s) = \int_0^\infty e^{-sv}\, dG(v) \tag{10.5.1}$$

Let t_n, $n = 1, 2, 3, \ldots$ ($t_0 = 0$) be the nth departure epoch, i.e. the instant at which the nth unit completes his service and leaves the system. These points t_n are the regeneration points of the process $\{Q(t)\}$, $Q(t)$ being the number of customers in the system. The sequence of points $\{t_n\}$ forms a renewal process. $Q(t_n + 0)$, the number in the system immediately after the nth departure has a denumberable state space $0, 1, 2, \ldots$ Write $Q(t_n + 0) = Q_n$, $n = 0, 1, 2, \ldots$ and then $\{Q_n\}$ is a M.C. (which will be demonstrated shortly).

Let X_n be the r.v. giving the number of customers that arrive during the service time of the nth customer whose service time is denoted by V_n.

Then
$$Q_{n+1} = \begin{cases} Q_n - 1 + X_{n+1} & \text{if } Q_n \geq 1 \\ X_{n+1} & \text{if } Q_n = 0 \end{cases}$$

i.e. $Q_{n+1} = Q_n - H(Q_n) + X_{n+1}$, H is the heavyside function given by

$$H(x) = \begin{cases} 1 \text{ if } x > 0 \\ 0 \text{ if } x \leq 0 \end{cases}$$

Now the service times of all the customers have the same distribution so that $X_n \stackrel{d}{=} X$ for all $n \geq 1$.

Then we have $P\{X = r \mid \text{service time of a customer is } v\}$

$$= \frac{e^{-\lambda v}(\lambda v)^r}{r!} = \int_0^\infty P(X_n = r \mid V_n)\, P(v < V_n \leq v + dv)$$

and so
$$k_r = P(X = r) = \int_0^\infty \frac{e^{-\lambda v}(\lambda v)^r}{r!}\, dG(v),\ r = 0, 1, 2, \ldots \tag{10.5.2}$$

gives the number of arrivals during the service time of a customer. The transition probabilities $p_{ij} = P(Q_{n+1} = j \mid Q_n = i)$ are given by

$$\left.\begin{array}{l} p_{ij} = \begin{cases} k_{j-i+1}, & i \geq 1, j \geq i - 1 \\ 0 & i \geq 1, j \leq i - 2 \end{cases} \\ p_{oj} = p_{ij} = k_j, j \geq 0 \end{array}\right] \tag{10.5.3}$$

Equation (10.5.3) shows that $\{Q_n, n \geq 0\}$ is a M.C. with transition matrix

$$P = (p_{ij}) \begin{bmatrix} k_0 & k_1 & k_2 & k_3 & \ldots \\ k_0 & k_1 & k_2 & k_3 & \ldots \\ 0 & k_0 & k_1 & k_2 & \ldots \\ 0 & 0 & k_0 & k_1 & \ldots \\ . & . & . & . & \end{bmatrix}$$

As every state can be reached from every other other state, the M.C. $\{Q_n\}$ is irreducible and also since $p_{ii} > 0$, the chain is aperiodic. It can be shown that if the traffic intensity $\rho = \dfrac{\lambda}{\mu} < 1$, the M.C. is recurrent, Nonnull and hence Ergodic. By Ergodic theorem of M.C. the limiting probabilities $\pi_j = \lim_{n\to\infty} p_{ij}^{(n)}$, $j = 0, 1, 2, 3, \ldots$ exist and are independent of the initial state i. The probabilities $\pi = (\pi_0, \pi_1 \ldots)$, $\sum_j \pi_j = 1$, are given as the unique solution of the equations

$$\pi = \pi P \tag{10.5}$$

Let $K(s) = \sum_{j=0}^{\infty} k_j s^j$ and $\pi(s) = \sum_{j=0}^{\infty} \pi_j s^j$.

Then
$$K(s) = \sum_{j=0}^{\infty} s^j \left\{ \int_0^{\infty} \frac{e^{-\lambda v}(\lambda v)^j}{j!} dG(v) \right\} \tag{10.5.5}$$

$$= \int_0^{\infty} e^{-(\lambda - \lambda s)v} dG(v) = G^*(\lambda - \lambda s)(|s| < 1) \tag{10.5.6}$$

Hence $E(X_n) = K'(1) = -\lambda G^{*}(0) = \lambda/\mu = \rho$.

Multiply the $(k + 1)$th equation by s^k in the infinite system of equations in (10.5.4), $k = 0, 1, 2, \ldots$ and adding over k we obtain, on simplification,

$$\pi(s) = \frac{\{1 - K'(1)\}(1 - s)K(s)}{K(s) - s}$$

$$= \frac{(1 - \rho)(1 - s)(s)}{K(s) - s} \; (0 < \rho = K'(1) < 1) \tag{10.5.7}$$

$$= \frac{(1 - \rho)(1 - s)G^*(\lambda - \lambda s)}{G^*(\lambda - \lambda s) - s}$$

which is the Pollaczek (1930)-Khintchine (1932) (P.K.) formula.

Note

(1) In the $M/G/1$ Queue by extracting a process at departure points t_n we get an associated process $Q_n = Q(t_n + 0)$ which is a M.C. Hence the transition probabilities p_{ij} of the M.C. give departure point probabilities of the Queue-length. We have just seen that when $\rho < 1$, the limiting probabilities $\pi_n = \lim_{m\to\infty} p_{ij}^{(m)}$ (which are also stationary or steady state probabilities) exist and can be obtained by applying ergodic theorem of M.C. Also in this case $p_{in} = \lim_{t\to\infty} p_{in}^{(t)}$ exist and one can show that $\pi_n = p_n$ for all n i.e. the limiting departure point probabilities are equal to the corresponding steady state probabilities of the queue-length (system size).

(2) Another interesting process can be extracted. The process $Y(t) = Q_n$, $t_n \le t \le t_{n+1}$ gives the number of units left behind in the system by the most recent departure. Then $\{Q_n, t_n\}$ is a Markov renewal process and $\{Y(t), t \ge 0\}$ is a semi-Markov process. Q_n being its imbedded M.C., and the waiting time in a state being distributed as the service time V.

10.5.3 Examples

Example 1 The Model $M/E_k/1$: Here the service time V is Erlang (k) with p.d.f.

$$dF(V) = g(v) = \frac{(\mu k)^k v^{k-1} e^{-k\mu v}}{(k-1)!}, 0 < v < \infty$$

Its Laplace Transform is $G^*(s) = \left(\frac{\mu k}{s + \mu k}\right)^k$

From (10.5.6) and (10.5.7) we get

$$K(s) = G^*(\lambda - \lambda s) = \left(\frac{\mu k}{\lambda - \lambda s + \mu k}\right)^k = \left(1 + \frac{\rho(1-s)}{k}\right)^{-k},$$

where $\rho = \lambda/\mu$ and $\pi(s) = \dfrac{(1-\rho)(1-s)}{1 - s(1 + \rho(1-s)/k)^k}$.

Expanding in powers of s and comparing coefficients of s^n on both sides, we get π_n, and equivalently p_n.

We can find the moments of p_n from $\pi(s)$. This model could also be considered as non-birth and death Markovian, by considering that the service comprises of k exponential phases each with mean $1/(k\mu)$ and the Chapman-Kolmogorov equations can be written and solution obtained there from.

Special case: $M/M/1$ Queue: putting $k = 1$, we get (for $M/M/1$ Queue),

$$\pi(s) = \frac{(1-\rho)(1-s)}{(1 - s - s\rho(1-s))} = \frac{1-\rho}{1 - s\rho}.$$

Example 2 The Model $M/D/1$: As $k \to \infty$ the Erlang E_k distribution tends to $1/\mu$ i.e. $E_k \to D$, as $k \to \infty$.

Therefore $G^*(s)$, the Laplace transform of $E_k \to e^{-s/\mu}$, so that

$$K(s) = G^*(\lambda - \lambda s) \to e^{-\rho(1-s)} \text{ and } \pi(s) = \frac{(1-\rho)(1-s)}{1 - se^{\rho(1-s)}}.$$

Expanding right hand side in powers of s, the coefficient of s^n gives

$$\pi_0 = 1 - \rho,\ \pi_1 = (1-\rho)(e^{\rho} - 1), \ldots \tag{10.5.8}$$

$$\pi_n = (1-\rho)\left[e^{n\rho} + \sum_{k=1}^{n-1} (-1)^{n-k} e^{k\rho} \left\{\frac{(k\rho)^{n-k}}{(n-k)!} + \frac{(k\rho)^{n-k-1}}{(n-k-1)!}\right\}\right], n \geq 2.$$

10.5.4 Pollaczek and Khintchine results waiting time: Pollaczek-Khintchine formula

Assume that the steady state exists ($\rho < 1$).

Let W and W_Q denote the waiting times in the system and in the Queue respecively and $W(t)$ and $W_Q(t)$ denote their d.f.s Then

$$p_n = \pi_n = P\{\text{departing customer leaves } n \text{ in the system}\}$$

$$= \int_0^{\infty} P\{\text{departing customer leaves } n \text{ in the system} \backslash t \leq W < t + dt\}$$
$$P(t \leq W < t + dt)$$

$$\simeq \int_0^\infty \frac{e^{-\lambda t}(\lambda t)^n}{n!} dW(t), n \geq 0, \tag{10.5.9}$$

Thus
$$\pi(s) = \sum_0^\infty \pi_n s^n = \int_0^\infty \{e^{-\lambda t} dW(t)\}\left\{\sum_{n=0}^\infty \frac{(\lambda t)^n}{n!} s^n\right\}$$

$$= \int_0^\infty e^{-\lambda t(1-s)} dW(t) = W^*(\lambda - \lambda s) \tag{10.5.10}$$

where $W^*(\theta)$ is the L.T. of $W(t)$. Now

$$P(s) = \sum_{n=0}^\infty p_n s^n = \sum_{n=0}^\infty \pi_n s^n = \pi(s)$$

$$= \frac{(1-\rho)(1-s)G^*(\lambda - \lambda s)}{G^*(\lambda - \lambda s) - s} \text{ by (10.5.6), (10.5.7) and (10.5.10).}$$

Hence $W^*(\alpha) = \dfrac{(1-\rho)(\alpha/\lambda)G^*(\alpha)}{G^*(\alpha) - (1-\alpha/\lambda)}$, where $\alpha = \lambda - \lambda s$

$$= \frac{\alpha(1-\rho)G^*(\alpha)}{\alpha - \lambda(1 - G^*(\alpha))}, \text{ or } W^*(s) = \frac{s(1-\rho)G^*(s)}{s - \lambda(1 - G^*(s))} \tag{10.5.11}$$

This connects the Laplace transform of the distribution of the service time V and waiting time W in the system.

Now $\qquad W = W_Q + V$, so that $W^*(s) = W_Q^*(s)G^*(s)$

(W_Q and V are independent), where W_Q^* is the Laplace transform of W_Q. Hence from (10.5.11) we get

$$W_Q^* = \frac{s(1-\rho)}{s - \lambda(1 - G^*(\alpha))}$$

The relations (10.5.11) and (10.5.12) are also known as *Pollaczek-Khintchine* (P.K.) formula. To distinguish it from the other P.K. formula (10.5.7) it is called *P.K. transform formula.*

10.5.5 Expected waiting time and expected number in the system

Using (10.5.7) and (10.5.12), moments of $Q(t)$ and W_Q can be obtained. We have

$$G^{*(k)}(0) = \frac{d^k}{ds^k} G^*(s)\,|_{s=0} = (-1)^k E(V^k), k \geq 1.$$

From (10.5.12)

$$W_Q^{*(1)}(s)/(1-\rho) = \frac{s - \lambda + \lambda G^*(s) - s\{1 + \lambda G^{*(1)}(s)\}}{[s - \lambda - \lambda G^*(s)]^2} \tag{10.5.13}$$

Since $W_Q^{(1)}(0)$ is 0/0 form as $s \to 0$, by L'Hospital's rule as $s \to 0$, the right hand side of (10.5.13) becomes

$$\left.\frac{(1-\lambda G^{*(1)}(s))-(1+\lambda G^{*(1)}(s))-s(\lambda G^{*(2)}(s))}{2(s-\lambda+\lambda G^{*}(s))(1+\lambda G^{*(1)}(s))}\right|_{s=0}=\frac{-\lambda G^{*(2)}(0)}{2(1+\lambda G^{*(1)}(0))^{2}}$$

$$=\frac{\lambda}{2(1-\rho)}(\sigma_v^2+1/\mu^2)$$

Using Little's formula or otherwise

$$E(W_Q)=W^{*(1)}(0)=\frac{\lambda}{2(1-\rho)}E(V^2)=\frac{\lambda E(V^2)}{2(1-\lambda/\mu)^2}.$$

$$E(W)=E(W_Q+V)=\frac{\lambda}{2(1-\rho)}E(V^2)+\frac{1}{\mu},$$

$$EQ(t)=\lambda E(W)=\rho+\frac{\lambda^2}{2(1-\rho)}E(V^2)$$

and $$E(Q^*(t))=\lambda(EW_Q)=\lambda^2/[2(1-p)]E(V^2).$$

Kleinrock (1976) called this *P.K. Mean value formula*.

By differentiating $W_Q^*(s)$ k times with respect to s and putting $s = 0$, we can find the kth moment of W_Q. It can be proved that

$$\text{Var}(W_Q)=\frac{\lambda}{12(1-\rho)^2}[4(1-\rho)E(V^3)+3\lambda\{E(V^2)\}^2]$$

and $C^2 \geq 1$, where $C^2 = \text{Var}(W_Q)/E^2(W_Q)$.

10.6 System Where Customers Arrive in Stages: $E_k/M/1$ Model

Here we have Frlangian arrivals, and the service time distribution $\mu e^{-\mu v}dv$ $(0 < v < \infty)$ with mean inter-arrival time λ^{-1} and traffic intensity $p = \lambda/\mu$.

[The Erlangian input $dA(u)=\frac{(\lambda k)^k}{(k-1)!}e^{-\lambda k u}u^{k-1}du$, $1 < k < \infty$ is a k-fold convolution of the negative exponential distribution and can be considered as representing an arrival process which occurs in k successive stages such that the time required for each stage has the negative exponential distribution independently of the others, and the arrival takes place only at the completion of these k-stages, when this happens, the process starts with the first stage of the next arrival.]

Let $Q_1(t)$ be the number of stage completed at time t, so that if $Q_1(t) = jk + v$ $(0 \leq k < k-1)$, then the Queue length is $Q(t) = j$, thus $Q(t)=\left[\frac{Q_1(t)}{k}\right]$.

The successive stages enter the system at a Poisson rate $k\lambda$, whereas the departure of each customer removes k stages from it. Thus $Q_1(t)$ is a denumerable Markov process whose transitions during $(t, t + dt]$ are given by

$$p_{ii}(dt) = 1 - k\lambda\, dt + o(dt) \quad (0 \leq i \leq k-1),$$

$$p_{ii}(dt) = 1 - (k\lambda + \mu)\, dt + o(dt) \quad (i \geq k),$$

$$p_{ii+1}(dt) = k\lambda dt + o(dt),$$

$$p_{ii-k}(dt) = \mu dt + o(dt) \ (i \geq k),$$

$$p_{ij}(dt) = o(dt) \ (i \geq 0, j \neq i, \neq i \neq i - k).$$

The forward Kolmogorov equations are

$$p'_{io}(t) = -k\lambda p_{io}(t) + \mu p_{ik}(t),$$

$$p'_{ij}(t) = -k\lambda p_{ij}(t) + k\lambda p_{ij-1}(t) + \mu p_{i,j+1}(t) \ (1 \leq j \leq k - 1) \tag{10.6.1}$$

$$p'_{ij}(t) = -k\lambda p_{ij}(t) + k\lambda p_{ij-1}(t) + \mu p_{i,j+k}(t) \ (1 \leq j \leq k - 1),$$

$$p'_{ij}(t) = -(k\lambda + \mu)\, p_{ij}(t) + k\lambda p_{ij-1}(t) + \mu p_{i,j+k}(t) \ (j \geq k).$$

From these equations, an expression for the transform $G^*(Z, \theta)$ can then be derived exactly as $M \backslash E_k \backslash 1$ Model. However, Jackson and Vickols (1956) obtain the limiting probability distribution u_j of $Q_1(t)$ by solving these equations after setting $p'_{ij}(t) = 0$. Thus u_j's satisfy the equations.

$$\left.\begin{aligned} k\lambda u_0 &= \mu u_k \\ k\lambda u_j &= k\lambda u_{j-1} + \mu u_{j+k} \qquad (1 \leq j \leq k - 1) \end{aligned}\right] \tag{10.6.3}$$

$$(k\lambda + \mu)u_j = k\lambda\, u_{j+1} + \mu u_{j+k} \qquad (j \geq k)$$

To solve (10.6.3), multiplying (10.6.3) successively by 1, z, z^2, . . . and summing, we get

$$k\lambda U(z) + \mu\left[U(z) - \sum_{j}^{k-1} u_j z^j\right] = k\lambda \sum_{j=1}^{\infty} u_{j-1} z^j + \mu \sum_{0}^{\infty} u_{j+k} z^j$$

$$= k\lambda z U(z) + \mu z^{-k}\left[U(z) - \sum_{o}^{k-1} u_j z^j\right],$$

which gives

$$U(z) = \frac{(1 - z^k) \sum_{0}^{k-1} u_j z^j}{k\rho z^{k+1} - (1 + k\rho) z^k + 1} \tag{10.6.4}$$

where $U(z) = \sum_{0}^{\infty} u_j z^j \ (|z| \leq 1)$, the G.F. of $\{u_j\}$.

Now for $|z| = 1 + \partial$, $\partial > 0$ and small, we have

$$\left|\frac{1 + k\rho z^{k+1}}{(1 + k\rho) z^k}\right| \leq \frac{(1 + \partial)^{-k} + k\rho(1 + \partial)}{1 + k\rho} = 1 + \frac{(1 + \partial)^{-k} + (1 - k\rho\partial)}{1 + k\rho}$$

$$= 1 + \frac{[1 - k\partial + 1/2k(k + 1)\partial^2 + \ldots] - (1 - k\rho\partial)}{1 + k\rho}$$

$$= 1 + \frac{k(\rho - 1)\partial + \ldots}{1 + k\rho} < 1 \ (\because \rho < 1).$$

By Rouche's theorem it follows that the equation $k\rho z^{k+1} - (1 + k\rho)z^k + 1 = 0$ has exactly k roots within the circle $| z | = 1 + \partial$; the remaining root Z_0 (say) is such that $| z_0 | > 1$. Now, z_0^{-1} is the only root within the unit circle of the equation

$$z^{k+1} - (1 + k\rho)z + k\rho = 0 \qquad (10.6.5)$$

or $z = K(z)$, where $K(z) = \dfrac{(z^{k+1} + k\rho)}{1 + k\rho}$. Here $K(z)$ is a p.g.f. with the mean $K'(1) = \dfrac{k + 1}{1 + k\rho} > 1$ since $\rho < 1$.

From the theory of Branching process it is known that $0 < z_0^{-1} = \mathscr{L}_0$ (say) < 1 iff $K'(1) > 1$. Since $\mathscr{L}_0$ satisfies equations (10.6.5), the relation $k\rho(1 - \mathscr{L}_0) = \zeta_0 (1 - \mathscr{L}_0^k)$ holds (10.6.6)

Since the p.g.f. $U(z)$ is analytic in the region $| z | \le 1$, it is clear that $k - 1$ zeros of the polynomial $\sum_{j=0}^{k-1} u_j z^j$ and the zeros of the denominator on the right-hand side of (10.6.4) must coincide in this region, we must thus have

$$A[k\rho z^{k+1} - (1 + k\rho)z^k + 1] = (z - 1)(z - z_0) \sum_0^{k-1} u_j z^j \qquad (10.6.7)$$

where A is a constant. Substituting (10.6.7) in (10.6.4) and simplyfying $U(z) = A \dfrac{\sum_0^{k-1} z^j}{z_0 - z}$. The condition $U(1) = 1$ gives $A = (z_0 - 1)/k$, so that finally

$$U(z) = \frac{1}{k}\left(\sum_{j=0}^{k-1} z^j\right) \frac{1 - \mathscr{L}_0}{1 - z\mathscr{L}_0} \quad (| z | \le 1)$$

$$= \frac{1}{k}\left(\sum_0^{k-1} z^j\right)(1 - \mathscr{L}_0)\left(\sum_{j=0}^{\infty} (z\mathscr{L}_0)^j\right)$$

$$= \frac{1 - \mathscr{L}_0}{k}\left[\sum_0^{k-1} z^j \sum_{v=0}^{j} \mathscr{L}_0^{j-v} + \sum_{j=k}^{\infty} z^j \sum_{v=0}^{k-1} \mathscr{L}_0^{j-v}\right]$$

$$= 1/k \sum_{j=0}^{k-1} z^j (1 - \mathscr{L}_0^{j+1}) + 1/k \sum_{j=k}^{\infty} z^j \mathscr{L}_0^{j-k+1}(1 - \mathscr{L}_0^k)$$

$$= 1/k \sum_{j=0}^{k-1} z^j (1 - \mathscr{L}_0^{j+1}) + 1/k \sum_{j=k}^{\infty} z^j \mathscr{L}_0^{j-k+1}(1 - \mathscr{L}_0^k) \text{ (by 10.6.6)}$$

Hence, the limiting distribution of $Q_1(t)$ is given by

$$u_j = \begin{cases} 1/k(1 - \mathscr{L}^{j+1}) & (0 \le j \le k - 1) \\ \rho(1 - \mathscr{L}_0)\mathscr{L}_o^{j-k} & (j \ge k). \end{cases}$$

Using the relation $Q_1(t) = [Q_1(t)/k]$, we get the limiting distribution of the Queue-length

$$v_0 = \sum_{v=0}^{k-1} u_v = 1 - \frac{\mathscr{L}_0(1 - \mathscr{L}_0^k)}{k(1 - \mathscr{L}_0} = 1 - \rho \qquad \text{(by (10.6.6))},$$

$$v_j = \sum_{jk}^{jk+k-1} u_v = \rho(1 - \mathscr{L}_0^k)\mathscr{L}_0^{k(j-1)} \qquad (j \geq 1).$$

Limiting distribution of waiting time is given by

$$dF(0) = V_0 = 1 - \rho$$

$$dF(x) = \sum_1^{\infty} v_j e^{-\mu x} \mu^j x^{j-1} dx/(j-1)!$$

$$= \lambda(1 - \mathscr{L}_0^k)\exp[-\mu x(1 - \mathscr{L}_0^k)]\,dx \quad (0 < x < \infty).$$

10.7 Machine Interference Problem

10.7.1 The model with finite input source

In this world of machine and automation, employing the right number of repairmen to look after a certain number of machines is a problem of some significance. The number of repairmen should be decided so as to minimize the loss due to idle machines and repairmen. We shall consider only a simplified model of such a situation.

Suppose that there are $M < \infty$ machines, each of which calls for service after being in use for an interval of time that has a negative exponential distribution with mean λ^{-1}. When there are m machines working at time t, the probability that any one of these machines would call for service in the infinitesimal interval $(t, t + \Delta t]$ is $m\lambda\Delta t + o(\Delta t)$. Let the service times of these machines be negative exponential r.v.s with mean μ^{-1}. Suppose there are *S repairmen* (called *operatives*). We shall call $\rho = \lambda/\mu$, the servicing factor. First, we shall derive some behaviourial results of this process which can be used in decision problems.

Let $N(t)$ be the number of machines not working at time t. Define

$$p_n(t) = P[N(t) = n \mid N(0) = i].$$

The stochastic process $N(t)$ can now be modelled as a birth and death process with parameters:

$$\lambda_n = (M - n)\lambda \text{ and } \mu_n = \begin{cases} n\mu & \text{if } 1 \leq n \leq s \\ s\mu & \text{if } n \geq s \end{cases}$$

The forward Kolmogorov's D.E.s of the process take the form

$$p_0'(t) = -M\lambda p_0(t) + \mu p_1(t),$$

$$p_n'(t) = -[(M - n)\lambda + n\mu]p_n(t) + (M - n + 1)\lambda p_{n-1}(t)$$
$$+ (n + 1)\mu p_{n+1}(t),\ 1 \leq n < s$$

$$p_n'(t) = -[(M - n)\lambda + s\mu]p_n(t) + (M - n + 1)\lambda p_{n-1}(t)$$
$$+ s\mu p_{n+1}(t),\ s \leq n < M$$

$P'_M(t) = -s\mu p_M(t) + \lambda p_{M-1}(t)$. As $t \to \infty$, $M\lambda\pi_0 = \mu\pi_1$,

$$[(M-n)\lambda + n\mu]\pi_n = (M-n+1)\lambda\,\pi_{n-1} + (n+1)\,\mu\pi_{n-1},\ 1 \le n < s,$$

$$[(M-n)\lambda + n\mu]\pi_n = (M-n+1)\lambda\,\pi_{n-1} + s\mu\,\pi_{n-1},\ s \le n < M$$

$s\mu\pi_M = \pi_{M-1}$, where $\pi_n = \lim_{t\to\infty} p_n(t)$, $n \ge 0$.

Solving recursively, we get

$$\pi_n = \begin{cases} \binom{M}{n}\rho^n\pi_0 & \text{if } 0 \le n \le s \\ \binom{M}{n}\dfrac{n!}{s!\,s^{n-s}}\rho^n\pi_0 & \text{if } s \le n \le M \end{cases}$$

and π_0 can be obtained from $\sum_{n=0}^{M}\pi_n = 1$. We get

$$\pi_0 = \left[\sum_{n=0}^{s-1}\binom{M}{n}\rho^n + \sum_{n=s}^{M}\frac{M!}{(M-n)!\,s!\,s^{n-s}}\rho_n\right]^{-1}$$

When $s = 1$, we get

$$\pi_n = \frac{M!}{(M-n)!}\rho^n\pi_0$$

and $$\pi_0 = \left[1 + \frac{M!}{(M-1)!}\rho + \frac{M!}{(M-2)!}\rho^2 + \ldots + M!\,\rho^M\right]^{-1}$$

The probability that the number of machines in working order is r ($\le M$) is given by π_{M-r}.

Two measures of effectiveness for the system can be defined as follows:

$$\text{Machine availability} = 1 - E(N)/M$$

$$\text{Operative utilization} = \sum_{n=0}^{s}\frac{n\pi_n}{s} + \sum_{n=s+1}^{M}\pi_n.$$

Clearly, the *machine availability* represents the fraction of total running time on all machines and the *operative utilization* represents the fraction of time any operative would be working. When $s = 1$, the operative utilization has the simple form $1 - \pi_0$.

Other measures of effectiveness of the system are coefficient of loss for machines, defined as

$$\frac{\text{Average number of machines actually waiting}}{\text{Total number of machines}}$$

and

$$\text{Coefficient of loss for operatives} = \frac{\text{Average number of idle operatives}}{\text{Total number of operatives}}.$$

10.8 Bulk Queues-Elementary Results (Non-Birth and Death Queueing Processes)

Bulk Queues: This section some non-birth and death processes which are Markovian. Namely discusses we shall consider *Group Arrival System* $M^{(x)}/M/1$ and processes with *Bulk Service Rules.* Consider $M^{(x)}/M/1$ queue with Poisson input with parameter λ. In this model the number of units arriving at an instant is a r.v. X with p.f. $c_k = P(X = k)$, $k = 1, 2, \ldots$ This is same as saying that in an infinitesimal interval $(t, t + h)$ the probability that a group of k arrives is $\lambda c_k h + o(h)$. Service takes place singly and independently with exponential distribution having mean $1/\mu$.

It is still Markovian but is a non-birth and death process, as transition may occur to states not necessarily neighbouring states. We denote this model by $M^{(x)}/M/1$. We obtain

$$p_0'(t) = -\lambda p_0(t) + \mu p_1(t)$$

$$p_n'(t) = -(\lambda + \mu)p_n(t) + \mu p_{n+1}(t) + \sum_{k=1}^{n} \lambda c_k p_{n-k}(t), n \geq 1$$

Then the P.G.F. of X is $C(s) = \sum_{k=1}^{\infty} c_k s^k$ with mean arrival rate $\lambda EX = \lambda C'(1)$, and the traffic intensity $\rho = \dfrac{\lambda EX}{\mu} = \lambda C'(1)/\mu$. Assuming steady state solution exists, we get

$$0 = -\lambda p_0 + \lambda p_1 \tag{10.8.1}$$

$$0 = -(\lambda + \mu)p_n + \mu p_{n+1} + \lambda \sum_k c_k p_{n-k}, n \geq 1 \tag{10.8.2}$$

Let $P(s)$ be the G.F. of $\{p_n\} = \sum_0^{\infty} p_n s^n$.

Multiplying (10.8.2) by s^n for $n = 1, 2, 3 \ldots$ and adding to (10.8.1) we get

$$0 = \lambda P(s) - \mu(P(s) - p_o) + (\mu/s)(P(s) - p_0) + \lambda \sum_{n=1}^{\infty} \sum_{k=1}^{n} c_k p_{n-k} s^n$$

$$= -\lambda(s) - \mu(P(s) - p_0) + (\mu/s)(P(s) - p_0) + \lambda \sum_{k=1}^{n} c_k s^k \left(\sum_k^{\infty} p_{n-k}\, s^{s-k} \right)$$

$$= -\lambda P(s) - \mu(P(s) - p_0) + (\mu/s)(P(s) - p_0) + \lambda c(s)P(s).$$

Therefore $$P(s) = \frac{\mu(1 - s)p_0}{\mu(1 - s) - \lambda s(1 - C(s))}.$$

To get p_0 we use $P(1) = 1$. By L'Hospital's rule

$$1 = \lim_{s \to 1} P(s) = \frac{-\mu p_0}{-\mu + \lambda C'(1)} \text{ or } p_0 = 1 - \rho.$$

Therefore $$P(s) = \frac{\mu(1 - s)(1 - \rho)}{\mu(1 - s) - \lambda s(1 - C(s))}.$$

Particular case: M/M/1 Since $c_1 = 1$, $c_k = 0$, $k > 1$ and $P(s) = (1 - \rho)/(1 - \rho s)$.

10.8.1 Bulk service rules

There are a number of policies/rules according to which batches for bulk service may be formed. The bulk service frequently used in the literature are as follows:

1. Bailey (1954) considers that units are served in batches of not more than b (say). If after the completion of a service, the server finds more than b units waiting he takes a batch of b for service while other wait. If he finds r units $(0 \le r \le b)$, he takes all the r units in a batch for service. Jaiswal (1960) and Neuts (1967) consider the same rule with the restriction that $r \ne 0$ $(1 \le b)$ i.e. the service facility stops until a unit arrives. This rule is called *usual bulk service rule* while Bailey's rule is called *modified type rule.*

2. The rule with a fixed batch size k has been considered by Fabens (1961) and Takacs (1962). In this the server waits until there are k units.

3. Neuts (1967) considers that if, after completion of a service, the server finds less than a units, he waits for service, if he finds a or more but at most b, he takes in the batch for service b units, while others wait. The batch takes a minimum of a units and maximum of b units. Since rules (1) and (2) can be considered as special cases of this rule, this is called *general bulk service rule.*

4. U.N. Bhat (1963) considers that the number taken in a batch is a r.v.X. The corresponding Markov model is denoted by $M^{(x)}/M/1$.

Exercises and Complements

Exercise 10.1 Find the expected number L in the system $M/M/1/k$ and expected Queue length L_q.

Exercise 10.2 Find the steady-state probabilities π_n for the finite-waiting-space multichannel queueing system $M/M/c/k$, where the number of channels is c and the maximum queue allowed in the system is $k(k > c)$. Also deduce expressions for L, L_q, $F(W)$, W_q by using Little's formula. Find the expected number of busy channels.

Exercise 10.3 Use Bayes theorem to find arrival point probabilities q_n in the system $M/M/c/k$. Find also the waiting time distribution function in terms of q_n's Show that in $M/M/c/\infty$, $q_n = \pi_n$ for $n \ge 1$.

Exercise 10.4 (Erlang's loss system) Find the steady-state probability π_n for the queueing system $M/M/c/c$ and also find the probability that all the channels are busy in this system. Describe a practical situation represented by the above model.

Exercise 10.5 Find the steady-state probability π_n, given that

$$\lambda_n = \begin{cases} (N-n)\lambda, & 0 \le n \le N \\ 0 & n \ge n \end{cases}$$

$$\mu_n = \begin{cases} n\mu, & 1 \le n < c \\ c\mu & n \le c \end{cases}$$

Describe a practical situation represented by the above model.

Exercise 10.6 A machine shop company has N machines. These machines break down periodically and the company has repairmen to service the machines when they break down. When a machine is fixed, the time until the next breakdown is exponentially distributed with mean $\frac{1}{\lambda}$. The shop always has enough of a work backlog to ensures that all machines in operating condition will be working. The repair time for each repairman is exponentially distributed with a mean $\frac{1}{\mu}$. Find the average number of machines operational at any given time, the expected "downtime" of a machine requires repair, and the expected percent idle time of each repairman.

Exercise 10.7 Show that the two expressions for the effective mean arrival rate, in the $M/M1/k$ model are equal, that is, $\lambda' (1 - \pi_n) = \mu(L - L_q)$.

Exercise 10.8 In the $M/G/1$ queueing system, let Q_n be the number of customers left behind in the system by the nth departing customer. Using the fact that in the steady-state the expectation $E(Q_{n+1}) = E(Q_n)$ and $E(Q_{n+1}^2) = E(Q_n^2)$, deduce that the steady-state expectation $E(Q)$ is given by

$$E(Q) = \rho + \frac{\rho^2 (1 + C_v^2)}{2(1 + \rho)},$$

where C_v is the coefficient of variation of service times, and ρ is the traffic intensity.

Exercise 10.9 (*Takacs integral equation*) In $M/G/1$ Queueing system, show that the busy period distribution function $G(x)$ is given by

$$G(x) = \int_0^x \sum_{n=0}^{\infty} \{(\lambda v)^n e^{-\lambda v} / n!\} G^{(n)} (x - v) dV(v)$$

where $V(v)$ is the service time distribution and $G_{(x)}^{(n)}$ denotes the n-fold convolution of $G(x)$ with itself. Hence deduce the functional equation

$$G^*(s) = V^*(s + \lambda - \lambda G^*(s))$$

where $G^*(s)$ and $V^*(s)$ are the Laplace-Stietijes transforms of $G(x)$ and $V(v)$, respectively. Also verify the expression for $G^*(s)$ obtained in the text for the $M/M/1$ system.

Exercise 10.10 Show that, for a $M/G/1$ queue, the sequence of times at which the server passes from being busy to being free constitutes a renewal process.

APPENDIX I

Sample Function of Brownian Motion

Let $\{X(t), t \in T\}$ be a real valued stochastic process on a complete probability space $\{\Omega, \mathscr{F}, P\}$ and $\mathscr{B}$ is the Borel $\sigma-$ field of the real line. From the definition of a stochastic process, it follows that the mapping $X(t,.)$: $\Omega \to R$ is $\mathscr{B}$-measurable for every $t \in T$. In other words, $\{\omega: X(t, \omega) \in B\} \in \mathscr{B}$ for each Borel set $B \subset R$. However,

$$\{\omega: X(t, \omega) \in B, t \in T\} = \bigcap_{t \in T} \{\omega: X(t, \omega) \in B\}$$

need not be in $\mathscr{B}$ in general, unless T is countable. This then implies that functions $\sup_{t \in T} X(t)$ and $\inf_{t \in T} X(t)$ may not be $\mathscr{B}$-measurable because they involve uncountable intersections and unions like $\{\sup_{t \in T} X(t) \le x\} = \bigcap_{t \in T} \{X(t) \le x\}$.

Therefore, a large number of important functionals of a continuous parameter stochastic process may not be random variables. A remedy for this situation is the separability concept introduced by J.L. Doob.

Definition I.1 The Process $\{X(t), t \in T\}$ is said to be *separable* iff there is a countable dense subset T_0 of T, called the *separating set* and a set A of probability zero called the negligible set, such that if $\omega \notin A$ and $t \in T$, there is a sequence $t_n \in T_0$, $t_n \to t$, with $X(t_n, \omega) \to X(t, \omega)$. Following results are well-known and easy to prove.

1. If a stochastic process $\{X(t), t \in T\}$ is *separable* and continuous in probability (i.e. $X(t) \xrightarrow{P} X(t_0)$ as $t \to t_0$), then any countable dense set $T_0 \subset T$ can be used as a separating set.

2. Let A be the negligible set and T_0 be the separating set of a separable process $\{X(t), t \in T\}$. If $\omega \notin A$, t_0 is a cluster point of T, and $\lim_{t \to t_0, t \in T_0} X(t, \omega)$ exists, then $\lim_{t \to t_0, t \in T} X(t, \omega)$ exists and the two limits are equal.

3. If $\{X(t), t \in T\}$ be stochastic process which is continuous in probability is separable, then every countable set dense in T is a separating set.

A basic property of separable processes is that many sets whose definitions involve uncountably many values of the parameter t become measurable. So separability allows us to handle events and random variables that involves uncountably many values of the parameter t of a stochastic process. However, if we wish to integrate the sample functions, a new concept is needed. Suppose that for $\{X(t), t \in I\}$, I an interval of R we ask the question whether almost every sample function is Lebesgue integrable on I. Using Fubini's Theorem

$$\int_{\Omega}\int_{I} | X(t, \omega) | \, dtdP = \int_{I}\int_{\Omega} | X(t, \omega) | \, dPdt = \int_{I} E[\, | X(t) | \,]dt \text{ and that}$$

implies $\int_{I} | X(t, \omega) | \, dt < \infty$ for almost every ω provided $\int_{I} E[\, | X(t) | \,]dt < \infty$ and $X(t, \omega)$ should be measurable in both variables t and ω.

Definition I.2 A stochastic process $\{X(t), t \geq 0\}$ is said to *adapted to the family* of σ fields $\mathcal{F}(t), t \geq 0$ if for $s < t$ $\mathcal{F}(s) \subset \mathcal{F}(t) \subset \mathcal{F}$ and $X(t)$ is $\mathcal{F}(t)$ – measurable for each t.

The stochastic process is said to be *progressively measurable* iff for all $t > 0$ the map $(s, \omega) \rightarrow X(s, \omega)$ of $[0, t] \, x\Omega \rightarrow \overline{R}$ is measurable relative to $\mathcal{B}[0, t] \, x \, \mathcal{F}(t)$ and $\mathcal{B}(\overline{R})$.

A Brownian motion $\{W(t), t \geq 0\}$ has the property $E(W(t) \mid W(\tau), 0 \leq \tau \leq s) = W(s)$ a.s. for any $t \geq s$. A process having this property is called a *continuous parameter martingale.* For discrete parameter martingale the reader may refer An introduction to measure theory and probability, A. Basu Prentice Hall, India. For Brownian motion we define a continuous parameter martingale a little more generally as follows.

Definition I.3 Let $\{X(t), t \in T\}$ be a stochastic process, and let $\{\mathcal{F}(t), t \in T\}$ be an increasing family of σ-algebras such that for each t, $X(t)$ is $\mathcal{F}(t)$ – measurable. $\{X(t), \mathcal{F}(t), t \in T\}$ is said to be a *martingale* if $i > s$ implies $E(X(t) \mid \mathcal{F}(s)) = X(s)$ *a.s.* and $E(X(t)) < \infty$. The process is said to be a *submartingale* (*supermartingale*) if the last equality is replaced by $\geq$ (respectively, $\leq$).

For any Brownion motion $\{X(t), t \geq 0\}$, we can take $\mathcal{F}(t)$ to be the smallest σ-algebra with respect to which $\{X(s), s \leq t\}$ is a martingale. To prove this merely note that for $t > s$,

$$E(X(t) \mid \mathcal{F}(s)) = E\{(X(s) + X(t) - X(s)) \mid \mathcal{F}(s)]$$

$$= X(s) + E(X(t) - X(s)) = X(s) \text{ a.s.}$$

If $\{X(t), t \geq 0\}$ is a Brownian motion, then

$$E(X(t) - X(s))^2 = | t - s |.$$

Therefore, by Chebychev's inequality, we have

$$P(\, | X(t + h) - X(t) | \geq \varepsilon) \leq \frac{| h |}{\varepsilon^2} \rightarrow 0 \text{ as } h \rightarrow 0 \text{ for every } \varepsilon > 0.$$

Thus, a Brownian motion is continuous in probability. It follows that every Brownian motion has a separable and measurable modification.

A separable Brownian motion has some important sample function properties. The most important is the following.

Theorem I.1 With probability one, every sample function of a separable Brownian motion is uniformly continuous on every finite interval.

To prove theorem I.1, let us first state and prove an extremely useful condition due to Kolmogorov which guarantees almost sure sample continuity of a separable process. (see Definition 1.1, Appendix I).

Theorem I.2 (Kolmogorov) Let $\{X(t), t \in T\}$ be a separable process and let T be a finite interval. If there exists $\alpha, \beta, c > 0$ such that $E \mid X(t+h) - X(t) \mid^{\alpha} \leq c\, h^{1+\beta}$ then

$$\sup_{\substack{t,s \in S \\ |t-s|<2^{n}}} \mid X(t) - X(s) \mid \to 0 \text{ a.s. as } h \to 0$$

so that almost every sample function is uniformly continuous on T.

Proof From the Markov inequality, we get

$$P(\mid X(t+h) - X(t) \mid \geq \varepsilon) \leq \frac{E \mid X(t+h) - X(t) \mid^{\alpha}}{\varepsilon^{\alpha}}, \ \varepsilon > 0.$$

Therefore, $P(\mid X(t+h) - X(t) \mid \geq h^{r}) \leq ch^{1+\beta-\alpha r}$

Let $0 \leq r < \beta/\alpha$, and set $\delta = \beta - \alpha r > 0$. Then

$$P(\mid X(t+h) - X(t) \mid \geq h^{r}) \leq ch^{1+\delta}$$

Let $$Z_{\upsilon} = \sup_{0 \leq k \leq 2^{\upsilon}-1} (\mid X((k+1)/2^{\upsilon}) - X(k/2^{\upsilon}) \mid)$$

Hence $$P(Z_{\upsilon} \geq (1/2^{\upsilon})^{r}) \leq \sum_{k=0}^{2^{\upsilon}-1} P(\mid X((k+1)/2^{\upsilon}) - X(k/2^{\upsilon}) \mid \geq (1/2^{\upsilon})^{r})$$

$$\leq c2^{\upsilon}(1/2^{\upsilon})^{1+\delta} = c2^{-\delta\upsilon}$$

Since $\sum_{\upsilon=0}^{\infty} 2^{-\delta\upsilon} < \infty$ we get $\sum_{\upsilon=0}^{\infty} P(Z_{\upsilon} \geq (1/2^{\upsilon})^{r}) < \infty$.

By Borel-Cantell Lemma, there exists $N(\omega)$ almost surely finite such that $Z_{\upsilon}(\omega) < 1/2^{\upsilon r}$ for all $\upsilon \geq N(\omega)$ and

$$\lim_{n \to \infty} \sum_{\upsilon=n+1}^{\infty} Z_{\upsilon} = 0 \text{ a.s.} \tag{A.I.1}$$

If $\mid t - s \mid < 2^{-n}$, then we can find k such that $0 < k < 2^{n}$ and $\mid t - k/2^{n} \mid < 2^{-n}$, $\mid s - k/2^{n} \mid < 2^{-n}$.

If $t \in S = \{k/2^{n}, k = 0, 1, \ldots, 2^{n} - 1; n = 0, 1, 2, \ldots\}$ and $|t - k/2^{n}| < 2^{-n}$, t must be of the form

$$t = k/2^{n} \pm \sum_{\upsilon=1}^{m} t_{\upsilon} 2^{-(n+\upsilon)} \ (t_{\upsilon} = 0, 1)$$

Thus,

$$\mid X(t) - X(k/2^{n}) \mid \leq \sum_{\upsilon=n+1}^{n+m} Z_{\upsilon} \leq \sum_{\upsilon=n+1}^{\infty} Z_{\upsilon}$$

Since, $\mid X(t) - X(s) \mid \leq \mid X(t) - X(k/2^{n}) \mid + \mid X(s) - X(k/2^{n}) \mid$, we have

$$\sup_{\substack{t,s \in S \\ |t-s|<2^{-n}}} \mid X(t) - X(s) \mid \leq 2 \sum_{\upsilon=n+1}^{\infty} Z_{\upsilon} \tag{A.I.2}$$

From (A.I.1) and (A.I.2), we get

$$\sup_{\substack{t,s\in S\\ |t-s|<2^{-n}}} |X(t)-X(s)| \to 0 \text{ a.s. as } n\to\infty$$

Proof of Theorem I.1

For a Brownian motion $W(t)$

$$E\,|\,W(t)-W(s)\,|^4 = 3\,|\,t-s\,|^2 \text{ and } E\,|\,W(t+h)-W(t)\,|^4 = 3h^2$$

which varifies Kolmogorov's condition of Theorem I.2 with $c = 3$, $\alpha = 4$, $\beta = 1$.

Since Brownian motion is separable and continuous in probability, the diadic rational set S is dense in $T = [0, 1]$.

Hence $\sup_{\substack{t,s\in S\\ |t-s|<h}} |\,W(t)-W(s)\,| = \sup_{\substack{t,s\in T\\ |t-s|<h}} |\,W(t)-W(s)\,|$ a.s.

Hence every sample function of a separable Brownian motion is uniformly continuous on every finite interval T.

Although the sample functions of a separable Brownian motion are almost surely continuous, they are highly irregular. For example, with probability one, the sample functions of a separable Brownian motion are nowhere differentiable and are of unbounded variation on every interval. Roughly, $(W(t+h)-W(t))$ is of order $O(\sqrt{h})$, which is incompatible with either differentiability or bounded variation. This behaviour of the Brownian motion is made precise by the following theorem.

Theorems I.3 (Quadratic variation) Let $T = [a, b]$ be a closed subinterval of $(0, \infty)$.

Let $\pi_n = [a = t_0^{(n)} < t_1^{(n)} < \ldots < t_{N(n)}^{(n)} = b]$, $n = 1, 2, 3, \ldots$, be a sequence of partitions of T such that

$$\Delta_n = \max_{1\le k\le N(n)} (t_k^{(n)} - t_{k-1}^{(n)}) \to 0 \text{ as } n\to\infty.$$

Let, $\{W(t), t\ge 0\}$ be a Brownian motion, then

$$\sum_{k=1}^{N(n)} (W(t_k^{(n)}) - W(t_{k-1}^{(n)}))^2 \xrightarrow{L_2} b-a \text{ as } n\to\infty$$

If $\sum_{n=1}^{\infty} \Delta_n < \infty$ then the convergence is also almost sure.

Proof To prove quadratic mean convergence (L_2), we write

$$S_n = \sum_{k=1}^{N(n)} (W(t_k^{(n)}) - W(t_{k-1}^{(n)}))^2 - (b-a)$$

$$= \sum_{k=1}^{N(n)} [(W(t_k^{(n)}) - W(t_{k-1}^{(n)}))^2 - (t_k^{(n)} - t_{k-1}^{(n)})]$$

Since S_n is a sum of independent r.v.s each with zero mean, we have

$$ES_n = 0,\ ES_n^2 = \sum_{k=1}^{N(n)} E[(W(t_k^{(n)}) - W(t_{k-1}^{(n)}))^2 - (t_k^{(n)} - t_{k-1}^{(n)})]^2$$

$$= 2\sum_{k=1}^{N(n)} (t_k^{(n)} - t_{k-1}^{(n)})^2 \le 2\Delta_n(b-a)$$

[Since $E(W(t) - W(s))^4 = 3\,|\,t-s\,|^2$ and

$$E(W(t) - W(s))^2 = (t-s)]$$

Since $\Delta_n \to 0$ as $n \to \infty$, $S_n \xrightarrow{L_2} 0$ as $n \to \infty$.
By Chebychev's inequality we find

$$P(\,|\,S_n\,| > c) \le ES_n^2/\varepsilon^2 \le 2(b-a)\Delta_n/\varepsilon^2$$

Since $\sum_{n=1}^{\infty} \Delta_n < \infty$, we have for $\varepsilon > 0$,

$$\sum_n P(|\,S_n\,| > \varepsilon) < \infty.$$

By Borel-Cantelli lemma, $|\,S_n\,| > \varepsilon$ i.o. has probability zero and hence $S_n \to 0$ a.s. as $n \to \infty$.

Note

1. The condition $\sum_n \Delta_n < \infty$ for a.s. convergence can be replaced by the condition that T_n is nested, i.e. $T_{n+1} \supset T_n$ for each n (see Doob, p. 395).

2. Taking $t_k^{(n)} = k/2^n, k = 0, 1, \ldots, 2^n$, $\Delta_n = \max_k\left(k/2^n - \frac{k-1}{2^n}\right) = 1/2^n \to 0$

and $\sum_n 1/2^n < \infty$. therefore

$$\lim_{n\to\infty} \sum_{k=1}^{2^n}\left[W\left(\frac{k}{2^n}t\right) - W\left(\frac{k-1}{2^n}t\right)\right]^2 = t \text{ a.s. and in } L_2,$$

$$\sum_{k=1}^{2^n}\left|W\left(\frac{k}{2^n}t\right) - W\left(\frac{k-1}{2^n}t\right)\right| \ge \frac{\sum_{k=1}^{n}\left[W\left(\frac{k}{2^n}t\right) - W\left(\frac{k-1}{n}t\right)\right]^2}{\max_{1\le j\le 2^n}\left|W\left(\frac{j}{2^n}t\right) - W\left(\frac{j-1}{2^n}t\right)\right|}.$$

The numerator on the right converges a.s. to t, while the denominator tends to zero because Brownian motion is uniformly continuous over bounded intervals. Therefore

$$\lim_{n\to\infty} \sum_{k=1}^{2^n}\left|W\left(\frac{k}{2^n}t\right) - W\left(\frac{k-1}{2^n}t\right)\right| = \infty$$

i.e. the *total variation of Brownian motion* path is infinite with probability one.

Theorem I.4 Almost every Brownian path is nowhere differentiable.

Proof We follow Dvoretzky, Erdös, and Kakutani (1961, 4th Berkely symposium) and shall prove stronger result that $W(t)$ has (a.s.) no point of increase or decrease.

Fix $\beta > 0$ and suppose that $W(t)$ has derivative $W'(s)$, $|W'(s)| < \beta$, at some point $s \in [0, 1]$; then there is an n_0 such that for $n > n_0$

$$|W(t) - W(s)| \leq 2\beta |t - s| \text{ if } |t - s| \leq 2/n, \tag{A.I.3}$$

Let $W(.)$ denote functions on [0, 1].

$A_n = \{W(.)$; there exists s such that

$$|W(t) - W(s)| \leq 2\beta |t - s|, \text{ if } |t - s| \leq 2/n\}$$

Then $A_n \uparrow A \subset$ {set of all sample paths of $W(t)$ on [0, 1] having a derivative at any point which is less than β in absolute value}.

If (A.I.3) holds, and let k be the largest integer such that $k/n \leq s$, then the following is implied

$$Y_n = \max\left(\left|W\left(\frac{k+2}{n}\right) - W\left(\frac{k+1}{n}\right)\right|, \left|W\left(\frac{k+1}{n}\right) - W\left(\frac{k}{n}\right)\right|, \left|W\left(\frac{k}{n}\right) - W\left(\frac{k-1}{n}\right)\right|\right) \leq 6\beta/n$$

Therefore, if

$$B_n = \{W(\cdot); \text{ at least one } Y_k \leq 6\beta/n\},$$

Then $A_n \subset B_n = \bigcup_{k=1}^{n-2} \{W(\cdot); Y_k \leq 6\beta/n\}$,

$$P(B_n) \leq \sum_{k=1}^{n-2} P(\max)\left|W\left(\frac{k+2}{n}\right) - W\left(\frac{k+1}{n}\right)\right|, \left|W\left(\frac{k+1}{n}\right) - W\left(\frac{k}{n}\right)\right|,$$

$$\left|W\left(\frac{k}{n}\right) - W\left(\frac{k-1}{n}\right)\right| \leq \left(\frac{6\beta}{n}\right)$$

$$\leq nP(\max)\left|W\left(\frac{3}{n}\right) - W\left(\frac{2}{n}\right)\right|, \left|W\left(\frac{2}{n}\right) - W\left(\frac{1}{n}\right)\right|, \left|W\left(\frac{1}{n}\right)\right| \leq \left(\frac{6\beta}{n}\right)$$

$$= nP^3\left(\left|X\left(\frac{1}{n}\right)\right| \leq \frac{6\beta}{n}\right) = n\left[\sqrt{\frac{n}{2\pi}}\int_{-6\beta/n}^{6\beta/n} e^{-nx^2/2}dx\right]^3$$

$$= n\left[\frac{1}{\sqrt{2\pi n}}\int_{6\beta}^{6\beta} e^{-y^2/(2n)}dy\right]^3 \qquad (y = nx)$$

$$\to 0.$$

Therefore $P(A) = \lim_{n\to\infty} P(A_n) \leq \lim_{n\to\infty} P(B_n) = 0$, which implies the theorem.

The law of the iterated logarithm (P. Levy)

It is one of the most precise and well-known theorems regarding oscillations of a Brownian motion. It has gone through many refinement and generalizations since its proof by Khintchine (1924).

Theorem I.5 If $\{X(t), 0 \leq t \leq \infty\}$ is a separable Brownian motion with $\text{var}(X(t)) = t$, then

$$P\left[\limsup_{t\to\infty} \frac{X(t)}{\sqrt{2t \log\log t}} = 1\right] = 1 \tag{A.I.4}$$

and

$$P\left[\limsup_{t\to 0} \frac{X(t)}{\sqrt{2t \log\log (1/t)}} = 1\right] = 1 \tag{A.I.5}$$

(A.I.5) is called the local law.

Proof We first prove that the probabilities given in the statement of the theorem are well defined. By separability the sample functions of $\{X(t), t \geq 0\}$ are continuous, since $\sqrt{2t \log\log t}$ is continuous and bounded away from zero for $t \geq 3$, then for any interval $(a, b) \subset (3, \infty)$

$\sup\limits_{t\in(a,b)} \dfrac{X(t)}{\sqrt{2t \log\log t}} = \sup\limits_{n} \dfrac{X(t_n)}{\sqrt{2t_n \log\log t_n}}$ a.s. for any sequence $\{t_n\}$ which is dense in (a, b). So the supermums are r.v.s.

To prove (A.I.4) it is sufficient to prove that for every $\varepsilon > 0$,

$$P\left[\limsup_{t\to\infty} \frac{X(t)}{\sqrt{2t \log\log t}} > 1 + \varepsilon\right] = 0 \tag{A.I.6}$$

and

$$P\left[\limsup_{t\to\infty} \frac{X(t)}{\sqrt{2t \log\log t}} > 1 - \varepsilon\right] = 1 \tag{A.I.7}$$

We first prove (A.I.6). It is sufficient to show that for $\varepsilon > 0$ there exists t_0 such that $P[X(t) \leq (1 + \varepsilon) \sqrt{2t \log\log t}$ for all $t \geq t_0] = 1$

Let $a > 1$ and let us denote

$$A_k = [\sup \{X(t) \mid 0 \leq t \leq a^k\} > (1 + \varepsilon/2) \sqrt{2a^k \log\log a^k}]$$

By Lemma 8.1

$$P(A_k) = \sqrt{2/(\pi a^k)} \int_{(1+\varepsilon/2\sqrt{2a^k \log\log a^k}}^{\infty} \exp(-y^2/(2a^k))dy$$

$$\leq \sqrt{2/(\pi a^k)} \int_{(1+\varepsilon/2)\sqrt{2a^k \log\log a^k}}^{\infty} \left\{\frac{y}{(1+\varepsilon/2)\sqrt{2a^k \log\log a^k}} \times \exp(-y^2/(2a^k))\right\} dy$$

$$= \frac{1}{(1 + \varepsilon/2)\sqrt{\pi \log\log a^k}} \exp\{-(1 + \varepsilon/2)^2 (\log\log a^k)\}$$

$$= \frac{1}{(1 + \varepsilon/2)\{\pi(\log k + \log\log a)\}^{1/2}} = \frac{1}{(k \log a)^{1+\varepsilon/2}}$$

Hence $\sum_k P(A_k) < \infty$ so by Borel cantelli Lemma

$$P(A_k \text{ i.o.}) = 0.$$

Hence with probability one, for all sufficiently large k, ($k \geq k_0(\omega)$), and for all $t \in [a^{k-1}, a^k]$,

$$X(t) \leq (1 + \varepsilon/2) \sqrt{2a^k \log \log a^k}$$

$$\leq (1 + \varepsilon/2) \sqrt{2t \log \log t} \sqrt{(2a^k \log \log a^k)/(2a^{k-1} \log \log a^{k-1})}.$$

Choose $a > 1$ such that

$$(1 + \varepsilon/2)\sqrt{a} \left\{ \frac{\log k + \log \log a}{\log (k-1) + \log \log a} \right\}^{1/2} < 1 + \frac{3\varepsilon}{4} \text{ for all } k \geq 2.$$

Then for t sufficiently large (depending on ω)

$$X(t) \leq (1 + \varepsilon)\sqrt{2t \log \text{lot}\, t} \text{ a.s.}$$

[**Note:** If we had taken $a < 1$ but sufficiently close to 1 we could have proved in the same manner the first half of the local law, namely

$$P\left[\limsup_{t \to 0} \frac{X(t)}{\sqrt{2t \log \log t}} > 1 + \varepsilon \right] = 0 \text{ for every } \varepsilon > 0.]$$

To prove (A.I.6), let us take $a > 1$ (but this time very large).

Let $B_k = [X(a^k) - X(a^{k-1}) \geq (1 - \varepsilon/2) \sqrt{2a^k \log \log a^k}\,]$, $0 < \varepsilon < 1$.

Since $X(t)$ has independent increments, the events B_k are independent. Since $X(a^k) - X(a^{k-1}) \underset{=}{d} N(0, a^k - a^{k-1})$,

$$P(B_k) = (2\pi(a^k - a^{k-1}))^{-1/2} \int_{(1-\varepsilon/2)\sqrt{2a^k \log \log a^k}}^{\infty} \exp[-y^2/(2(a^k - a^{k-1}))]dy$$

Let $z = y/(a^k - a^{k-1})^{1/2} = \dfrac{y}{\sqrt{a^k(1 + 1/a)}}$,

$$c_k = (1 - \varepsilon/2) \sqrt{(2 \log \log a^k)(1 - 1/a)} \text{ and } d_k = c_k + 1$$

then $P(B_k) = (2\pi)^{-1/2} \displaystyle\int_{c_k}^{\infty} e^{-z^2/2} dz > (2\pi)^{-1/2} \int_{c_k}^{d_k} \exp(-z^2/2)dz$

$$> (2\pi)^{-1/2} \exp(-d_k^2/2).$$

Let $\tau = (1 - \varepsilon/2)/\sqrt{1 - 1/a}$. If a is large $1 - 1/a$ is close to 1 and $\tau < 1 - \varepsilon/4$.

Now $1/2\{\tau\sqrt{2 \log \log a^k} + 1\}^2 = \tau^2(\log k + \log \log a) + \tau\sqrt{2 \log \log a^k} + 1/2$.

Since for large x, $k\sqrt{x} < \theta x$ for any $k > 0$, $\theta > 0$, we have for large k

$$\tau\sqrt{2 \log \log a^k} < \frac{1}{2}(1 - \tau^2) \log \log a^k$$

Hence for large k,

$$\theta^{1/2} d_k^2 = \exp\{1/2\{\tau\sqrt{2\log\log a^k} + 1\}^2$$
$$< \exp\{((\tau^2+1)/2)\{\tau\sqrt{2\log\log a^k} + 1\}^2$$
$$(\log k + \log\log a))\} = (\log a)^{(\tau^2+1)/2}\, k^{(\tau^2+1)/2}$$

Hence, $P(B_k) > \left(\dfrac{\log a)^{\tau^2+1}}{2\pi}\right)^{1/2} k^{-\beta}$, where $0 < \beta = (\tau^2+1)/2 < 1$. Therefore, $\sum_k P(B_k) = \infty$ and by Borel's zero-one law $P(B_k\ i.o.) = 1$. Since the distribution of process $\{X(t),\ t \geq 0\}$ is synmetric, we obtain from (A.I.6)

$$P[-X(a^{k-1}) \leq (1+\delta)\sqrt{2a^{k-1}\log\log a^{k-1}} \text{ for all large } k] = 1 \text{ for any } \delta > 0.$$

Hence $P[X(a^k) > (1-\varepsilon/2)\sqrt{2a^k\log\log a^k} - (1+\delta)\sqrt{2a^{k-1}\log\log a^{k-1}}$ for infinitely many $k] = 1$.
But

$$(1-\varepsilon/2)\sqrt{2a^k\log\log a^k} - (1-\delta)\sqrt{2a^{k-1}\log\log a^{k-1}}$$
$$= \sqrt{2a^k\log\log a^k}\,\{1 - \varepsilon/2 - A\},$$

where $A = \dfrac{1+\delta}{\sqrt{a}}\sqrt{\dfrac{\log\log a^{k-1}}{\log\log a^k}} < \varepsilon/2$ if a is large for all $k \geq 2$.

Then $P[X(a^k) > (1-\varepsilon)\sqrt{2a^k\log\log a^k}\ i.o.] = 1$

or
$$P\left[\limsup_{k\to\infty}\frac{X(a^k)}{\sqrt{2a^k\log\log a^k}} > 1-\varepsilon\right] = 1$$

Since

$$\left[\limsup_{k\to\infty}\frac{X(a^k)}{\sqrt{2a^k\log\log a^k}} > 1-\varepsilon\right] \subset \left[\limsup_{k\to\infty}\frac{X(t)}{\sqrt{2t\log\log t}} > 1-\varepsilon\right]$$

(A.I.7) is proved.

Note: If one select $0 < a < 1$, with a close to zero, then similarly we can prove

$$P\left[\limsup_{t\to 0}\frac{X(t)}{\sqrt{2t\log\log|1/t|}} > 1-\varepsilon\right] = 1. \quad \text{(A.I.8)}$$

Corollary I.5.1 The function $X(t)$ has (a.s.) arbitrarily large zeros.

Proof The r.v.s $\{-X(t)\}$ are again Brownian motion to which the law of iterated logaritham (A.I.6) applies. Hence (A.I.6) holds for $-X(t)$ as well as $X(t)$. This means that $X(t)$ has both positive and negative values for arbitrarily large t, and continuity does the rest.

Corollary I.5.2 For each t_0, the function $X(t) - X(t_0)$ has (a.s.) a sequence of zeros converging to t_0 from above. In particular $t_0 = 0$ is not an isolated root of $X(t)$.

Proof follows from local law of iterated logarithm for Brownian motion.

The Zeros of $X(t)$

For any continuous function, the zero set is a closed set. Let $T(\omega) = \{t \in [0, 1]; X(t, \omega) = 0\}$ be an infinite set a.s.

But Lebesgue measure of $T(\omega)$ is a.s. zero because

$$m_L(T(\omega)) = m_L\{t: X(t) = 0\} = \int_0^1 I_{\{0\}}(X(t))dt,$$

where $m_L(A)$ is the Lebesgue measure of A.

So that $$E(m_L(T(\omega))) = E\left(\int_0^1 I_{\{0\}}(X(t))dt\right) = \int_0^1 P[X(t) = 0]dt = 0$$

where interchange of E and m_L is justified by (Fubini theorem) and joint measurability of $X(t, \omega)$, hence of $I_{\{0\}}(X(t, \omega))$.

Theorem I.6 (Breiman) For almost all ω, $T(\omega)$ is a closed, perfect set of Lebesgue measure zero (therefore, uncountable).

Corollary I.6.1 For any value a, the set $T^{(a)} = \{t: X(t) = a, 0 \le t \le 1\}$ is, for almost all ω, either empty or a perfect closed set of Lebesgue measure zero.

Definition I.4 A real valued function f on R is locally Holder continuous of order α if, for every $L > 0$,

$$\sup \{ | F(t) - f(s) |/| t - s |^{\alpha}; | t |, | s | \le L, t \ne s\} < \infty \tag{A.I.9}$$

Theorem A.I.7 The Brownian motion is locally Holder continuous of order α for every $\alpha < 1/2$.

Proof We shall first show that if a stochastic process $\{X(t)\}$ be such that

$$E | X(t) - X(s) |^{\gamma} \le c | t - s |^{1+\eta}, \eta > 0, \gamma > 0 \tag{A.I.10}$$

then, there is a modification $\tilde{X}$ *of* X such that

$$E[\{\sup_{s \ne t} (| \tilde{X}(t) - \tilde{X}(s) |/| t - s |^{\alpha} \}^{\gamma}] < \infty$$

for every $\alpha \in [0, \eta/\gamma]$. In particular, the paths of $\tilde{X}$ are Holder Continuous of order α. For each positive integer m, let D_m be the set $s_m = \{2^{-m_i}\}$, i is an integer in the interval $[0, 2^m)$. The set $D = \cup D_m$ is the set of dyadic numbers. Let further Δ_m be the set of pairs (s, t) in D_m such that $| t - s | = 2^{-m}$; there are 2^m such pairs. Now set $K_i = \sup_{(s,t)\in\Delta_i} | X(s) - X(t) |$. By assumption

$$E[K_i^{\gamma}] \le \sum_{(s,t)\in\Delta_i} E | X(s) - X(t) |^{\gamma} \le 2^i \cdot C2^{-i(1+\eta)} = c\, 2^{i\eta}$$

For a point s (t) in D, there is an increasing sequence $\{s_n\}$ ($\{t_n\}$) of points in D such that $s_n(t_n)$ is in D_n, $s_n \le s(t_n \le t)$ and $s_n = s(t_n = t)$ for some n onward.

Let $s, t \in D$ and $|s - t| = 2^{-m}$; either $s_m = t_m$ or $(s_m, t_m) \in \Delta_m$, and in any case

$$X(s) - X(t) = \sum_{i=m}^{\infty} [X(s_{i+1}) - X(s_i)] + X(s_m) - X(t_m)$$

$$+ \sum_{i=m}^{\infty} [X(t_i) - X(t_{i+1})]$$

Here series are actually finite sums. It follows that

$$|X(s) - X(t)| \le K_m + 2 \sum_{i=m+1}^{\infty} K_i \le 2 \sum_{m}^{\infty} K_i$$

Set $M_\alpha = \sup_{s \ne t}\{(|X(t) - X(s)|/|t - s|^\alpha, s, t \in D\}$.

Then, we have

$$M_\alpha \le \sup_m \{2^{(m+1)\alpha} \sup_{|t-s| \le 2^{-m}} |X(t) - X(s)|; s, t \in D, s \ne t\}$$

$$\le \sup_m \left\{ \left(2.2^{(m+1)\alpha} \left(\sum_{i=m}^{\infty} K_i \right) \right) \right\}$$

$$\le 2^{\alpha+1} \sum_{i=0}^{\infty} 2^{i\alpha} K_i$$

For $\gamma \ge 1$ and $\alpha < \eta/\gamma$ we get, with $C_1 = 2^{\alpha+1} C$,

$$(E|M_\alpha|^\gamma)^{1/\gamma} \le C_1 \sum_{i=0}^{\infty} 2^{i\alpha} (EK_i^\gamma)^{1/\gamma} \le C_1 \sum_{i=0}^{\infty} 2^{i(\alpha - (\eta/\gamma))} < \infty.$$

For $\gamma < 1$, the same arguments holds to EM_α^γ instead of $(EM_\alpha^\gamma)^{1/\gamma}$. It follows, in particular that for almost every ω, X is uniformly continuous on D and it makes sense to set

$$\tilde{X}(t, \omega) = \lim_{\substack{s \to t \\ s \in D}} X(x, \omega)$$

By Fatou's lemma and the hypothesis, $\tilde{X}(t) = X(t)$ a.s. and $\tilde{X}$ is clearly the desired modification.

Now back to Brownian motion (increments being normal), for every $p > 0$,

$$E|W(t) - W(s)|^{2p} = C_p |t - s|^p.$$

So Brownian motion has a modification which is Holder continuous of order α for $\alpha < (p-1)/2p = 1/2 - 1/2p$. Since p can be taken arbitrarily large, the result follows.

Note Condition A.I.10 is the same condition which appears in theorem A.I.2 to prove much weaker property of Brownian motion.

Corollary I.7.1 The Brownian paths are a.s. nowhere locally holder continuous of order α for $\alpha > 1/2$.

Proof If $|W(t, \omega) - W(s, \omega)| \le C|t - s|^\alpha$ for $0 \le s, t \le 1$ and $\alpha > 1/2$ then

$$\sum_i (W(t_{i+1}, \omega) - W(t_i, \omega))^2 \le C^2 \sup_i | t_{i+1} - t_i |^{2\alpha-1} \quad \text{(A.I.10)}$$

Now by theorem A.I.3 and the note after that

if $\sup_n | \Delta_n | \to 0$ where $\{\Delta_n\} = \{t_{n+1} - t_n\}$ is a subdivision of $[a, b]$,

$$\sum_i (W(t_{i+1}) - W(t_i))^2 \to (b - a) \text{ a.s. as } n \to \infty. \quad \text{(A.I.11)}$$

Hence the right hand side of (A.I.11) would converge to 0 as $n \to \infty$ if $W(t)$ is locally Holder continuous which is contradictory. Note Theorem A.I.7 and its corollary leave open the case $\alpha = 1/2$. It can be proved by the law of iterated logarithm for Brownian motion (Theorem A.I.5) that the Brownian paths are not Holder continuous of order 1/2.

APPENDIX II

Second Order Stochastic Analysis

II(A) Second Order Calculus

A second-order process $\{X(t), t \in T\}$ is said to be continuous in quadratic mean (q.m.) or mean square (m.s.) at $t = t_0$ if $E \mid X(t_0 + h) - X(t_0) \mid^2 \to 0$ as $h \to 0$. If a process is q.m. continuous at every $t \in T$, we say that it is a q.m. continuous process.

From general property of m.s. convergence $E \mid X(t_0)|^2 < \infty$ and moreover $\lim_{t \to t_0} EX(t) = E(\lim_{t \to t_0} X(t))$ (m.s) if $E \mid X(t) - X(t_0) \mid^2 \to 0$ as $t \to t_0$. The q.m. continuity of a second order process must, of course, be closely related to questions of continuity of the covariance function of the process.

The following theorem summarizes succinctly the relationship between q.m. continuity and continuity of the covariance function.

Theorem II.1 Let $\{X(t), t \in T\}$ be a second-order process on an interval T, and let $\Gamma(t, s) = E(X_t \overline{X}_s)$ denote its covariance function.

(a) $\{X(t), t \in T\}$ is q.m. continuous at every $t \in T$ iff $\Gamma(. . .)$ is continuous at the diagonal point (t, t).

(b) If $\{X(t), t \in T\}$ is q.m. continuous at every $t \in T$, then $\Gamma(. . .)$ is continuous at every point of the square $T \times T$.

(c) If a non-negative definite function on $T \times T$ is continuous at every diagonal point, then it is continuous everywhere on $T \times T$.

Proof If Γ is continuous at (t, t), then by Hermition symmetry of Γ

$$
\begin{aligned}
E \mid X(t+ h) - X(t) \mid^2 &= \Gamma(t + h, t + h) - \Gamma(t, t + h) - \Gamma(t + h, t) + \Gamma(t, t) \\
&= [\Gamma(t + h, t + h) - \Gamma(t, t)] - [\Gamma(t, t + h) - \Gamma(t, t)] \\
&\quad - [\Gamma(t + h, t) - \Gamma(t, t)] \to 0 \text{ as } h \to 0.
\end{aligned}
$$

Conversely, if $\{X(t), t \in T\}$ is q.m. continuous at t, then

$$
\begin{aligned}
\Gamma(t + h, t + h') - \Gamma(t, t) &= E(X(t + h)\overline{X}(t + h')) - EX(t)\overline{X}(t) \\
&= E((X(t + h) - X(t))\overline{X}(t + h')) + EX_t(\overline{X_{t+h'} - X_t})
\end{aligned}
$$

From Cauchy-Schwartz inequality we have

$$
\begin{aligned}
|\Gamma(t + h, t + h') - \Gamma(t, t)| &\le [E|X(t + h) - X(t)|^2 \, E|\overline{X}(t + h')|^2]^{1/2} \\
&\quad + [E \mid X(t_1) \mid^2 E \mid X(t + h') - X_t|^2]^{1/2} \to 0 \text{ as } h, h' \to 0.
\end{aligned}
$$

(b) If $\{X(t), t \in T\}$ is q.m. continuous at every $t \in T$, then

$$|\Gamma(t+h, s+h') - \Gamma(t,s)| = |EX(t+h)\overline{X}(s+h') - EX(t)\overline{X}(s)|$$

$$= |\{E(X(t+h) - X(t))\overline{X}(s+h')\} + E\{X(t)(\overline{X(s+h') - X(s)})\}|$$

$$\leq \sqrt{\{E|X(t+h) - X(t)|^2 \, E|X(s+h')|^2\}}$$

$$+ \sqrt{\{E|X(t)|^2 \, E|X(s+h') - X(s)|^2\}} \to 0 \text{ as } h, h' \to 0.$$

(c) Since every nonnegative definite function on $T \times T$ is the covariance function of some second-order process on T, part (c) follows immediately from (a) and (b).

Corollary II.1 Let $X(t)$ and $Y(t)$ be two second order stochastic process ($t \in T$), Let t_0, s_0 be two limit points of T. Then if

$X(t) \to X$ (m.s.) as $t \to t_0$ and $Y(s) \to Y$ (m.s.) as $s \to s_0$, then

$$EX(t)\overline{Y}(s) \to EX\overline{Y} \text{ as } t \to t_0, s \to s_0.$$

Definition II.1 A Second order process $\{X(t), t \in T\}$ is said to have a derivative $X(t)$ in q.m. (m.s.) at $t \in$ T if

$$\lim_{h\to 0} \frac{1}{h}(X(t+h) - X(t)) \text{ in q.m.} = \dot{X}(t) \qquad \text{(A.II.1)}$$

Theorem II.2 Let $\{X(t), t \in T\}$ be a second order process. If $X(t)$ has a derivative at all $t \in T$, then the derivatives

$\frac{\partial}{\partial t}\Gamma(t,s), \frac{\partial}{\partial s}\Gamma(t,s), \frac{\partial^2}{\partial t \partial s}\Gamma(t,s)$ of the covariance function $\Gamma(t, s)$ exist and are finite on $T \times T$.

Proof If $X(t)$ has the derivative $X(t)$ (q.m.) at $t \in T$, Then

$$\frac{X(t+u) - X(t)}{u} \to \dot{X}(t) \text{ as } u \to 0.$$

Therefore $\lim_{u\to 0}\left(\frac{\Gamma(t+u,s) - \Gamma(t,s)}{u}\right) = \lim_{u\to 0} E\left(\frac{X(t+u) - X(t)}{u}\overline{X}(s)\right)$

$$= E\left(\lim_{u\to 0}\frac{X(t+u) - X(t)}{u}\overline{X}(s)\right) = EX(t)\overline{X}(s). \qquad \text{(A.II.2)}$$

Hence the derivative $\frac{\partial}{\partial t}\Gamma(t,s)$ exists and

$$\frac{\partial}{\partial t}\Gamma(t,s) = E\dot{X}(t)\overline{X}(s) < \infty.$$

Similarly $$\frac{\partial}{\partial s}\Gamma(t,s) = EX(t)\dot{\overline{X}}(s) \qquad \text{(A.II.3)}$$

Finally $$\lim_{u\to 0}\frac{1}{u}\left\{\frac{\partial}{\partial t}\Gamma(t,s+u) - \frac{\partial}{\partial t}\Gamma(t,s)\right\}$$

$$= \lim_{u \to 0} E(\dot{X}(t)\left(\frac{\overline{X}(s+u) - \overline{X}(s)}{u}\right) \quad \text{(from A.II.2)}$$

$$= E\dot{X}(t)\overline{\dot{X}}(s), \tag{A.II.4}$$

which shows that $\frac{\partial^2}{\partial t \partial s}\Gamma(t, s) = E\dot{X}(t)\overline{\dot{X}}(s)$. (A.II.5)

Let $\{X(t), t \in T\}$ be second order process with mean function $m(t) = EX(t)$ and covariance function $\Gamma(t, s)$. Suppose that $\dot{X}(t)$ has derivative $X(t)$ on T. Then

$$\frac{d}{dt}m(t) = \lim_{h \to 0}\frac{m(t+h) - m(t)}{h} = \lim_{h \to 0} E\left(\frac{X(t+h) - X(t)}{h}\right) = E\left(\frac{d}{dt}X(t)\right)$$

which exhibits validity of interchange of the operations of differentiation and taking expectation. Further (A.II.5) can be written as $\frac{\partial^2}{\partial t \partial s}\Gamma_x(t, s) = \Gamma_{\dot{x}}(t, s)$. Next

$$\frac{\Gamma(t+h, t') - \Gamma(t, t')}{h} = E\left(\frac{X(t+h) - X(t)}{h}\overline{X}(t')\right)$$

shows that the existence of the partial derivative $\frac{\partial}{\partial t}\Gamma(t, t')$ depends on that of the derivative of $X(t)$. We shall need the concept of the second generalized derivative of $\Gamma(t, t')$, defined as the limit (if it exists) as $h, h' \to 0$ of the quotient

$$\frac{1}{hh'}\{\Gamma(t+h, t'+h') - \Gamma(t+h, t') - \Gamma(t, t'+h') + \Gamma(t, t')\}.$$

Above concepts are due to Slutsky (1928) and the following results are due to Loeve (1945–48, 63).

Theorem II.3 A second order process $\{X(t), t \in T\}$ has a derivative at $t \in T$ iff, the second generalized derivative of $\Gamma(t, t)')$ exists and is finite at (t, t).

Proof For fixed $t \in T$, let us put

$$Y_s = \frac{X(t+s) - X(t)}{s}.$$

Then $Y_s \to \dot{X}_t$ (q.m.) as $s \to 0$, where $\dot{X}(t)$ is a stochastic process iff $E(Y_s\overline{Y}_{s'})$ converges to a limit as $s, s' \to 0$ (by Theorem A II.1). But

$$E(Y_s\overline{Y}_{s'}) = \frac{1}{ss'}E\{(X(t+s) - X(t))(\overline{X}(t+s') - \overline{X}(t))\}$$

$$= \frac{1}{ss'}\{\Gamma(t+s, t+s') - \Gamma(t+s, t) - \Gamma(t, t+s') + \Gamma(t, t)\}; \tag{A.II.6}$$

the limit of which (if it exists) as $s, s' \to 0$ is the second generalized derivative of $\Gamma(t, t')$ at (t, t).

The Stochastic Integral

Quadratic-mean integrals arise even more frequently than q.m. derivatives.

Let $\{X(t), a \leq t \leq b\}$ be a second order stochastic process.

Let $$a = t_0 < t_1 < \ldots < t_n = b$$

be a subdivision of $[a, b]$,

$$t_{k-1} \leq \xi_k \leq t_k (k = 1, 2, \ldots, n),$$

and consider the Riemann sum

$$S_n = \sum_{k=1}^{n} (t_k - t_{k-1}) X(\xi_k) \tag{A.II.7}$$

If, as $n \to \infty$ in such a way that $\max_{1 \leq k \leq n} |t_k - t_{k-1}| \to 0$, S_n converges (q.m.) to a limit, we say that this limit is the Riemann Integral of $X(t)$ over [a, b] and denote it by $I = \int_a^b X(t)dt$. Clearly, I is a r.v. If I exists, we say that $X(t)$ is Riemann integrable over $[a, b]$. A simple and typical use of this integral is the estimate $\mu = \dfrac{1}{b-a}\int_a^b X(t)dt$ of the constant but unknown mean of the second order stationary process $\{X(t), a \leq t \leq b\}$.

Theorem II.4 A second order process $X(t)$ with covariance function $\Gamma(t, s)$ is Riemann integrable over $[a, b]$ iff the double integral

$$\int_a^b \int_a^b \Gamma(t, s)\, dt\, ds \text{ exists.} \tag{A.II.8}$$

If this is the case, then $$E\left\{\int_a^b X(t)dt \int_a^b \overline{X}(s)ds\right\} = \int_a^b \int_a^b \Gamma(t, s)\, dtds$$

Proof Now S_n defined by (A.II.7) converges to I (q.m.) iff

$$E(S_n \overline{S}_m) \to E(I\overline{I}), \tag{A.II.9}$$

where S_m corresponds to a subdivision $a = s_0 < s_1 < \ldots < s_m = b$ of $[a, b]$ and $s_{j-1} \leq \eta_j \leq s_j$ $(j = 1, 2, \ldots, m)$. Then we have

$$ES_n\overline{S}_m = \sum_{k=1}^{n} \sum_{j=1}^{m} (t_k - t_{k-1})(s_j - s_{j-1})\Gamma(\xi_k, \eta_j)$$

and if the limit of this double sum exists as

$$n \to \infty, m \to \infty, \max_{1 \leq k \leq n} |t_k - t_{k-1}| \to 0 \text{ and } \max_{1 \leq j \leq m} |s_j - s_{j-1}| \to 0,$$

then it is the double integral (A.II.8). This proves the first part of the theorem; the second part follows from (A.II.9).

Remarks

1. In the last theorem instead of Riemann integral Lebesgue and Stieltjes integrals can be considered along similar lines.
2. If $m(t) = EX(t)$, then

$$E\int_a^b X(t)dt = \int_a^b EX(t)dt = \int_a^b m(t)dt.$$

Let $\{X(t), t \in T\}$ be a second-order process. A r.v. Y is said to be derived from a linear operation on $\{X(t), t \in T\}$ if either

(a) $Y(\omega) = \sum_{k=1}^{n} C_k X(t_k, \omega)$ or

(b) Y is the q.m. limit of a sequence of such finite linear combinations.

Denote the collection of all such r.v.s derived from a given process $\{X(t), t \in T\}$ by χ_x. Two elements Y and Z are not distinguished whenever $E \mid Y - Z \mid^2 = 0$. Equality between second-order r.v.s will always be understood to be up to zero q.m. differences. The space χ_x becomes an inner product space if we define the inner product $\langle Y, Z\rangle = EY\bar{Z}$. The inner product automatically defines a norm $\| Y \| = \sqrt{\langle Y, Y\rangle}$ and a distance function $d(y, z) = \| Y - Z \|$. A sequence $\{Y_n\}$ in χ_x is said to be a Cauchy sequence if $\|Y_m - Y_n\| \to 0$ as $n, m \to \infty$. Since $\| Y_n - Y_m \|^2 = E \mid Y_n - Y_m \mid^2$, a Cauchy sequence in χ_x is a mutually q.m. convergent sequence, which by completeness must converge in q.m. to some r.v. Y in χ_x. By definition χ_x is a Hilbert space.

Let $f(t)$ be a complex valued piecewise continuous function defined on a sub-interval $[a, b]$ of T. If for each $\omega \in \Omega$ the stochastic process $X(t, \omega)$, $a \le t \le b$ is piecewise continuous, then $f(t)X(t, \omega)$, $a \le t \le b$ is also piecewise continuous. Since the ordinary Riemann integrals of calculus are well defined for piecewise continuous functions, $\int_a^b f(t)X(t, \omega)dt$ is well defined for each $\omega \in \Omega$. We define the q.m. integral $\int_a^b f(t)X(t)dt$ as an element of χ_x whenever $\lim_{n\to\infty}$ q.m. $\sum_{k=0}^{n-1} f(t'_{k_n})X(t'_{k_n})(t_k^{(n)} \quad t_{k-1}^{(n)})$, where t'_{k_n} are any sequence of points satisfying $t_{k-1}^{(n)} \le t'_{k_n} < t_k^{(n)}$ exists and is independant of the choice of the partition points t'_{k_n}.

Theorem II.5 The quadratic mean (q.m.) interal $\int_a^b f(t)X(t)dt$ exists if and only if $\int_a^b\int_a^b f(t)\ \bar{f}(s)\Gamma(t, s)dt\,ds$ exists as a Riemann integral.

Proof is similar to that of Theorem II.4 and hence is omitted.

Remarks

1. The q.m. integral can easily be generalized to include infinite integrals by taking q.m. limit as one or both of the end points goes to infinity.

2. A common way in which q.m. integrals arise is when a process $\{Y_t, t \in T'\}$ is generated from a second process $\{X(t), t \in T\}$ according to the formula.

$$Y(t) = \int_T h(t, s)X(s)ds, t \in T' \tag{A.II.10}$$

Then (AII.10) often admits the interpretation of a linear system with input $X(t)$, output $Y(t)$, and impulse response $h(t, s)$.

3. From applications point of view one might prefer $\int_T f(t)X(t,\omega)dt$ to be defined as the Lebesgue integral of sample function $\{X(t), t \in T\}$. The existence of such a Lebesque integral is insured if the process is a measurable process and if $\int_T |f(t)|E|X(t)|dt < \infty$.

II(B) Continuity, Integration and Differentiation of Second Order Process

Definition II.2 A stochastic process $\{X(t), t \in T)$ is said to be *sample continuous* (or $\{X(t), t \in T\}$ has continuous sample path), if for each $\omega \in \Omega$, $X(t, \omega)$, $t \in T$ is a continuous function of t. For example Brownian motion is a sample continuous stoohastic process but Poisson process is not. Let $\{X(t), t \in T\}$ be a continuous parmater second order process satisfying (i) the mean function $m_x(t) = E(X(t))$, $t \in T$ is a continuous function of t and (ii) $\Gamma_x(s, t)$, $s \in T$, and $t \in T$, is jointly continuous in s and t. We say that this process is differentiable if there is a second order process $\{Y(t), t \in T\}$, satisfying assumptions (i), (ii), (iii) $\lim_{s \uparrow t} X(s, \omega)$ in finite, (iv) $X(s, \omega) \to X(t, \omega)$ as $s \downarrow t$ and (v) the function $X(t, \omega)$, $t \in T$ has only finite number of points of discontinuity on any closed bounded subinterval of T such that for $t \in T$

$$X(t) - X(t_0) = \int_{t_0}^{t} Y(s)ds, t \in T$$

then the process $Y(t)$ is called the derivative of $X(t)$ and is denoted by $X'(t)$, $t \in T$. Thus

$$X(t) - X(t_0) = \int_{t_0}^{t} X'(s)ds, t \in T. \tag{A.II.11}$$

It can be proved that if a second order process is differentiable then it is differentiable in q.m.

It follows now that $X(t)$ has continuous sample functions and that $\frac{d}{dt}X(t) = X'(t)$ holds excepts at point of dicontinuity of the $X'(t)$ process.

It also follows that $m_x(t) - m_x(t_0) = \int_{t_0}^{t} m_x'(s)ds$ and hence $m_x'(t) = \frac{d}{dt}m_x(t), t \in T$.

To find the covariance function of $X'(t)$ process it is necessary to consider cross covariance functions of two second order proceses $\{X(t), t \in T\}$ and $\{Y(t), t \in T\}$ defined by $E(X(t)\overline{Y}(s)) = \Gamma_{xy} = \Gamma_{xy}(t, s)$. Suppose that $X(t)$ is differentiable and satisfies conditions (i) and (ii). We shall show that

$$\Gamma_{yx'}(s,t) = \frac{\partial}{\partial t}\Gamma_{yx}(s,t), s \in T \text{ and } t \in T \tag{A.II.12}$$

and

$$\Gamma_{x'y}(s,t) = \frac{\partial}{\partial s}\Gamma_{xy}(s,t), s \in T \text{ and } t \in T \tag{A.II.13}$$

If moreover $\{X_t\}$ is second order stationary and twice differentible then $\Gamma_{x'}(s,t) = -\Gamma_{x''}(t-s)$.

Proof Choose

$$t_0 \in T \text{ and write } X(t_0) = \int_{t_0}^{t} X'(u)\,du.$$

Then we have $\quad Y(s)(X(t) - X(t_0)) = \int_{t_0}^{t} X'(u)\,du\,Y(s).$

Thus
$$E[Y(s)\{X(t) - X(t_0)\}] = \int_{t_0}^{t} E[Y(s)X'(u)]\,du \tag{A.II.14}$$

Now $\quad m_x(t) - m_x(t_0) = \int_{t_0}^{t} m_{x'}(u)\,du,$

and hence
$$m_y(t)(m_y(t) - m_x(t_0)) = \int_{t_0}^{t} m_y(s)m_{x'}(u)\,du \tag{A.II.15}$$

Subtracting (A.II.15) from (A.II.14) and rewriting the resulting expression in terms of coveriance function, we conclude that

$$\Gamma_{yx}(s,t) - \Gamma_{yx}(s,t_0) = \int_{t_0}^{t} \Gamma_{yx'}(s,u)\,du \tag{A.II.16}$$

Now the cross covariance function $\Gamma_{yx'}(s,t)$ is jointly continuous in s and t if the second order processes $\{X(t), Y(t), t \in T\}$ are q.m. continuous. Thus for each fixed s we can differentiate (A.II. 16) with respect to t and obtain (A.II.12) as desired. Now (A.II.13) follows from (A.II.12) by symmetry.

Now (A.II.12) and (A.II.13) imply $\Gamma_{xx'}(s,t) = \frac{\partial}{\partial t}\Gamma_{xx}(s,t) = \frac{\partial}{\partial t}\Gamma_x(s,t)$ and $\Gamma_{x'x'}(s,t) = \frac{\partial^2}{\partial s \partial t}\Gamma_{xx}(s,t)$, $s \in T$ and $t \in T$.

Since $\{X_t, t \in T\}$ is second order stationary, $m_{x'}(t) = 0$ and

$$\Gamma_{x'}(s,t) = \frac{\partial^2}{\partial s \partial t}\Gamma_x(s,t) = \frac{\partial^2}{\partial s \partial t}\Gamma_x(t-s) = -\Gamma_{x''}(t-s).$$

Orthogonal Expansions

As on page 275 the Hilbert space generated by the class of second order r.v.s $X(t), t \in T$ will be denoted by $\chi_x = \chi$. A family $\mathscr{F}$ of elements (random Variables) in χ_x is said to be orthonormal $(0 - N)$ family if any two distinct elements Y and Z of $\mathscr{F}$ satisfy

$$E\|Y\| = (E|Y|^2)^{1/2} = 1 = (E|Z|^2)^{1/2} \quad \text{and} \quad \langle Y, Z\rangle = E(Y\bar{Z}) = 0 \tag{A.II. 17}$$

Suppose that $\{X(t), t \in T\}$ is a q.m. continuous process, where T is an interval, finite or infinite. Let T' be the set of all rational points in T. For every $t \in Y$, there exists a sequence $\{t_n\}$ in T' such that $t_n \to t$ as $n \to \infty$.

Since $\{X(t), t \in T\}$ is q.m. continuous, we have for every t, $X(t) = \lim_{n\to\infty}$ q.m. $X(t_n)$.

If follows that every element in χ_x is a linear combination of $\{X(t), t \in T'\}$ or the q.m. limit of a sequence of such linear combinations. In short, the countable family $\{X_t, t \in T'\}$ is dense in χ_x. It follows that every O-N family in χ_x is at most countable. For a q.m. continuous process $\{X(t), t \in T\}$, let $\{Z_j, j \geq 1\}$ be an O-N family in χ_x. If $Y \in \chi_x$, then

$$E|Y - \sum_{j=1}^{n} < Y, Z_j > Z_j|^2 = E|Y|^2 - \sum_{j=1}^{n} | < Y, Z_j >|^2 \geq 0 \text{ for every } n \geq 1.$$

Therefore, $\infty > E|Y|^2 \geq \sum_{j=1}^{\infty} |\langle Y, Z_j \rangle|^2$.

Hence $\sum_{j=1}^{\infty} \langle Y, Z_j\rangle Z_j$ is well defined and $Y - \sum_{j=1}^{\infty} \langle Y, Z_j\rangle Z_j$ is orthogonal to every Z_j. So by completeness of $\{Z_j\}$ in χ_x every

$Y \in \chi_x$ has the representation $\quad Y = \sum_{j=1}^{\infty} \langle Y, Z_j\rangle Z_j \qquad$ (A.II.18)

Suppose that $\{Z_j, j \geq 1\}$ is a given complete O-N family in χ_x and set

$$\phi_j(t) = \langle X(t), Z_j\rangle. \tag{A.II.19}$$

Then from (A.II.18) we get

$$X(t) = \sum_{j=1}^{\infty} \phi_j(t) Z_j, t \in T \tag{A.II.20}$$

since $\phi_r(t)$'s are the cross covariance function of the q.m. continuous $X(t)$ and the r.v. Z_j, are continuous functions.

Further, the set of functions $\{\phi_j(t), j \geq 1\}$ is also linearly independent, i.e. for every n

$$\sum_{j=1}^{n} c_j \phi_j(t) = 0 \text{ for all } t \in T \text{ implies } c_j = 0, j = 1, 2, \ldots, n.$$

The linear independence of $\{\phi_j(t), j \geq 1\}$ is due to the fact that $\sum_{j=1}^{n} c_j \phi_j(t) = 0$ for all $t \in T$ implies $\sum_{j=1}^{n} c_j \phi_j$ is orthogonal to $X(t)$ for every $t \in T$, Hence, also orthogonal to every ϕ_j which implies $c_j = 0$ for every j. It follows from (A.II.20) that

$$E(X(t)\overline{X}(s)) = \Gamma(t, s) = \sum_{j=1}^{\infty} \phi_j(t)\overline{\phi}(s)\, t, s \in T \tag{A.II.21}$$

Conversely, Suppose that $\{\phi_j(t), t \in T, j \geq 1\}$ is a linerly independent family of continuous functions such that

$$\Gamma(r, s) = \sum_{j=1}^{\infty} \phi_j(t)\overline{\phi}_j(s)\, t, s \in T.$$

Then by Karhunen general representation theorem (1947) there exists complete O-N family of r.vs. $\{Z_j, j \geq 1\}$ in χ_x such that

$$X(t, \omega) = \sum_{j=1}^{\infty} \phi_j(t) Z_j(\omega), t \in T.$$

Thus (A.II.20) and (A.II.21) represent each other. The representation (A.II.20) permit uncountable set of r.v.s $\{X(t), t \in T\}$ to be presented by a countable number of orthonormal r.vs. $\{\phi_j(t)\}$.

Though it is usually difficult to express the r.v.s $\{Z_j\}$ explicitly in terms of $\{X(t), t \in T\}$ but in case when $\{\phi_j\}$ are orthogonal, i.e. $\int_T \phi_i(t)\phi_j(t)\,dt = 0$ whenever $i \neq j$ the task is easier and that leads to well known Karhunen-Loeve expansion (1963).

Consider a q.m. continuous process $\{X(t), a \le t \le b\}$. Suppose that there exists an expansion of the form

$$X(t, \omega) = \sum_{j=1}^{\infty} \phi_j(t) Z_j(\omega), a \le t \le b \tag{A.II.22}$$

where $\{Z_j\}$ and $\{\phi_j\}$ satisfy $\quad E(Z_j \bar{Z}_j) = \delta_{ij}$ (A.II.23)

$$\int_a^b \Gamma_1(t, s)\phi_1(s)\,ds = \lambda_1\phi_1(t), a \le t \le b.$$

Therefore $\displaystyle\int_a^b \phi_1(t)\bar{\phi}_0(t)\,dt = \int_a^b \bar{\phi}_0(t)\left[\frac{1}{\lambda_1}\int_a^b \Gamma_1(t, s)\phi_1(s)\,ds\right]dt$

and

$$\int_a^h \phi_i(t)\phi_j(t)\,dt = \lambda_j\delta_{ij} \tag{A.II.24}$$

Now, from (A.II.21), the covariance function $\Gamma(.)$ must satisfy

$$\Gamma(t, s) = \sum_{j=1}^{\infty} \phi_i(t)\bar{\phi}_j(s) \text{ for each } (t, s) \in [a, b] \times [a, b] \tag{A.II.25}$$

Now from Cauchy-Schwartz inequality and the fact that Γ is continuous on $[a, b] \times [a, b]$ we have

$$\sup_{a \le t, s \le b}\left|\sum_{j=1}^{n} \phi_j(t)\bar{\phi}_j(s)\right| \le \sup_{a \le t, s \le b} \sum_{j=1}^{n} |\phi_j(t)|^2$$

$$\le \sup_{a \le t \le b} \Gamma(t) < \infty, (\Gamma(t, t) = \Gamma(t)) \tag{A.II.26}$$

Therefore, the convergence in (A.II.25) is bounded and

$$\int_a^b \Gamma(t, s)\phi_j(s)\,ds = \lambda_j(t)\phi_j(t), a \le t \le b \tag{A.II.27}$$

This proves that if an expansion (A.II.22), satisfying (A.II.23) and (A.II.24) exists, then $\{\phi_j\}$ must be solutions to the integral equation (A.II.27). We shall show that under our assumptions, such an expansion always exists.

The integral equations of the form (A.II.27) are well known and we shall now summarize some of the elementary facts about integral equation of the form (A.II.27) which are necessary for Loeve-Karhunen expansion of second order process.

Rudiments of Integral equation

Consider the integral equation

$$\int_a^b \Gamma(t, s)\phi(s)\,ds = \lambda\phi(t),\ a \le t \le b \qquad \text{(A.II.28)}$$

where $[a, b]$ is finite closed interval and $\Gamma(t, s)$ is continuous on $[a, b] \times [a, b]$. The $\lambda \neq 0$ for which there exists ϕ satisfying both (A.II.28) and the condition $\int_a^b |\phi(t)|^2\,dt < \infty$, is called an *eigen value* of the integral equation. The corresponding ϕ is called an *eigen function*. Let us list the important facts.

1. Any eigen value of (A.II.28) must be real and positive. The fact that λ is real follows from the Hermition symmetry $\Gamma(t, s) = \overline{\Gamma(s, t)}$. The fact that λ is positive follows from the non-negative definitness of Γ.

2. There is at least one eigen value for (A.II.28), if Γ is not identically zero. The largest eigen value λ_0 is given by

$$\lambda_0 = \max_{\|\phi\|=1} \int_a^b \int_a^b \Gamma(t, s)\phi(s)\phi(t)\,ds\,dt, \qquad \text{(A.II.29)}$$

where $$\|\phi\| = \left[\int_a^b |\phi(t)|^2\,dt\right]^{1/2}.$$

Note: This fact is not easily proved. It depends on both the nonnegative definiteness of Γ and its continuity (see Taylor 1961, p 334–336).

3. Let $\phi_0(t)$ denote the normalized eigen function corresponding to λ_0, then $\phi_0(t)$ is continuous on $[a, b]$, since

$$\phi_0(t) = \frac{1}{\lambda_0}\int_a^b \Gamma(t, s)\phi_0(s)\,ds,\ a \le t \le b \qquad \text{(A.II.30)}$$

and continuity of ϕ_0 follows from continuity of $\Gamma(t, s)$ is t.

4. Let $\Gamma_1(t, s) = \Gamma(t, s) - \lambda_0\phi_0(t)\overline{\phi}_0(s)$. Then $\Gamma_1(t, s)$ is both continuous and nonnegative definite. The continuity of Γ_1 is obvious from 3.

To show nonnegative definiteness, let

$$Y(t) = X(t) - \phi_0(t)\int_a^b X(\tau)\phi_0(\tau)\,d\tau \qquad \text{(A.II.31)}$$

then $$E(Y(t)\overline{Y}(s)) = E(X(t)\overline{X}(s)) - E\left[X(t)\,\overline{\phi}_0(s)\int_a^b \overline{X}(u)\,\phi_0(u)\,du\right]$$

$$- E\left[\overline{X}(s)\,\phi_0(t)\int_a^b X(\tau)\,d\tau\right] + E\left[\phi_0(t)\overline{\phi}_0(s)\int_a^b\int_a^b X(\tau)\phi_0(u)\,d\tau\,du\right]$$

$$= \Gamma(t, s) - \lambda_0 \phi_0(t) \bar{\phi}_0(s) = \Gamma_1(t, s) \tag{A.II.32}$$

Hence, $\Gamma_1(t, s)$, being a covariance function, must be nonnegative definite.

5. We observe that $\int_a^b \Gamma_1(t, s) \phi_0(s)\, ds = 0$. Therefore, if we repeat step 2 and obtain λ_1, ϕ_1 then

$$\int_a^b \Gamma_1(t, s)\, \phi_1(s)\, ds = \lambda_1 \phi_1(t),\ a \le t \le b.$$

Therefore $\int_a^b \phi_1(t) \bar{\phi}_0(t)\, dt = \int_a^b \bar{\phi}_0(t) \left[\frac{1}{\lambda_1} \int_a^b \Gamma_1(t, s) \phi_1(s)\, ds \right] dt$

$$= \frac{1}{\lambda_1} \int_a^b \phi_1(s) \left[\int_a^b \overline{\Gamma_1(s, t) \bar{\phi}_0(t)\, dt} \right] ds$$

$$= \frac{1}{\lambda_1} \int_a^b \phi_1(s) \left[\int_a^b \overline{\Gamma_1(s, t) \phi_0(t)\, dt} \right] ds = 0$$

Hence, ϕ_1 is orthogonal to ϕ_0. Also

$$\int_a^b \Gamma(t, s) \phi_1(s)\, ds = \int_a^b \Gamma_1(t, s) \phi_1(s)\, ds - \lambda_0 \phi_0(t) \int_a^b \phi_1(s) \bar{\phi}_0(s)\, ds$$

$$= \lambda_1 \phi_1(t),\ a \le t \le b \tag{A.II.33}$$

i.e. λ_1 and ϕ_1 are eigen value and eigen function of (A.II.28).

6. Repeating the above procedure we get a nonincreasing sequence of eigen values $\lambda_0, \lambda_1, \ldots$ and a corresponding sequence of eigen functions $\phi_0, \phi_1, \ldots$ which are orthonormal, i.e.

$$\int_a^b \phi_i(t) \bar{\phi}_j(t)\, dt = \delta_{ij} \tag{A.II.34}$$

7. The sequence $\lambda_0, \lambda_1, \ldots$ may terminate after a finite number of terms, in which case we have

$$\Gamma(t, s) = \sum_{n=0}^{N} \lambda_n \phi_n(t) \bar{\phi}_n(s) \tag{A.II.35}$$

If the sequence $\lambda_0, \lambda_1, \ldots$ does not terminate,

$$\infty > \int_a^b \Gamma(t, t)\, dt \ge \sum_{n=0}^{N} \lambda_n \int_a^b |\phi_n(t)|^2\, dt \text{ for every } N$$

$$= \sum_{n=0}^{N} \lambda_n \quad \text{for every } N \text{ (by (A.II.34)}$$

and hence as $\lambda_n \to 0$ as $n \to \infty$.

8. If the number of eigen values is infinite, then

$$\lim_{n \to \infty} \sup_{s \le t \le b} \left| \Gamma(t, s) - \sum_{k=0}^{n} \lambda_k \phi_k(t) \bar{\phi}_k(s) \right| = 0 \tag{A.II.36}$$

i.e. $$\lim_{n\to\infty} \sum_{k=1}^{n} \lambda_k \phi_k(t) \bar{\phi}_k(s) = \Gamma(t, s) \quad \text{(A.II.37)}$$

uniformly on $[a, b]^2$.

This result is known as *Mercer's theorem* (see M.Loeve 1963). Since uniform convergence implies convergence in mean square,

$$\lim_{n\to\infty} \int_a^b \int_a^b \left| \Gamma(t, s) - \sum_{k=0}^{n} \lambda_k \phi_k(t) \bar{\phi}_k(s) \right|^2 dt ds = 0 \quad \text{(A.II.38)}$$

9. In general, the O-N family $\{\phi_k\}$ is not complete in $L_2[a, b]$.

At the most we can say that, given $\int_a^b |f(t)|^2 dt < \infty$, we can write

$$f(t) = f_0(t) + \lim_{n\to\infty} \sum_{k=0}^{n} \phi_k(t) \int_a^b f(s) \bar{\phi}_k(s) ds \text{ in q.m.} \quad \text{(A.II.39)}$$

and f_0 satisfies $$\int_a^b \Gamma(t, s) f_0(s) ds = 0 \quad \text{for all} \quad t \in [a, b] \quad \text{(A.II.40)}$$

If follows that $\{\phi_k\}$ is complete in $L_2[a, b]$, iff (A.II.40) implies

$$\int_a^b |f_0(t)|^2 dt = 0.$$

Now we are in a position to state and prove the celebrated.

Theorem (Karhunen-Loeve expansion, Loeve 1963, pp. 478-479)
Let $\{X(t), t \in [a, b]\}$ be q.m. continuous second-order process with covariance function $\Gamma(t, s)$.

(a) If $\{\phi_k\}$ are the orthonormal eigen functions of

$$\int_a^b \Gamma(t, s) \phi(s) ds = \lambda \phi(t), a \le t \le b \quad \text{(A.II.41)}$$

and $\{\phi_k\}$ the eigen-values, then $X(t) = \lim_{n\to\infty} \sum_{k=0}^{n} \sqrt{\lambda_k} \phi_k(t) A_k$ uniformly for

$$a \le t \le b \quad \text{(A.II.42)}$$

where $\{A_k\}$ is a sequence of r.v.s satisfying

$$\sqrt{\lambda_k} A_k = \int_a^b \bar{\phi}_k(t) X(t) dt \text{ (q.m.)} \quad \text{(A.II.43)}$$

and $$E(A_k \bar{A}_j) = \delta_{kj} \quad \text{(A.II.44)}$$

(b) Conversely, if $X(t)$ has an expansion of the form (A.II.42) with

$$\int_a^b \phi_k(t) \bar{\phi}_j(t) = \delta_{kj} = E(A_k \bar{A}_j),$$

$\{\phi_k\}$ and $\{\lambda_k\}$ must be eigen functions and eigenvalues of (A.II.41).

Proof (a) First note that (A.II.4A) is well difined, because $X(t)$ being q.m. continuous and hence $\phi_n(t)$ is continuous on $[a, b]$ by result (3).

Now $$\sqrt{\lambda_k}\, A_k = \int_a^b X(t)\bar{\phi}_k(t)\,dt,$$

$$E(X(t)\bar{A}_k) = \frac{1}{\sqrt{\lambda_k}}\int_a^b E(X(t)\bar{X}(s))\phi_k(s)\,ds,$$

$$= \frac{1}{\sqrt{\lambda_k}}\int_a^b \Gamma(t,s)\phi_k(s)\,ds = \frac{1}{\sqrt{\lambda_k}}\lambda_k\phi_k(t) = \sqrt{\lambda_k}\,\phi_k(t).$$

Hence $$E\left\{X(t)\sum_{k=0}^{n}\sqrt{\lambda_k}\;\bar{\phi}_k(t)\bar{A}_k\right\} = \sum_{k=0}^{n}\sqrt{\lambda_k}\;\bar{\phi}_k(t)E(X(t)\bar{A}_k)$$

$$= \sum_{k=0}^{n}\sqrt{\lambda_k}\;\bar{\phi}_k(t)\phi_k(t)\sqrt{\lambda_k} = \sum_{k=0}^{n}\lambda_k|\phi_k(t)|^2.$$

Therefore, $E|X(t) - \sum_{k=0}^{n}\sqrt{\lambda_k}\,\phi_k(t)A_k|^2 = \Gamma(t,t) - \sum_{k=0}^{n}\lambda_k|\phi_k(t)|^2 \to 0$ as $n \to \infty$ uniformly in t by virtue of Mercer's theorem (Result 8, equation (A.II.36)).

(b) Suppose $X(t)$ has the expansion (A.II.43). Then we have

$$\Gamma(t,s) = E(X(t)\;\bar{X}(s)) = \sum_{k=0}^{\infty}\lambda_k\phi_k(t)\bar{\phi}_k(s)$$

$$= \lim_{n\to\infty} E(X_n(t)\bar{X}_n(s)),$$

where $X_n(t) = \sum_{k=0}^{n}\sqrt{\lambda_k}\,\phi_k(t)A_k$.

Hence $$\int_a^b \Gamma(t,s)\phi_j(s)\,ds = \int_a^b \sum_{k=0}^{\infty}\lambda_k\phi_k(t)\bar{\phi}_k(s)\phi_j(s)\,ds$$

$$= \lambda_j\phi_j(t),\; a \le t \le b.$$

Let us give an example of how an integral equation of the form (A.II.41) might be solved.

Let $\Gamma(t, s) = \min(t, s)$, the covariance function of Brownian motion and consider

$$\int_0^T \min(t,s)\phi(s)\,ds = \lambda\phi(t),\; 0 \le t \le T \qquad \text{(A.II.45)}$$

or $$\int_0^t s\phi(s)\,ds + t\int_t^T \phi(s)\,ds = \lambda\phi(t)$$

Differentiating with respect to t yields

$$\int_t^T \phi(s)\,ds = \lambda\phi'(t),\; 0 < t < T \qquad \text{(A.II.46)}$$

Differentiating again, we get

$$-\phi(t) = \lambda\,\phi''(t),\ 0 < t < T \tag{A.II.47}$$

Now the boundary conditions are $\phi(0) = 0$ and $\phi'(T) = 0$.

Equation (A.II.48) with boundary condition $\phi(0) = 0$ yields

$$\phi(t) = A \sin \frac{1}{\sqrt{\lambda}}\, t$$

Applying the boundary $\phi'(T) = 0$ in (A.II.47), we find $\cos\left(\frac{1}{\sqrt{\lambda}}\,T\right) = 0$. Hence the eigen values are given by

$$\lambda_k = \frac{T^2}{(k + 1/2)^2\,\pi^2},\ k = 0, 1, 2, \ldots \tag{A.II.48}$$

The normalized eigen functions are given by

$$\phi_k(t) = \sqrt{\frac{2}{T}}\,\sin\,[k + 1/2)\,\pi\,(t/T)] \tag{A.II.49}$$

By Mercer's Theorem we have

$$\Gamma(t, s) = \min\,(t, s)$$

$$= \frac{2}{T}\sum_{k=0}^{\infty} \frac{T^2}{\pi^2\,(k + 1/2)^2}\,\sin[(k + 1/2)\,\pi(s/T)]\ \sin\,[(k + 1/2)\,\pi(t/T)]$$

uniformly on $[0.\ T]^2$.

Wold Decomposition Theorem (1938)

Any stationary process $X(t)$ can be expressed in the form $X(t) = U(t) + V(t)$,

(a) where $U(t)$ and $V(t)$ are uncorrelated proces.

(b) $U(t)$ is regular (or indeterministic) with a one-sided linear representation

$$U(t) = \sum_{i=0}^{\infty} c_i\,\varepsilon_{t-i} \text{ with } c_0 = 1,\ \sum_{i=0}^{\infty} c_i^2 < \infty$$

and ε_t is white noise W.N. $(0, \sigma^2)$ and uncorrelated with $V(t)$, i.e. $E(\varepsilon_t V(s)) = 0$ for all t, s.

The sequence $\{c_i\}$ and the process $\{\varepsilon_t\}$ are uniquely determined.

(c) $V(t)$ is singular (or determinstic) i.e. can be predicted from its own past with zero prediction variance (See Note 1, page 286).

Proof Let $\hat{X}(t)$ denote the projection of $X(t)$ on χ_{t-1} (i.e. the linear subspace of χ spanned by the r.vs. $X(s)$, $s \le t - 1$ i.e. limits of Cauchy sequences of linear combinations of $X_{t-1}, X_{t-2}, \ldots$ and the all such linear combinations. In fact $\hat{X}(t)$ is the linear least square predictor of $X(t)$ given $X(t-1), X(t-2), \ldots$). Define the process $\{\varepsilon_t\}$ by $\varepsilon_t = X(t) - \hat{X}(t)$. Then $\varepsilon_t \perp \chi_{t-1}$ i.e. ε_t is orthogonal to every element of χ_{t-1}. For any pair, $\{\varepsilon_s, \varepsilon_t\}$ with, say $s < t$, We have $\varepsilon_t \perp \chi_s \subset \chi_{t-1}$, and $\varepsilon_s \in \chi_s$ (since both $X(t)$ and $\hat{X}(s)$ are in χ_s).

Hence ε_s and ε_t are orthogonal i.e. $E(\varepsilon_s\ \varepsilon_t) = 0$ for $s \ne t$. Now consider the projection of $X(t)$ on $\chi_t(\varepsilon)$, the subspace spanned by $\varepsilon_t, \varepsilon_{t-1}, \ldots$ and this projection can be written as

$$U(t) = \sum_{i=0}^{\infty} c_i \varepsilon_{t-i} \tag{A.II.50}$$

where ε_t's are orthogal,

$$c_i = E[X(t)\varepsilon_{t-i}]/\sigma_\varepsilon^2 \tag{A.II.51}$$

with $\sigma_\varepsilon^2 = \| \varepsilon_t \|^2 = E\varepsilon_t^2$.

Thus $\quad c_0 = E[X(t)\varepsilon_t]/\sigma_\varepsilon^2 = E[(\hat{X}(t) + \varepsilon_t)\varepsilon_t]/\sigma_\varepsilon^2 = E(\varepsilon_t^2/\sigma_\varepsilon^2) = 1$

$$(\because \varepsilon_t \perp \hat{X}(t), \varepsilon_t = X(t) - \hat{X}(t))$$

Now write $\quad V(t) = X(t) - U(t) = X(t) - \sum_{j=0}^{\infty} c_j \varepsilon_{t-j} \quad$ (A.II.52)

Thus for $s \le t$

$$E(V(t)\varepsilon_s) = E[(X(t) - U(t))\varepsilon_s] = E[X(t)\varepsilon_s] - E[U(t)\varepsilon_s] = 0$$

(since $E[U(t)\varepsilon_s] = \sigma_\varepsilon^2 c_{t-s}$ and by (A.II.51)

$$E[X(t)\varepsilon_s] = \sigma_\varepsilon^2 c_{t-s}).$$

For $s > t$,

$V(t) \in \chi_t$ and $\varepsilon_s \perp \chi_t$ $(\because \varepsilon_t \in \chi_{s-1}^{\perp} \subset \chi_t^{\perp}$ for $s > t)$

Hence ε_s is orthogonal to $V(t)$ for all s, t, and $U(s)$ is orthogonal to $V(t)$ for all $s, t, U(s)$ being linear combination of ε_s for all s.

Therefore $\quad \| X(t) \|^2 = \| U(t) \|^2 + \| V(t) \|^2$

and we then obtain

$$\sigma_\varepsilon^2 \left(\sum_{i=0}^{\infty} c_i^2 \right) = \|U(t)\|^2 \le \| X(t)\|^2 = E(X^2(t)) < \infty.$$

For orthogonal subspaces $£_1$ and $£_2$ we define the direct sum $£_1 \oplus £_2 = \{x + y : x \in £_1, y \in £_2\}$. Let $\chi_t(v)$ be the space spanned by $V(t)$, $V(t-1)$, . . . Then $V(t) \in \chi_t = \chi_{t-1} \oplus \overline{sp}\{\varepsilon_t\}$, where $\overline{sp}\{\varepsilon_t\}$ is the space spanned by ε_t. Thus $V_t \in \chi_{t-1}$. Since $V_t \perp \varepsilon_t$, and by repeating the same argument we find that $V_t \in \chi_{t-s}$, for all $s \ge 0$ $(\because V_t \in \chi_{t-1} \oplus \overline{sp}\{\varepsilon_t\}, < V_t, \varepsilon_{t-1} > = 0 \Rightarrow V_t \in \chi_{t-2})$.

Hence $V_t \in \chi_{-\infty} = \bigcap_{t=-\infty}^{\infty} \chi_t$, and is therefore determined by the "remote past" of $\{X_t\}$. But the projection of V_t on χ_{t-1} is the same as the projection of V_t on $\chi_{t-1}(V)$, since we may write $\chi_{t-1}(V) \oplus \chi_{t-1}(\varepsilon)$ (by A.II.52).

Hence $V_t \in \chi_{t-1}(V)$ i.e. $\chi_t(V) = \chi_{t-1}(V)$ for all t and clearly $V_t \in \chi_{t-s}(V)$ for all $s \ge 0$.

We can write $V_t \in \chi_{-\infty}(V) = \bigcap_{t=-\infty}^{\infty} \chi_t(V)$, i.e. V_t is determined by its own "remote past" and can be predicted with zero prediction variance. To establish uniqueness we observed from

$$X(t) = \sum_{j=0}^{\infty} c_j \varepsilon_{t-j} + V_t \tag{A.II.53}$$

that if $\{\varepsilon_t\}$ and $\{V_t\}$ are any sequences satisfying (A.II.53) and having the desired properties proved above, then $\chi_{t-1} \subset \overline{sp}\,\{\varepsilon_s, s \le t-1\} \oplus \overline{sp}\,\{V_e, s \le t-1\}$ from which it follows, $\{\varepsilon_t\}$ being white noise (orthogonal) and $E(\varepsilon_t V_s) = 0$, that ε_t is orthogonal to χ_{t-1}. Projecting each side of (A.II.53) onto χ_{t-1} and subtracting the resulting equation from (A.II.53), we find that the process $\{\varepsilon_t\}$ must satisfy $\varepsilon_t = X_t - \hat{X}_t = X_t - P\chi_{t-1}X_t$. By taking inner products of each side of (A.II.53) with ε_{t-j} we see that c_j must also satisfy (A.II.51). Finally, if (A.II.53) is to hold, it is obviously necessary that V_t must be defined as in (A.II.52).

Note

1. For a general stationary discrete parameter process $\{X_t\}$ (whose spectrum need not necessarily be continuous) let us consider the sequence of prediction variances $\sigma_m^2 = E[X_{t+m} - \hat{X}_{t+m}]^2$, $m \ge 1$. By stationarity of X_t, σ_m^2 does not depend on t. Clearly $\chi_s \subset \chi_t$, $s \le t$ and hence $\sigma_1^2 \le \sigma_2^2 \le \ldots \le \sigma_m^2 \le \ldots$ (obviously we would expect that variances to increase as we predict further ahead).

If $\sigma_1^2 > 0$ the process is called regular or *nondeterministic* where as if $\sigma_1^2 = 0$ process is called *singular* or *deterministic*. If $\sigma_1^2 > 0$, it follows that $\sigma_m^2 > 0$ for all m. On the other hand if $\sigma_1^2 > 0$ then X_{t+1} is its projection on χ_t i.e. $\chi_{t+1} = \chi_t$ for all t and all the χ_t spaces are same. Thus, X_{t+m} is its own projection on χ_t for all $m \ge 1$ and hence $0 = \sigma_1^2 = \sigma_2^2 = \ldots \sigma_m^2 = \ldots$ We therefore see that the $\{\sigma_m^2\}$ are either all positive (corresponding to X_t regular) or all zero (corresponding to X_t singular). In the later case X_t can be predicted "prefectly" i.e. with zero prediction variance, from its past values.

2. If $\chi_{-\infty} = \{0\}$, then $V_t = 0$ in Wold's decomposition and $X_t = U_t^{'}$ is "purely nondeterministic". If $\sigma_\varepsilon^2 = 0$, then $U_t = 0$, and $X_t = V_t$ is then a deterministic process.

Appendix III
Stochastic Integral (Ito Integral)

Introduction

In recent years, it has become apparent that physical systems, classically modelled by deterministic differential equations, can be more satisfactorily modelled by certain stochastic counterparts if random effects in the physical phenomena as well as measuring devices are to be taken into account. In this context, an ordinary differential equation $\frac{dx}{dt} = f(t, x)$ would be replaced by a random differential

$$\frac{dX(t)}{dt} = F(t, X(t), Y(t)) \tag{A.III.1}$$

where $Y(t)$ represent some stochastic input process explicitly. It is often seen that it is not possible to interpret (A.III.1) as an ordinary differential equation along each sample path. For example

$$\frac{dX(t)}{dt} = f(t, X(t)) + g((t, X(t))N(t) \tag{A.III.2}$$

with $N(t)$ being a Gaussion white noise process. Equation (A.III.2) has been popular in the engineering literature especially, since Gaussion white noise approximates the effect of the superposition of a large number of small random disturbances, a situation encountered in engineering systems. However, the irregularity of the sample paths of N makes (A.III.2) intractable mathematically. Just as a solution of (A.III.1) satisfies the deterministically integral equation $X(t) = X(t_0) + \int_{t_0}^{t} f(s, X(s))ds$ a solution of (A.III.2) should be a solution of the random equation

$$X(t) = X(t_0) + \int_{t_0}^{t} f(s, X(s))ds + \int_{t_0}^{t} g(s, X(s))N(s)ds \tag{A.III.3}$$

but unfortunately the last integral in (A.III.3) cannot be defined in any meaningful way. To deal with this difficulty, the integral in question is replaced by an integral of the form

$$\int_{t_0}^{t} g(s, X(s))dW(s). \tag{A.III.4}$$

where W is the Wiener process and with the motivation being that, at least formally, $\dot{W}(t) = N(t)$ and so $dW(t) = N(t)dt$. However, even equation (A.II.3) with the second integral replaced by (A.III.4) is not unquestionable. The natural interpretation of (A.III.4) would be as a Stieltjes integral along sample paths. But this is not possible, since the sample paths of the Wiener process are not of bounded variation. Furthermore, if different choices of the τ_i are made from the subinterval $[t_{i-1}, t_i]$ of the partition $t_0 < t_1 < t_2 < \ldots < t_n = t$, the natural approximating sums

$$\sum_{i=1}^{n} g(\tau_i, X(\tau_i))[W(t_i) - W(t_{i-1})]. \tag{A.III.5}$$

converge in the q.m. sense to different values of the integral. If, in (A.III.5), τ_i is taken as t_{i-1}, the Ito integral results. The impact is that the integral inherits many of the probabilistic properties of the Wiener process at the expense of the corresponding calculus differing from what one would expect in the Stieltjes case. The main instrument of this calculus is Ito's formula, which can be thought of as a probabilistic chain rule; it yields a stochastic equation for any sufficiently smooth function of a solution. Stochastic integral $\int_a^b g(t)dW(t)$ of Ito satisfies

$$E\left(\int_a^b g(t)dW(t)\right) = 0 \tag{A.III.6}$$

and

$$E\left|\int_a^b g(t)dW\right|^2 = \int_a^b E\,|\,g(t)\,|^2\,dt \tag{A.III.7}$$

and viewed as a function of the upper limit of integration, forms a martingale. Note that $W(t)$ is a martingale with $E[W(b) - W(c)] = 0$ and $E|\,W(b) - W(a)\,|^2 = b - a$. This illustrate tracability that motivates the choice of the Ito integral from a mathematical point of view.

The difficulty alluded here from $g(\tau_i)$ being a random variable which may not be measurable with respect to the σ-algebra generated by $W(t_{i-1})$. From our discussion it seems reasonable to start by defining stochastic integral for random step functions. However, there are infinitely many (indexed by parameter λ) such definitions for random step functions leading to distinct q.m. stochastic integrals. Specifically, consider the integral $\int_a^b W(t)d\,W(t)$ and for any fixed λ, $0 \le \lambda \le 1$, the random step function approximation of $W(t)$

$$f_n^{\lambda}(t) = \lambda W(t_k^{(n)}) + (1 - \lambda)W(t_{k-1}^{(n)}),\ t_{k-1}^{(n)} \le t < t_k^{(n)} \tag{A.III.8}$$

for which

$$\int_a^b f_n^{\lambda}(t)dW(t) = \sum_{k=1}^{n} f_n^{\lambda}(t_{k-1})\Delta W_k^{(n)} \tag{A.III.9}$$

where $0 \le a = t_o^{(n)} < \ldots < t_n^{(n)} = b$ is a sequence of partitions of the interval $[a, b]$, $\Delta t_k^{(n)} = t_k^{(n)} - t_{k-1}^{(n)}$, $\partial_n = \max_k \Delta t_k^{(n)}$, and $\Delta W_k^{(n)} = W(t_k^{(n)}) - W(t_{k-1}^{(n)})$ results in the stochastic integral (by taking q.m. limit as $\partial_n \to 0$)

$$\int_a^b W(t)dW(t) = \frac{1}{2}[W^2(b) - W^2(a)] + (\lambda - 1/2)(b - a) \tag{A.III.10}$$

For example, when $\lambda = 1/2$ the Stratonovich stochastic integral exists. In this case the form of (A.III.10) indicates that the corresponding stochastic calculus will be analogous to the usual Riemann-Stieltjes Calculus. However, a disadvantage of this and any $\lambda > 0$ case is that expansion

$$\lambda W(t_k^{(n)}) + (1 - \lambda)W(t_{k-1}^{(n)}) \tag{A.III.11}$$

is not measurable with respect to the σ-algebra generated by $W(t_{k-1}^{(n)})$, a situation which prevents the martingale property as well as (A.III.6) and (A.III.7) from holding. The corresponding step functions $f_n^\lambda(t)$ are said to be *anticipating* in this case. The *nonanticipating* case, namely, when (A.III.11) is measurable with respect to the σ-algebra generated by $W(t_{k-1}^{(n)})$ and hence independent of the increments $\Delta W_k^{(n)}$, occurs only when $\lambda = 0$. This leads to the Ito case and with the random step functions $f_n^0(t) = W(t_{k-1}^{(n)}), t_{k-1}^{(n)} \le t < t_k^{(n)}$ in (A.III.9), one obtains the special case of (A.III.10),

$$\int_a^b W(t)dW(t) = 1/2\{[W^2(b) - W^2(a)] - (b - a)\}$$

and it can be shown that the integral interpreted in the Ito sense satisfies

(a) $E\left(\int_a^b W(t)dW(t)\right) = 0$ and

(b) $E\left|\int_a^b W(t)dW(t)\right|^2 - 1/2(b^2 - a^2)$.

Definition of Ito Integral

Let $W(t)$ be a standard Wiener process defined on the probability space $(\Omega, \mathcal{F}, P)$ and let $\{\mathcal{F}_t, -\infty < t < \infty\}$ be an increasing family of sub σ-algebras of $\mathcal{F}$ satisfying the following conditions (i) $\mathcal{F}_{t_1} \subset \mathcal{F}_{t_2}$ if $t_1 < t_2$, (ii) $W(t)$ is $\mathcal{F}_t$- measurable for each t and (iii) For $s > 0$, $W(t + s) - W(t)$ is independent of $\mathcal{F}_t$. If the integrand f does not depend on ω, then the integral $I(f) = \int_a^b f(t)dW(\omega, t)$ can be treated as second order (q.m.) integral as defined before (see Appendix II, theorem II.5).

Interpreted in that way, only the property of a Brownian motion as a process with orthogonal increaments is made use of, and not its other properties. If f is random i.e. it depends on ω and, then the integral has to be defined anew. It turns out that its definition now depends in a crucial way on the martingale property and

$$E(W_{t+s} \mid \mathcal{F}_t) = W_t \text{ and } E(W_{t+s} - W_t)^2 = s \text{ a.s., } s \ge 0, \text{ of } W_t \tag{A.III.12}$$

We make the following assumptions on the integrand f.

(a) The function f is jointly measurable in (ω, t) (with respect to $\mathscr{F}$ in ω and the Lebesgue measureable sets in t) (A.III.13)

(b) For each t, $f_t = f(., t)$ is measurable with respect to $\mathscr{F}_t$ and f satisfies

$$\int_a^b E|f(t)|^2 \, dt < \infty \tag{A.III.14}$$

Denote by ε the class of random step functions satisfying (A.III.12), (A.III.13) and (A.III.14).

The stochastic Integral is now defined as follows:

1. If $a = t_0 < t_1 < \ldots < t_n = b$ be partition points of the interval $[a, b]$ are independent of ω and

$$f_i(\omega) = f(t_i, .) = f(\omega, t) \text{ if } t_i \leq t < t_{i+1} \text{ and } i = 0, \ldots, n-1 \tag{A.III.15}$$

satisfy conditions (A.III.13) and (A.III.14), then we define the stochastic Integral by

$$\int_a^b f(\omega, t) dW(\omega, t) = \sum_{i=0}^{n-1} f_i(\omega)[W(\omega, t_{i+1}) - W(\omega, t_i)] \tag{A.III.16}$$

2. If f satisfy (A.III.13) and (A.III.14), then we shall show that three exists a sequence of random-step functions $\{f_n(\omega, t)\}$ satisfying (A.III.13) and (A.III.14) such that

$$\int_a^b E|f(., t) - f_n(., t)|^2 \, dt \to 0 \text{ as } n \to \infty \tag{A.III.17}$$

It will then follow that $\int_a^b f_n(\omega, t) dW(\omega, t)$ converges in mean square as $n \to \infty$, and this q.m. limit is the same as any sequence of random step functions which satisfy (A.III.17).

Therefore, we define

$$\int_a^b f(\omega, t) dW(\omega, t) = \lim_{n \to \infty} \int_a^b f_n(\omega, t) dW(\omega, t) \text{ in q.m.,} \tag{A.III.18}$$

where $\{f_n(\omega, t)\}$ is any sequence of random step functions satisfying (A.III.17). We assume that both a and b are finite. If not, the stochastic Integral is defined as the limit in q.m. as $a \to -\infty$ and $b \to \infty$, or both.

Theorem III.1 Let $\{W_t, t \in T\}$ be a Brownian motion, and let $f(\omega, t)$ satisfy (A.III.13) and (A.III.14). Then,

(a) there exists a sequence of random step functions $\{f_n\}$ satisfying (A.III.13) and (A.III.14) such that

$$\|f - f_n\|^2 = \int_a^b E|f(., t) - f_n(., t)|^2 \, dt \to 0 \text{ as } n \to \infty.$$

(b) For each $n, I(f_n) = \int_a^b f(\omega,t)dW(\omega,t)$ is well defined by (A.III.16), and $\{I(f_n)\}$ converges in q.m.

(c) If $\{f_n\}$ and $\{f_n^*\}$ are two sequences of random step-functions satisfying (A.III.13) and (A.III.14) such that $\| f - f_n \|$ and $\| f - f_n^* \|$ both go to zero as $n \to \infty$, then $\lim_{n\to\infty} I(f_n) = \lim_{n\to\infty} I(f_n^*)$ in q.m.

Proof (a) Suppose $E[f(.,t)\bar{f}(.,s)]$ is continuous on $[a, b] \times [a, b]$, that is, f_t is q.m. continuous on $[a, b]$, then an approximating sequence of random step functions $\{f_n\}$ can be constructed by partitioning $[a, b]$, sampling $f(\omega, t)$ at partition points $t_j(n)$, defining $f_n(\omega,t) = f(\omega, t_j^{(n)}), t_j^{(n)} \le t < t_{j+1}^{(n)}$, and refining the partitions to zero (taking $\max_j (t_{j+1}^{(n)} - t_j^{(n)}) \to 0$ as $n \to \infty$). Since f_t is q.m. continuous, $E\,|f(.,t) - f_n(.,t)|^2 \to 0$ as $n \to \infty$ for every $t \in [a, b]$. By dominated convergence theorem, we have $\int_a^b E|f(.,t) - f_n(.,t)|^2 dt \to 0$ as $n \to \infty$.

More generally; if f merely satisfies (A.III.13) and (A.III.14) and is not necessarily q.m. continuous on $[a, b]$, we construct a sequence of approximating random step functions in the following manner:

By setting $f(t) = 0$, $t \notin [a, b]$ f can be considered defined on the entire line $(-\infty, \infty)$. Then by (A.III.14) and Fubini Theorem

$$E\int_{-\infty}^{\infty} |f|^2\, dt = \int_{-\infty}^{\infty} E\,|f|^2\, dt < \infty \qquad \text{(A.III.19)}$$

so that,

$$\int_{-\infty}^{\infty} |f(t)|^2\, dt < \infty \text{ a.s.} \qquad \text{(A.III.20)}$$

Now

$$\int_{-\infty}^{\infty} |f(t+h) - f(t)|^2 dt \le 4 \int_{-\infty}^{\infty} |f(t)|^2 dt; \qquad \text{(A.III.21)}$$

hence (A.III.20) and (A.III.21) imply

$$\int_{-\infty}^{\infty} |f(t+b) - f(t)|^2 dt \to 0 \text{ a.s. as } h \to 0 \qquad \text{(A.III.22)}$$

by the dominated convengence theorem for Lebesgue integration along sample paths. Now from (A.III.20),

$$E\int_{-\infty}^{\infty} |f(t+h) - f(t)|^2 dt \le 4E\int_{-\infty}^{\infty} |f(t)|^2 dt,$$

So by (A.III.19) applying dominated convengence theorem in the probability space we get

$$E\int_{-\infty}^{\infty} |f(t+h) - f(t)|^2 dt \to 0 \text{ a.s. } h \to 0 \qquad \text{(A.III.23)}$$

Define the function $\phi_j(t) = [jt]/j, j$ an integer $\geq$ where [ ] denotes the "greatest integer" function.

Since $\phi_j(t) \to t$ as $j \to \infty$, (A.III.23) can be replaced by

$$E\int_{-\infty}^{\infty} |f(s+\phi_j(t)) - f(s+t)|^2 ds \to 0 \text{ as } j \to \infty \text{ for any fixed } t \quad \text{(A.III.24)}$$

As this expression is bounded by $4E\int_{-\infty}^{\infty} |f(s)|^2\, ds < \infty$, the dominated convengence theorem together with Fubini theorem, yields

$$\int_{-\infty}^{\infty}\int_{a-1}^{b} E\,|f(s+\phi_j(t)) - f(s+t)|^2\, dt\, ds = \int_{a-1}^{b} E\int_{-\infty}^{\infty} |f(s+\phi_j(t)) - f(s+t)|^2\, ds dt \to 0 \text{ as } j \to \infty \quad \text{(A.III.25)}$$

Therefore there exists a subsequence $\{j_n\}$ such that

$$\int_{a-1}^{b} E\,|f(s+\phi_{j_n}(t)) - f(s+t)|^2\, dt \to 0 \text{ as } j_n \to \infty, \text{ for almost all } s \quad \text{(A.III.26)}$$

Fixing such $s \in [0, 1]$ and replacing t by $t - s$ in (A.III.26), we have

$$\int_{a-1+s}^{b+s} E\,|f(s+\phi_{j_n}(t-s)) - f(t)|^2\, dt \to 0 \quad \text{(A.III.27)}$$

Therefore, since $0 \leq s \leq 1$, (A.III.27) implies

$$\int_{a}^{b} E\,|f(s+\phi_{j_n}(t-s)) - f(t)|^2\, dt \to 0 \quad \text{(A.III.28)}$$

Taking $f_n(t) = f(s+\phi_{j_n}(t-s))$, we have $f_n \in \varepsilon$ and (A.III.28) states that $\| f_n - f \| \to 0$ as $n \to \infty$.

(b) Assume that $\{f_n\}$ is a sequence of random step functions such that $\| f - f_n \| \to 0$ as $n \to \infty$ and define $I(f_n)$ by (A.III.16) as

$$I(f_n) = \sum_j f_{n_j}(\omega)[W(\omega, t_{j+1}^{(n)}) - W(\omega, t_j^{(n)})] \quad \text{(A.III.29)}$$

Writing $\Delta_j W^{(n)} = W(\omega, t_{j+1}^{(n)}) - W(\omega, t_j^{(n)})$, we have

$$E|\,I(f_n)\,|^2 = \sum_j \sum_k E(f_{n_j} \bar{f}_{n_k} \Delta_j W^{(n)} \Delta_k W^{(n)})$$

If $k > j$, then $\Delta_k W^{(n)}$ is independent of $\mathscr{F}_{t_j}(n)$ while $(\phi_{n_j}\phi_{n_k}\Delta_j W^{(n)})$ is $\mathscr{F}_{t_j}(n)$-measurable. Hence $E(\phi_{nj}\bar{\phi}_{nk}\Delta_j W^{(n)}\Delta_k W^{(n)}) = 0$ if $k \neq j$

and
$$E|\,I(f_n)\,|^2 = \sum_j E[|f_{n_j}|^2 (\Delta_j W^{(n)})^2]$$

$$= \sum_j E\,|f_{n_j}|^2 E[(\Delta_j W^{(n)})^2 \mid \mathscr{F}_{t_j}(n)]$$

$$\sum_j E|f_{n_j}|^2 (t_{j+1}^{(n)} - t_j^{(n)}) = \int_a^b E|f_n(.,t)|^2\, dt \qquad \text{(A.III.30)}$$

Now, $I(f_{m+n}) - I(f_n) = I(f_{m+n} - f_n)$ and $f_{m+n} - f_n$ is again a random step function. Therefore,

$$E|I(f_{m+n}) - I(f_n)|^2 = \int_a^b E|f_{m+n}(.,t) - f_n(.,t)|^2 dt$$

$$\leq 2\int_a^b E|f_{m+n}(.,t) - f(.,t)|^2 dt + 2\int_a^b E|f_n(.,t) - f(.,t)|^2 dt$$

$$\to 0 \text{ as } n \to \infty \qquad \text{(A.III.31)}$$

Hence $\{I(f_n)\}$ is a mutually q.m. convergent sequence, and there exists a second-order r.v. $I(f)$ such that $E|I(f_n) - I(f)|^2 \to 0$ as $n \to \infty$.

(c) Suppose $\{f_n)\}$ and $\{f_n'\}$ are random step functions such that

$$\|f_n - f\| \to 0 \text{ as } n \to \infty \text{ and } \|f_n' - f\| \to 0 \text{ as } n \to \infty.$$

Then, $\|f_n - f_n'\| \leq \sqrt{2}(\|f_n - f\|^2 + \|f_n' - f\|^2)^{1/2} \to 0$ as $n \to \infty$

Therefore, $E|I(f_n) - I(f_n')|^2 \to 0$ as $n \to \infty$ and

$$\lim_{n\to\infty} I(f_n) = \lim_{n\to\infty} I(f_n') \text{ in q.m.}$$

Theorem III.2 Let $f(\omega, t)$ and $g(\omega, t)$ satisfy (A.III.13) and (A.III.14).

Then
$$E\left[\int_a^b f(.,t)dW(t) \int_a^b g(.,t)dW(t)\right] = \int_a^b E(f\bar{g})dt \qquad \text{(A.III.32)}$$

Proof It is enough to prove (A.III.32\0 for $f = g$, because

$$E[I(f)\overline{I(g)}] = \frac{1}{4}[E|I(f) + I(g)|^2 - E|I(f) \quad I(g)|^2]$$

$$+ i.\frac{1}{4}[E|I(-if) + I(g)|^2 - E|I(-if) - I(g)|^2]$$

$$= \frac{1}{4}[E|I(f+g)|^2 - E|I(f-g)|^2]$$

$$+ \frac{i}{4}[E|I(-if+g)|^2 - E|I(-if-g)|^2 \qquad \text{(A.III.33)}$$

(A.III.32) follows from the special case $E|I(f)|^2 = \int_a^b E|g|^2 dt$. Now if f is a random step function, we have already proved the last equality in (A.III.30) that $E|I(f)|^2 = \int_a^b E|f(.,t)|^2 dt$.

If f is not a random step function, let $\{f_n\}$ be an approximating sequence of random step functions, then

$$E|I(f)|^2 = E|I(f-f_n)+I(f_n)|^2$$
$$= E|I(f_n)|^2 + 2\text{Re}(E[I(f-f_n\overline{I(f_n)})])$$
$$+ E|I(f-f_n)|^2.$$

Since $E|I(f-f_n)|^2 \to 0$ as $n \to \infty$, we have

$$E|I(f)|^2 = \lim_{n\to\infty} E|I(f_n)|^2 = \lim_{n\to\infty} \int_a^b E|f_n(.,t)|^2 dt$$
$$= \int_a^b E|f(.,t)|^2 dt.$$

Now we want to define definite Ito Integral and in this vain let us state and prove the following theorem.

Theorem III.3 Let $\{W(t), \mathcal{F}_t\}$ be a Brownian motion and let f satisfy (A.III.13) and (A.III.14). Define a process $\{X_t, a \le t \le b\}$ by

$$X(\omega,t) = \int_a^t f(\omega,s)dW(\omega,s) \tag{A.III.34}$$

Then $\{X_t, \mathcal{F}_t, a \le t \le b\}$ is a martingale, i.e.

a.s. $\quad t > s, E(X_t|\mathcal{F}_s) = X_s$ a.s.

Proof First suppose that the integrand f in (A.III.34) is a random step function. For $t > s$, let $t_1, t_2, \ldots, t_n$ be jump points of f between s and t. Then we can write

$$X(\omega,t) - X(\omega,s) = \int_s^t f(\omega,s)dW(\omega,t)$$
$$= f_0(\omega)[W(\omega,t_1) - W(\omega,s)] + f_1(\omega)[W(\omega,t_2) - W(\omega,t_1)] + \ldots$$
$$+ f_n(\omega)[W(\omega,t) - W(\omega,t_n)] \tag{A.III.35}$$

where f_0 is $\mathcal{F}_s$ – measurable and f_k is $\mathcal{F}_k$-measurable, $k = 1, 2, ..., n$.

By successively taking conditional expectation with respect to $\mathcal{F}_{t_n}$, $\mathcal{F}_{t_{n-1}}, \ldots, \mathcal{F}_n$ we find $E[X_t - X_s|\mathcal{F}_s] = 0$. Since X_s is obviously $\mathcal{F}_s$-measurable, this proves the theorem when f is a step function. If f is not a step function, $\{f_n\}$ be a sequence of step functions convenging to f in q.m.

Define $$X_n(\omega,t) = \int_s^t f_n(\omega,t)dW(\omega,t) \tag{A.III.36}$$

Then for each n, $\{X_n(\omega,t), a \le t \le b\}$ is a martinglae and hence

$$E[X_t - X_s|\mathcal{F}_s] = E[X(.,t) - X_n(.,t)|\mathcal{F}_s] - E[X(.,s) - X_n(.,s)|\mathcal{F}_s].$$

Since $E|X(.,t) - X_n(.,t)|^2 \to 0$ as $n \to \infty$, applying Jenson's inequality for conditional expectation, $E[X(.,t) - X_n(.,t)] - E[X(.,s) - X_n(.,s)] \to 0$ in q.m. as $n \to \infty$.

Hence, $E[X_t - X_s|\mathcal{F}_s] = 0$ a.s.

One immediate consequence of the martingale property is that a stochastic integral does not behave like an ordinary integral. Note that if the stochastic Integral, $\int_0^t W(s)dW(s)$ behave like an ordinary Riemann Integral, surely it must be equal to $\frac{1}{2}(W(t)^2 - W(0)^2) = \frac{1}{2}W^2(t)$. However, $\frac{1}{2}W^2(t)$ is not a martingale. What $\int_0^t W(s)dW(s)$ will be clarified by the so called Ito's differentiation rule.

Ito Formula

Our aim here is to give a feel for simple ITO formula and not to give the most general formula. In usual calculus we know that for a continuously differentiable function $f(t)$ on $(0, \infty)$.

$$f(t) = \int_0^t f'(s)ds + f(0)$$

or

$$f(t) - f(0) = \int_0^t f'(s)ds.$$

Can we replace t by the r.v. $W(t)$ and we can write

$$f(W(t)) - f(W(0)) = \int_0^t f'(W(s))dW(s)?$$

The answer is no, but some interesting and useful facts can be proved. Why does not the formula work in ordinary calculus? This is because of Taylor's expansion which says $f(t + \Delta t) - f(t) = f'(t)\, \Delta t + f''(t)\, (\Delta t)^2$ and $(\Delta t)^2$ term is too small and can be ignored. To derive the calculus formula we assume that f has continuous second derivatives (i.e. $f \in C^2$) and $\xi \in (t, t + \Delta t)$. Take a sequence of partitions Π_n of $[0, t]$ with $\| \Pi_n \| = \max_{0 \le j \le k-1} (t_{j+1} - t_j) \to 0$, $\Pi_n = \{0 = t_0 < t_1 < \ldots < t_k = t\}$. Of course the points t_k depend on n.

Then

$$f(t) - f(0) = \sum_{i=0}^{n-1} [f(t_{i+1}) - f(t_i)] \text{ (Talescopic sum)}$$

$$= \sum_{i=0}^{n-1} f'(t_i)[t_{i+1} - t_i] + \sum_{i=0}^{n-1} 1/2 f''(\xi_i)(t_{i+1} - t_i)^2 \quad \text{(by Taylor's expansion)}$$

As $n \to \infty$, the first term $\to \int_0^t f'(s)ds$ by definition of Riemann integral and

$$|\text{ Second term }| \le 1/2\, C \sum_{i=0}^{n-1} (t_{i+1} - t_i)^2, \text{ where } C = \sup_{0 \le s \le t} |f''(s)|$$

$$\le 1/2\, Ct \,\| \Pi_n \| \to 0 \text{ as } n \to \infty.$$

For Brownian motion on the average $(W(t + \Delta t) - W(t))^2$ behaves like Δt (not like $(\Delta t)^2$) and hence can not be ignored.

Ito formula (1) Let f be a real valued function with continuous second derivatives. Then

$$f(W(t)) = f(W(0)) + \int_0^t f'(W(s))dW(s) + \frac{1}{2}\int_0^t f''(W(s))ds.$$

Remark Left side is the process whose value at (ω, t) is $f(W(\omega, t))$ and $f(W(0))$ is a r.v. and $f'(W(s))$ is the process $\int_0^t f(\omega, s)dW(s)$, where $f(\omega, s) = f'(W(\omega, s))$. This Ito integral makes sense because $s \to W(\omega, s) \to f'(W(\omega, s))$ is continuous function of s and hence $\int_0^t f^2(\omega, s)ds < \infty$. Further it is adopted. It is jointly measurable, because $(\omega, s) \to W(\omega, s)$ is so. The last term is usual pathwise integral $\frac{1}{2}\int_0^t g(\omega, s)ds$ where $g(\omega, s) = f''(W(\omega, s))$ i.e. first keep $f''(x)$ ready, given (ω, s) calculate $W(\omega, s)$ and evaluate f'' at the point $x = W(\omega, s)$.

Proof of Ito formula (1)

Fix a sequence $\{\Pi_n\}$ of partitions $[0 = t_o^{(n)} < t_1^{(n)} < \ldots < t_k^{(n)} = t] = \Pi_n$ so that $\| \Pi_n \| \to 0$. First fix a $t > 0$ and then fix Π_n. Of course the $t_i^{(n)}$ depends on n. Then

$$\begin{aligned} f(W(t)) - f(W(0)) &= \sum_{i=0}^{n-1} [f(W(t_{i+1}^{(n)}) - f(W(t_i^{(n)}))] \\ &= \sum_{i=0}^{n-1} f'(W(t_i^{(n)}))\,[W(t_{i+1}^{(n)}) - W(t_i^{(n)})] \\ &\quad + \sum_{i=0}^{n-1} 1/2\, f''(\xi_i)[W(t_{i+1}^{(n)}) - W(t_1^{(n)})]^2 = I + II. \quad \text{(A.III.37)} \end{aligned}$$

where ξ_i lies between points $W(\omega, t_i^{(n)})$ and $W(\omega, t_{i+1}^{(n)})$. Note that first term I of (A.III.37) is equal to $\int_0^t f_n(s)dW(s)$, where $f_n(\omega, s) = f'(W(\omega, t_i^{(n)}))$ if $t_i^{(n)} \le s < t_{i+1}^{(n)}$ is a random step function for each n. Note that $f_n(\omega, s) \to f(\omega, s)$ $= f''(W(\omega, s))$ for each ω and each s. Then the first term converges in probability to the r.v. $\int_0^t f'(W(s)dW(s)$. This is obtained under weaker condition $\int_a^b |f(\omega, t)|^2 dt < \infty$ a.s. instead of the condition $\int_a^b E|f(\omega, t)|^2 dt < \infty$. Here the stochastic integral is defined by

$$\int_a^b f(\omega, t)dW(\omega, t) = \lim_{n\to\infty} \int_a^b f_n(\omega, t)dW(\omega, t) \text{ in probability sense,}$$

where $f_n(\omega, t)$ is defined by

$$f_n(\omega, t) = \begin{cases} f(\omega, t) \text{ if } \int_a^b |f(\omega, t)|^2 dt < n \\ 0 \text{ otherwise} \end{cases}$$

This can be proved as follows :

Now $I(f_n) = \int_a^b f_n(\omega, t) dW(\omega, t)$ is jointly measurable, for fixed ω it is $\mathscr{F}_t$-measurable and $\int_a^b E|f_n(\omega, t)|^2 dt < \infty$, so that $I(f_n)$ is well defined. Now, for any ω such that $\int_a^b |f(\omega, t)|^2 dt \leq \min(m, n)$, we have

$$\sup_t |f_m(\omega, t) - f_n(\omega, t)| = 0$$

and that implies

$$\int_a^b f_n(\omega, t) dW(\omega, t) = \int_a^b f_m(\omega, t) dW(\omega, t).$$

It follows that for every $\varepsilon > 0$,

$$P(|I(f_m) - I(f_n)| \geq \varepsilon) \leq P\left(\int_a^b |f(\omega, t)|^2 dt > \min(m, n)\right) \to 0$$

as $m, n \to \infty$ which proves that $\{I(f_n)\}$ converges in probability to $I(f)$, say.

Back to the Proof of Ito formula (1),

$$II = -\sum_{i=0}^{n-1} 1/2[f''(W(t_i)) - f''(\xi_i)][W(t_{i+1}) - W|(t_i)]^2 + 1/2$$

$$\sum_{i=0}^{n-1} f''(W(t_i))[W(t_{i+1}) - W(t_i)]^2 \qquad \text{(A.III 38)}$$

The first term of (A.III.38) in modulus $\leq c_n(\omega) \sum_{i=0}^{n-1} [W(t_{i+1}) - W(t_i)]^2$, where $c_n(\omega) = 1/2 \sup_i \sup_x |f''(W(t_i)) - f''(x)|$, between $W(\omega, t_i)$ and $W(\omega, t_{i+1})$. Note that by continuity of $f''(x)$ and $W(t)$ paths it is easy to see that $c_n(\omega) \to 0$ a.s. as $\| \Pi_n \| \to 0$ and $\sum_{i=0}^{n-1} [W(t_{i+1}) - W(t_i)]^2 \to t$ in q.m. The product

$$c_n(\omega) \sum_{i=0}^{n-1} [W(t_{i+1}) - W(t_i)]^2 \xrightarrow{P} 0.$$

The second term of (A.III.38) $= \frac{1}{2} \sum_{i=0}^{n-1} f''(W(t_i))\{[W(t_{i+1}) - W(t_i)]^2 - [t_{i+1} - t_i]\}$

$$+ \frac{1}{2} \sum_{i=0}^{n-1} f''(W(t_i)) [t_{i+1} - t_i] \qquad \text{(A.III.39)}$$

The first term of (A.III.39) in modulus $\leq c_1(\omega)\left|\sum_{i=0}^{n-1}(W(t_{i+1})-W(t_i))^2-t\right|$

where $c_1(\omega)=[\sup f''(x),\ x\in\{W(\omega,s),\ 0\leq s\leq t\}]$.

$$\left|\sum_{i=0}^{n-1}(W(t_{i+1})-W(t_i))^2-t\right|\to 0 \text{ in q.m.}$$

Therefore the first term in (A.III.39) $\xrightarrow{P} 0$.

The second term of (A.III.39) (being the Riemann sum) converges to $\int_0^t f''(W(\omega,s))ds$ in probability. Now for each fixed t the two random variables on either side of ITO formula (1) are equal a.s. So the two processes are equal a.s. because both sides are obviously continuous path processes.

ITO formula (2) Let $f(x, t)$ be a real valued function on $R\times(0,\infty)$ with continuous second partial derivative with respect to x and has continuous first derivative with respect to t.

$$\text{Then } f(W(t),t)=f(W(0),t)+\int_0^t\frac{\partial f}{\partial x}(W(s),s)\,dW(s)$$

$$+\int_0^t\frac{\partial f(W(s),s)}{\partial x}ds+\frac{1}{2}\int_0^t\frac{\partial^2 f(W(s),s)ds}{\partial x^2}$$

Proof Proof is exactly the same as formula (1)—only use Taylor expansions for functions of two variables and a little more messy computation than before. Hence the details of proof is left to the reader.

APPENDIX IV
Some Important Martingale Theorems

Lemma A.I V.1 Let $\{X_n\}$ be a (super) martingale with respect to $\mathscr{F}_n = \mathscr{B}(Y_0, \ldots, Y_n)$ and T is a stopping time with respect to $\mathscr{F}_n$.

Then for all $n \geq k$,

$$E(X_n I_{\{T=k\}})(\leq) = E[X_k I_{\{T=k\}}] \tag{A.IV.1}$$

and

$$E(X_0)(\geq) = E(X_{T\wedge n})(\geq) = E(X_n),\ n \geq 1 \tag{A.IV.2}$$

Proof

$$\begin{aligned} E(X_n I_{[T=k]}) &= E\{E[X_n I_{[T=k]} \mid \mathscr{F}_n]\} \\ &= E[I_{[T=k]} E(X_n \mid \mathscr{F}_n)]\ (\leq) \\ &- E[I_{[T=k]} X_k]. \end{aligned}$$

$$\begin{aligned} E(X_{T\wedge n}) &= \sum_{k=0}^{n-1} E[X_T I_{[T=k]}] + E[X_n I_{[T\geq n]}] \\ &= \sum_{k=0}^{n=1} E[X_k I_{[T=k]}] + E[X_n I_{[T\geq n]}] \\ (\geq) &= \sum_{k=0}^{n-1} E[X_n I_{[T=k]}] + E[X_n I_{[T>n]}] \\ &= E(X_n). \end{aligned}$$

For a martingale, $EX_n = EX_0$, which completes the proof of (A.IV.2) in this case.

For a supermartingale we have to prove $E(X_0) \geq E(X_{T\wedge n})$. We first assume $E|X_n| < \infty$, for all $n \geq 1$.

Define $X_n^* = \sum_{k=1}^{n} \{X_k - E(X_k \mid \mathscr{F}_{k-1})\}$. Then $\{X_n^*, \mathscr{F}_n, n \geq 0\}$ is a martingale with $X_0 = 0$. Then by (A.2)

$$\begin{aligned} 0 = E[X_{T\wedge n}^*] &= E\left[\sum_{1}^{T\wedge n} \{X_k - E(X_k \mid \mathscr{F}_{k-1})\}\right] \\ &\geq E\left[\sum_{k=1}^{T\wedge n} \{X_k - X_{k-1}\}\right] = E[X_{T\wedge n}] - E(X_0) \end{aligned}$$

Therefore $E[X_0] \geq E[X_{T\wedge n}]$.

The general case can be obtained by truncation considering the new super-martingale $\tilde{X}_n = \begin{cases} X_n \text{ if } X_n < c \\ c \text{ if } X_n \geq c \end{cases}$, where c is a constant.

Theorem (Optional stopping theorem) A.IV.1

Let $\{X_n\}$ be a martingale and T a stopping time if

(i) $P[T < \infty] = 1$

(ii) $E[|X_T|] < \infty$

(iii) $\lim_{n\to\infty} E[X_n I_{(T>n)}] = 0.$

Then $E(X_T) = E(X_0)$.

Proof For all $n \geq 1$,

$$E(X_T) = E[X_T I_{(T\leq n)}] + E[X_T I_{(T>n)}]$$
$$= E[X_{T\wedge n}] - E[X_n I_{(T\geq n)}] + E[X_T I_{(T>n)}]$$

Now $E[X_{T\wedge n}] = E(X_0)$ by lemma A.IV.1 and $\lim_{n\to\infty} E[X_n I_{(T>n)}] = 0$ by assumption (iii)

$$0 \leq | E(X_T) - E[X_T I_{(T\leq n)}] |$$
$$= | E[X_T I_{(T>n)}] |$$
$$\leq E[| X_T | I_{(T>n)}] \to 0 \quad \text{by assumption (ii)}$$

Thus $E(X_T) = \lim_{n\to\infty} E[X_{T\wedge n}] = E[X_0]$.

Let us state and prove the other optional theorem.

Theorem (Optional sampling theorem) A.IV.2

Suppose $\{X_n\}$ is a martingale and T is a stopping time. If $P(T < \infty) = 1$ and $E[\sup_{n\geq 0} |X_{T\wedge n}|] < \infty$, then $E(X_T) = E(X_0)$.

Proof Let $W = \sup_{n\geq 0} | X_{T\wedge n} |$. Since $P(T < \infty)$,

$$X_T = \sum_{j=0}^{\infty} X_j I_{(T=j)} = \sum_{j=0}^{\infty} X_{T\wedge j} I_{(T=j)}$$

and that implies $| X_T | \leq W$, and therefore $E | X_T | \leq E(W) < \infty$.

By lemma A.IV 1 we need only to show $\lim_{n\to\infty} E[X_{T\wedge n}] = E(X_T)$. We have

$$| E[X_{T\wedge n}] - E[X_T] | \leq E [| (X_{T\wedge n} - X_T) | I_{(T>n)}] \leq 2E[W I_{(T>n)}].$$

Now

$$E[| W |] \geq E[W I_{(T\leq n)}] = \sum_{j=0}^{n} E[| W | | T = j] P[T = j]$$
$$\to \sum_{j=0}^{\infty} E[| W || T = j] P[T = j] \text{ as } n \to \infty$$
$$= [E | W |].$$

Hence $\lim_{n \to \infty} E[|W| I_{(T \le n)}] = E|W| < \infty$ and therefore $\lim_{n \to \infty} E[W I_{(T>n)}] = 0$. By A.IV.2 the proof is complete.

Under the assumption that the second moments of $\{X_n\}$ are uniformly bounded let us state without proof the martingale mean square convergence theorem.

Theorem (Martingale convergence theorem) A.IV.3
Let $\{X_n\}$ be a martingale with respect to $\sigma-$fields $\{\mathcal{F}_n\}$ satisfying $E[X_n^2] \le c < \infty$ for all $n \ge 1$. Then $\{X_n\}$ converges as $n \to \infty$ to a limit r.v. X_∞ with probability one and in mean square, that is

$$P(\lim_{n \to \infty} X_n = X_\infty) = 1 \text{ and } E[|X_n - X_\infty|^2] = 0$$

Also $E(X_0) = E(X_n) = E(X_\infty)$ for all $n \ge 1$.

Let $\{X_t, t \ge 0\}$ be a continuous parameter martingale adapted to the increasing sequence of sub σ-fields $\mathcal{F}_t$, i.e. $\mathcal{F}_s \subset \mathcal{F}_t$ if $s \le t$, $\mathcal{F}_t \subset \mathcal{F}$ for all $t \ge 0$.

A r.v. T having possible values in $[0, \infty]$, is called a Markov time (Stopping time) relative to $\{\mathcal{F}_t\}$ if, for every $t \ge 0$, the event $\{T \le t\}$ is in $\mathcal{F}_t$ or equivalently $\{T > t\} \in \mathcal{F}_t$, $t > 0$. For continuous parameter process, it is not sufficient to require $\{T = t\} \in \mathcal{F}_t$ for each t.

Let A be a closed set and define $T(A)$, the entry time to A, to be the random time

$$T(A) = \inf [t \ge 0; X(t) \in A].$$

Then $T(A)$ is a Markov time with respect to $\mathcal{F}_t = \mathcal{B}(X(u); 0 \le u \le t)$, provided $X(t)$ is a continuous function of t. But unfortunately the entry time to a set A is not necessarily a Markov time with respect to $\{\mathcal{F}_t\}$ if $X(t)$ is not continuous or A is not closed. We can rectify this defect by enlarging the σ-fields $\mathcal{F}_t$. Suppose $X(t)$ is a right continuous (i.e. $\lim_{s \downarrow t} X(s)$ for all $t \ge 0$) process with left hand limit (i.e. $X(t-) = \lim_{s \downarrow t} X(s)$ exists for all $t > 0$).

Define $\mathcal{F}_{t^+} = \bigcap_{\varepsilon > 0} \mathcal{F}_{t+\varepsilon}$. Then each $X(t)$ is $\mathcal{F}_{t^+}$-measurable, and $\mathcal{F}_{s^+} \subset \mathcal{F}_{t^+}$, if $s < t$. Also let $\overline{\mathcal{F}_{t^+}}$ be the completion of $\mathcal{F}_{t^+}$ (i.e. the smallest σ-field containing $\mathcal{F}_{t^+}$ and every subset A of Ω for which $A \in \mathcal{F}$ and $P(A) = 0$).

Then for every Borel set A, the entry time

$$T(A) = \begin{cases} \inf t \ge 0: X(t) \in A, & \text{if } X(t) \in A \text{ for some } t \ge 0 \\ \infty, & \text{if } X(t) \notin A \text{ for all } t \ge 0. \end{cases}$$

is a Markov time with respect to $\{\overline{\mathcal{F}_{t^+}}\}$.

Both the martingale optional sampling and convergence theorems are valid in continuous time. If $\{X(t), t \ge 0\}$ is a supermartingale with respect to $\{\mathcal{F}_t\}$, then $E(X(0)) \ge E[X(T \wedge t)] \ge E[X(t)]$, $t \ge 0$, for all Markov times. T. The inequalities are reversed for a submartingale and equality is achieved for a martingale. If $P(T < \infty) = 1$, then $X(T \wedge t) \to X(T)$ as $t \to \infty$.

Like discrete parameter case we obtain the optional sampling theorem for continuous parameter martingale.

Theorem (Optional sampling theorem) A.IV.4
Let $\{X(t): \geq 0\}$ be a martingale and T is a Markov time. If

$$P(T < \infty) = 1 \text{ and } E[\sup_{t \geq 0} | X(t) |] < \infty, \text{ then}$$

$$E(X(0)) = E(X(T)).$$

For more discussions about both discrete and continuous parameter martingales readers are advised to consult excellent books by Doob (1953). Brieman (1998), Chow and Tiecher (1990) and many others.

Note: Martingale Convergence theorem A.IV.3 is valid under much less restrictive condition $\sup_n E | X_n | < \infty$ (see J.L. Doob).

Solution to Exercises

Chapter 1

Exercise 1.1 Let $X = 4ac$ and $Y = b^2$, $0 \le Y \le 1$.
Then $F_Y(U) = P(b^2 \le u) = P(b \le \sqrt{u}) = \sqrt{u}$ for $0 \le u \le 1$.

and
$$F_X(u) = P(4ac \le u) = \int_0^1 P(ac \le u \mid v < c < v + dv) df_c(v)$$

But
$$P\left(a \le \frac{u}{4v}\right) = \begin{cases} 0 & \text{if } v < 0 \\ \dfrac{4}{4v} & \text{if } 0 \le \dfrac{u}{4v} \le 1 \text{ or if } \dfrac{u}{4} \le v \\ 1 & \text{if } \dfrac{u}{4v} \ge 1 \text{ or } \dfrac{u}{4} \ge v \end{cases}$$

Hence
$$F_x(u) = \int_0^{u/4} 1 \cdot dv + \int_{u/4}^1 \frac{1}{v} dv = \frac{u}{4}\left[1 - \log\frac{u}{4}\right], 0 \le u \le 4$$

$$P(b^2 - 4ac \ge 0) = P(X \le Y) = \int_0^1 P(X - Y \le 0 \mid u < Y \le u + du] \, dF_Y(u)$$

$$= \int_0^1 u/4\left[1 - \log\frac{u}{4}\right]\frac{1}{2\sqrt{u}} du = \frac{5 + 6\log 2}{36}$$

Exercise 1.2
$$P(X = k) = \int_0^\infty P(X = k \mid u < \lambda \le 4 + du) \, P(u < \lambda \le u + du)$$

$$= \int_0^\infty e^{-u}\frac{u^k}{k!}\frac{1}{\sqrt{n}} u^{n-1} e^{-u} du = \frac{\sqrt{k+n}}{\sqrt{n}\sqrt{k+1}} (1/2)^{k+n}.$$

Exercise 1.3 Joint probability distribution of X and Y is

$$\frac{e^{-\lambda}\lambda^x}{x!}\binom{x}{y} p^y (1-p)^{x-y} = \frac{e^{-\lambda}}{y!}\left(\frac{p}{1-p}\right)^y \frac{\lambda^x (1-p)^x}{(x-y)!}$$

Therefore
$$P(Y = y) = \frac{e^{-\lambda}}{y!}\left(\frac{p}{1-p}\right)^y \sum_{x=y}^{\infty} \frac{\lambda^x (1-p)^x}{(x-y)!}$$

$$= \frac{e^{-\lambda}}{y!}\left(\frac{p}{1-p}\right)^y \sum_{x=0}^{\infty} \lambda^{x+y}(1-p)^{x+y}/x!$$

$$= \frac{e^{-\lambda p}}{y!}(\lambda p)^y, y = 0, 1, 2, \ldots$$

Exercise 1.4 The joint Ch.f. of $X_{t_1}, \ldots X_{t_n} = E\left[\exp\left(i \sum_{j=1}^{n} \theta_j X_{t_j}\right)\right]$

Now $\sum_{j=1}^{n} \theta_j X_{t_j} = X_{t_1}(\theta_1 + \theta_2 + \ldots + \theta_n) + (X_{t_2} - X_{t_1})(\theta_2 + \ldots + \theta_n)$

$$+ \ldots + (X_{t_n} - X_{t_{n-1}})\,\theta_n$$

Since $\{X(t)\}$ has independent increments,

$$E\left[\exp\left(i \sum_{j=1}^{n} \theta_j X_{t_j}\right)\right] = E\left[\exp\left(i \sum_{j=1}^{n} \theta_j X_{t_1}\right)\right]$$

$$\times E\left[\exp\left(i \sum_{j=2}^{n} \theta_j (X_{t_2} - X_{t_1})\right)\right] \ldots E\left[\exp\left(i\theta_n (X_{t_n} - X_{t_{n-1}})\right)\right]$$

$$= \phi\left(\sum_{j=1}^{n} \theta_j, t_1\right)\phi\left(\sum_{j=2}^{n} \theta_j, (t_2 - t_1)\right) \ldots \phi(\theta_n, (t_n - t_{n-1})).$$

Exercise 1.5 Let $t_0 < t_1 < \ldots < t_n < t_{n+1} \in T$, the index set and let A be an arbitrary subset of the real line. Let $\{a_n\}$ denote the set consisting of the single point a_n

$$P[X_{t_{n+1}} \in A \mid X_{t_n} = a_n, X_{t_{n-1}} = a_{n-1}, \ldots, X_{t_0} = a_0]$$

$$= P[(X_{t_{n+1}} - X_{t_n}) \in A - \{a_n\} \mid X_{t_n} = a_n, X_{t_n} - X_{t_{n-1}} =$$

$$a_n - a_{n-1}, \ldots, X_{t_1} - X_{t_0} = a_1 - a_0]$$

$$= P[(X_{t_{n+1}} - X_{t_n}) \in A - \{a_n\} \mid X_{t_n} = a_n]$$

(due to independent increments)

$$= P[X_{t_{n+1}} \in A \mid X_{t_n} = a_n]$$

Exercise 1.6 Suppose T has a geometric distribution given by

$$P[T = k] = pq^{k-1}, k = 1, 2, \ldots, 0 < p < 1, q = 1 - p$$

Then $$P[T > x + y] = \sum_{k=x+y+1}^{\infty} P[T = k] = q^{x+y}$$

So $$P[T > x]P[T > y] = q^x q^y = q^{x+y} = P[T > x + y]$$

Conversely let $$p_n = P[T = n],\ q_n = P[T > n]$$

then $q_{n+1} = P[T > n + 1] = P[T > 1]P[T > n] = q_1 q_n = \ldots = q_1^{n+1}$.

So $$p_n = P[T > n - 1] - P[T > n] = q_{n-1} - q_n$$

$$= q_1^{n-1} - q_1^n = q_1^{n-1}(1 - q_1)$$

which is geometric.

Exercise 1.7 Suppose that T has a exponential distribution given by $P[T > x] = e^{-\lambda x}$, $x > 0$, $\lambda > 0$.

Then $$P(T > x)P(T > y)e^{-\lambda x} \cdot e^{-\lambda y} = e^{-\lambda(x+y)} = P(T > x + y)$$

Conversely let $g(x) = P[T > x]$ then $g(x + y) = g(x)g(y)$, T is continuous with $g(0) = 1$. This is a Cauchy functional equation and the solution is given by $g(x) = e^{-\lambda x}$ for some $\lambda > 0$.

Exercise 1.8(a) $E[e^{-ex(T)}] = E[E(e^{-ex(T)} \mid T = t)]$

$$= \int_0^\infty E[\theta^{-sx(T)} \mid T = t]\, dF(t)$$

but $$E[\theta^{-sx(T)} \mid T = t] = \int_0^\infty \theta^{-su} \frac{1}{\sqrt{\alpha t}} (\lambda u)^{\alpha t-1} \theta^{-\lambda u}\, du = \left(\frac{\lambda}{\lambda + s}\right)^{\alpha t}$$

Hence, $$E[\theta^{-sx(T)}] = \int_0^\infty \left(\frac{\lambda}{\lambda + s}\right)^{\alpha t} dF(t) \tag{1}$$

(b) If T is negative Binomial then

$$E[e^{-sx(T)}] = \sum_{k=0}^{\infty} \left(\frac{\lambda}{\lambda + s}\right)^{\left(\frac{\beta+k}{\alpha}\right)} \binom{\beta + k - 1}{k} p^\beta (1 - p)^k = \left(\frac{\lambda p}{\lambda p + s}\right)^\beta$$

which is the Laplace transform of a gamma distribution with parameters β and λp.

Conversely, if $X(T)$ has a gamma density with Laplace transform

$$E[e^{-sx(T)}] = \left(\frac{\mu}{\mu + s}\right)^\beta$$

$$= \int_0^\infty \left(\frac{\lambda}{\lambda + s}\right)^{\alpha t} dF(t) \quad \text{(by (1))} \tag{2}$$

Put $\left(\frac{\lambda}{\lambda + s}\right)^\alpha = e^{-u}$. Then $s = \lambda[e^{u/\alpha} - 1]$.

So $$\int_0^\infty e^{-u} dF(t) = \left\{ \frac{\mu}{\mu + \lambda[e^{u/\alpha} - 1]} \right\}^\beta \qquad \text{(by (2))}$$

$$= \left[\frac{pe^{-u/\alpha}}{1 - qe^{-u/\alpha}} \right]^\beta \text{ where } 0 < p = \frac{\mu}{\lambda} < 1.$$

This is the Laplace transform of a geometric distribution given by

$$P(\alpha T = \beta + k) = \binom{\beta + k - 1}{k} p^\beta (1 - p)^k \text{ where } p = \frac{\mu}{\lambda}.$$

Exercise 1.9 $P[Y \le y] = \int_0^\infty P[Y \le y \mid X = x] f(x)\, dx$, (since $Y \mid x$ is uniform)

$$= \int_0^y f(x)\, dx + \int_y^\infty \frac{y}{x} f(x)\, dx$$

Hence $$f_Y(y) = \begin{cases} 0 & \text{if } y \le 0 \\ \displaystyle\int_y^\infty \frac{f(x)}{x} dx & \text{if } y > 0 \end{cases} \qquad (1)$$

$$P(X - Y \le u) = \int_u^\infty P[X - Y \le u \mid x = x]\, dx = \int_u^\infty \frac{u}{x} f(x)\, dx.$$

Similarly $$f_{x-y}(u) = \begin{cases} 0 & \text{if } u \le 0 \\ \displaystyle\int_u^\infty \frac{f(x)}{x} dx & \text{if } u > 0 \end{cases} \qquad (2)$$

Thus Y and $X - Y$ are independent and identically distributed. But $X = Y + (X - Y)$. Let $\phi(t)$ be the characteristic function then

$$\phi_X(t) = \phi_Y(t)\, \phi_{X-Y}(t) = [\phi_Y(t)]^2 \qquad (3)$$

We have to find $\phi_Y(t)$.

$$\phi_Y(t) = \int_0^\infty e^{ity} \left(\int_y^\infty \frac{f(x)}{x} dx \right) dy \text{ (from (1))}$$

$$= \int_0^\infty \frac{f(x)}{x} \left[\int_0^x e^{ity} dy \right] dx$$

$$= \frac{1}{it} \int_0^\infty \frac{e^{itx} - 1}{x} f(x)\, dx \text{ (by Fubini Theorem)} \qquad (4)$$

So *it* $\phi_Y'(t) + i\phi_Y(t) = i\int_0^\infty \frac{f(x)}{x}[xe^{itx}]\,dx$ (differenting the expression in (4)).

Hence $$t\phi_Y'(t) + \phi_Y(t) = \int_0^\infty f(x)[e^{itx}]\,dx = \phi_x(t) \tag{5}$$

Therefore, by (3) and (5) we get $t\phi_Y'(t) + \phi_Y(t) = \phi_Y^2(t)$ which is a linear differential equation. The solution is $\phi_Y(t) = \frac{1}{1 - At}$ with initial condition

$$\phi_Y'(0) = i\mu \Rightarrow A = i/a. \text{ So } \phi_x(t) = \frac{1}{(1 - i/at)^2},\ a = \frac{1}{\mu}$$

which is the ch.f. of gamma distribution given by

$$f(x) = a^2xe^{-ex},\ a > 0.$$

Exercise 1.10 $EY(t) = EZ(t) = 0$, and since $\sin^2 t + \cos^2 t = 1$ for all $t \in T$, $EY^2(t) = EZ^2(t) = 1$. Since $Y(t)$ and $Z(t)$ have a joint normal distribution, it is enough to show that $E(Y(t)Z(t)) = 0$. $Y(t)$ and $Z(t)$ are independent normal having zero mean and variance unity for all $t \in T$. Therefore $X(t)$ has same distribution as $X(t + h)$ for any h. Here the bivariate process $X(t) = (Y(t), Z(t))$ is strictly stationary.

Exercise 1.11 $EX(t) = 0$, $\text{Cov}\,(X(t), X(s)) = E(X(t)X(s))$

$$= E(A^2)\cos\lambda t\cos(\lambda s) + E(B^2)\sin(\lambda t)\sin(\lambda s)$$
$$+ E(AB)\cos\lambda t\sin(\lambda s) + E(AB)\cos(\lambda s)\sin(\lambda t)$$
$$= \sigma^2(\cos(\lambda t)\cos(\lambda s) + \sin(\lambda t)\sin(\lambda s)) = \sigma^2\cos\lambda(t - s)$$

Obviously $\{X(t)\}$ is a covariance stationary process.

Exercise 1.12 $EZ_t = 0$, $\cos(Z_t, Z_s) = \sum_{j=1}^{n}\sigma_j^2\cos\lambda_j(t - s)$, $t \neq s$.

Therefore $\{Z_t, t \geq 0\}$ is covariance stationary. Since given upto second order moments is not enough to determine joint distribution of a stochastic process, we can not say whether or not the process is strictly stationary.

Exercise 1.13 $E\{X(t)\} = \frac{1}{2\pi}\int_0^{2\pi}\sin\omega t\,d\omega = 1/2\pi\cos\omega t/t\Big|_0^{2\pi} = 0$

$$E\{X(t + s)X(s)\} = \frac{1}{2\pi}\int_0^{2\pi}\sin\omega(t + s)\sin\omega s\,d\omega$$
$$= \frac{1}{4\pi}\int_0^{2\pi}[\cos\omega t - \cos\omega(t + 2s)]d\omega$$
$$= \frac{1}{4\pi}\left[\frac{1}{t}\sin\omega t\Big|_0^{2\pi} - \frac{1}{t + 2s}\sin\omega(t + 2s)\Big|_0^{2\pi}\right]$$
$$= 0 \text{ if } t, s = 1, 2, 3, \ldots$$

But $E(X(t+s)X(s)) = f(t, s) \neq 0$ if t, s are not integers, where f is a function of t and s but not of t only.

So $\{X(t), t = 1, 2, 3, \ldots\}$ is covariance stationary.

Now $\text{Cov}(X(s), X(t+s))$ depends on t and s not on $|t|$ only if we consider the process $\{X(t), t \geq 0\}$. Hence it is not even covariance stationary and therefore can not be strictly stationary.

Exercise 1.14 Use trigonometric identity to evaluate

$$E(X_n) = E(\cos nU) = \begin{cases} 1 & \text{if } n = 0 \\ 0 & \text{if } n \geq 1 \end{cases}$$

$$E(X_n X_{n+v}) = \int_{-\pi}^{\pi} \cos nu \cos (n+v)\, u du$$

$$= \int_{0}^{\pi} [\cos (2n+v)\, u + \cos vu]\, du$$

$$= 0 \quad \text{if } n \geq 1.$$

Therefore, $\{X_n\}$ is covariance stationary.

Similarly calculate the third moment to determine

$$E(X_n X_{n+v} X_{n+v+l}) \text{ depends on } n.$$

Hence $\{X_n)$ is not strictly stationary.

Exercise 1.15 Now $Y(t) = Y_n$ for $(n-1)T < t < nT$

Where Y_n's are i.i.d. with $P(Y_n = \pm 1) = 1/2$

Since U and $Y(t)$ are independent.

$$E(Y(t-U) \mid U = u) = E(Y(t-u)) = 0.$$

Consequently, $m(t) = EX(t) = E[Y(t-u) \mid U = u] = 0$

Therefore $\text{Cov}(X(s), X(t)) = E(X(s)\, X(t))$.

The time axis divided into successive intervals, it follows from the independence assumption on the r.v.s Y_n that $\text{Cov}(X(s), X(t)) = 0$ if $|t-s| > T$. Let $|t-s| > T$. Since U is uniform in (O, T), it follows that $E(X(s), X(t)) = 0$ and consequently $E(X(s)X(t)) = E(E[X(s)X(t) \mid u]] = 0$. Now first $|t-s| < T$ and $s = nT$. Then t and s will lie in the same interval if $U + [t-s| < T$ that is $U < T - |t-s|$ and this event occurs with probability $(T - |t-s|)/T$. Therefore

$$E(X(s)X(t)) = \begin{cases} 1 - |t-s|/T & \text{if } |t-s| < T \\ 0 & \text{if } |t-s| > T \end{cases}$$

Hence $X(t)$ is wide sense stationary.

Exercise 1.16 $E(X(t)) = E(Y)E(Z(t))$ where $Z(t) = (-1)^{N(t)}$

$$E(X(t)X(s)) = E(Y^2)E(Z(t)Z(s)) = E(Z(t)Z(s))$$

Now $Z(t)$ takes values $+1$ and -1 with probabilities given by

$$P(Z(t) = 1) = P(N(t) \text{ is even}) = e^{-\lambda t} \sum_{n=0}^{\infty} \frac{(\lambda t)^{2n}}{(2n)!} = e^{-\lambda t} \cos ht\lambda$$

and $\quad P(Z(t) = -1) = P(N(t) \text{ is odd}) = e^{-\lambda t} \sin ht\lambda$

Therefore $EZ(t) = e^{-2\lambda t}$ and hence

$$E(X(t)) = 0 \cdot e^{-2\lambda t} = 0. \text{ For } t > s$$

$$P(Z(t) = 1, Z(s) = 1) = e^{-\lambda s} \cos hs\lambda\, e^{-\lambda(t-s)} \cos h\lambda(t-s)$$

$$= P(N(s) \text{ is even and } N(t-s) \text{ is even})$$

$$P(Z(t) = -1, Z(s) = -1) = e^{-\lambda s} \sin ht\lambda\, e^{-\lambda(t-s)} \cos h\lambda(t-s)$$

$$= P(N(s) \text{ is odd and } N(t-s) \text{ is even})$$

$$P(Z(t) = -1, Z(s) = -1) = P(N(s) \text{ is even and } N(t-s) \text{ is odd})$$

$$= e^{-\lambda s} \cos hs\lambda\, e^{-\lambda(t-s)} \sin h\lambda(t-s)$$

$$P(Z(t) = 1, Z(s) = -1) = P(N(s) \text{ is odd and } n(t-s) \text{ is odd})$$

$$= e^{-\lambda s} \sin h(s\lambda) e^{-\lambda(t-s)} \sin h\lambda(t-s)$$

Therefore $E(Z(t)Z(s)) = e^{-2\lambda(t-s)}$ when $t < s$ and

$E(Z(t)Z(s)) = e^{2\lambda(t-s)}$ when $t > s$.

Hence Cov $(X(s), X(t)) = E(X(s)\, X(t)) = e^{-2\lambda|t-s|}$.
Thus, $X(t)$ is covariance stationary.

Exercise 1.17 If $H(t)$ is known then we have

$$P(X_t = k) = \binom{N}{k} (H(t))^k \{1 - H(t)\}^{N-k}.$$

To find $H(t)$, partition $(0, t]$ into n subintervals of equal length $\Delta t = t/n$. By assumption, the average number of germination in $(0, \Delta t] \cong N(1-p)\lambda\Delta t$. So the number of nongerminating seeds at time Δt is $N(1-p)(1-\lambda\Delta t)$. Thus the number of nondegerminating seeds at time t is approximately $N(1-p)(1-\lambda\Delta t)^n$. Hence for sufficiently large n,

$$P(T \le t) \cong \{1 - (1 - \lambda\Delta t)^n\}(1-p) = \{1 - (1 - \lambda t/n)^n\}(1-p)$$

By letting $\Delta t \to 0$, we obtain

$$H(t) = (1-p)(1-e^{-\lambda t}) \text{ and } P(X_t = k) = \binom{N}{k} [(1-p)(1-e^{-\lambda t})]^k$$

$$\{1 - (1-p)(1-e^{-\lambda t})\}^{n-k}.$$

Chapter 2

Exercise 2.1 From the directed graph we see that the state space decomposes into two classes $C_1 = \{1, 2, 3\}$ and $C_2 = \{4, 5\}$. Also for $i \in C_1$ and $j \in C_2$ we see that $i \nrightarrow j$ or $j \nrightarrow i$. Therefore C_1 and C_2 are closed sets of states.

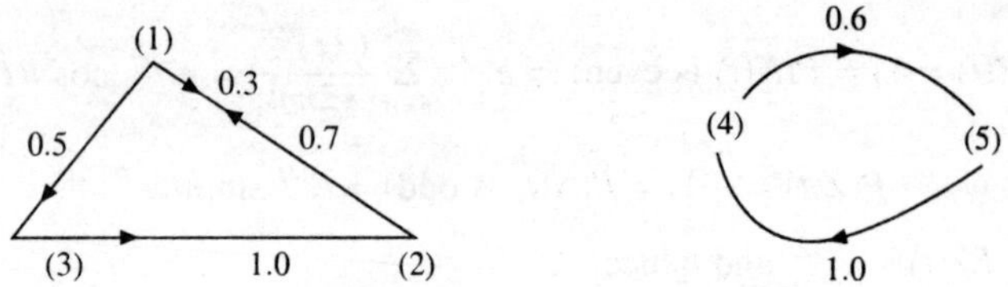

The state spaces decomposes into $C = \{1, 2, 3\}$ and $A = \{4\}$. Hence C is closed and A is not.

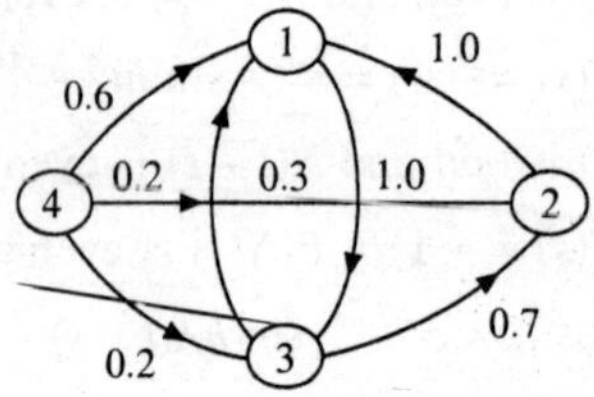

Exercise 2.2 State space of the M.C. is given by $S = \{0, 1, 2, \ldots N\}$. $X_{n+1} = 0$ if $X_n = 0$, $X_{n+1} = \min\{N, X_n + J_{n+1} - 1\}$ if $X_n > 0$, where J_{n+1} = number of new jobs arriving during nth service period. Then the transition matrix in its cannonical form is given by

$$P = \begin{bmatrix} 1 & 0 & 0 & \cdots & 0 & 0 \\ p_0 & p_1 & p_2 & \cdots & p_{N-1} & p_N \\ 0 & p_0 & p_1 & \cdots & p_{N-2} & p_{N-1} \\ 0 & 0 & 0 & \cdots & p_0 & p_1 \end{bmatrix}$$

State "0" is absorbing and others are transient.

Exercise 2.3 (Bartky's sampling inspection scheme)

$$P(X_k = j) = p_j = \binom{N}{j+1} p^{j+1}(1-p)^{N-j-1}.$$

Obviously the conditional distribution of S_{n+1} given $S_1, S_2, \ldots, S_n$ is a Markov chain with states space $\{0, 1, \ldots, a, \text{"}> a\text{"}\}$ and the transition matrix is given by

$$P = \begin{bmatrix} 1 & 0 & 0 & \cdots & 0 & 0 & 0 & 0 \\ p_{-1} & p_0 & p_1 & \cdots & p_{N-1} & 0 & 0 & 0 \\ . & . & .. & \cdots & \cdots & \cdots & \cdots & \cdots \\ 0 & 0 & 0 & & 0 & p_{-1} & p_0 & 1 - p_1 - p_{-1} \\ 0 & 0 & 0 & \cdots & 0 & 0 & 0 & 1 \end{bmatrix}$$

States "0" and "> a" are absorbing and other states are transient.

Exercise 2.4 Let X_n be the number of customers present at time n. Since the number of customes present at time $n + 1$ equals the number of customers present at time n less the one served at time n (if there is any), plus the number of customers arriving in time interval $(n, n + 1)$ if the total does not exceed m and equals m otherwise, we have $X_{n+1} = \min\{X_n + \delta_{0,x_n} - 1 + \xi_n, m\}$, where $\delta_{i,j}$ is the Kronecker's delta. Since X_{n+1} depends only on X_n and ξ_n, and X_n and ξ_n are independent, $\{X_n\}_{n\geq 0}$ is a M.C. with state space $\{0, 1, \ldots, m]$. Put $P_l + p_l = p_{l+1} + \ldots, 1 \leq l \leq m$. Then

$$p_{oj} = \begin{cases} P(X_{n+1} = j \mid X_n = 0) = P(\xi_n = j \mid X_n = 0) = p_j & \text{if } 0 \leq j \leq m - 1 \\ P(\xi_n \geq m) = p_m + p_{m+1} + \ldots = P_m & \text{if } j = m. \end{cases}$$

$$\text{For } 1 \leq j \leq m, p_{ij} = \begin{cases} P(X_{n+1} = j \mid X_n = i) = P(\xi_n = j + 1 - i) \\ \qquad = p_{j-i+1} \quad \text{if } i - 1 \leq j \leq m - 1 \\ P(x_n > m + 1 - j) = P_{m+1-j} \quad \text{if } j = m \\ 0 \quad \text{otherwise} \end{cases}$$

So the transition matrix is given by

$$\begin{matrix} 0 \\ 1 \\ 2 \\ \vdots \\ m-1 \\ m \end{matrix} \begin{bmatrix} 0 & 1 & 2 & \ldots & m-1 & m \\ p_0 & p_1 & p_2 & \ldots & p_{m-1} & p_m \\ p_0 & p_1 & p_2 & \ldots & p_{m-1} & p_m \\ 0 & p_0 & p_1 & \ldots & p_{m-2} & p_{m-1} \\ 0 & 0 & \ldots & \ldots & p_1 & p_2 \\ 0 & 0 & \ldots & \ldots & p_0 & p_1 \end{bmatrix}$$

Exercise 2.5 (*Storage mode 1*)
Like the previous exercise (Queueing) it is easy to see that

$$X''_{n+1} = \text{Min}\,(X''_n + \eta_{n+1} + \delta_{0,x''_n} + \eta_{n+1} - 1, m - 1).$$

Hence X''_n is a M.C. with state space $\{0, 1, \ldots, m - 1\}$.

$$p_{00} = P(X''_{n+1} = 0 \mid X''_n = 0) = P[(\eta_{n+1} = 0) \cup (\eta_{n+1} = 1) \mid X''_n = 0]$$

(since X''_n and η_{n+1} are independent) $= p_0 + p_1$.

Similarly

$$p_{oj} = P(X''_{n+1} = j \mid X''_n = 0) = P(\eta''_{n+1} = j + 1 \mid X''_n = 0) = p_{j+1},\ 1 \leq j \leq m - 2$$

$$p_{0,m-1} = P(\eta_{n+1} \geq m \mid X''_n = 0) = P_m, \text{ where } P_m = \sum_{k=m}^{\infty} p_k\ p_{ij},\ 1 \leq i \leq m - 1$$

are calculated like Exercise 2.4 (Queueing). Hence the transition matrix is given by

$$P = \begin{array}{c} \\ 0 \\ 1 \\ \vdots \\ m-1 \\ m \end{array} \begin{array}{c} \begin{array}{cccccc} 0 & 1 & 2 & \ldots & m-1 & m \end{array} \\ \begin{bmatrix} p_0 + p_1 & p_2 & p_3 & \cdots & p_{m-1} & P_m \\ p_0 & p_1 & p_2 & \cdots & p_{m-2} & P_{m-1} \\ 0 & p_0 & p_1 & \cdots & p_{m-3} & P_{m-2} \\ 0 & 0 & 0 & \ldots\ldots & p_1 & P_2 \\ 0 & 0 & 0 & \ldots\ldots & p_0 & p_1 \end{bmatrix} \end{array}$$

Exercise 2.6

$P(X_{n-1} = i_{n-1} \mid X_n = i_n, X_{n+1} = i_{n+1}, \ldots)$

$$= \frac{P(X_{n-1} = i_{n-1}, X_n = i_n, X_{n+1} = i_{n+1}, \ldots)}{P(X_n = i_n, X_{n+1} = i_{n+1}, \ldots)}$$

$$\frac{P(X_{n+1} = i_{n+1}, X_{n+2} = i_{n+2}, \ldots \mid X_n = i_n, X_{n-1} = i_{n-1})P(X_n = i_n, X_{n-1} = i_{n-1})}{P(X_{n+1} = i_{n+1}, X_{n+2} = i_{n+2}, \ldots, \mid X_n = i_n)P(X_n = i_n)}$$

$= P(X_{n-1} = i_{n-1} \mid X_n = i_n)$ [since $P(X_{n-1} = i_{n-1}, X_{n+2} = i_{n+2}, \ldots \mid$

$X_n = i_n, X_{n-1} = i_{n-1}) = P(X_{n+1} = i_{n+1}, X_{n+2} = i_{n+2}, \ldots \mid X_n = i_n)$].

Exercise 2.7

$$P^{k\times k}\begin{pmatrix}1\\1\\\vdots\\1\end{pmatrix}^{k\times 1} = \begin{pmatrix}1\\1\\\vdots\\1\end{pmatrix}^{k\times 1} u^{k\times 1} \Rightarrow \lambda = 1$$

is an eigen value of the stochastic matrix P of order k.

If λ is an eigen value of the stochastic matrix $P^{k\times k} = (p_{ij})$ and $u^{k\times 1} = \begin{pmatrix}u_1\\\vdots\\u_k\end{pmatrix}$ is a left eigenvector corresponding to λ, we have $\lambda u_j = \sum_{i=1}^{k} u_i p_{ij}$, $1 \le j \le k$, whence

$$\sum_{j=1}^{k} |\lambda u_j| = \sum_{j=1}^{k} \left| \sum_{i=1}^{k} u_i p_i \right| \le \sum_{i=1}^{k} |u_1| \sum_{j=1}^{k} p_{ij}$$

$$= \sum_{i=1}^{k} |u_i| \Rightarrow |\lambda| \sum_{j=1}^{k} |u_j| \le \sum_{i=1}^{k} |u_i| \Rightarrow |\lambda| \le 1.$$

Exercise 2.8

Clearly $$X_0 = r,\ X_n = \xi_{n+1} + X_{n+1},\ n \ge 0 \ldots \tag{1}$$

The probability that among i susceptables k of them ($k \le i$) have contacts with infectious individuals equals $\binom{i}{k} p^k q^{i-k}$ ($q = 1 - p$). Hence

$$P(X_{n+1} = j \mid X_n = i) = P(\xi_{n+1} = X_n - j \mid X_n = i) =$$

$$P(\xi_{n+1} = i - j \mid X_n = i) = \begin{cases} \binom{i}{j} p^{i-j} q^j & \text{if } i \geq j \\ 0 & \text{if } i < j \end{cases}$$

and obviously, the distribution of X_{n+1} is independent of the values assumed by the r.v.s X_m, $m < n$. Therefore $\{X_n, n \geq 0\}$ is a Markov chain with states 0, 1, 2, . . . , r and transition matrix

$$\begin{bmatrix} 0 & 1 & 2 & 0 & \dots & k \\ 1 & 0 & 0 & & \dots & 0 \\ p & q & 0 & 0 & \dots & 0 \\ p^2 & 2pq & q^2 & 0 & \dots & 0 \\ p^r & rpq^{r-1} & \binom{r}{2} p^{r-2} q^2 & & \dots & q^r \end{bmatrix}$$

The duration of the epidemic is the smallest number τ such that $\xi_\tau = 0$ i.e. the moment τ when there are no further infectives i.e. the smallest number τ such that $X_{\tau-1} = X_\tau$ using the equality $X_n = \xi_{n+1} + X_{n+1}$. Hence the number of cases in the epidemic is $\xi_0 + X_0 - X_\tau = \xi_0 + r - X_\tau$.

Exercise 2.9 By chain rule of conditional probabilities

$$P(X_n = i_n, X_k = i_k \mid X_m = i_m) = \frac{P(X_m = i_n, X_k = i_k, X_m = i_m)}{P(X_m = i_m)}$$

$$= \frac{P(X_n = i_n \mid X_k = i_k, X_m = i_m)\, P(X_m = i_m \mid X_k = i_k)}{P(X_m = i_m)} \cdot P(X_k = i_k)$$

$$= \frac{P(X_n = i_n \mid X_m = i_m) P(X_m = i_m \mid X_k = i_k)\, P(X_k = i_k)}{P(X_m = i_m)}$$

(by Markov property)

$$= P(X_n = i_n \mid X_m = i_m) P(X_m = i_m, X_k = i_k) / P(X_m = i_m)$$

$$= P(X_n = i_n \mid X_m = i_m) P(X_k = i_k \mid X_m = i_m).$$

Exercise 2.10 (a) $P(X_n = i_n \mid X_{n+1} = i_{n+1}, \dots, X_{n+k} = i_{n+k})$

$= P(X_n = i_n, X_{n+1} = i_{n+1}, \dots, X_{n+k} = i_{n+k}) / P(X_{n+1} = i_{n+1}, \dots, X_{n+k} = i_{n+k})$

$= [P(X_{n+k} = i_{n+k} \mid X_n = i_n, \dots, X_{n+k-1} = i_{n+k-1}) P(X_{n+k-1} = i_{n+k-1} \mid$

$X_n = i_n, \dots, X_{n+k-2} = i_{n+k-2}) \dots P(X_{n+1} = i_{n+1} \mid X_n = i_n)$

$\times P(X_n = i_n)] / [P(X_{n+k} = i_{n+k} \mid X_{n+1} = i_{n+1}, \dots, X_{n+k-1} = i_{n+k-1}) P(X_{n+k-1}$

$= i_{n+k-1} \mid X_{n+1} = i_{n+1}, \dots, X_{n+k-2} = i_{n+k-2}) \dots P(X_{n+1} = i_{n+1})]$

By Markov property and cancellation, the last expression

$$= P(X_{n+1} = i_{n+1} \mid X_n = i_n)\, P(X_n = i_n)/P(X_{n+1} = i_{n+1})$$

$$= P(X_{n+1} = i_{n+1}, X_n = i_n)/P(X_{n+1} = i_{n+1})$$

$$= P(X_n = i_n \mid X_{n+1} = i_{n+1}).$$

(b) We shall prove this for $m = 2$ only. Since $P(AB \mid C) = P(A \mid BC)P(B \mid C)$,

$$P(X_{n_{k+2}} = x_{n_{k+2}}, X_{n_{k+1}} = x_{n_{k+1}} \mid X_{n_j} = x_j, 1 \le j \le k)$$

$$= P(X_{n_{k+2}} = x_{n_{k+2}} \mid X_{n_{k+1}} = x_{n_{k+1}}, X_{n_j} = x_j, 1 \le j \le k)$$

$$\times P(X_{n_{k+1}} = x_{n_{k+1}} \mid X_{n_j} = x_j, 1 \le j \le k)$$

By Markov property the last expression

$$= P(X_{n_{k+2}} = x_{n_{k+2}} \mid X_{n_{k+1}} = x_{n_{k+1}})P(X_{n_{k+1}} = x_{n_{k+1}} \mid X_{n_k} = x_k)$$

$$= P(X_{n_{k+2}} = x_{n_{k+2}}, X_{n_{k+1}} = x_{n_{k+1}} \mid X_{n_k} = x_k)$$

The proof for general m is exactly similar and hence is omitted.

Exercise 2.11 It is easy to see that

$$\frac{1}{n} \sum_{m=1}^{n} P^m, \; n = 1, 2, \ldots$$

are stochastic matrices. We have to prove that

$$\lim_{n\to\infty} \frac{1}{n} \sum_{m=1}^{n} P^m = Q \text{ (say), exists.} \tag{0}$$

Since the elements of the stochastic matrices are bounded by 1, every sequence of Stochastic matrices contains a convergent sub-sequence. To show existence of Q it is sufficient to show that the convergent subsequences of the averages on the left all have the same limit.

Suppose for some convergent subsequence $n_1 < n_2 < \ldots$.

$$\lim_{k\to\infty} \frac{1}{n} \sum_{m=1}^{n_k} \rho^m = A \tag{1}$$

Multiplying both sides on the left and right by P we get

$$\lim_{k\to\infty} \frac{1}{n_k} \sum_{m=2}^{n_k+1} P^m = AP = PA \tag{2}$$

Since $\lim_{k\to\infty} P^{n_k+1}/n_k = 0 = \lim_{k\to\infty} P/n_k$ (elements of powers of P are bounded by 1).

By (1) and (2) $$A = AP = PA \tag{3}$$

Therefore $$A = AP^n = P^n A = A\left(\frac{1}{n} \sum_{m=1}^{n} P^m\right) = \left(\frac{1}{n} \sum_{m=1}^{n} P^m\right) A \tag{4}$$

for all $n \ge 1$.

Hence, if B is any limit matrix of a subsequence of averages in (0),

$$A = AB = BA \tag{5}$$

Since A and B are arbitrary limit matrices here, they can be interchanged giving

$$B = BA = AB. \tag{6}$$

By (5) and (6) $A = B$. So there is only one limit matrix.
Thus, by (3), $A = QP = PQ$ and taking limits $n \to \infty$ in (4), we get $Q = Q^2$.

Exercise 2.12 Suppose that C is recurrent but is not closed, then there exists $i \in C$ such that $\sum_{j\in c} p_{ij} < 1$ and hence there exists $k \notin C$ such that $p_{ik} > 0$, i.e. $i \to k$. State i being recurrent we have $i \leftrightarrow k$, hence $k \in C$. This is a contradiction. Hence C is closed. Conversely, suppose that C be a closed class of states. Since the state space of a finite M.C. contains at least one recurrent state, the sub-Markov chain with state space C has atleast one recurrent state. But recurrence is a class property, hence C is a recurrent class.

****Exercise 2.13** (a) We proceed by induction on r. For $r = 1$ this is true by the very definition of f_{ii}. Assume it holds for $r = m-1$.

Write $P_i(A) = P(A \mid X_0 = i)$ and $T_i = \min(n \geq 1 : X_n = i)$

$P_i(X_n = i$ for at least $m-1$ values of $n \geq 1)$

$$= \sum_{k\geq 1} P_i(T_i = k, X_{T_i+j} = i \text{ for at least } m-1 \text{ values of } j \geq 1)$$

$$= \sum_{k\geq 1} P_i(T_i = k) P_i(X_{T_i+j} = i \text{ for at least } m-1 \text{ values of } j \geq 1 \mid T_i = k)$$

Now $P_i(X_{T_i+j} = i$ for at least $m-1$ values of $j \geq 1 \mid T_i = k)$

$$= P_i(X_{T_i+j} = i \text{ for at least } m-1 \text{ values of } j \geq 1 \mid X_{T_i} = i, T_i = k)$$

$$= P_i(X_{T_i+j} = i \text{ for at least } m-1 \text{ values of } j \geq 1 \mid X_{T_i} = i)$$

(by strong Markov property of M.C.)

$$= P_i(X_j = i \text{ for at least } m-1 \text{ values of } j \geq 1)$$

$$= [f_{ii}]^{m-1} \text{ (by induction hypothesis).}$$

Therefore $P_i(X_n = i$ for at least m values of $n \geq 1)$

$$= \sum_{k\geq 1} f_{ii}^{(k)} [f_{ii}]^{m-1} = [f_{ii}][f_{ii}]^{m-1} = [f_{ii}]^m$$

(b) Proof is entirely similar to that of (a) using T_{ij} instead of T_i when $T_{ij} = \min\{n \geq 1; X_n = j$ given that $X_0 = i\}$.

(c) Let $A = \{$The Markov chain returns to i infinitely often$\}$

$$= \bigcap_{r=1}^{\infty} \{\text{The Markov chain returns to } i \text{ at least } r \text{ times}\}$$

$$= \lim_{r\to\infty} \{\text{The Markov chain returns to } i \text{ at least } r \text{ times}\}$$

By (a)

$$P(A) = \lim_{r\to\infty} [f_{ii}]^r = \begin{cases} 1 & \text{if } f_{ii} = 1 \\ 0 & \text{if } f_{ii} < 1 \end{cases}$$

Exercise 2.14 (a) First prove that (by induction on n)

$$P_i[v(j) \geq n] = f_{ij}[f_{jj}]^{n-1} \tag{1}$$

For $n = 1$, $P_i[v(j) \geq 1] = f_{ij}$ (by definition)
For $n = 2$, $P_i[v(j) \geq 2] = f_{ij}f_{jj}$
Since $P_i[v(j) \geq n+1] = f_{ij}[f_{jj}]^n$, inductively we obtain (1). We have

$$\begin{aligned} P_i[v(j) = n] &= \sum_{k\geq n} P_i[v(j) = k] - \sum_{k \geq n+1} P_i[v(j) \geq k] \\ &= P_i[v(j) \geq n] - P_i[v(j) \geq n+1] \\ &= f_{ij}[f_{jj}]^{n-1}[1 - f_{jj}]. \end{aligned} \tag{2}$$

(b) $$\mu(k, i) = \sum_{n=1}^{\infty} P_k[v(i) \geq n] = \sum_{n=1}^{\infty} f_{ki}[f_{ii}]^{n-1}$$

$$= f_{ki}\frac{1}{1 - f_{ii}} < \infty \quad \text{if } 0 \leq f_{ii} < 1, \text{ i.e. if } i \text{ is transient.}$$

From equation (2)

$$P_k(v(i) < \infty) = f_{ki}(1 - f_{ii}) \sum_{n=1}^{\infty} [f_{ii}]^{n-1}$$

$$= f_{ki}\frac{1 - f_{ij}}{1 - f_{ij}} = 1 \quad \text{if } f_{ij} < 1 \text{ and } f_{ki} = 1.$$

(c) Obvious from (b).

Exercise 2.15

$$\sum_{n=1}^{m} p_{ij}^{(n)} = \sum_{n=1}^{m}\sum_{k=1}^{n} f_{ij}^{(k)} p_{ji}^{(n-k)} = \sum_{k=1}^{m}\left(f_{ij}^{(k)} \sum_{n=k}^{m} p_{jj}^{(n-k)} \right)$$

Hence $$\left(1 + \sum_{k=1}^{m} p_{ij}^{(k)}\right)\left(\sum_{k=1}^{m} f_{ij}^{(k)}\right) \geq \sum_{n=1}^{m} p_{ij}^{(n)} \geq \left(1 + \sum_{n=1}^{m-m'} p_{jj}^{(n)}\right)\left(\sum_{k=1}^{m'} f_{ij}^{(k)}\right) \tag{1}$$

for all $m' < m$. The last inequalities hold because

$$1 + \sum_{n=1}^{m} p_{jj}^{(n)} \geq \sum_{n=1}^{m} P_{jj}^{(n-k)} = 1 + \sum_{n=1}^{m-k} p_{jj}^{(n)}$$

and $$\sum_{k=1}^{m} (f_{ij}^{(k)})\left(\sum_{n=k}^{m} p_{jj}^{(n-k)}\right) \geq \sum_{k=1}^{m'}\left((f_{ij}^{(k)}) \sum_{m=k}^{m-m'+k} p_{jj}^{(n-k)}\right)$$

$$= \left(1 + \sum_{n=k}^{m-m'} p_{ij}^{(n)}\right)\left(\sum_{k=1}^{m'} f_{ij}^{(k)}\right)$$

Dividing both sides of (1) by $1 + \sum_{n=1}^{m} p_{ij}^{(n)}$ and letting first $m \to \infty$ and then $m' \to \infty$ yields the result.

The last part is trivial and is omitted.

Exercise 2.16

$$p_{ii} = \frac{i+1}{k+1} \quad \text{for } i = 0, 1, \ldots, k$$

$$p_{ji} = \frac{i+1}{k+1} \quad \text{if } j < i$$

$$= 0 \quad \text{if } j > i$$

From the relation $X_n = X_{n+1} + (Y_n - X_{n-1})^+$ it is easy to see that X_n is a Markov chain. Since $p_{kk} = 1$, k is an absorbing state.

From the transition matrix

$$\begin{pmatrix} \frac{1}{k+1} & \frac{1}{k+1} & \frac{1}{k+1} & \cdots & \frac{1}{k+1} \\ 0 & \frac{2}{k+1} & \frac{1}{k+1} & \cdots & \frac{1}{k+1} \\ 0 & 0 & \frac{3}{k+1} & \cdots & \frac{1}{k+1} \\ 0 & 0 & \cdots & \cdots & \cdots \\ \cdots & \cdots & \cdots & \cdots & \cdots \\ 0 & 0 & . & . & 1 \end{pmatrix}$$

It is clear that $j \to k$ but $k \not\to j$ for $j < k$ and the chain is finite. Hence all the states $0, 1, \ldots, k-1$ are transient and k is positive recurrent.

Since $p_{ii} > 0$ for all $i = 0, 1, 2, \ldots, k$ all the states $\{0, 1, 2, \ldots, k\}$ have period one though the chain is not aperiodic (since not irreducible).

Exercise 2.17 *(Polya urn Scheme)*
Here $P(S_0 = a) = 1$. After the nth drawing there are a total of $a + b + n$ balls in the box. If i of these are white, then the probability of drawing another white ball is $\frac{i}{a+b+n}$. Hence ${}_n p_{i,\, j+1} = \frac{i}{a+b+n}$. Also if $S_n = i$ then $S_{n+1} = i$ means a black ball will be drawn in the next n+1th drawing. Since there will be $a + b + n - i$ black balls in the box

$${}_n p_{i,j} = P(S_{n+1} = i \mid S_n = i) = \frac{a+b+n-i}{a+b+n}.$$

It is obvious from the nature of transition probabilities that the past history of values of S_m for $m < n$ does not influences the conditional distribution but does depend on n. Obviously ${}_n p_{ij} = 0$ if $j \neq i, i + 1$ and it is an nonhomogeneous chain since transition probabilities depend on n.

Exercise 2.18 The states space of the M.C. is {1, 2, 3, 4, 5, 6}. The transition matrix is

$$P = \begin{matrix} 1 \\ 2 \\ 3 \\ 4 \\ 5 \\ 6 \end{matrix} \begin{pmatrix} \frac{1}{6} & \frac{1}{6} & \frac{1}{6} & \frac{1}{6} & \frac{1}{6} & \frac{1}{6} \\ 0 & \frac{2}{6} & \frac{1}{6} & \frac{1}{6} & \frac{1}{6} & \frac{1}{6} \\ 0 & 0 & \frac{3}{6} & \frac{1}{6} & \frac{1}{6} & \frac{1}{6} \\ 0 & 0 & 0 & \frac{4}{6} & \frac{1}{6} & \frac{1}{6} \\ 0 & 0 & 0 & 0 & \frac{5}{6} & \frac{1}{6} \\ 0 & 0 & 0 & 0 & 0 & 1 \end{pmatrix}$$

We obtain, for example, the third row of the matrix as follows. Suppose the system is in state 3, i.e. the maximum of the numbers occurring on the nth toss is 3. Then the system remains in state 3 if a 1, 2 or 3 occurs on the $(n + 1)$th toss; hence $p_{33} = \frac{3}{6}$. On the other hand, the system moves to state 4, 5 or 6 respectively, if a 4, 5 or 6 occurs on the $(n + 1)$ the toss; hence $p_{34} = p_{35} = p_{36} = \frac{1}{6}$. The system can never move to 1 or 2 since a 3 has occurred on one of the tosses; hence $p_{31} = p_{32} = 0$. The other rows are obtained similarly. The matrix is not regular since state 6 is absorbing and there is a 1 on the main diagonal in row 6.

Exercise 2.19 (*See Exercise 2.26*)

Let Y_n be the content of the dam at time n and X_n is the input at the nth period. Then

$$Y_{n+1} = \min (k, Y_n + X_n) - 1$$

Hence, clearly $\{Y_n\}$ is a Markov chain. The transition matrix is

$$\begin{matrix} 0 \\ 1 \\ 2 \\ \\ k-1 \end{matrix} \begin{bmatrix} q + qp & qp^2 & qp^3 & \dots & qp^{k-1} & p^k \\ q & qp & qp^2 & \dots & qp^{k-2} & p^{k-1} \\ 0 & q & qp & \dots & qp^{k-3} & p^{k-2} \\ \vdots & & & & & \\ 0 & 0 & \dots & \dots & q & p \end{bmatrix}$$

Exercise 2.20 The transition matrix of the M.C. is given by

$$P = \begin{pmatrix} 0 & 0 & 0 & 0 & 0 & \dots & 1 \\ 0 & 0 & 0 & 0 & 0 \dots & \frac{1}{2} & \frac{1}{2} \\ 0 & 0 & 0 & 0 & \frac{1}{3} & \frac{1}{3} & \frac{1}{3} \\ \dots & \dots & \dots & \dots & \dots & \dots & \dots \\ 0 & \frac{1}{M-1} & & \dots & & \frac{1}{M-1} & \frac{1}{M-1} \\ \frac{1}{M} & \frac{1}{M} & \frac{1}{M} & \dots & & & \frac{1}{M} \end{pmatrix}$$

This is a finite irreducible chain and the chain is positive recurrent. Hence a stationary distribution exist.

The stationary distribution is given by

$$(\pi_1, \pi_2, \dots, \pi_M)$$

$$\times \begin{pmatrix} 0 & 0 & 0 & 0 & 0 & \dots & & 0 & 1 \\ 0 & 0 & 0 & 0 & 0 & \dots & & \frac{1}{2} & \frac{1}{2} \\ 0 & 0 & 0 & 0 & 0 & \dots & \frac{1}{3} & \frac{1}{3} & \frac{1}{3} \\ & & & & \dots & & & & \\ 0 & \frac{1}{M-1} & \frac{1}{M-1} & & \dots & & & \frac{1}{M-1} & \frac{1}{M-1} \\ \frac{1}{M} & & & & \dots & & & & \frac{1}{M} \end{pmatrix}$$

$$= \begin{pmatrix} \pi_1 \\ \pi_2 \\ \vdots \\ \pi_M \end{pmatrix}$$

Thus, $\pi_1 = \pi_M / M, \pi_2 = \dfrac{\pi_{M-1}}{M-1} + \dfrac{\pi_M}{M}, \dots$

$$\pi_{M-1} = \pi_2/2 + \pi_3/3 + \dots + \pi_{M-1}/M - 1 + \pi_M/M,$$

$$\pi_M = \pi_1 + \pi_2/2 + \pi_3/3 + \dots + \pi_M/M.$$

Hence

$$\pi_1 = \pi_M/M,\ \pi_2 = \pi_1 + \pi_{M-1}/(M-1),\ \pi_3 = \pi_2 + \pi_{M-2}/(M-2), \dots, \pi_{M-1} = \pi_{M-2} + \pi_2/2,\ \pi_M = \pi_{M-1} + \pi_1.$$

Therefore taking $\pi_M = 1$,

$\pi_1 = \dfrac{1}{M}, \pi_{M-1} = \dfrac{M-1}{M}, \pi_2 = 2/M, \pi_{M-2} = \dfrac{M-2}{M}, \dots$ and the stationary distribution is given by

$$\pi_1 = \frac{2}{M(M+1)}, \pi_2 = \frac{4}{(M+1)M}, \pi_3 = \frac{6}{M(M+1)}, \dots, \pi_M = \frac{2}{(M+1)}.$$

Exercise 2.21

$$P = \begin{pmatrix} 0 & 1 & 0 \\ 1/2 & 0 & 1/2 \\ 0 & 1 & 0 \end{pmatrix}.$$

The chain is irreducible as the transition matrix P is so. We have

$$P^2 = \begin{pmatrix} 1/2 & 0 & 1/2 \\ 0 & 1 & 0 \\ 1/2 & 0 & 1/2 \end{pmatrix}, p^3 = P.$$

In general $P^{2n} = P^2$, $P^{2n+1} = P$, so that $p_{ii}^{(2n)} > 0, p_{ii}^{(2n+1)} = 0$ for each i. So the states are periodic with period 2.

We find that $f_{11}^{(1)} = 0, f_{11}^{(2)} = 1$ so that $f_{11} = \sum_n f_{11}^{(n)} = 1$, i.e. the state 1 is recurrent. Since the chain is irreducible, the other states 0 and 2 are also recurrent.

Being a finite chain all the states are positive recurrent.

Exercise 2.22 Here $\{X_n\}$ is a homogeneous two-state Markov chain with transition matrix $P = \begin{pmatrix} q & p \\ p & q \end{pmatrix}$. It can be shown by induction or otherwise that

$$P^m = \begin{bmatrix} 1/2 + 1/2(q-p)^m & 1/2 - 1/2(q-p)^m \\ 1/2 - 1/2(q-p)^m & 1/2 + 1/2(q-p)^m \end{bmatrix}$$

and
$$\lim_{m\to\infty} P^m = \begin{bmatrix} 1/2 & 1/2 \\ 1/2 & 1/2 \end{bmatrix}.$$

Since the initial distribution is given by

$$P(X_0 = 0) = a = 1 - P(X_0 = 1),$$

$$P(X_m = 0, X_0 = 0) = P(X_m = 0 \mid X_0 = 0)\ P(X_0 = 0). = a p_{00}^{(m)}$$

and $P(X_m = 0, X_0 = 1) = (1-a) p_{10}^{(m)}$.

By Bayes Theorem

$P(X_0 = 0 \mid X_m = 0)$

$$= \frac{P(X_m = 0 \mid X_0 = 0) P(X_0 = 0)}{P(X_m = 0 \mid X_0 = 0) P(X_0 = 0) + P(X_m = 0 \mid X_0 = 1) P(X_0 = 1)}$$

$$= \frac{a p_{00}^{(m)}}{a p_{00}^{(m)} + (1-a) p_{10}^{(m)}}$$

$$= \frac{a\{1/2 + 1/2(q-p)^m\}}{a\{1/2 + 1/2(q-p)^m\} + (1-a)\{1/2 - 1/2(q-p)^m\}}$$

$$= \frac{a\{1 + (q-p)^m\}}{1 + (2a-1)(p-q)^m}.$$

Exercise 2.23 Let $I_y(z)$, $z \in S$, denote the indicator function of the set $\{y\}$, defined by

$$I_y(z) = \begin{cases} 1 & \text{if } z = y \\ 0 & \text{if } z \neq y \end{cases}$$

Since $I_y(X_n) = 1$ if the chain is in state y at time n and $I_y(X_n) = 0$ otherwise,

$$N(y) = \sum_{n=1}^{\infty} I_y(X_n) \qquad (1)$$

Now $P(N(y) \geq 1 \mid X_0 = x) = \rho_{xy}$. Let T_y be the first passage time to the state y. By Markov property $P(X_{n+1} = x_{n+1}, \ldots, X_{n+m} = x_{n+m} \mid X_0 = x, \ldots, X_n = x_n) = p_{x_n, x_{n+1}}, \ldots p_{x_{n+m-1}, x_{n+m}}$. Hence the probability that a Markov chain starting at x first visit y at time m and next visit yn units of time later is

$$P(T_y = m \mid X_0 = x)\, P(T_y = n \mid X_0 = y).$$

Thus, $P(N(y) \geq 2 \mid X_0 = x) = \sum_{m=1}^{\infty} \sum_{n=1}^{\infty} P(T_y = m \mid X_0 = x) P(T_y = n \mid X_0 = y)$

$$= \sum_{m=1}^{\infty} P(T_y = m \mid X_0 + x) \sum_{n=1}^{\infty} P(T_y = n \mid X_0 = y) = \rho_{xy}\rho_{yy}.$$

By induction and Markov property

$$P(N(y) \geq m + 1 \mid X_0 = x) = P(N(y) \geq m \mid X_0 = x)\, P(N(y) \geq 1 \mid X_0 = y)$$

$$= \rho_{xy}\rho_{yy}^{m-1} p_{yy} = p_{xy} p_{yy}^{m} \qquad (2)$$

Now $\qquad E(I_y(X_n) \mid X_0 = x) = P(X_n = y \mid X_0 = x) = p_{xy}^{(n)}.$

Hence by (1) $\qquad E(N(y) \mid X_0 = x) = E\left(\sum_{n=1}^{\infty} I_y(X_n) \mid X_0 - x \right)$

(Monotone Convergence Theorem)

$$= \sum_{n=1}^{\infty} E(I_y(X_n) \mid X_0 = x) = \sum_{n=1}^{\infty} p_{xy}^{(n)} \qquad (3)$$

Since $P(N(y) = m \mid X_0 = x) = P(N(y) \geq m \mid X_0 = x) - P(N(y) \geq m + 1 \mid X_0 = x)$. It follows from (2) that

$$P(N)(y) = m \mid X_0 = x) = \rho_{xy}\rho_{yy}^{m-1}(1 - \rho_{yy}),\ m \geq 1.$$

Let $\qquad G(X, y) = E(N(y) \mid X_0 = x) = \sum_{m=1}^{\infty} m P(N(y) = m \mid X_0 = x)$

$$= \sum_{m=1}^{\infty} m \rho_{xy}\rho_{yy}^{m-1}(1 - \rho_{yy})$$

Substituting $t = \rho_{yy}$ in power series

$$\sum_{m=1}^{\infty} mt^{m-1} = \frac{1}{(1-t)^2},$$

we conclude that

$$G(x, y) = \frac{\rho_{xy}}{1-\rho_{yy}} < \infty \quad \text{if } 0 \le \rho_{yy} < 1, \text{ i.e. } y \text{ is a transient state.}$$

Hence by (3)

$$\lim_{n\to\infty} p_{xy}^{(n)} = 0 \quad \text{if } x \in S \text{ and } y \text{ is transient} \tag{4}$$

If S is finite and all states are transient, then by (4)

$$0 = \sum_{y\in s} \lim_{n\to\infty} \sum_{y\in S} p_{xy}^{(n)} = \lim_{n\to\infty} \sum_{y\in S} p_{xy}^{(n)} = \lim_{n\to\infty} P(X_n \in S \mid X_0 = x)$$

$$= \lim_{n\to\infty} 1 = 1 \text{ which is a contradiction.}$$

Hence the Markov chain must have at least one recurrent state.

Exercise 2.24 By Chapman-Kolmogorov equation

$$p_{jj}^{(n+m)} = \sum_{j\in s} p_{jj}^{(m)} p_{jj}^{(n)} \ge p_{jj}^{(n)} p_{jj}^{(n)}$$

$$= p_{jj}^{(m)} p_{jj}^{(n)} + p_{jj}^{(n)} \sum_{\substack{j\in s \\ k\ne j}} p_{kj}^{(m)} \tag{1}$$

$$= p_{jj}^{(m)} p_{jj}^{(n)} + p_{jj}(1 - p_{jj}^{(m)}) \left(\because \sum_{j\in s} p_{kj}^{(m)} = 1 \right)$$

$$\le p_{jj}^{(m)} p_{jj}^{(n)} + (1 - p_{jj}^{(m)}).$$

Exercise 2.25

Parent mating	$aa \times aa$ $AA \times AA$	$aa \times Aa$ $AA \times Aa$	$Aa \times Aa$	$AA \times aa$
Progeny mating				
$aa \times aa$, $AA \times AA$	1	1/4	1/8	0
$aa \times Aa$, $AA \times Aa$	0	1/2	1/2	0
$Aa \times Aa$	0	1/4	1/4	1
$AA \times aa$	0	0	1/8	0

The latent roots of the matrix P, i.e. the solution of

$$|P = \lambda I| = 0 \text{ are given by } (\lambda - 1)\left(\lambda - \frac{1}{4}\right)(\lambda^2 - \lambda/2 - 1/4) = 0$$

or $\lambda_1 = 1, \lambda_2 = 1/4 + 1/4\sqrt{5}, \lambda_3 = 1/4 - 1/4\sqrt{5}, \lambda_4 = 1/4.$

From spectral decomposition theorem $p^r = \sum_{i=1}^{4} \lambda_i^r A_i$.

So the proportion of heterozygotes in rth generation is given by

$$h_r = A(1/4 + 1/4\sqrt{5})^r + B(1/4 - 1/4\sqrt{5})^r + C(1/4)^r$$

and determine, A, B and C from the values of h_0, h_1, and h_2 which are 1, 1/2 and 1/2. This gives

$$A = 2/\sqrt{5}(1/4 + 1/4\sqrt{5}),$$

$$B = -2/\sqrt{5}(1/4 - 1/4\sqrt{5}) \text{ and } C = 0.$$

Thus we found

$$h_r = 2/\sqrt{5}\{(1/4 + 1/4\sqrt{5})^{r+1} - (1/4 - 1/4\sqrt{5})^{r+1}\}.$$

Exercise 2.26 Let k be the finite capacity of the dam and Z_n be the storage at time n i.e. the amount of water in the dam before the input X_n flows into it. Then if $Z_n + X_n > k$, an amount $Z_n + X_n - k$ will overflow, but if $Z_n + X_n \le k$, there will be no overflow, the dam now contains a quantity min $(k, Z_n + X_n)$. At time $n + 1$, an amount of water $m(< k)$ if $Z_n + X_n \ge m$ or $Z_n + X_n$ if $Z_n + X_n < m$ is released from the dam. The release is thus $Y_n = \min(m, Z_n + X_n)$.

Hence Z_n satisfies the recurrence relation

$$Z_{n+1} = \min(k, Z_n + X_n) - \min(m, Z_n + X_n) \; (n = 0, 1, 2, \ldots) \text{ if } m > 1$$

$$= \min(k, Z_n + X_n) - 1 \text{ if } m = 1.$$

Thus, $\{Z_n\}$ is clearly a Markov chain. Let $\{g_j\}$ be the probability distribution of X_n with P.G.F

$$G(z) = \sum_{j=0}^{\infty} g_j z^j, |z| < 1$$

and transition matrix of $\{Z_n\}$ is denoted by $P = (p_{ij}) = P(Z_{n+1} = j \mid Z_n = i)$. $\{Z_n\}$ is a finite state M.C. with states 0, 1, 2, . . . , $k - m$ and the transition matrix $(m > 1)$

$P =$		0	1	2		$k-m-1$	$k-m$
	0	G_m	g_{m+1}	g_{m+2}	$\cdots$	g_{k-1}	h_k
	1	G_{m-1}	g_m	g_{m+1}	$\cdots$	g_{k-2}	h_{k-1}
	$\vdots$	$\vdots$	$\cdots$				
	m	G_0	g_1	g_2	$\cdots$	g_{k-m-1}	h_{k-m}
	$m+1$	0	g_0	g_1	$\cdots$	g_{k-m-2}	h_{k-m-1}
	$\vdots$	$\vdots$	$\cdots$				
	$k-m$	0	0	0	$\cdots$	g_{m-1}	h_m

(Assume $m < k/2$) where $G_i = g_0 + g_1 + \ldots + g_i$, $h_i = g_i + g_{i+1} + \ldots$ $(i \geq 0)$. If $m = 1$.

	0	1	2	...	k–2	k – 1
0	$g_0 + g_1$	g_2	...		g_{k-1}	h_k
1	G_0	g_1	...		g_{k-2}	h_{k-1}
⋮ $P =$			...			
2	0	g_0	...		g_{k-3}	h_{k-2}
$k-1$	0	0	...		g_0	h_1

where $h_i = \sum_{j=i}^{\infty} g_j$. Let us assume that $g_j > 0$ for all j, the chain is then irreducible and aperiodic, so that the stationary distribution $\pi_i (i = 0, 1, \ldots, k-1)$ exists where $\pi_i = \lim_{n\to\infty} P(Z_n = i)$. The π_i is the unique solution of the equations

$$\pi_j = \sum_{i=o}^{k-1} \pi_i p_{ij} \quad (j = 0, 1, \ldots, k-1) \text{ together with}$$

$$\pi_0 + \pi_1 + \ldots + \pi_{k-1} = 1.$$

In case of geometric input $P(X_n = j) = qp^j = g_j$, $0 < q < 1$, $p = 1 - qj = 0, 1, 2, \ldots$ when $m = 1$, the transition matrix is given by

	0	1	2	...	$k-2$	$k-1$
0	$q + qp$	qp^2	qp^3	...	qp^{k-1}	p^k
1	q	qp	qp^2	...	qp^{k-2}	p^{k-1}
$P =$ 2	0	q	qp	...	qp^{k-3}	p^{k-2}
⋮						
$k-1$	0	0	...	...	q	p

Exercise 2.27 Loss suffered in the first year is $Y \cdot R/100 = Y_1 \cdot U_1$. Thus $X_1 = Y_1 - Y_1U_1 = Y_1(1 - U_1)$. Loss suffered in the second year is $Y_2(1 - U_1)R_2/100 = Y_2(1 - U_1)U_2$. Therefore $X_2 = Y_2(1 - U_1) - Y_2(1 - U_1)U_2 = Y_1(1 - U_1)(1 - U_2)$. Hence by induction on n, the crop yield in the nth year is $X_n = Y_n \prod_{i=1}^{n} Z_i$, $Z_i = 1 - U_i$ and $X_n = Y_nL_n$. Since $L_n = \prod_{i=1}^{n} Z_i = L_{n-1}Z_n$, L_n is obviously a Markov chain.

Exercise 2.28 Let $p_{jk} = P\{k$ white balls in first urn after $n + 1$ interchanges $| j$ white balls in first urn after n interchanges$\}$.

Note that if there are j white balls in the first urn after n interchanges then except for $j - 1, j, J + 1$ states other states are impossible after $n + 1$ interchanges.

Therefore $p_{j,j-1} = P$ (Replacing a white ball in the first urn by a black ball in the 2nd urn) $= j/N \cdot j/N = j^2/N^2$.

$p_{jj} = P$ (Replacing white ball in the 2nd urn or replacing a black ball in the 1st urn by a black ball in the 2nd urn)

$$= j/N \cdot \frac{N-j}{N} + \frac{N-j}{N} \cdot \frac{j}{N} = 2\left(\frac{j}{N}\right)\left(\frac{N-j}{N}\right).$$

$p_{jj+1} = P$ (Replacing a black ball in the first urn by a white ball in the 2nd urn)

$$= \frac{N-j}{N} \cdot \frac{N-j}{N} = \left(\frac{N-j}{N}\right)^2$$

Here
$$p_{jk} = \begin{cases} \left(\frac{j}{N}\right)^2 & \text{if } k = j-1,\ j = 2, \ldots N \\ 2(j/N)\left(\frac{N-j}{N}\right) & \text{if } k = j,\ j = 0, 1, 2, \ldots N \\ (1 - j/N)^2 & \text{if } k = j+1,\ j = 0, 1, \ldots N-1 \\ 0 & \text{otherwise} \end{cases}$$

$p_{jj}^{(N)} > 0$ for all $i, j \in S$ so that there are no transient states, the chain is irreducible and aperiodic finite chain.

Hence
$$\lim_{n\to\infty} p_{ij}^{(N)} = \pi_j > 0 \text{ for all } i, j = 0, 1, \ldots, N.$$

π_j can be solved by solving the equations $\sum_{i=0}^{N} \pi_i p_{ij} = \pi_j$ and $\sum_{j=0}^{N} \pi_j = 1$.

The solution is given by $\pi_j = \dfrac{(N!)^4}{(j!)^2\,((N-j)!)^2\,(2N!)}$.

Exercise 2.29

(a)
$$p_{ik} = \begin{cases} (i/N)q & \text{if } k = i+1 \\ \frac{i}{N}p + \frac{N-i}{N}q & \text{if } k = i,\ i = 1, 2, \ldots, N-1 \\ \frac{N-i}{N}p & \text{if } k = i-1 \\ 0 & \text{otherwise} \end{cases}$$

$p_{ii} = 1$ if $i = 0$ and $i = N$. Equivalence classes are $\{0\}, \{N\}, \{1, 2, \ldots, N-1\}$. Hence $\{0\}$ and $\{N\}$ are absorbing states.

(b)
$$p_{ik} = \begin{cases} i^2/N^2 + (N-i)^2/N^2 & \text{if } k = i \\ \frac{i(N-i)}{N^2} & \text{if } k = i+1, i-1 \quad i = 1, 2, \ldots N \\ 0 & \text{otherwise} \end{cases}$$

(c) Equivalence classes are $\{0\}, \{1, 2, \ldots, N-1\}, \{N\}$

Exercise 2.30 In fact we shall show that a 2×2 stochastic matrix is the two-step transition matrix of a Markov chain iff the sum of its diagonal terms is greater than or equal to 1.

Let $P = \begin{pmatrix} p & 1-p \\ 1-q & q \end{pmatrix}$ and $A = \begin{pmatrix} \alpha & 1-\alpha \\ 1-\beta & \beta \end{pmatrix}$ be two stochastic matrices such that $P = A^2$. We shall prove that $p + q \geq 1$.

Proof $p = \alpha^2 + (1-\alpha)(1-\beta)$, $q = \beta^2 + (1-\alpha)(1-\beta)$.

So $p + q = \alpha^2 + 2(1-\alpha)(1-\beta) + \beta^2 = (\alpha + \beta - 1)^2 + 1 \geq 1$. Conversely if $p + q \geq 1$, then there exists a stochastic matrix A such that $P = A^2$.

1st Proof We have to show that there exists a solution of

$$\left.\begin{aligned} \alpha^2 + (1-\alpha)(1-\beta) &= p && \ldots (1) \\ (1-\alpha)(\alpha+\beta) &= 1-p && \ldots (2) \\ (1-\beta)(\alpha+\beta) &= 1-q && \ldots (3) \\ \beta^2 + (1-\alpha)(1-\beta) &= q && \ldots (4) \end{aligned}\right\} \text{if } p+q \geq 1.$$

From (3) and (2) $1 - \beta = (1-\alpha)\dfrac{1-q}{1-p}$. Use this in (1) to get

$$\alpha^2\{2 - p - q\} - 2\alpha\{1 - q\} + \{1 - p - q + p^2\} = 0.$$

The solution of this quadratic equation in α is

$$\begin{aligned}(2 - p - q)\alpha &= 1 - q \pm \sqrt{(1-q)^2 + (2-p-q)(q+p-1-p^2)} \\ &= 1 - q \pm (1-p)\sqrt{p+q-1}\end{aligned}$$

Since $p + q - 1 \geq 0$, this solution is real. Choose the positive root, then

$$\alpha = \frac{1-q}{2-p-q} + \frac{(1-p)\sqrt{p+q-1}}{2-p-q} \geq 0.$$

Similarly $$\beta = \frac{1-p}{2-p-q} + \frac{(1-q)\sqrt{p+q-1}}{2-p-q} \geq 0.$$

2nd proof Let P and A be as above. Now if $P = A^2$ then the eigen values of P and A^2 coincide. Since A and P are stochastic, 1 is an eigen value of P and A. The other eigen value of P is $p + q - 1$ and that of A is $\alpha + \beta - 1$. So $p + q - 1 = (\alpha + \beta - 1)^2 \geq 0$. Now since $1 \geq 0$ and $p + q - 1 \geq 0$, $P = A^2$ has a solution $A = \sqrt{P}$ (all eigen values are non-negative).

Exercise 2.31 **1st Solution** Since the M.C. is finite and irreducible, it is positive recurrent. Since it is aperiodic $p_{ij}^{(n)} \to \pi_j > 0$ as $n \to \infty$. Hence there exists a $N(i, j)$ such that $p_{ij}^{(n)} > 0$ for all $n \geq N(i, j)$. Since the M.C. is finite, $N = \max_{i,j} N(i \cdot j) < \infty$.

Hence for all $\quad n \geq N \quad p_{ij}^{(n)} > 0$ for all $i, j \in S$.

Second Solution If $p_{ij}^{(m)} > 0$, then $p^{(m+nd(j))} > 0$ for all sufficiently large positive integer n and where $d(j)$ is the period of state j. Now same type of arguments as the first solution applies.

Exercise 2.32 We have $p_{jj}^{(n)} = \sum_{k=0}^{n} f_{ij}^{(k)} p_{jj}^{(n-k)}$.

$$\text{So} \quad \sum_{n=0}^{\infty} p_{ij}^{(n)} = \sum_{n=0}^{\infty} \sum_{k=0}^{n} f_{ij}^{(k)} p_{jj}^{(n-k)} = \sum_{k=0}^{\infty} \sum_{n=k}^{\infty} f_{ij}^{(k)} p_{jj}^{(n-k)} \text{ (by Fubini Theorem)}$$

$$= \sum_{k=0}^{\infty} f_{ij}^{(k)} \sum_{l=0}^{\infty} p_{jj}^{(l)} = f_{ij} \sum_{l=0}^{\infty} p_{jj}^{(l)} < \infty$$

(Since j is transient and $f_{ij} \leq 1$). Also $p_{jj}^{(0)} = 0$ if $i \neq j, p_{jj}^{(0)} = 1$, and $\sum_{n=1}^{\infty} p_{ij}^{(n)} < \infty$.

Exercise 2.33 Since P is doubly stochastic $1 = \sum_{i=0}^{\infty} p_{ij}^{(n)}$ for all $n \geq 1$ (by induction). Fix $N > 0$.

Then $1 = \sum_{i=0}^{\infty} p_{ij}^{(n)} \geq \sum_{i=0}^{N} p_{ij}^{(n)} \geq N \min_{0 \leq i \leq N} p_{ij}^{(n)}$. Since chain is irreducible, $0 \leq \lim_{n \to \infty} [\min_{0 \leq i \leq N} p_{ij}^{(n)}]$ does not depend on j and is $\leq \frac{1}{N}$ for all n. Take $N \to \infty$, then $\lim_{n \to \infty} p_{ij}^{(n)} = 0$ exists for all $i, j \in S$ and hence the chain is either transient or null recurrent.

Exercise 2.34 Let $P_{ij}(z)$ be the generating function of $\{p_{ij}^{(n)}\}$,

(a) $\pi_{ij} = \lim_{z \to 1-} (1 - z) P_{ij}(z)$ (by Cesaro-Taubor Theorem)

$$= \lim_{z \to 1-} \frac{F_{ij}(z)(1 - z)}{1 - F_{ij}(z)}$$

and this limit exists always (may be zero) where $F_{ij}(z)$ is the generating function of $\{f_{ij}^{(n)}\}$.

(b) By Fatau's lemma

$$\sum_j \pi_{ij} = \sum_j \lim_{n \to \infty} \frac{1}{n} \sum_{m=1}^{n} p_{ij}^{(m)} \leq \lim_{n \to \infty} \sum_{m=1}^{n} \sum_j p_{ij}^{(m)} = 1.$$

(c) $$\sum_k \pi_{ik} p_{kj} = \sum_k \lim_{n \to \infty} \frac{1}{n} \sum_{m=1}^{n} p_{ik}^{(m)} p_{kj} = \lim_{n \to \infty} \frac{1}{n} \sum_{m=1}^{n} \sum_k p_{ik}^{(m)} p_{kj}$$

(by Dominated Convergence Theorem and Fubini Theorem)

$$= \lim_{n\to\infty} \frac{1}{n} \sum_{m=1}^{n} p_{ij}^{(m+1)} = \lim_{n\to\infty} \sum_{m=1}^{n} \frac{1}{n} p_{ij}^{(m)} + \lim_{n\to\infty} \frac{1}{n} [p_{ij}^{(n+1)} - p_{ij}] = \pi_{ij}$$

Similiarly $\sum_k p_{ik}\pi_{kj} = \pi_{ij}$.

By induction $\pi_{ij} = \sum_k \pi_{ik} p_{kj}^{(m)}$. Sum this for $m = 1, 2, \ldots, n$.

Then $\pi_{ij} = \frac{1}{n} \sum_k \sum_{m=1}^{n} \pi_{ik} p_{kj}^{(m)}$. Taking $n \to \infty$ and appling Dominated Convergence Theorem $\pi_{ij} = \sum_k \pi_{ik}\pi_{kj}$.

(d) If j is positive recurrent then $F_{ij}(1) = f_{ij}$, $\sum_n n f_{jj}^{(n)} = F'_{jj}(1) < \infty$.

$$\text{By (a) } \pi_{ij} = f_{ij} \frac{1}{\sum_n n f_{jj}^{(n)}} = f_{ij}\pi_j, \quad \text{where } \pi_j = \frac{1}{\sum_n n f_{jj}^{(n)}}.$$

Exercise 2.35 Let 0, 1, 2, . . . , N are the states of the M.C. and the M.C. is irreducible. Therefore the chain is positive recurrent and hence has a stationary distribution. But $\pi_j = \frac{1}{N+1}$ for all j is a solution of the equations $x_j = \sum_{i=0}^{N} x_i p_{ij}$ since $\sum_{i=0}^{N} p_{ij} = 1$. Moreover $\pi_j > 0$ and $\sum_{j=0}^{N} \pi_j = 1$. By uniqueness $\pi_j = \frac{1}{N+1}$ is the only stationary distribution.

Exercise 2.36 We apply induction on r. For $r = 0$, $P[N_i = 0] = 1 - f_{ii}$. Assume that

$P[N_i = r] = (1 - f_{ii})(f_{ii})^r$. $P[N_i = r + 1] = \sum_n P[N_i = r + 1, X_n = i \text{ for the first time}]$

(omitting the condition $X_0 = i$)

$$= \sum_n P[N_i = r + 1 \mid X_n = i]\, P[X_n = i \text{ for the first time}]$$

$$= \sum_n P[N_i = r] f_{ii}^{(n)} = (1 - f_{ii})(f_{ii})^r \sum_n f_{ii}^{(n)}$$

$$= (1 - f_{ii}) f_{ii}^r f_{ii} = (1 - f_{ii}) f_{ii}^{r+1} \left(\because \sum_n f_{ii}^{(n)} = f_{ii}\right).$$

$$E(N_i) = \sum_{r=1}^{\infty} r P(N_i = r) = (1 - f_{ii}) \sum_{r=1}^{\infty} r f_{ii}^r$$

$$= (1 - f_{ii}) \frac{f_{ii}}{(1 - f_{ii})^2} = \frac{f_{ii}}{1 - f_{ii}} \quad (\because f_{ii} < 1).$$

Exercise 2.37 First Solution By induction (or by first entrance) it follows that $f_{ij}^{(n+1)} = \sum_{k \neq j} p_{ik}^{(n)}$ for all $n \geq 1$. Hence

$$\mu_{ij} = \sum_{n=0}^{\infty} (n+1) f_{ij}^{(n+1)} = f_{ij} + \sum_{n=0}^{\infty} n f_{ij}^{(n+1)} = f_{ij} + \sum_n n \sum_{k \neq j} p_{ik} f_{kj}^{(n)}$$

and since the chain is positive and irreducible

$$f_{ij} = 1.$$

Hence
$$\mu_{ij} = 1 + \sum_{k\neq j} p_{ik} \sum_{n=0}^{\infty} n f_{kj}^{(n)} \text{ (Fubini Theorem)}$$

$$= 1 + \sum_{k\neq j} p_{ik}\, \mu_{kj} = \sum_{k\in s} p_{ik} + \sum_{k\neq j} p_{ik}\mu_{kj} \qquad (1)$$

$$= p_{ij} + \sum_{k\neq j} p_{ik}\, [1 + \mu_{kj}].$$

Second Solution Given a state j, the mean first passage times $\{\mu_{ij}, i \neq j\}$ can be regarded as mean absorption times in a new Markov chain obtained from the original Markov chain by making j an absorbing chain. If $P = (P_{ij})$ is the transition matrix of the original M.C., define a new transition matrix $P' = (p'_{ki})$ as follows: $p'_{ki} = 1$ if $k = j, i = j$, $p'_{ki} = 0$ if $k = j, i \neq j$, $p'_{ki} = p_{ki}$ if $k \neq j$, i is any state. The resulting chain contains a single absorbing state j, while all other states are non-recurrent since they are inessential. The behaviour of the new chain before absorption is the same as behaviour of the original chain before visiting state j for the first time. In particular, the mean first passage time from i to j in the original chain is the same as the mean time to absorption.

Exercise 2.38

(a) $p_{jk}^{(n)} = \sum_{m=0}^{n} f_{jk}^{(m)} p_{kk}^{(n-m)} \leq \sum_{m=0}^{n} f_{jk}^{(m)} \leq f_{jk}$ for all $n \geq 0$.

Therefore
$$\sup_n p_{jk}^{(n)} \leq f_{jk} \qquad (1)$$

$$f_{jk}^{(n)} = p_{jk}^{(n)} - \sum_{m=0}^{n-1} f_{jk}^{(m)}\, p_{kk}^{(n-m)} \leq p_{jk}^{(n)}.$$

Hence
$$f_{jk} \leq \sum_{n=1}^{\infty} p_{jk}^{(n)} \qquad (2)$$

Combining (1) and (2) we get (a).

(b) If $j \to k$, then $p_{jk}^{(n)} > 0$ for some n. Hence by (1)

$$f_{jk} \geq \sup_n p_{jk}^{(n)} \geq p_{jk}^{(n)} > 0 \quad \text{i.e. } f_{jk} > 0.$$

Alternately, if $f_{jk} > 0$, then by (2) $\sum_{n=1}^{\infty} p_{jk}^{(n)} > 0 \Rightarrow$ for some n, $p_{jk}^{(n)} > 0$. So $j \to k$.

(c) Similarly, $k \to j$ iff $f_{kj} > 0$. So $j \leftrightarrow k$ iff $f_{jk} f_{kj} > 0$.

(d) $\sum_{n=1}^{\infty} p_{jk}^{(n)} = \sum_{n=1}^{\infty} \sum_{m=1}^{n} f_{jk}^{(m)} p_{kk}^{(n-m)} = \sum_{m=1}^{\infty} f_{jk}^{(m)} \sum_{n=m}^{\infty} p_{kk}^{(n-m)}$ (by Fubini)

$$= f_{jk} \sum_{m=0}^{\infty} p_{kk}^{(m)} = f_{jk}\left[\sum_{m=1}^{\infty} p_{kk}^{(m)} + 1\right].$$

(e) From (d) taking $j = k$

$$\sum_{n=1}^{\infty} p_{kk}^{(n)} = f_{kk}\left[\sum_{n=1}^{\infty} p_{kk}^{(n)} + 1\right]$$

Thus, if $f_{kk} < 1$ then $\sum_{n=1}^{\infty} p_{kk}^{(n)} = \dfrac{f_{kk}}{1 - f_{kk}}$.

Exercise 2.39

(a) $p_j^{(n)} = P[X_n = j] = \sum_{i \in S} P[X_n = j \mid X_0 = i]P[X_0 = i] = \sum_{i \in S} p_{ij}^{(n)} \alpha_i$

Thus, $\sum_{j \in S} p_j^{(n)} = \sum_{j \in S} \sum_{i \in S} p_{ij}^{(n)} \alpha_i = \sum_{i \in S} \alpha_i \sum_{j \in S} p_{ij}^{(n)}$ (by Fubini Theorem)

$$= \sum_{j \in S} \alpha_i = 1.$$

(b) By dominated convergence theorem

$$\lim_{n \to \infty} p_j^{(n)} = \sum_{j \in S} (\lim_{n \to \infty} p_{ij}^{(n)})\alpha_i$$

$$= \pi_j \sum_{i \in S} \alpha_i = \pi_j,$$

where $\pi_j = 0$ if j is transient or null recurrent and > 0 if j is positive recurrent.
(c) A stationary distribution is such that

$$\pi_j = \sum_{i \in S} \pi_i\, p_{ij}^{(n)} \quad \text{for all } n \geq 1.$$

Hence $\qquad p_j^{(n)} = \sum_{i \in S} \alpha_i p_{ij}^{(n)} = \sum_{i \in S} \pi_i p_{ij}^{(n)} = \pi_j.$

Therefore the n-step unconditional probabilities are the same for all $n \geq 1$ and the M.C. is "Stationary" i.e. the joint distribution of $\{X_{t_1+h}, X_{t_2+h}, \ldots, X_{t_n+h}\}$ for positive integers $t_1 < t_2 \ldots < t_n$ is independent of h, if h is a positive integer.

Exercise 2.40 First Solution Check that

$$x_k = e^{-\lambda/q}\left(\frac{\lambda}{q}\right)^k (k!)^{-1},\ k = 0, \ldots$$

is a solution of $x_k = \sum_j x_k p_{jk}$ with

$$p_{jk} = e^{-\lambda} \sum_{n=0}^{k} \binom{j}{n} p^n q^{j-n} \lambda^{k-n}/(k-n)!$$

Now $p_{00} = e^{-\lambda} > 0$, $p_{0i} = e^{-\lambda}\lambda > 0$ if $0 < p, q < 1$ and $0 < \lambda < \infty$.

Similarly every other $p_{jk} > 0$ for all $j, k \in S$.

Hence, the M.C. is irreducible.

Since $p_{jk} > 0, p_{jk}^{(2)} > 0, \ldots, p_{jk}^{(n)} > 0$ for all $n \geq 1$. So the M.C. is aperiodic.

Since $0 < \sum_{k=0}^{\infty} |x_k| < \infty$, the M.C. is positive recurrent. Moreover $x_k > 0$ and $\sum_{k=0}^{\infty} x_k = 1$. Hence it is the stationary distribution of the M.C.

Second Solution Again irreducibility and aperiodicity of the chain is obvious.

Let $g(z) = \sum_{k=0}^{\infty} x_k z^k$ be the generating function of $\{x_k\}$. We try to find a solution of $x_k = \sum_{j=0} x_j p_{jk}$. Then

$$g(z) = \sum_{k=0}^{\infty} \left\{ \sum_{j=0}^{\infty} x_j e^{-\lambda} \sum_{n=0}^{k} \binom{j}{n} p^n q^{j-n} \lambda^{k-n} / (k-n)! \right\} z^k, \; p + q = 1$$

$$= e^{-\lambda} \sum_{n=0}^{\infty} \sum_{j=0}^{\infty} x_j \binom{j}{n} p^n q^{j-n} z^n \sum_{k=n}^{\infty} (z\lambda)^{k-n} / (k-n)! \text{ (by Fubini)}$$

$$= e^{-(1-z)\lambda} \sum_{j=0}^{\infty} x_j \sum_{n=0}^{\infty} \binom{j}{n} (zp)^n q^{j-n}$$

$$\text{(by Fubini and } \sum_{k=n}^{\infty} (z\lambda)^{k-n} / (k-n)! = e^{-z\lambda} \text{)}$$

$$= e^{\lambda(z-1)} \sum_{j=0}^{\infty} x_j (zp+q)^j = e^{\lambda(z-1)} \sum_{j=0}^{\infty} x_j \, [1 + p(z-1)]^j .$$

So $g(z) = e^{-\lambda(1-z)} g[1 + p\,(z-1)]$. By induction

$$g(z) = e^{-\lambda(1-z)} e^{-\lambda(1-z)p}\, g[1 + p^2(z-1)]$$

$$= e^{-\lambda(1-z)(1+p+p^2+\ldots+p^n)} g[1 + p^n (z-1)].$$

Now $1 + p + \ldots + p^n \to \frac{1}{q}$ and $1 + p^n(z-1) \to 1$ as $n \to \infty$ for $|z| < 1$.

Therefore

$$g(z) = \exp\,[-\lambda(1-z)/q]\; g(1),$$

where $0 < |g(1)| < \infty$ since otherwise $|g(z)| = \infty$ or 0.

We choose $g(1) = 1$. Then

$$g(z) = e^{-\lambda/q\{1-z\}} = e^{-\lambda/q} e^{(\lambda/q)z} = \sum_{j=0}^{\infty} e^{-\lambda/q} (\lambda/q)^j / j! z^j .$$

Therefore, $$x_j = e^{-\lambda/q} (\lambda/q)^j / j!$$

is a solution of the given system which is accidently also a stationary distribution. Hence the M.C. is positive recurrent.

Exercise 2.41

$P[Y_n = j] = q_{n-j} \sum_{m=0}^{j} P[S_m = j]$, where $q_n = P(W_1 > n)$ and $S_m = W_1 + \ldots + W_m$. Hence

$$\sum_{n=0}^{\infty} t^n \sum_{j=0}^{n} x^j P[Y_n = j] = \sum_{j=0}^{\infty} x^j P_{00}^{(j)} \sum_{n=j}^{\infty} q_{n-j} t^{n-j} t^j \quad (1)$$

$$= P_{00}(xt) \sum_{m=0}^{\infty} q_m t^m$$

where

$$p_{00}^{(j)} = \sum_{m=0}^{j} P[S_m = j] \text{ and } P_{00}(xt) = \sum_{j=0}^{\infty} p_{00}^{(j)} (xt)^j$$

is the generating function of $\{p_{00}^{(j)}\}$.

But $$P_{00}(xt) = [1 - G(xt)]^{-1} \tag{2}$$

and $$\sum_{m=0}^{\infty} q_m t^m = \sum_{m=0}^{\infty} \left[1 - \sum_{n=0}^{\infty} f_{00}^{(n)}\right] t^m = \frac{1 - G(t)}{1 - t} \tag{3}$$

In fact $$P[W_1 > j] = P[W_1 > j - 1] - P[W_1 > j] = q_{j-1} - q_j.$$

Hence $$G(t) = \sum_{m=1}^{\infty} t^m (q_{m-1} - q_m) = \sum_{m=1}^{\infty} t^m q_{m-1} - \sum_{m=1}^{\infty} t^m q_m$$

$$= t \sum_{m=0}^{\infty} t^m q_m - \sum_{m=1}^{\infty} t^m q_m + 1.$$

Thus, $$1 - G(t) = (1 - t)\left(\sum_{n=0}^{\infty} t^m q_m\right) \Rightarrow \sum_{m=0}^{\infty} t^m q_m = \frac{1 - G(t)}{1 - t}.$$

Since W_n's are i.i.d. r.v.s, $G_j^m(t) = \sum_{k=0}^{\infty} t^k P[S_m = k]$.

$$P_{00}(xt) = \sum_{j=0}^{\infty} p_{00}^{(j)} (xt)^j = \sum_{j=0}^{\infty} \sum_{m=0}^{j} P[S_m = j](xt)^j \text{ (by Fubini)}$$

$$= \sum_{m=0}^{\infty} \sum_{j=0}^{\infty} P[S_m = j](xt)^j = \sum_{m=0}^{\infty} \sum_{j=1}^{\infty} (xt)^j P[S_m = j]$$

$$= \sum_{m=0}^{\infty} G^m(xt) = \frac{1}{1 - G(xt)} \quad (\because G(t) < 1, |t| < 1).$$

Therefore, $$\sum_{n=0}^{\infty} t^n \sum_{j=0}^{n} x^j P[Y_n = j] = \frac{1 - G(t)}{(1 - t)(1 - G(xt))}.$$

Exercise 2.42 Let α_k be the probability of absorption in state 0, starting at k. Then $\alpha_0 = 1$, $\alpha_N = 0$ (since N is also absorbing)

$$\alpha_k = \mu_k \alpha_{k-1} + (1 - \lambda_k - \mu_k) \alpha_k + \lambda_k \alpha_{k+1}, \ 0 < k < N \tag{1}$$

$$\Rightarrow \quad \mu_k/\lambda_k[\alpha_{k-1} - \alpha_k] = [\alpha_k - \alpha_{k+1}], \ 0 < k < N$$

So $$\alpha_1 - \alpha_2 = \rho_1(1 - \alpha_1), \ \alpha_2 - \alpha_3 = \rho_2(1 - \alpha_1), \ldots$$

$$\alpha_{N-1} - 0 = \rho_{N-1}(1 - \alpha_1) \tag{2}$$

where $\rho_k = \dfrac{\mu_1 \ldots \mu_k}{\lambda_1 \ldots \lambda_k}$, $k \geq 1$, $\rho_0 = 1$.

Summing over (2) gives $\alpha_1 = (\rho_1 + \ldots + \rho_{N-1})(1 - \alpha_1)$

or $$\alpha_1 = \left(\sum_{i=1}^{N-1} \rho_i\right)\left(\sum_{i=1}^{N-1} \rho_i\right)^{-1} = \frac{\sum_{i=0}^{N-1} \rho_i - 1}{\sum_{i=0}^{N-1} \rho_i}$$

From (2) $$\alpha_i - 1 = \left(\frac{\sum_{j=1}^{N-1} \rho_j}{\sum_{j=0}^{N-1} \rho_j} - 1\right) \sum_{j=0}^{i-1} \rho_j$$

$\Rightarrow$ $$\alpha_i = 1 - \frac{\sum_{0}^{i=1} \rho_i}{\sum_{j=0}^{N-1} \rho_j} \quad (\because \rho_0 = 1)$$

Putting $$i = k,\ \alpha_k = \sum_{j=k}^{N-1} \rho_j \Big/ \sum_{j=0}^{N-1} \rho_j$$

By summetry, if β_k is the probability of absorption at N, starting at k, then $\beta_0 = 1$, $\beta_N = 1$, and the same reasoning works. From the explicit formulas one shows that P(absorption at 0)+P (absorption at N) = 1 for every k.

Hence, $$\beta_k = \left(\sum_{0}^{k-1} \rho_j\right) \Big/ \left(\sum_{j=0}^{N-1} \rho_j\right).$$

Exercise 2.43 First Solution Define $Z_n = 1$ if $X_n = i$ and 0 otherwise. Then $E(Z_n) = P(Z_n = 1) = P(X_n = i \mid X_0 = i) = p_{ii}^{(n)}$. Since i is positive recurrent $f_{ii} = 1$. So

$$E\frac{[N_n(i)]}{n} = \frac{1}{n} E\left[\sum_{m=1}^{n} Z_m\right] = \frac{1}{n}\sum_{m=1}^{n} p_{ii}^{(m)} \to \frac{1}{\mu_{ii}} = \frac{1}{\mu_i} = \frac{1}{\sum_{1}^{\infty} n f_{ii}^{(n)}}$$

Second Solution Define a renewal process as follows: Let $\{W_n, n \geq 1\}$ be the inter recurrence times to state i. By Markov property $\{W_n, n \geq 1\}$ are i.i.d. non-negative r.v.s with distribution $P[W_n = k] = f_{ii}^{(k)}$. Now since i is positive recurrent,

$$E(W_n) = \mu_i = \sum_{n=1}^{\infty} n f_{ii}^{(n)} < \infty.$$

Let $N_n^{(i)}$ be the renewal counting function for state i, that is i.e. $N_n(i) = \max\,[m : W_1 + \ldots + W_m \leq n]$. Then by discrete renewal Theorem

$$\lim_{n\to\infty} \frac{E(N_n(i))}{n} = \frac{1}{\mu_i}.$$

Exercise 2.44 First Solution Assume that the M.C. is transient and put $p_{ij}^* = \sum_{n=0}^{\infty} p_{ij}^{(n)}$. Since j is transient, $p_{ij}^* < \infty$.

By Problem 2.38

1. $f_{00} > 0$ since the M.C. is irreducible.
2. $f^*_{k0} \le p^*_{k0}$ by 2.38 (a) for $k \ne 0$.
3. $p^*_{k0} = f_{k0}\, p^*_{00}$ by 2.38 (e) and 2.38 (d), and $p^*_{00} = \dfrac{1}{1 - f_{00}} > 0$.

Let $k \ne 0$. Then $0 < \alpha \le f_{k0} \le p^*_{k0}$ (by 2)

$$= f_{k0}\, p^*_{00} \text{ (by 3) } = f_{k0} f_{00}\, p^*_{00} = \ldots = f_{k0} [f_{00}]^n\, p^*_{00} \text{ for all } n \ge 1.$$

Since $p^*_{00} = \dfrac{1}{1 - f_{00}}$ (by 3) > 0 and $f_{00} < 1$ by assumption, we can choose n so large that $[f_{00}]^n < \dfrac{\alpha}{2 f_{k0}\, p^*_{00}}$.

Hence $0 < \alpha \le f_{k0} [f_{00}]^n\, p^*_{00} < \alpha/2$ which is a contradiction. Therefore, the M.C. is recurrent.

Second Solution

Let $Q_{ij} = P$ [visit j i.o. $| X_0 = i$] $= P\left\{ \bigcap_{m=1}^{\infty} \bigcap_{n=m}^{\infty} [X_n = j \mid X_0 = i] \right\}$

(by Monotone Convergence Theorem)

$$= \lim_{m \to \infty} P\left\{ \bigcup_{n=m}^{\infty} [X_n = j \mid X_0 = i] \right\}$$

$$= \lim_{m \to \infty} \sum_k P\{X_n = j, X_m = k \text{ for some } n \ge m \mid X_0 = i\} \text{ (by Markov property)}$$

$$= \lim_{m \to \infty} \sum_k P\{X_n = j \text{ for some } n \ge m \mid X_m = k\}\, P\{X_m = k \mid X_0 = i\}$$

$$= \lim_{m \to \infty} \sum_k P\{\text{visit } j \text{ in sometime} \mid X_m = k\}\, p_{ik}^{(m)} = \lim_{m \to \infty} \sum_k f_{kj}\, p_{ik}^{(m)}$$

Hence
$$Q_{ij} = \lim_{m \to \infty} \sum_k p_{ik}^{(m)} f_{kj} \tag{1}$$

It is easy to see that

$$Q_{ij} = f_{ij}\, Q_{ij} \tag{2}$$

Given $0 < \alpha \le f_{i0}$ for all $i \ne 0 \Rightarrow$ from (1) with $k \ne 0, j = 0$,

Hence $\quad Q_{i0} \ge \lim_{m \to \infty} \sum_{k \ne 0} p_{ik}^{(m)} f_{k0} \ge \alpha \lim_{m \to \infty} \sum_{k \ne 0} p_{ik}^{(m)} = \alpha > 0.$

$$Q_{00} = \frac{Q_{i0}}{f_{i0}} > 0 \text{ since } Q_{i0} > 0 \text{ and } f_{i0} > 0. \tag{3}$$

Since Q_{00} is either 0 or 1, by (3) $Q_{00} = 1 \Rightarrow 0$ is recurrent and the chain being irreducible, it is recurrent.

Exercise 2.45 (a) Assume that j is transient. Let S^* be the Communicating class of j. If there is no k such that $j \not\rightarrow k$ and $k + j$ then all states of S are in S^* so that S is positive recurrent (since the chain is finite). Since j is transient, there will be a k such that $j \not\rightarrow k$ and $k + j$.

Conversely if $j \to k$ and $k \not\to j$, then $p_{jk}^{(n)} > 0$ for some n and $p_{jk}^{(m)} = 0$ for all m.

Hence $1 - f_{jj} = P(X_m \neq j \text{ for all } m > 0 \mid X_0 = j) \geq p_{jk}^{(n)} > 0$ for some n and some $k \neq j$. Hence $f_{jj} < 1$, i.e. j is transient.

Exercise 2.46

Solution (a) If $j \in S$ is absorbing, then $p_{jj} = 1$ and $p_{ji} = 0$ for all $i \neq j$. Since S is denumerable, record S such that the absorbing states are numbered 1, 2, ... The remaining states are supposed to be transient. Clearly

$$P = \begin{array}{c} \quad E \quad E^c \\ \begin{pmatrix} I & O \\ E & Q \end{pmatrix} \begin{array}{c} E \\ E^c \end{array} \end{array}.$$

(b) If $i, j \in E^c$, let m_{ij} = number of visits to j starting from i.

Then let $Z_k = \begin{cases} 1 & \text{if } X_k = j \\ 0 & \text{if } X_k \neq j \end{cases}$

Clearly $m_{ij} = \sum_{k=0}^{\infty} Z_k$ and $N_{ij} = E(m_{ij}) = \sum_{k=0}^{\infty} E(Z_k) = \sum_{k=0}^{\infty} p_{ij}^{(k)} = \sum_{k=0}^{\infty} Q_{ij}^{(k)}$ since $i, j \in E^c$.

(c)
$$B_{ij} = f_{ij} = \sum_{n=1}^{\infty} f_{ij}^{(n)} = \sum_{n=0}^{\infty} f_{ij}^{(n+1)} = \sum_{n=0}^{\infty} \sum_{k \in E^c} p_{ik}^{(n)} p_{kj} \tag{1}$$

$$= \sum_{k \in E^c} \sum_{n=0}^{\infty} p_{ik}^{(n)} p_{kj} = \sum_{k \in E^c} N_{ik} p_{kj} = (NR)_{i,j} \text{ (by Fubini)}$$

If $j \in E$, then

$$p_{ij}^{(n)} = \sum_{m=1}^{n} f_{ij}^{(m)} p_{jj}^{(n-m)} = \sum_{m=1}^{n} f_{ij}^{(m)} \cdot 1 \text{ so that } f_{ij}^{(n+1)} = p_{ij}^{(n+1)} - p_{ij}^{(n)}, j \in E.$$

$$p_{ij}^{(n+1)} = \sum_k p_{ik}^{(n)} p_{kj} = \sum_{k \in E^c} p_{ik}^{(n)} p_{kj} + \sum_{k \in E} p_{ik}^{(n)} p_{kj}$$

$$= p_{ij}^{(n)} + \sum_{k \in E^c} p_{ik}^{(n)} p_{kj}. \text{ Hence } f_{ij}^{(n+1)} = \sum_{k \in E^c} p_{ik}^{(n)} p_{kj}.$$

(d)
$$P = \begin{pmatrix} I & O \\ R & Q \end{pmatrix} \Rightarrow P^n = \begin{pmatrix} I & O \\ R(I + Q + \ldots + Q^{n-1}) & Q^n \end{pmatrix}$$

$$\to \begin{pmatrix} I & O \\ RN & O \end{pmatrix} = \begin{pmatrix} I & O \\ B & O \end{pmatrix},$$

since $RN = B$ and $Q^n \to 0$ (since j is transient if $j \in E^c$).

Exercise 2.47 Since P is symmetric $p_{ij} = p_{ji}$. By Chapman-Kolmogorov equation

$$P_{ij}^{(2)} = \sum_{k\in S} p_{ik}p_{kj} = \sum_{k\in S} p_{ki}p_{jk} = p_{ji}^{(2)}.$$

Hence, $P^{(2)}$ is symmetric. Similarly, by induction on n $P^{(n)}$ is symmetric for all $n \geq 1$.

Now $\sum_{k\in S} p_{ij} = \sum_{j\in S} p_{ji} = 1$ so that P is douldy stochastic. If the M.C. is irreducible and infinite then it is not positive (by Ex. 2.33). Since $p_{ij} > 0 \Rightarrow p_{ji} = p_{ij} > 0$. So the M.C. consists of closed classes only. Every closed class can be treated as a separate M.C. So if every closed class is infinite then by previous argument the original M.C. can not be positive. But if the M.C. consists of finite number of states only then there will be no transient state, not all states are null recurrent and therefore there will be some positive states. But if the M.C. is irreducible and has finite number of states then it will be positive recurrent and same thing is true for every commucating class.

Exercise 2.48 Let the initial distribution be $\{\pi_i, i \in S\}$, then by Ex. 2.13. (1) $\pi_j = \sum_{i\in S} \pi_i p_{ij}^{(n)}$ for all $n \geq 1$, (2) $p_j^{(n)} = P(X_n = j) = \pi_j$, for all $j \in S$, $n \geq 1$.

Hence

$$R_{ji}^{(m)} = \frac{P(X_{n+m} = j, X_n = i)}{P(X_{n+m} = j)} = \frac{\pi_i}{\pi_j} p_{ij}^{(m)} \qquad \text{(by (2))}$$

independent of n ($\Rightarrow$ the new chain is stationary).

(a) $R_{ji}^{(m)} \geq 0$ is obvious.

$$\sum_{i\in S} R_{ji}^{(m)} = \frac{1}{\pi_j} \sum_{i\in S} \pi_i p_{ij}^{(m)} = \frac{\pi_j}{\pi_j} = 1 \qquad \text{(by (1))}$$

$$R_{ij}^{(m+n)} = \frac{\pi_i}{\pi_j} p_{ij}^{(m+n)} = \sum_{k\in S} \left(\frac{\pi_i}{\pi_k} p_{ik}^{(m)}\right)\left(\frac{\pi_k}{\pi_j} p_{kj}^{(n)}\right) \quad \text{(Chapman-Kolmogorov)}$$

$$= \sum_{k\in S} R_{ki}^{(m)} R_{jk}^{(n)} \quad \text{for all } m, n \geq 0.$$

Hence R_{ji} is a transition matrix and since the initial distribution is given, $R^{(m)}$ is the m-step transition matrix of a discrete time Markov Chain.

Note Let N be any large integer. Define $Y_n = X_{N-n}$ for $n = 0, 1, 2, \ldots, N$ then $\{Y_n, n = 0, 1, \ldots, N\}$ is a Markov chain with transition matrix R. It is called the *reverse* or *dual* chain of $\{X_n\}$.

(b) Since the M.C. is irreducible, so is its dual. Moreover

$$\lim_{m\to\infty} R_{ij}^{(m)} = \frac{\pi_i}{\pi_j} \lim_{m\to\infty} P_{ji}^{(m)} = \frac{\pi_j}{\pi_i} \pi_i = \pi_j > 0.$$

Thus the chain is positive and the stationary distribution is given by the initial distribution $\{\pi_i, i \in S\}$.

Exercise 2.49

$$p_n = p_{n-1}((1-c)p + c) + q_{n-1}((1-c)p), \text{ where } q_{n-1} = 1 - p_{n-1}$$
$$= cp_{n-1} + (1-c)p.$$

Therefore $p_n = Ac^{n-1}p_1 + \dfrac{(1-c)p}{1-c} = Ac^{n-1}p + p, n \geq 1$ (since $p_1 = p$).

From initial condition $p = Ap + p$ so that $A = 0$. Therefore

$$p_n = p = P(X_0 = 1) \text{ for all } n \geq 1.$$

Therefore $E(X_n) = p$, Var $(X_n) = p - p^2 = p(1-p)$.

$$E(X_{n-1}X_n) = 1.1\ ((1-c)p + c)p = ((1-c)p + c)p = p^2 - cp^2 + cp$$

and $\quad$ Cov $(X_n - 1, X_n) = (p^2 - cp^2 + cp) - p^2 = cp)(1-p)$.

Similarly $\quad$ Cov $(X_{n-2}, X_n) = c^2p\ (1-p)$

and in general

Cov $(X_{n-k}, X_n) = c^kp(1-p$ for $n \geq 1$. Hence $E(S_n) = np \to \lambda$.

$$\sum_{k=1}^{n} \text{Var}(X_n) = np(1-p),\ \sum_{j<k} \text{Cov}(X_j, X_k)$$
$$= p\,(1-p)\{(c + c^2 + \ldots + c^{n-1})$$
$$+ (c + c^2 + \ldots + c^{n-2}) + \ldots + c\}$$
$$= p(1-p)\frac{c}{1-c}\left[(n-1) - c\frac{1-c^{n-1}}{1-c}\right]$$

Hence, $\quad \text{Var}(S_n) = \sum_{k=1}^{n} \text{Var}(X_k) + 2\sum_{j<k} \text{Cov}(X_j, X_k)$

$$= np(1-p) + 2p(1-p)\left[\frac{c(n-1)}{1-c} - c^2(1-c^{n-1})/(1-c)^2\right]$$
$$\to \lambda + \frac{2\lambda c}{1-c} \quad (\because p < c < 1).$$

Exercise 2.50

First Solution If $P = P^2$ then $p_{ij}^{(n)} = p_{ij}$ for all $n \geq 1$. Hence $p_{ij} = \dfrac{1}{n}\sum_{m=0}^{n} p_{ij}^{(m)} \to \pi_j$

and $p_{ij} = \dfrac{1}{n}\sum_{m=0}^{n} p_{ij}^{(m)} \to \pi_j$ (since the chain is irreducible). So $p_{ij} = p_{jj} = \pi_j$.

Assume $p_{00} = 0$, then $p_{i0} = 0$ for all $i \in S$ and hence $\{0\}$ can not be reached, so the M.C. is not irreducible.

Hence $p_{00} > 0$ and $\{0\}$ is aperiodic. Since the chain is irreducible, all states are aperiodic.

Second Solution If $P^2 = P$, the $p_{ij}^{(n)} = p_{ij}$ for all $i \in S$ and hence $\lim_{n\to\infty} p_{ij}^{(n)} = p_{ij}$. Assume j as transient or null recurrent then $p_{ij} = 0$ for all i and j and so P is not a stochastic matrix. Hence j is positive and the chain being irreducible, the M.C. is positive recurrent. Hence $p_{ij}^{(n)} \to \pi_j$ and $p_{jj}^{(n)} > 0$. So the M.C. is aperiodic and $p_{ij} = p_{jj}$.

Chapter 3

Exercise 3.1

$$\pi_k(n) = p\pi_{k+1}(n-1) + q\pi_{k-1}(n-1),\ 1 < k < a-1$$

$$\pi_1(n) = p\pi_2(n-1) + q\pi_0(n-1)\delta_{n1} \tag{1}$$

$\pi_0(n) = \delta_{n0}$, $\pi_a(n) = 0$ where δ_{ni}, $i = 1, 0$ are Kronecker's deltas. Multiplying both sides of the above equations and summing over all values of n we obtain

$$G_k(s) = ps\ G_{k+1}(s) + qs\ G_{k-1}(s),\ 1 \le k \le a-1 \tag{2}$$

$$G_0(s) = 1,\ G_a(s) = 0. \tag{3}$$

Since (2) is a linear difference equation with one variable k, taking $G_k(s) = \lambda^k(s)$ as a trial solution we find from (2)

$$\lambda^k = ps\lambda^{k+1} + qs\lambda^{k-1}$$

or $ps\lambda^2 - \lambda + qs = 0$. The roots of the quadratic equation are

$$\lambda_1(s), \lambda_2(s) = \frac{1 \pm \phi(s)}{2ps}, \text{ where } \phi(s) = (1 - 4pqs^2)^{1/2} \tag{4}$$

Since $1 - 4pqs^2 > 0$ or $s < \dfrac{1}{2\sqrt{pq}}$, we assume here $0 < s < \dfrac{1}{2\sqrt{pq}}$ for two real roots $\lambda_1(s)$ and $\lambda_2(s)$.

The general solution is given by

$$G_k(s) = A[\lambda_1(s)]^k + B[\lambda_2(s)]^k$$

where A and B are functions of s and must be deduced from the boundary conditions (3). Hence we get

$$G_k(s) = (q/p)^k \frac{\lambda_1^{a-k}(s) - \lambda_2^{a-k}(s)}{\lambda_1^a(s) - \lambda_2^a(s)} \tag{5}$$

$$= (q/p)^k (2ps)^k \frac{[1+\phi(s)]^{a-k} - [1-\phi(s)]^{a-k}}{[1+\phi(s)]^a - [1-\phi(s)]^a}$$

(substituting $\lambda_1(s)$ and $\lambda_2(s)$ from (4))

$$= (q/p)^k (2ps)^k \frac{\sum_{r=0}^{a-k} \binom{a-k}{r} [\phi(s)]^r \{1-(-1)^r\}}{\sum_{r=0}^{a} \binom{a}{r} [\phi(s)]^r \{1-(-1)^r\}}$$

Since no contribution comes from even r, setting $r = 2j + 1$, we get

$$G_k(s) = (q/p)^k (2ps)^k \frac{\sum_{j=0}^{(a-k-1)/2} \binom{a-k}{2j+1} [\phi(s)]^{2j}}{\sum_{j=0}^{(a-1)/2} \binom{a}{2j+1} [\phi(s)]^{2j}} \quad \text{if } a-k \text{ and } a \text{ are odd.}$$

$$= \left(\frac{q}{p}\right)^k (2ps)^k \frac{\sum_{j=0}^{(a-k-2)/2} \binom{a-k}{2j+1} [\phi(s)]^{2j}}{\sum_{j=0}^{(a-2)/2} \binom{a}{2j+1} [\phi(s)]^{2j}} \quad \text{if } a-k \text{ and } a \text{ are even.}$$

Now $G_k(s) = \dfrac{U(s)}{V(s)}$ where $U(s)$ and $V(s)$ are polynomials in s with degree of $U(s)$ does not exceed that of $V(s)$ by more than one. So we assume the degree of $U(s)$ is less than the degree of $V(s)$ by m (say). Let the equation $V(s) = 0$ has m distinct roots $s_1, s_2, \ldots, s_m$. Write $G_k(s)$ in partial fraction as

$$G_k(s) = \rho_1/(s_1 - s) + \frac{\rho_2}{(s_2 - s)} + \ldots + \frac{\rho_m}{(s_m - s)}$$

with coefficients ρ_k are given by

$$\rho_k = \lim_{s \to s_k} (s_k - s) G_k(s) = \frac{U(s_k)}{V'(s_k)}, \quad k = 1, 2, \ldots m.$$

Hence the coefficients of s^n is power series expansion of $G_k(s)$ are given by

$$\pi_k(n) = \frac{\rho_1}{s_1^{n+1}} + \ldots + \frac{\rho_m}{s_m^{n+1}} \tag{7}$$

Making auxillary variable $\dfrac{1}{\cos\theta} = 2(pq)^{1/2} s$ we then have $\lambda_1(s), \lambda_2(s) = (q/p)^k e^{\pm i\theta}$, so that equation (5) yields

$$G_k(s) = (q/p)^{k/2} \frac{\sin(a-k)\theta}{\sin a\theta} \tag{8}$$

The roots of the denominator are $\theta = 0, \dfrac{\pi}{a}, \ldots, \dfrac{2\pi}{a}$. These roots are related to the roots of $V(s) = 0$ by the relation

$$s_j = \frac{1}{2(pq)^{1/2} \cos\left(\frac{j\pi}{a}\right)}, \quad j = 0, 1, \ldots, a.$$

For $j = 0$ and $j = a$, we get the corresponding values for θ as $\theta = 0$ and $\theta = \pi$ which are also roots of the numerator of (8). If a is even, there is no s_j corresponding to $j = a/2$. So s_j, $j = 1, 2, \ldots, a-1$ are real and distinct.

So $\rho_j = \lim_{s\to s_j} (p/q)^{k/2} \dfrac{\sin(a-k)\theta}{\sin(a\theta)}(s_j - s)$

$$= -(q/p)^{k/2} \frac{\sin\{(a-k)j\pi/a\}}{a\cos j\pi\left(\dfrac{d\theta}{ds}\right)_{s=s_j}} = (q/p)^{k/2} \frac{\sin(kj\pi/a)\sin(j\pi/a)}{2a(pq)^{1/2}\cos^2\left(\dfrac{j\pi}{a}\right)}$$

Now from (7) we get

$$\pi_k(n) = a^{-1}2^n p^{(n-k)/2} q^{(n+k)/2} \sum_{j=1}^{a-1} \int_0^1 \cos^{n-1}(j\pi/a)\sin(j\pi/a)\sin(kj\pi/a)$$

Letting $a \to \infty$, we get the probability $\pi_k(n)$ that a player with initial capital $k > 0$ will be ruined in exactly at the nth step against an infinitely rich opponent

$$\pi_k(n) = 2^n p^{(n-k)/2} q^{(n+k)/2} \int_0^1 \cos^{n-1}(\pi x)\sin(\pi x)\sin(k\pi x)\,dx$$

By an elementary way this is also

$$k/n \binom{n}{(n-k)/2} p^{(n-1-k)/2}\, q^{(n+k)/2}.$$

Exercise 3.2 If the particle starts at a, absorption occurs at time 0 with probability 1, and if it starts at any other point, absorption cannot occur at time 0. So

$$f_{ja}^{(n)} = \begin{cases} 1 & \text{if } j = a \\ 0 & \text{if } j \neq a \end{cases} \tag{1}$$

Since the random walk takes positive jump or negative or no jump at all at the first step

$$f_{ja}^{(n)} = pf_{j+1,a}^{(n)} + (1-p-q)f_{ja}^{(n-1)} + qf_{j-1,a}^{(n)} \tag{2}$$

$$(j = -b+1, \ldots, a-1,\ n = 0, 1, 2, \ldots)$$

The boundary conditions are given by

$$f_{aa}^{(n)} = 1 \text{ and } f_{-b,a}^{(n)} = 0, \quad n = 1, 2, \ldots \tag{3}$$

(2) is a difference equation of the first order in n and second order in j with initial conditions and boundary conditions (3).

Multiplying equation (2) by s^n and summing over $n = 1, 2, \ldots$ we obtain

$$F_{ja}(s) = s[\,pF_{j+1,a}(s) + (1-p-q)F_{j,a}(s) + qF_{j-1,a}(s)] \tag{4}$$

with boundary conditions

$$F_{aa}(s) = 1 \text{ and } F_{-b,a}(s) = 0 \tag{5}$$

(4) is a difference equation in one variable j.

Substituting a trial solution $F_{ja}(s) = \lambda^j(s)$ we get from (4)

$$\lambda^j = s[p\lambda^{j+1} + (1-p-q)\lambda^j + q\lambda^{j-1}]$$

or
$$ps\lambda^2 - \lambda[1 - s(1-p-q)] + qs = 0 \tag{6}$$

The roots of the quadratic equation (6) in λ are given by

$$\lambda_1(s), \lambda_2(s) = \frac{1 - s(1-p-q) \pm [1 - s(1-p-q)^2 - 4pqs^2]^{1/2}}{2ps} \tag{7}$$

Assuming s is real and positive and the function under the radical sign is also positive leads to

$$[1 - s(1-p-q)]^2 > 4pqs^2 \quad \text{or} \quad 0 < s < \frac{1}{1 - (\sqrt{p} - \sqrt{q})^2}.$$

The general solution of (4) is given by

$$F_{ja}(s) = A[\lambda_1(s)]^j + B[\lambda_2(s)]^j$$

where A and B are functions of s and are to be determined from the boundary conditions (5).

This gives
$$F_{ja}(s) = \frac{[\lambda_1(s)]^{j+b} - [\lambda_2(s)]^{j+b}}{[\lambda_1(s)]^{a+b} - [\lambda_2(s)]^{a+b}}. \tag{8}$$

putting $j = 0$ in (8)

$$F_{0,s}(s) = \frac{[\lambda_1(s)]^{b} - [\lambda_2(s)]^{b}}{[\lambda_1(s)]^{a+b} - [\lambda_2(s)]^{a+b}}.$$

Exercise 3.3 Since the jumps are independent, the particle's position at time n depends only on its position at time $(n-1)$ and the nth jump. During the interval $(n-1, n)$, the state j can be reached by one of three mutually exclusive and exhaustive ways, viz. a jump of +1, 0 or –1 from states $j-1, j, j+1$, respectively.

$$p_{i,j}^{(n)} = pp_{i,j-1}^{(n-1)} + (1-p-q)\,p_{i,j}^{(n-1)} + qp_{i,j+1}^{(n-1)} \quad (0 < j < a)$$

At the barrier states we have

$$p_{i,a}^{(n)} = pp_{i,a-1}^{(n-1)} + (1-q)\,p_{i,a}^{(n-1)}$$

$$p_{i,0}^{(n)} = (1-p)\,p_{i,0}^{(n)} + qp_{i,1}^{(n-1)}$$

Since the chain is finite irreducible the stationary distribution exists and are the limiting values of occupation probabilities (see theorem 2.9). Then π_k satisfies the equations

$$\pi_j = p\pi_{j-1} + (1-p-q)\,\pi_j + q\pi_{j-1} \quad (k = 1, 2, \ldots,)$$

$$\pi_a = p\pi_{a-1} + (1-q)\,\pi_a$$

$$\pi_0 = (1-p)\,\pi_0 + q\pi_1$$

Solving these equations we get

$\pi_j = (p/q)^j\ \pi_0 (j = 0, 1, \ldots, a)$ and π_0 can be calculated from $\sum_{j=0}^{a} \pi_j = 1$.

Hence, we obtain the stationary distribution

$$\pi_j = \frac{1 - p/q}{1 - (p/q)^{a+1}} (p/q)^j \quad (j = 0, 1, \ldots, a).$$

Writing $\pi_j(x) = \dfrac{1-x}{1-x^{a+1}} x^j$ where $x = p/q$, by L'Hospital's rule

$$\lim_{x \to 1} \pi_j = \lim_{x \to 1} \pi_j(x) = \frac{1}{a+1} \quad \text{for all } j = 0, 1, \ldots, a.$$

Exercise 3.4 For $\theta > 0$ we have

$$\phi(\theta) > P(e^x > 1 + \delta)(1 + \delta)^\theta$$

Hence, since $P(e^X > 1 + \delta) > 0$, $\lim_{\theta \to \infty} \phi(\theta) = \infty$. (1)

Similarly for $\theta < 0$, $\phi(\theta) > P(e^x > 1 - \delta)(1 - \delta)^\theta$

Hence, since $P(e^X < 1 - \delta) > 0$, $\lim_{\theta \to -\infty} \phi(\theta) = \infty$. (2)

Since $\phi''(\theta) = E(X^2 e^{\theta X})$ (since m.g.f. exists for real θ differentiation under expectation is valid), it follows that

$$\phi''(\theta) > 0 \text{ for all real } \theta. \quad (3)$$

Now (1), (2) and (3) imply that $\phi(\theta)$ has a unique minimum at $\theta = \theta''$ say. Since $\phi'(0) = E(X) \neq 0$, $\theta^* \neq 0$ and $\phi(\theta^*) < \phi(0) = 1$. Also $\phi(\theta)$ is strictly decreasing over the interval $(-\infty, \theta^*)$ and is strictly increasing over the interval (θ^*, ∞). Since $\phi(0) = 1$ and $\phi(\theta^*) < 1$, there exists an unique real $\theta_0 \neq 0$ such that $\phi(\theta_0) = 1$.

Exercise 3.5 Under the conditions of Exercise 3.4 there exists real $\theta_0 \neq 0$ that $\phi(\theta_0) = 1$. Substituting θ_0 for θ in Wald's Identity, we obtain $E(e^{\theta_0 S_N}) = 1$. Therefore

$$P(S_N \le -b) E(e^{\theta_0 S_N} \mid S_N \le -b) + P(S_N \ge a) E(e^{\theta_0 S_N} \mid S_N \ge a)$$

$$= E(e^{\theta_0 S_N}) = 1.$$

Since $P(S_n \le a) = 1 - P(S_n \le -b)$, solving we get $P(S_N \le -b) = \dfrac{E_1 - 1}{E_1 - E_0}$.

Exercise 3.6 Let $E^* = E(e^{\theta S_N} \mid N = n)$. We shall first find an upper bound of E^*. If

$$\theta > 0, E^* \le E(e^{\theta S_N} \mid S_N \ge a, N = n) \quad (1)$$

and $E(e^{\theta S_N} \mid S_N \ge a, N = n, e^{\theta S_{N-1}} = \rho e^{a\theta}) = e^{a\theta} \rho E\left(e^{\theta X_n} \mid e^{X_n \theta} \ge \dfrac{1}{\rho}\right)$ (2)

$$\leq l(\theta) \text{ where } l(\theta) = \sup_{\rho \in (e^{-(a+b)|\theta|}, 1)} \rho E(e^{X\theta} \mid e^{X\theta} \geq 1/\rho)$$

Now since $\phi(\theta)$ exists, $l(\theta) > \infty$. From (1) and (2) we get

$$E^* \leq e^{a\theta}\, l(\theta) \quad \text{if } \theta > 0 \tag{3}$$

and hence $$E(e^{\theta S_N}) \leq e^{a\theta} l(\theta) \quad \text{if } \theta > 0$$

Similarly we get $$E^* \leq e^{-b\theta} l(\theta) \quad \text{if } \theta < 0 \tag{4}$$

and hence $$E(e^{\theta S_N}) \leq e^{-b\theta} l(\theta) \quad \text{if } \theta < 0$$

Now to prove that Wald's identity can be differentiated under the expectation sign any number of times with respect to θ at any value θ in the domain $|\phi(\theta)| \geq 1$, it is sufficient to show the following two results hold.

Result 1 All derivatives of $(e^{\theta S_N})[\phi(\theta)]^{-N}$ with respect to θ exists in the domain $|\phi(\theta)| \geq 1$.

Result 2 For any positive integer r and any finite interval I in which $|\phi(\theta) \geq 1$, there exists a function $H(S_N, N)$ such that

$$H(S_N, N) \geq \left| \frac{d^r}{d\theta^r} [e^{\theta S_N} \{\phi(\theta)\}^{-N}] \right|$$

for all values of θ in I and $E(H(S_N, N)) < \infty$.

Now existence of $\phi(\theta)$ for all θ implies the existence of all derivative of $\phi(\theta)$ (Due of analyticity of $\phi(\theta)$). This implies result 1.

Since $\frac{d^r}{d\theta^r}\{e^{\theta S_N}[\phi(\theta)]^{-N}\}$ is equal to the sum of a finite number of terms of the type $S_N^{r_1} N^{r_2} e^{\theta S_N}[\phi(\theta)]^{-N}$, Result 2 is proved if we can show that any given integers r_1 and r_2 there exists a function $H_{r_1 r_2}(S_N, N) \geq |S_N^{r_1} N^{r_2} e^{\theta S_N}[\phi(\theta)]^{-N}|$ for all $\theta \in I$ and $E\{H_{r_1 r_2}(S_N, N)\} < \infty$.

Since $$|\phi(\theta)| \geq 1 \text{ in } I, |S_N^{r_1} N^{r_2} e^{\theta S_N}[\phi(\theta)]^{-N}| \leq S_N^{r_1} N^{r_2} e^{|S_N|\theta_0} \tag{5}$$

where $\theta_0 \geq |\theta|$ in I. Let $\theta_1 > \theta_0$. Then for $c \geq \text{Max}(a, b)$ we have

$$|S_N^{r_1}| e^{|S_N|\theta_0} < Ce^{|S_N|\theta_1}. \tag{6}$$

Hence it follows from (5) and (6) that

$$|S_N^{r_1} N^{r_2} e^{\theta S_N}[\phi(\theta)]^{-N}| \leq CN^{r_2} e^{\theta_1 |S_N|} \leq CN^{r_2}(e^{S_N\theta_1} + e^{-S_N\theta_1})$$

for all $\theta \in I$.

Put $$H_{r_1 r_2}(S_N, N) = CN^{r_2}(e^{S_N\theta_1} + e^{-S_N\theta_1}).$$

Then $$E(H_{r_1 r_2}(S_N, N)) = C \sum_{n=1}^{\infty} P(N = n) N^{r_2} [E(e^{S_N\theta_1} \mid N = n)$$

$$+ E(e^{-S_N\theta_1} \mid N = n)] \tag{7}$$

Hence by (7), (3) and (4) we obtain

$$E[H_{r_1 r_2}(S_N, N)] \le C[e^{\theta_1 a} l(\theta_1) + e^{-b\theta_1} 1(-\theta_1)]\, E(N^{r_2})$$

Since $E(N^r) < \infty$ for all integers $r > 0$ (by Theorem 3.1) the Result 2 holds. Differentiating the Wald's identity at $\theta = 0$, the derivative of $e^{S_N \theta}[\phi(\theta)]^{-N}$ at $\theta = 0$ is given by

$$S_N - \frac{\phi'(0)\, N}{[\phi(0)]} = S_N - E(X)\, N \tag{8}$$

Taking expectation we get $E(S_N) = E(N)E(X)$ and hence

$$E(N) = \frac{ES_N}{E(X)} \quad \text{if } E(X) \ne 0.$$

Differentiating Wald's identity twice under the expectation sign at $\theta = 0$, we get

$$\left\{[S_N - N\phi'(\theta)/\phi(\theta)]^2 - N\frac{\phi''(\theta)\phi(\theta) - [\phi'(\theta)]^2}{[\phi(\theta)]^2}\right\} e^{\theta S_N}[\phi(\theta)]^{-N}. \tag{9}$$

Since $\phi(0) = 1$, $\phi'(0) = E(X) = 0$ and $\phi''(0) = EX^2$, putting $\theta = 0$, (9) becomes $S_N^2 - N\phi''(0) = S_N^2 - NEX^2$.

Taking expectation $E[S_N^2 - NE(X^2)] = 0$.

Therefore $EN = E(S_N^2)/E(X^2)$ if $EX = 0$.

Chapter 4

Exercise 4.1

Let C_1 be the first time a customer makes a purchase and $0 < S_1 < S_2 < \ldots$ be the arrival points (renewal points) of the customers. Then, by the total probability rule

$$P(C_1 \le x) = \sum_{k=1}^{\infty} P(C_1 \le x \mid C_1 = S_k)\, P(C_1 = S_k).$$

By the assumptions, $P(C_1 = S_k) = (1 - p)^{k-1} p$ and $P(C_1 \le x \mid C_1 = S_k) = F_{(x)}^{(k)}$. Thus, $P(C_1 \le x) = p \sum_{k=1}^{\infty} (1-p)^{k-1} F_{(x)}^{(k)} = F_1(x)$, say. Similarly, if C_m is the mth time that a purchase is made.

$$P(C_1 \le x, C_2 - C_1 \le y) = \sum_{k,m} P(C_1 \le x,\ C_2 - C_1 \le y \mid G_{k,m})\, P(G_{k,m}),$$

where $G_{k,m} = \{C_1 = S_k, C_2 = S_m\}$, $1 \le k \le m$. We can easily prove that the points C_j from a renewal process with interarrival distribution $F_1(x)$. If the time scale is changed by a factor $1/p$, the interarrival distribution becomes $F_1(x/p)$. The actual formula for $F_1(x)$ reduces to $F(x)$ if $F(x)$ is exponential.

Exercise 4.2 From exercise 4.6

$$W_t(x) = P(Y(t) \le x) = \int_t^{t+x} [1 - F(t + x - y)]\, dH(y).$$

Taking Laplace transform on both sides we get

$$e^{-st}\int_0^\infty e^{-sx}\,dW_t(x) = \int_0^\infty d\left[\int_t^{t+x} e^{-s(x+t-y)}\{1 - F(t+x-y)\}e^{-sy}\,dH(y)\right].$$

Differentiating under the integral sign and making change of variable along with substitution $z = x + t - y$ yields

$$e^{-st}L_w(s) = [1 - L_F(s)]\int_t^\infty e^{-sy}\,dH(y).$$

Exercise 4.3

(a) $$E(X - x \mid X > x) = \frac{\int_x^\infty \{1 - F(t)\}dt}{1 - F(x)}.$$

(b) $$E(Y_t + Z_t) = E(Y_t) + E(Z_t) = 1/\lambda + \int_0^t P(Z_t > x)\,dx$$

$$= 1/\lambda + \int_0^t e^{-\lambda x}\,dx = 1/\lambda + \frac{1}{\lambda}(1 - e^{-\lambda t}).$$

Therefore, $\lim_{t\to\infty} E(Y_t + Z_t) = 2/\lambda$.

Exercise 4.4

1. (a) By Fatou's Lemma $\lim_{t\to\infty} H(t) = \lim_{t\to\infty}\sum_{n=1}^{\infty} F^{(n)}(t) \ge \sum_{n=1}^{\infty} F^{(n)}(\infty) = \infty$.

(b) $0 \le H(t)$ is obvious, and since $F(t) \le 1$,

$$\frac{F(t)}{1 - F(t)} \le \frac{1}{1 - F(t)}.$$

To prove $H(t) \le \dfrac{F(t)}{1 - F(t)}$ if $F(t) = 1$ for $t \ge 0$, then the result is obvious. So let $F(t) < 1$ for all $t > 0$.

Proof 1. $$H(t) = F(t) + \int_0^t F(t - x)\,dH(x) \le F(t) + F(t)H(t)$$

$$(\text{since } F(t)\uparrow) \Rightarrow H(t) \le \frac{F(t)}{1 - F(t)}.$$

Proof 2. Since $F^{(n)}(t) = \int_0^t F^{(n-1)}(t - x)\,dF(x)$ and $F(t)\uparrow$, induction on n shows $F^{(n)}(t) \le [F(t)]^n$.

Therefore $H(t) \le \sum_{n=1}^{\infty} [F(t)]^n = \dfrac{F(t)}{1 - F(t)}$, since $F(t) < 1$ for $0 \le t < \infty$.

2(a) $$E[N^2(t)] = \sum_{n=1}^{\infty} n^2\{F^{(n)}(t) - F^{(n+1)}(t)\}$$

$$= \lim_{N\to\infty}\left\{\sum_{n=1}^{N}(2n-1)F^{(n)}(t) + N^2F^{(N)}(t)\right\} \geq 0$$

$$= \sum_{n=1}^{\infty}(2n-1)F^{(n)}(t) + \lim_{N\to\infty} N^2F^{(N)}(t)$$

Since $F^{(n)}(t) < F(t)F^{(n-1)}(t)$ (except if F is degenerate at 0), by applying the ratio test $\sum_{n=1}^{\infty} n^2F^{(n)}(t) < \infty$.

So $n^2F^{(n)}(t) \to 0$ as $n \to \infty$ for all finite $t \geq 0$

Therefore $$E[N^2(t)] = \sum_{n=1}^{\infty}(2n-1)F^{(n)}(t).$$

(b) $$H*H(t) = \int_0^t \sum_{n=1}^{\infty} F^{(n)}(t-u)\, d\sum_{n=1}^{\infty}\{F^{(n)}(t)\} \text{ (by Fubini's theorem)}$$

$$= \sum_{n=1}^{\infty}\sum_{m=1}^{\infty}\int_0^t F^{(n)}(t-u)\,dF^{(m)}(t) = \sum_{n=1}^{\infty}\sum_{m=1}^{\infty}F^{(n+m)}(t)$$

(by definition of convolution).

Puting $n + m = k$ and applying Fubini Theorem

$$H*H(t) = \sum_{n=1}^{\infty}\sum_{k=n+1}^{\infty}F^{(k)}(t) = \sum_{k=2}^{\infty}\sum_{n=1}^{k-1}F^{(k)}(t) = \sum_{k=2}^{\infty}(k-1)F^{(k)}(t)$$

So $$H(t) + 2H*H(t) = \sum_{k=1}^{\infty}F^{(k)}t + \sum_{k=1}^{\infty}2(k-1)F^{(k)}(t)$$

$$= \sum_{k=1}^{\infty}(2k-1)F^{(k)}(t) = H_2(t)$$

i.e. $$H_2(t) = H(t) + 2\int_0^t H(t-u)\,dH(u).$$

3. Note $F^*(t) = 0$ for $t \leq 0$ so that $X^* \geq 0$.

Since $$\frac{d}{dt}[F^*(t)] = \frac{1}{\mu}[1 - F(t)] \geq 0$$

by the Fundamental Theorem of Calculus, $F^*(t)$ is (absolutely) continuous and nondecreasing.

We have only to prove that $F^*(\infty) = 1$.

$$F^*(\infty) = \lim_{t\to\infty}F^*(t) = \frac{1}{\mu}\left\{\lim_{t\to\infty}t[1-F(t)] + \int_0^{\infty}x\,dF(x)\right\} \quad (1)$$

(integration by parts)

But $t[1 - F(t)] \to 0$, since

$$0 = \lim_{t\to\infty}\left[\mu - \int_0^t x\,dF(x)\right] = \lim_{t\to\infty}\int_t^\infty x\,dF(x)$$

$$\geq \lim_{t\to\infty} t\int_t^\infty dF(x) = \lim_{t\to\infty} t\,[1 - F(t)] \geq 0.$$

Hence (1) implies $F^*(\infty) = \int_0^\infty x\,dF(x)/\mu = 1.$

$F^*(t)$ is differentiable and hence continuous. So it is right continuous. Alternately

$$F^*(t + h) = \frac{1}{\mu}\int_0^{t+h} [1 - F(x)]\,dx = \frac{1}{\mu}C.\int_0^{t+h} dx$$

(by mean value theorem of integral calculus) $= C\cdot\dfrac{t + h}{\mu}$.

Therefore $\quad \lim_{h\downarrow 0} F^*(t + h) = Ct/\mu = \dfrac{1}{\mu}\displaystyle\int_0^t [1 - F(x)]dx = F^*(t)]$

(b) Let $f^*(x) = \int_0^\infty e^{-st}\,dF^*(t)$

$$= \frac{1}{\mu}\int_0^\infty e^{-st}[1 - F(t)]dt = \frac{1}{\mu s} - \frac{1}{\mu}\int_0^\infty e^{-st}F(t)\,dt \tag{2}$$

Integrating by parts for $s \geq 0$.

$$f(s) = \int_0^\infty e^{-st}\,dF(t) = e^{-st}F(t)\,|_0^\infty + s\int_0^\infty e^{-st}F(t)\,dt \tag{3}$$

(2) and (3) $\Rightarrow f^*(s) = \dfrac{1}{\mu s} - \dfrac{1}{\mu s}f(s) = \dfrac{1}{\mu s}[1 - f(s)].$

(c) Since F^* is the d.f. of $X^* \geq 0$, $E(X^*) = -f^{*\prime}(0)$.
But $E(X^2) = f''(0)$ also since $X \geq 0$, finite or not.

By (3) $\qquad E(X^*) = -f^{*\prime}(0) = \lim_{s\to 0+} - \dfrac{-sf'(s) - 1 + f(s)}{\mu s^2}$

$$= \lim_{s\to 0+}\frac{s\,f''(s)}{2\,\mu s} \text{ (by L'Hospital's rule)}$$

$$= \frac{E(X^2)}{2\,\mu} = \frac{\mu^2}{2\,\mu}.$$

4(a) *First Proof* Let $h = \dfrac{f}{1 - f}$ be the Laplace-Stieltjes transform of $H(t) = \sum_{n=1}^{\infty} F^{(n)}(t), f$ and f^* be as above. Then Laplace-Stieltjes transform of

$$[(1 - F^*) * H] = (1 - f^*)h = h - \frac{1-f}{\mu s} f/(1-f) \text{ (by 3(b))}$$

$$= h - \frac{1}{\mu s} + \frac{1-f}{\mu s} = \text{Laplace-Stieltjes transform of}$$

$[H - t/\mu + F^*]$ (by 3 (b)).
By uniqueness theorem of Laplace-Stieltjes transform

$$(1 - F^*) * H = H(t) - t/\mu + F^*(t)$$

i.e. $\displaystyle\int_0^t (1 - F^*(t-u))\,dH(u) = \int_0^t Q(t-u)\,dH(u) = H(t) - t/\mu + F^*(t).$

Second Proof

$$F^*(t) = \frac{1}{\mu}\int_0^t (1 - F(x))\,dx = \frac{1}{\mu}\int_t^0 (1 - F(t-u))\,d(t-u) \text{ (put } x = t-u)$$

$$= -\frac{1}{\mu}\int_t^0 (1 - F(t-u))\,du = \frac{1}{\mu}\int_0^t (1 - F(t-u))\,du = \frac{1}{\mu}(1-F)^*t$$

$$= t/\mu - \frac{1}{\mu} F^*t. \tag{4}$$

Let $\quad G(t) = H(t) - t/\mu$. Then $\quad H(t) = F(t) + F^*H(t)$

becomes by (4) $(1 - F)^*G(t) = F(t) - F^*(t)$. Convoluting with t/μ

$$F^* * G = \frac{t}{\mu} * (1-F)^*G = \frac{t}{\mu} * F - \frac{t}{\mu} * F^*$$

So $F^* * H = F^* * G + F^* * t/\mu = \dfrac{t}{\mu} * F$

or $\quad (1 - F^*) * H = H - F^* * H = H - t/\mu * F = H - t/\mu + F^*$ (by (4))

(b) $\quad 0 \le Q(t) = 1 - F^*(t) \downarrow$ and $\displaystyle\int_0^\infty Q(t)\,dt = E(X^*) = \mu_2/2\mu$

by (3c). By the Key renewal Theorem and part 4(a)

$$\lim_{t\to\infty} \{H(t) - t/\mu\} + \lim_{t\to\infty} F^*(t) = \frac{1}{\mu}\int_0^\infty Q(t)\,dt = \frac{\mu_2}{2\mu^2}$$

i.e.
$$\lim_{t\to\infty} \{H(t) - t/\mu\} + 1 = \mu_2/2\mu^2.$$

or
$$\lim_{t\to\infty} \{H(t) - t/\mu\} = \frac{\mu_2}{2\mu^2} - 1 = \frac{\mu_2 - 2\mu^2}{2\mu^2}.$$

(c) The above relation means that

$H(t)$ tends to its asymptotic value $t/\mu + \dfrac{\mu_2 - 2\mu^2}{2\mu^2} = t/\mu + c$

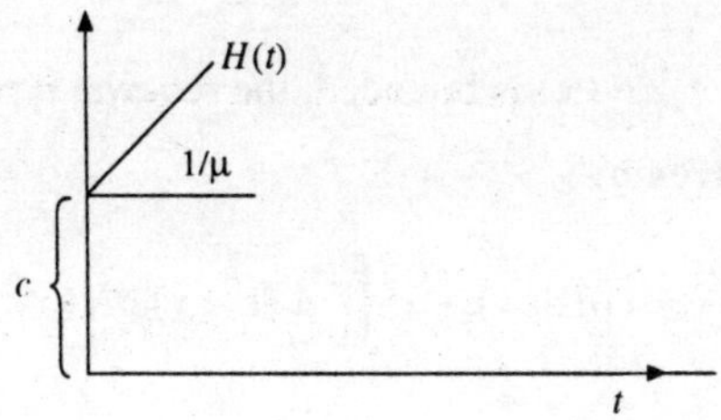

Proof of 4(b)

Define $$\begin{aligned} A(t) &= H(t) + 1 - \mu^{-1}t \\ &= E[N(t) + 1] - t/\mu \end{aligned} \tag{1}$$

Since $E(N(t) + 1) < \infty$, by Wald's equation

$$\begin{aligned} A(t) &= \mu^{-1}\{E[S_{N(t)+1}] - t\} \\ &= \mu^{-1}E(Y(t)) \text{ where } Y(t) = S_{N(t)+1} - t, \end{aligned}$$

the excess life time.

By the renewal argument, conditioning on the time $X_1 = x$ of the first renewal we shall get a renewal type equation for $A(t)$. Distinguishing two cases

we get $$E[Y(t) \mid X_1 = x] = \begin{cases} x - t & \text{if } x \geq t \\ \mu Y(t - x) & \text{if } x < t \end{cases}$$

Therefore $\mu A(t) = \int_0^\infty E[Y(t) \mid X_1 = x]\, dF(x)$

$$= \int_t^\infty (x - t)\, dF(x) + \mu \int_0^t A(t - x)\, dF(x) \tag{2}$$

Now $$\int_t^\infty (x - t)\, dF(x) = \int_0^\infty y\, dF(t + y) = \int_0^\infty \{1 - F(t + y)\}\, dy$$

$$= \int_0^\infty \int_{t+y}^\infty dF(u) \tag{3}$$

is a decreasing function of t.

Since the integrand is nonnegative, by Fubini theorem interchanging the orders of integration we get

$$\int_0^\infty \left\{ \int_0^\infty [1 - F(t + y)]\, dy \right\} dt = \int_0^\infty \int_t^\infty \int_{t+y}^\infty dF(u)\, dy\, dt$$

$$= \int_0^\infty \int_t^\infty \left\{ \int_0^{u-t} dy \right\} dF(y)\, dt = \int_0^\infty \int_t^\infty (u - t)\, dF(u)\, dt$$

$$= \int_0^\infty \int_0^v (u - t)\, dt\, dF(u) = 1/2 \int_0^\infty u^2\, dF(u) = 1/2(\sigma^2 + \mu^2) < \infty \tag{4}$$

Since $\int_0^{\infty} (x - t)\,dF(x)$ is bounded, the renewal type of equation (2) has an unique solution given by

$$\mu A(t) = \int_t^{\infty} (x - t)\,dF(x) + \mu \int_0^t A(t - x)\,dH(x),\ H(x) = \sum_{k=1}^{\infty} F^{(k)}(t).$$

Now taking $Q(t) = \int_t^{\infty} (x - t)\,dF(x)$ by Blackwell's renewal theorem and key renewal theorem and by equation (4)

$$\lim_{t \to \infty} \mu A(t) = \mu^{-1} 1/2(\sigma^2 + \mu^2).$$

or

$$\lim_{t \to \infty} \{H(t) - t/\mu\} = \lim_{t \to \infty} \{A(t) - 1\} \quad \text{(by (1))}$$

$$= \frac{\sigma^2 + \mu^2}{2\mu^2} - 1 = \frac{\sigma^2 - \mu^2}{2\mu^2}.$$

Exercise 4.5 $P(N(t) \geq n) = P(S_n \leq t) = F^{(n)}(t)$

$$EN^k(t) = \sum_{n=1}^{\infty} n^k [P(S_n \leq t) - P(S_{n+1} \leq t)]$$

$$= \sum_{n=1}^{\infty} n^k [F^{(n)}(t) - F^{(n+1)}(t)]$$

$$= \sum_{n=1}^{\infty} [(n-1)^k - n^k]\, F^{(n+1)}(t) \tag{1}$$

by rearrangement of terms provided we can prove that the above sum is convergent.

If $F(x)$ is degenerate at a point $c > 0$, then both sides of (1) are clearly finite for every fixed t. If $F(x)$ is not degenerate, then there is a $y > 0$ such that $0 < F(y) < 1$. Hence for every $m \geq 1$, $F^{(m)}(my) < 1$ because $1 - F^{(m)}(my) = P(S_m > my) \geq P(X_j > y,\ 1 \leq j \leq m) = [1 - F(y)]^m > 0$.

Furthermore, for any $z > 0$,

$$F^{(n)}(z) = \int_0^z F^{(k)}(z - x)\,dF^{(n-k)}(x) \leq F^{(k)}(z)\,F^{(n-k)}(z)$$

i.e. $F^{(n)}(z) \leq [F^{(m)}(z)]^d$, where $n \geq md$. Now if $x > 0$ is arbitrary we choose (and fix) $m \geq 1$ so that $my \geq x$. Then, with d the integer part of n/m

$$F^{(n)}(x) \leq F^{(n)}(my) \leq [F^{(m)}(my)]^d$$

Now (1) shows that $EN^k(t) < \infty$.

Exercise 4.6

$$P(t + x > S_n > t, S_{n+1} > t + x) = \int_t^{t+x} P(S_{n+1} > t + x \mid S_n = y)\, dF^{(n)}(y)$$

$$= \int_t^{t+x} P(X_{n+1} > t + x - y)\, dF^{(n)}(y) = \int_t^{t+x} [t - F(t + x - y)]\, dF^{(n)}(y)$$

Hence $P(Y(t) \le x) = \sum_{n=1}^{\infty} \int_t^{t+x} [1 - F(t + x - y)]\, dF^{(n)}(y)$

$$= \int_t^{t+x} [1 - F(t + x - y)]\, dH(y) \left(\text{Since } \sum_{n=1}^{\infty} F^{(n)}(y) = H(y) \text{ and } 0 < t \le y \right)$$

Exercise 4.7 Starting with total probability law, we first note that $P(Z(t) \le z, N(t) = 0) = 0$ for $z < t$.

Hence, $P(Y(t) > x, Z(t) \le z) = \sum_{n=1}^{\infty} P(Y(t) > x, Z(t) \le z, N(t) = n)$

$$= \sum_{n=1}^{\infty} \int P(Y(t) > x, Z(t) \le z, N(t) = n \mid S_n = y)\, dP(S_n \le y) \qquad (1)$$

Note that conditional probabilities are equal to zero if $y \notin (t - z, t)$, while for $y \in (t - z, t)$, the meaning of such conditional probabilities is that the process $N(t)$ has no points in the interval $(y, t + x)$, i.e. $S_{n+1} - S_n > t + x - y$. Therefore by (1)

$$P(Y(t) > x, Z(t) \le z) = \sum_{n=1}^{\infty} \int_{t-z}^{t} [1 - F(t + x - y)]\, dF^{(n)}_{(y)}$$

$$= \int_{t-z}^{t} [1 - F(t + x - y)]\, dH(y)$$

(interchanging the order of Summation and Integration by Fubini Theorem).

Exercise 4.8

$$X(s) X(s + t) = \sum_{i,j=1}^{\infty} V_i V_j I\{Y_{i-1} \le s < y_i\}\, I\{Y_{i-1} \le s + t < y_j\}$$

where $Y_i = T_1 + \ldots + T_i$. Let Z_s is the waiting time from s to the arrival of the next customer. If $i \ne j$ $E \mid V_i V_j \mid = E[\mid V_i \mid\mid V_j \mid] \le \sigma^2$ (by Cauchy-Schwarz) and

$$\sum_{i,j=1}^{\infty} E[I_{\{Y_{i-1} \le s < y_i\}} I_{\{Y_{j-1} \le s+t < y_j\}}] = 1.$$

Thus term-by-term integration is possible and we get

$$K(t) = E[X(s)X(s+t)] = \sigma^2 \sum_{i=1}^{\infty} P[Y_{i-1} \le s < y_i, Y_{i-1} \le s+t < y_i]$$

$$= \sigma^2 P\{\text{no transition in } (s, s+t]\}$$

$$= \sigma^2 P[Z_s > t] = \sigma^2(1 - F_1(t)).$$

Also $$X(t) = \sum_{i=1} V_i I_{\{Y_{i-1} \le t < y_i\}}$$

and hence $EX(t) = 0$. Since $K(-t) = K(t)$, we have $\sigma^2(1 - F_1(|t|)) = K(t)$.

Exercise 4.9

(a) We have $$Y_t = \sum_{n=1}^{\infty} I_{\{S_{n-1} \le t\}} \cdot (S_{n-1} + X_n - t)^+, t > 0$$

where all terms but one are equal to 0. We can thus raise the sum to any power termwise. Let $H(t) = \sum_{n=1}^{\infty} F^{(n)}(t)$ be the renewal function. Deviding the range of integration into $[0, t)$ and $[t, \infty]$, after an elementary computation we get

$$EY_t^{r-1} = E((X_1 - t)^+)^{r-1} + \int_0^r Q(t-s)\, dH(s),$$

where $Q(t) = \int_t^{\infty} (u-t)^{r-1} dF(u)$.

Since $\int_0^{\infty} Q(t)\, dt = \frac{1}{r} EX_1^r < \infty$, by the key renewal theorem

We get $EY_1^{r-1} \to \frac{1}{r\mu} EX_1^r$ as $t \to \infty$.

(b) Since $EX_1^2 = \sigma^2 + \mu^2$, we get (b) from (a) for $r = 2$.

Exercise 4.10 From the integral equation of renewal theory

$$H(t) = F(t) + (H*F)(t), \text{ where } H(t) = \sum_{n=1}^{\infty} F^{(n)}(t).$$

Also $H(nd) = u_0 + u_1 + \ldots + u_n (n \ge 0)$.

The actual number of partial sums equal to nd is $Y_n = \sum_{j=1}^{\infty} I(S_n = nd)$ and $u_n = EY_n$. From discrete convolution formula we get the result.

Exercise 4.11 From Doob's renewal theorem it is enough to show that $\left\{\left(\frac{N(t)}{t}\right)^r, t \ge 1\right\}$ is uniformly integrable for all $r > 0$. Let $N(t) + 1 = \upsilon(t) = \min\{n: S_n > t\}$. So it is enough to show that $\left\{\left(\frac{\upsilon(t)}{t}\right)^r, t \ge 1\right\}$ is uniformly integrable (u.i.) for all $r > 0$.

Consider $\upsilon(t+s)$, $t, s > 0$. In order to reach the level $t + s$ we must reach the level t. When this has been done the process starts afresh. Since $S_{\upsilon(t)} > t$ the remaining distance of the process to climb is at most equal to s, and thus, the required number of steps to achive this is majorized by a random variable distributed as

$$\upsilon(s), \text{ i.e. } \upsilon(t+s) \le \upsilon(t) + \min\{k - \upsilon(t): S_k - S_{\upsilon(t)} > s\}$$

$$= \upsilon(t) + \upsilon_1(s) \tag{1}$$

where $\upsilon_1(s)$ is distributed as $\upsilon(s)$.

By induction $\qquad \upsilon(n) \le \upsilon_1(1) + \ldots + \upsilon_n(1) \qquad (2)$

where $\{\upsilon_k(1), k \ge 1\}$ are distributed as $\upsilon(1)$.

Since $EN^r(t) < \infty$ for all $r > 0$ and $t > 0$, $E\upsilon^r(1) < \infty$ for $r > 0$. Hence by Minkowski's inequality and (2)

$$(E\upsilon^r(n))^{1/r} \le n\,(E\upsilon^r(1))^{1/r} \tag{3}$$

Since $\upsilon(t) \le \upsilon([t]+1)$, we have for $t \ge 1$,

$$\frac{\upsilon(t)}{t} \le 2\,\frac{\upsilon([t]+1)}{[t]+1}.$$

Therefore from (3) we get

$$(E(\upsilon(t)/t)^r)^{1/r} \le 2\,(E(\upsilon([t]+1))/([t]+1)^r)^{1/r}$$

$$\le 2\,(E(\upsilon(1))^r)^{1/r} < \infty.$$

Since the last bound is uniform in t $\{(\upsilon(t)/t)^p, t \ge t\}$ is uniformly integrable for all $p < r$. Since r is arbitrary $\{(\upsilon(t)/t)^r, t \ge 1\}$ is u.i. for all $r > 0$.

Exercise 4.12

(a) Since $\quad \dfrac{N(t)}{t}$ a.s. $\to \dfrac{1}{\mu}$ as $t \to \infty \qquad (1)$

and $\qquad E\left(\dfrac{N(t)}{t}\right)^r \to \dfrac{1}{\mu^r}$ as $t \to \infty \qquad (2)$

(by Exercise 4.11) for any $r > 0$, it is enough to show that $\left\{\left(\dfrac{N(t) - t/\mu}{\sqrt{t}}\right)^2, t \ge 1\right\}$

is uniformly integrable. Due to monotonicity of $\{N(t), t \ge 0\}$, it is enough to

show that $\left\{\left(\dfrac{N(n) - n/\mu}{\sqrt{n}}\right)^2, n \ge 1\right\}$ is uniformly integrable. Hence for $r > 2$,

(2) implies

$$\left\{\left(\frac{N(t) - t/\mu}{\sqrt{t}}\right)^2, t \geq 1\right\}$$ is uniformly integrable.

(b) can be proved in a similar way.

Exercise 4.13 Conditioning on the time $X_1 = x$ of first renewal we get

$$P(Y_t > y \mid X_1 = x) = \begin{cases} 1 & \text{if } x > t + y \\ 0 & \text{if } t + y \geq x > t \\ A_y(t - x) & \text{if } t \geq x > 0 \end{cases}$$

where $A_y(t) = P(Y_t > y)$.

Also
$$A_y(t) = \int_0^\infty P(Y_t > y \mid X_1 = x) dF(x)$$
$$= 1 - F(t + y) + \int_0^t A_y(t - x) dF(x)$$

is a renewal type equation and the solution is given by

$$A_y(t) = 1 - F(t + y) + \int_0^t \{1 - F(t + y - x)\} dH(x).$$

Since $\mu = E(X_1) = \int_0^\infty (1 - F(x))\, dx < \infty$,

$\int_0^\infty \{1 - F(t + y)\} dt = \int_y^\infty \{1 - F(y)\} dx < \infty$ and $\{1 - F(t + y)\}$ is nonincreasing function of t. By key renewal theorem we get

$$\lim_{t\to\infty} P(Y_t > y) = \lim_{t\to\infty} A_y(t) = \mu^{-1} \int_y^\infty \{1 - F(y)\} dy, y > 0$$

Now the set of events $\{Y_t \geq x \text{ and } Z_t \geq y\}$ is equivalent to $\{Y_{t-y} \geq x + y\}$. Therefore

$$\lim_{t\to\infty} P(Z_t \geq y, Y_t \geq x) = \lim_{t\to\infty} P(Y_{t-y} \geq x + y)$$
$$= \mu^{-1} \int_{x+y}^\infty \{1 - F(z)\} d(z).$$

In particular,

$$\lim_{t\to\infty} P\{Z_t \geq z\} = \lim_{t\to\infty} P(Z_t \geq z, Y_t \geq 0) = \mu^{-1} \int_z^\infty \{1 - F(x)\} dx$$

Define $K_x(t) = P\{S_t = Y_t + X_t > x\}$. Therefore

$$P\{S_t > x \mid X_1 = y\} = \begin{cases} 1 & \text{if } y > \max(x, t) \\ K_x(t-y) & \text{if } y \le t \\ 0 & \text{otherwise} \end{cases}$$

Then by the law of total probability we get the renewal type equation

$$K_x(t) = 1 - F(\max(x, t)) + \int_0^t K_x(t-y)\, dF(y).$$

By key renewal theorem and integration by parts

$$\lim_{t\to\infty} P(S_t > x) = \lim_{t\to\infty} K_x(t) = \frac{1}{\mu}\int_0^\infty [1 - F(\max(x, \tau))]d\tau$$

$$= \frac{1}{\mu}\int_x^\infty y\, dF(y).$$

Therefore $$\lim_{t\to\infty} P(S_t \le x) = \mu^{-1}\int_0^\infty y\, dF(y).$$

Exercise 4.14 Now Y_i is the cost incurred at the ith replacement cycle, then

$$Y_i = \begin{cases} C_1 \text{ with probability } 1 - F(T) \\ C_2 \text{ with probability } F(T) \end{cases}$$

and $E(Y_i) = C_1(1 - F(T)) + C_2F(T)$. Since the expected length of a replacement cycle is

$$E(\min(X_i, T)) = \int_0^T [1 - F(x)]\, dx.$$

We have the long run mean cost per unit is given by

$$\theta(T) = \frac{C_1[1 - F(t)] + C_2F(T)}{\int_0^T [1 - F(x)]\, dx}.$$

Differentiating $\theta(T)$ with respect to T and equating to zero we get

$$\frac{(-C_1 + (C_2)f(T)\left\{\int_0^T [1 - F(x)]\, dx\right\} - \{C_1(1 - F(T)) + (C_2 - C_1)F(T)(1 - F(T))\}}{\left\{\int_0^T [1 - F(x)]\, dx\right\}^2} = 0$$

which yields the result.

Exercise 4.15 Let $U(t)$ be the interoccurrence distribution for $N(t)$. Then

$$1 - U(x) = P(N(x) = 0) = P(N_1(x) = 0, N_2(x) = 0) = [1 - F(x)]^2 \quad (1)$$

Let $Y_1(t)$, $Y_2(t)$ and $Y(t)$ be the excess life at time t for the processes N_1, N_2 and N respectively. Then, because the processes N_1 and N_2 and N respectively. Then, because the processes N_1 and N_2 are composed and we must have

$$Y(t) = \min \{Y_2(t), Y_2(t)\}$$

and $$P(Y(t) > x) = [P(Y_1(t) > x)]^2$$

Letting $t \to \infty$ and by exercise 4.14, we obtain

$$\frac{1}{\nu}\int_x^\infty [1 - U(y)]\, dy = \frac{1}{\mu^2}\left\{\int_x^\infty [1 - F(y)]\, dy\right\}^2 \quad (2)$$

where $\nu = \int_0^\infty [1 - U(y)]\, dy$, the mean of $U(y)$.

Differentiating (2) and using (1) gives

$$\frac{1}{\nu}[1 - F(x)]^2 = \frac{1}{\nu}[1 - U(Y)] = \frac{2}{\mu^2}\left\{\int_x^\infty [1 - F(y)]\, dy\right\}[1 - F(x)]$$

i.e. $$1 - F(x) = \frac{2\upsilon}{\mu^2}\int_x^\infty [1 - F(y)]\, dy \quad (3)$$

Letting $G(x) = 1 - F(x)$ and differentiating (3), we get

$$\frac{dG(x)}{dx} = -\frac{2\upsilon}{\mu^2} G(x) \quad (4)$$

Since $G(0) = 1$, the solution of differential equation (4) gives

$$G(x) = 1 - F(x) = e^{-\lambda x}, \text{ where } \lambda = 2\upsilon/\mu^2.$$

Hence all of the $N_1(t)$, $N_2(t)$ and $N(t)$ are Poisson, because of the fact that sum of two Poisson processes is a Poisson process.

Chapter 5

Exercise 5.1
$f(s) = 1 - p(1 - s)^\beta$ is a Probability generating function iff the coefficients in the expansion of $f(s)$ are nonnegative and sum to 1, since $\beta > 0, f(1) = 1$.

By the binomial expansion $f(s) = 1 - p\left(\sum_{n=0}^{\infty}\binom{\beta}{n}(-s)^n\right)$

$$= 1 - p - p\left(\sum_{n=1}^{\infty}\binom{\beta}{n}(-s)^n\right),\ 1 - p > 0 \text{ (since } p < 1).$$

Also $\binom{\beta}{n} = \beta(\beta-1)\ldots(\beta-n+1)/n! = \frac{(-1)^{n-1}}{n!}\beta(1-\beta)(2-\beta)\ldots(n-1-\beta)$

and $(1-\beta)(2-\beta)\ldots(n-1-\beta) > 0$, since $\beta < 1$.

Coefficient of s^n is $-p\binom{\beta}{n}(-1)^n = p\frac{(-1)^{2n}}{n!}\beta(1-\beta)\ldots(n-1-\beta) > 0$, since $p > 0$, for $n \geq 2$.

Also for $n = 1$ we have $-p\binom{\beta}{1}(-1) = p\beta > 0$. Hence $f(s)$ is a probability generating function. To compute the iterates, we use induction.

For $n = 1$, $f_1(s) = f(s) = 1 - p(1-s)^{\beta}$.

Assume for $n = m$ $f_m(s) = 1 - p^{1+\beta+\ldots+\beta^{m-1}}(1-s)^{\beta^m}$, then

$$f_{m+1}(s) = f[f_m(s)] = 1 - p\{p^{1+\beta+\ldots+\beta^{m-1}}(1-s)^{\beta^m}\}^{\beta}$$

$$= 1 - p\{p^{(1+\beta+\ldots+\beta^{m-1}}\}(1-s)^{\beta^m\beta} = 1 - p^{1+\beta+\ldots+\beta}(1-s)^{\beta^{m+1}}$$

Exercise 5.2 $E\left[\sum_{n=1}^{\infty} X_n\right] = E[\lim_{N\to\infty} E\left[\sum_{n=1}^{N} X_n\right]$ (since $X_n \geq 0$, by monotone convergence theorem)

$$= \lim_{N\to\infty}\sum_{n=1}^{N} E(X_n) = \lim_{N\to\infty}\sum_{n=1}^{N}(EX_1)^n = \lim_{N\to\infty}\sum_{n=1}^{N} m^n$$

$$= \lim_{n\to\infty}\frac{m - m^{N+1}}{1-m} = \frac{m}{1-m}, \text{ since } m < 1.$$

Exercise 5.3

$$f(s) = as^2 + bs + c,\ a > 0,\ b > 0,\ c > 0 \text{ and } a + b + c = 1.$$

Hence $m = f'(1) = 2a + b = 2a + (1 - a - c) = a - c + 1$. (1)

Now only root of $f(s) = s$ are $s = 1$ and $s = \frac{1-a-b}{a} = \frac{c}{a}$

Since the extinction probability is less than 1, it is c/a, for otherwise from (1) $m - 1 > 0 \Rightarrow a > c \Rightarrow a/c > 1 \Rightarrow c/a < 1$. Therefore, $s = 1$ cannot be the probability of extinction and d must be c/a.

Exercise 5.4 $q_n = \phi_n(0) = \phi_{n-1}(\phi(0)) = \phi_{n-1}(p_0)$, $0 < p_0 < 1$.

Exercise 5.5 Part 1 can be proved by induction or by a careful argument on branching process with more than one ancestors.

$$\phi_{l,n}(u, x) = \sum_{k=0}^{\infty}\sum_{j=0}^{\infty} P[X_n = j, X_l = k]u^k x^j = \sum_{k=0}^{\infty}\sum_{j=0}^{\infty} p_{kj}^{(n-l)} P(X_l = k)u^k x^j$$

$$= \sum_{k=0}^{\infty} P[X_l = k]u^k\left[\sum_{j=0}^{\infty} p_{kj}^{(n-l)} x^j\right]$$

$$= \sum_{k=0}^{\infty} P[X_l = k][u\phi_{n-l}(x)]^k = \phi_l[u\phi_{n-l}(x)]$$

if we can show that $\sum_{j=0}^{\infty} p_{kj}^{(n-l)} x^j = [\phi_{n-l}(x)]^k$. (1)

Proof of (1): Writing $n - l = m$ (1) becomes $\sum_{j=0}^{\infty} p_{kj}^{(m)} = [\phi_m(x)]^k$. (2)

For $m = 0$ $p_{kj}^{(0)} = \delta_{kj} \Rightarrow \sum_{j=0}^{\infty} p_{kj}^{(0)} x^j = x^k = [\phi_0(x)]^k$ since $\phi_0(x) = x$.

Assume (2) is true for m, then

$$\sum_{j=0}^{\infty} p_{kj}^{(m+1)} x^j = \sum_{j=0}^{\infty} \sum_{l=0}^{\infty} P[X_{m+1} = j, X_m = l \mid X_0 = k] x^j$$

$$= \sum_{j=0}^{\infty} \sum_{l=0}^{\infty} P[X_{m+1} = j \mid X_m = l]\, P[X_m = l \mid X_0 = k] x^j$$

(by Markov property)

$$= \sum_{l=0}^{\infty} P[X_m = l \mid X_0 = k] \sum_{j=0}^{\infty} p_{lj} x^j$$

$$= \sum_{l=0}^{\infty} p_{kl}^{(m)} [\phi(x)]^l = [\phi_m(\phi(x))]^k = [\phi_{m+1}(x)]^k.$$

2. $E(X_n X_1) = \dfrac{\partial^2 \phi_n(u, x)}{\partial u\, dx} \Big|_{u=1, x=1}.$

Now, $\partial \phi_{l,n}(u, x) / \partial u \,|_{u=1} = \phi_l'[u\phi_{n-l}(x)]\phi_{n-l}(x) \,|_{u=1} = \phi_l'[\phi_{n-l}(x)]\phi_{n-l}(x)$

Hence $\dfrac{\partial \phi_{l,n}(u, x)}{\partial u\, \partial x} \Big|_{u=1, x=1} = \phi_l''(1)[\phi_{n-l}(x)]\phi_{n-l}'(x)\phi_{n-l}(x) \,|_{x=1}$

$$+ \phi_l'[\phi_{n-l}(x)]\phi_{n-l}'(x) \,|_{x=1}$$

$$= \phi_l''(1)\, \phi_{n-l}'(1)\, .1 + \phi_l'(1)\, \phi_{n-l}'(1)$$

$$= \phi_l''(1)\, m^{n-l} + m^{n-l}\phi_l'(1)$$

$$= m^{n-1}\, \{\phi_l''(1) + \phi_l'(1) = m^{n-l} E(X_l^2),\ n > l.$$

3. $$\text{Var}(X_n) = \begin{cases} \sigma^2 m^{n-1}(m^n - 1)/(m-1) & \text{if } m \neq 1 \\ n\sigma^2 & \text{if } m = 1 \end{cases}$$

$$\text{Var}(X_l) = \begin{cases} \sigma^2 m^{l-1}(m^l - 1)/(m-1) & \text{if } m \neq 1 \\ l\sigma^2 & \text{if } m = 1 \end{cases}$$

$$\text{Cov}(X_n, X_l) = \begin{cases} \sigma^2 m^{n-1}(m^l - 1)/(m-1) & \text{if } m \neq 1 \\ l\sigma^2 & \text{if } m = 1 \text{ and } n > l. \end{cases}$$

$$R_{nl} = \frac{\text{Cov}(X_n, X_l)}{\sqrt{\text{Var}(X_n)\text{Var}(X_l)}} = \frac{\sigma^2 m^{n-1}(m^l - 1)(m-1)}{(m-1)\sigma^2\sqrt{m^{n+l-2}(m^n-1)(m^l-1)}}$$

$$= \sqrt{m^{n-l}(m^l - 1)/(m^n - 1)} \quad \text{if } m \neq 1$$

$$= \sqrt{l/n} \quad \text{if } m = 1$$

$$\to 0 \quad \text{as } n \to \infty$$

Thus, $R_{nl} \to 0$ if $m < 1$ as $n \to \infty$. If $m > 1$, then $m^n \to \infty$, $m^{-n} \to 0$ and hence $R_{nl} \to \sqrt{1 - m^{-l}}$.

Exercise 5.6 (i) For Poisson distribution with parameter λ the generating function $\phi(z) = e^{\lambda(z-1)}$ satisfies the difference equation $\lambda\phi(z) = \phi'(z)$; (ii) Similarly for Binomial distribution with generating function $\phi(z) = (q + pz)^n$ where parameters are n and p satisfies $np\phi(z) = (q + pz)\phi'(z)$; and (iii) negative Binomial distribution with (Parameters n and p) with generating function $\phi(z) = (pz)^n/(1 - qz)^n$ also satisfies $np\phi(z) = (1 - qz)\phi'(z)$. All of them are desired form of differential equation. Now from the relation $\phi_n(z) = \phi(\phi_{n-1}(z))$ differentiating with respect to z, we get $\phi_n'(z)$

$$= \phi'(\phi_{n-1}(z))\,\phi_{n-1}'(z) \tag{1}$$

Apply the assumed differential equation $(c + dz)\phi'(z) = a + b\phi(z)$ to equation (1), yielding $\phi_n(z) = [(a + b\phi(\phi_{n-1}(z)))/(c + d\phi_{n-1}(z))]$. $\phi_{n-1}'(z) = [(a + b\phi_n(z))/(c + d\phi_{n-1}(z))]\,\phi_{n-1}'(z)$. Multiplying the last relation we get $\phi_n'(z)[c + d\,\phi_{n-1}(z)] = [a + b\phi_n(z)]\,\phi_{n-1}'(z)$ (2)

Now any probability generating function $G(z)$ is differentiable any number of times within the unit circle and $G^{(k)}(0) = k!\,p_k$, $k \geq 1$. Hence,

$$P(X_{n+1} = k + 1) = \frac{1}{(k+1)!}\,\phi_n^{(k+1)}(0) \tag{3}$$

Also note that $q_n = P(X_n = 0) = \phi_n(0)$. (4)

Now differentiating k times both sides of (2), using (3) and (4) and rearranging we get the desired recursive formula.

Exercise 5.7 Clearly $E(X_n/X_{n-1}) = mX_{n-1}$ a.s. Also $EX_n = m^n$. Hence $Y_n = X_n/EX_n$ is a martingale. Also Var $(X_n) \simeq \sigma^2/m^{2n-2}$ if $m \neq 1$

Hence Var $(Y_n) \simeq \sigma^2/m^2$ if $m \neq 1$

$\leq \sigma^2$ if $m > 1$

Thus, $\sup_n$ Var $(Y_n) < \infty$. Hence the martingale $\{Y_n\}$ converges a.s.

Exercise 5.8 Let p_k, $k = 0, 1, 2$ be the probability that one red cell produces k red cells in one minute span. It is given that $p_0 = 1/12$, $p_1 = 2/3$, $p_2 = 1/4$.

We note that the 0th generation starts with one red cell.

Then $P(s) = \frac{1}{12} + \frac{2}{3}s + \frac{1}{4}s^2$, where $P(s)$ is the probability generating function of the number of off springs per individual.

Now $P'(s) = \frac{2}{3} + \frac{1}{2}s,\ P'(1) = \frac{7}{6}\ (>1) \rightarrow m > 1.$

Hence probability of extinction is the root (less than 1) of the equation $s = P(s)$

or $$s = \frac{1}{12} + \frac{2}{3}s + \frac{1}{4}s^2$$

or $$\frac{s^2}{4} - \frac{1}{3}s + \frac{1}{12} = 0$$

or $$3s^2 - 4s + 1 = 0$$

or $$(3s - 1)(s - 1) = 0$$

or $$s = 1 \text{ or } 1/3$$

Hence probability that the blood culture will eventually die out is 1/3.

Chapter 6

Exercise 6.1

For $$t > 0,\ P_\pi(X(t) = 1) = 1/2p_{1,1}(t) + 1/2p_{-1,1}(t) \tag{1}$$

and $$P_\pi(X(t) = 1) = 1/2.$$

By Chapman-Kolmogorov's equation

$$p_{1,1}(t + s) = p_{1,1}(s)p_{1,1}(t) + p_{1,-1}(s)p_{-1,1}(t) \tag{2}$$

From (2), we get the functional equation

$$\alpha(t + s) = \alpha(s)\alpha(t) + (1 - \alpha(s))(1 - \alpha(t))$$
$$= 2\alpha(s)\alpha(t) + 1 - \alpha(s) - \alpha(t) \tag{3}$$

Put $\alpha(t) = 1/2[1 + h(t)]$. Then equation (3) becomes

$$h(t + s) = h(s)\, h(t) \tag{4}$$

This is the Cauchy-functional equation whose only continuous solution is

$$h(t) = e^{-\lambda t},\ \lambda > 0 \tag{5}$$

From (4) and (5) we obtain that $\alpha(t) = (1 + e^{-\lambda t})/2$.

Exercise 6.2 This is a two state birth and death process. The Chapman-Kolmogorov equation for a homogenous M.C. is given by

$$p_{ij}(t + h) = \sum_{k \in s} p_{ik}(t)\, p_{kj}(h),\ k = 0, 1$$

For $i = j = 0$, we get the equation

$$p_{00}(t + h) = p_{00}(t)\, p_{00}(h) + p_{01}(t)\, p_{10}(h)$$
$$= (1 - \alpha h + o(h))p_{00}(t) + p_{01}(t)\beta h + o(h)$$

Taking limit $h \rightarrow 0$, we get

$$\frac{dp_{00}(t)}{dt} = -\alpha p_{00}(t) + \beta p_{01}(t) \tag{1}$$

Since we have

$$p_{00}(t) + p_{01}(t) = 1, \tag{2}$$

equation (1) becomes

$$\frac{dp_{00}(t)}{dt} = \beta - (\alpha + \beta) p_{00}(t) \tag{3}$$

Solution of equation (3) subject to the initial condition $p_{00}(0) = 1$ is given by

$$p_{00}(t) = p_{00}(0)\, e^{-(\alpha+\beta)t} + \int_0^t e^{-(\alpha+\beta)(t-s)} \beta\, ds$$

$$= e^{-(\alpha+\beta)t} + \beta\, e^{-(\alpha+\beta)t}\, 1/(\alpha+\beta)\, (e^{(\alpha+\beta)t} - 1)$$

$$= e^{-(\alpha+\beta)t} + \frac{\beta}{\alpha+\beta} - \frac{\beta}{\alpha+\beta} e^{-(\alpha+\beta)t}, \quad t \ge 0$$

$$= \frac{\alpha}{\alpha+\beta} e^{-(\alpha+\beta)t} + \frac{\beta}{\alpha+\beta}, \quad t \ge 0$$

Similarly $p_{10}(t) = \dfrac{\beta}{\alpha+\beta} - \dfrac{\beta}{\alpha+\beta} e^{-(\alpha+\beta)t}$

Also $p_0(t) = p_0(0) p_{00}(t) + (1 - p_0(0)) p_{10}(t)$

$$= \frac{\beta}{\alpha+\beta} + (p_0(0) - \frac{\beta}{\alpha+\beta}\, e^{-(\alpha+\beta)t}$$

$$= \frac{\beta}{\alpha+\beta} + \frac{\alpha}{\alpha+\beta} e^{-(\alpha+\beta)t} \quad \text{if } p_0(0) = 1.$$

Exercise 6.3

$$p_{00}(t) = \frac{\beta}{\alpha+\beta} + \frac{\alpha}{\alpha+\beta} [e^{-(\alpha+\beta)t}]$$

Let

$$Y_{ij}(t) = \begin{cases} 1 & \text{if } X(t) = j \text{ given } X(0) = i \\ 0 & \text{if } X(t) \ne j \text{ given } X(0) = i \end{cases}$$

Then $E(Y_{ij}(t)) = p_{ij}(t)$. The expected time spent in state j is given by

$$\mu_{ij}(t) = E\left[\int_0^t Y_{ij}(\tau)\, d\tau\right] = \int_0^t E[Y_{ij}(\tau)\, d\tau] = \int_0^t p_{ij}(t)\, dt$$

(by Fubini Theorem)

Then $\mu_{00}(t) = \displaystyle\int_0^t p_{00}(t)\, dt = \int_0^t \left[\frac{\beta}{\alpha+\beta} + \frac{\alpha}{\alpha+\beta} e^{-(\alpha+\beta)t}\right] dt$

$$= \frac{\beta t}{\alpha+\beta} + \alpha/(\alpha+\beta)^2 [1 - e^{-(\alpha+\beta)t}]$$

In a similar way the other functions $\mu_{01}(t)$, $\mu_{10}(t)$ and $\mu_{11}(t)$ can be determined.

Exercise 6.4 Put $q_{ii} = (v + i\lambda)$, $q_{i,i+1} = v + i\lambda$, $q_{ik} = 0$ if $k < i$ or if $k > i + 1$ in Kolmogorov's forward equation.

Clearly $\quad p_{00}(t) = e^{-vt}$, $p'_{01}(t) = -(v + \lambda)p_{01}(t) + vp_{00}(t)$

$$= -(v + \lambda)p_{01}(t) + ve^{-vt}. \tag{1}$$

(1) Can be solved with the aid of an integrating factor. Let $h(t) = e^{(v+\lambda)t}p_{01}(t)$. Then (1) may be expressed as $\dfrac{dh(t)}{dt} = ve^{\lambda t}$, or $h(t)$

$$= h(0) + \int_0^t ve^{\lambda u}\,du = \frac{v}{\lambda}(e^{\lambda t} - 1).$$

Therefore, $\quad p_{01}(t) = e^{-(v+\lambda)t}h(t) = \dfrac{v}{\lambda}(e^{-vt} - e^{-(v+\lambda)t})$

$$= \frac{v}{\lambda}e^{-vt}(1 - e^{-\lambda t}). \tag{2}$$

Now $\quad p'_{02}(t) = -(v + 2\lambda)p_{02}(t) + (v + \lambda)p_{01}(t)$

$$= -(v + 2\lambda)p_{02}(t) + (v + \lambda)\frac{v}{\lambda}e^{-vt}(1 - e^{-\lambda t}) \tag{3}$$

Similarly, as (2) we get

$$p_{02}(t) = e^{-(v+2\lambda)t}\left[\int_0^t e^{(v+2\lambda)u}\frac{(v + \lambda)v}{\lambda}e^{-vu}(1 - e^{-\lambda u})\,du\right]$$

$$= e^{-(v+2\lambda)t}\left[\frac{v(v + \lambda)}{\lambda}\int_0^t (e^{2\lambda u} - e^{\lambda u})\,du\right]$$

$$= e^{-(v+2\lambda)t}\left[\frac{v(v + \lambda)}{\lambda}\right]\left[\frac{e^{2\lambda t} - 1}{2\lambda} - \frac{e^{\lambda t} - 1}{\lambda}\right]$$

$$= \frac{v(v + \lambda)}{2\lambda^2}[e^{-vt} - 2e^{-(v+\lambda)t} + e^{-(v+2\lambda)t}]$$

$$= \frac{v(v + \lambda)}{2\lambda^2}e^{-vt}[1 - e^{-\lambda t}]^2$$

We shall prove (3) by induction on n. Assume that

$$p_{on}(t) = \frac{v(v + \lambda)\dots(v + (n - 1)\lambda)}{n!\,\lambda^n}e^{-vt}(1 - e^{-\lambda t})^n \text{ hold} \tag{4}$$

Then,

$$p'_{on}(t) = -(v + (n + 1)\lambda)p_{on+1}(t) + (v + n\lambda)p_{on}(t) \text{ yields}$$

$$p_{o,n+1}(t) = e^{-(v+(n+1)\lambda)t} \int_0^t e^{(v+(v+1)\lambda)u} \; (v + n\lambda)p_{on}(u)\, du$$

$$= \frac{v(v+\lambda)\ldots(v+n\lambda)}{n!\,\lambda^n} e^{-(v+(n+1)\lambda)t} \int_0^t e^{(n+1)\lambda u} \; [1 - e^{-\lambda u}]^n \, du$$

(substituting the value of $p_{on}(u)$ from (4))

Letting $x = e^{\lambda u}$ yields

$$\int_0^t e^{(n+1)\lambda u} [1 - e^{-\lambda u}]^n \, du = \frac{1}{\lambda} \int_1^{e^{\lambda t}} x^n \left(1 - \frac{1}{x}\right)^n dx$$

$$= \frac{1}{\lambda} \int_1^{e^{\lambda t}} (x-1)^n \, dx = \frac{1}{\lambda} \left[\frac{(x-1)^{n+1}}{n+1} \right]_1^{e^{\lambda t}} = \frac{1}{\lambda} \frac{(e^{\lambda t} - 1)^{n+1}}{n+1}$$

Therefore $\quad p_{0,n+1}(t) = \dfrac{v(v+\lambda)\ldots(v+n\lambda)}{(n+1)!\,\lambda^{n+1}} e^{-vt} (1 - e^{-vt})^{n+1}.$

Exercise 6.5 Consider a branching process starting out with i particles. Let $\xi_1, \ldots, \xi_i$ be the times until these particles split apart or disappear. Then $\xi_1, \ldots, \xi_i$ each has exponential distribution with parameter θ and hence $\tau_1 = \min(\xi_1, \ldots, \xi_i)$ has an exponential distribution with parameter $q_i = i\theta$. Whichever particle acts first has probability p of splitting into two particles and probability $1 - p$ of disappearing. Let $p_{ij}(t)$ be the probability that starting at i $X(t)$ jumps to state j after waiting a random length of time. Thus for $i \geq 1$

$$p_{i,\,i+1}(t) = p \text{ and } p_{i,i-1}(t_{i-1}) = 1 - p$$

State 0 is an absorbing state. Since $\lambda_i = q_i p_{i,\,i+1}(t)$ and $\mu_i = q_i \cdot p_{i,i-1}(t)$, we conclude that the birth and death rates are $\lambda_i = i\theta p$, and $\mu_i = i\theta(1-p)$, $i \geq 0$.

Exercise 6.6 Suppose there are initially j particles present. Let $(\xi_1, \ldots, \xi_j)$ be the times at which these particles split apart or disappear and let η be the first time a new particle enters the system. We interpret the description of the systems as implying that η is independent of $\xi_1, \xi_2, \ldots, \xi_j$, η are independent exponentially distributed random variables having respective parameters $\theta, \theta, \ldots, \theta, \lambda$. Thus $\tau_1 = \min(\xi_1, \ldots, \xi_j, \eta)$ is exponentially distributed with parameter

$$q_j = j\theta + \lambda \quad \text{and} \quad P(\tau_1 = \eta) = \frac{\lambda}{j\theta + \lambda} \tag{1}$$

The event $[X(\tau_1) = j + 1]$ occurs if either $\tau_1 = \eta$ or $\tau_1 = \min(\xi_1, \ldots, \xi_j)$ and a particle splits into two new particles at time τ_1. Thus

$$p_{j,j-1}(t) = \frac{\lambda}{j\theta + \lambda}(1-p) \quad \text{and} \quad p_{j,j+1}(t) = \frac{\lambda}{j\theta + \lambda} + \frac{j\theta}{j\theta + \lambda} p.$$

Hence we conclude that the process $X(t)$ has birth and death rates

$$\lambda_j = q_j p_{j,j+1} = j\theta p + \lambda \quad \text{and} \quad \mu_j = q_j p_{j,\,j-1} = j\theta(1-p).$$

[Note that equation (1) follows from some standard properties of independent exponentially distributed r.vs: If $\xi_1, \ldots, \xi_n$ be independent r.v.s having exponential distributions with respective parameters $\theta_1, \ldots, \theta_n$. Then $\xi_{(1)} = \min(\xi_1, \ldots, \xi_n)$ has an exponential distribution with parameter $\theta_1 + \ldots + \theta_n$ and $P(\xi_k = \min(\xi_1, \ldots, \xi_n)) = \dfrac{\theta_k}{\theta_1 + \ldots + \theta_n}$, $k = 1, 2, \ldots, n$. These results can easily be verified in the following manner.

$P(\min(\xi_1, \ldots, \xi_n) > x) = P(\xi_1 > x, \ldots, \xi_n > x) = P(\xi_1 > x) \ldots P(\xi_n > x) = e^{-\theta_1 x} \ldots e^{-\theta_n x} = e^{-(\theta_1 + \ldots + \theta_n)x}$. Hence $\xi_{(1)}$ is exponentially distributed. Set $\eta_k = \min(\xi_j : j \neq k)$. Then η_k has an exponential distribution with parameter $\mu_k = \sum_{j \neq k} \theta_j$, and ξ_k and η_k are independent.

Thus $P(\xi_k = \min(\xi_1, \ldots, \xi_n)) = P(\xi_k \leq \eta_k)$

$$= \int_0^\infty \left(\left(\int_x^\infty \left(\left(\int_x^\infty \theta_k e^{-\theta_k x} \mu_k e^{-\mu_k y} dy \right) dx \right.\right.\right.\right.$$

$$= \int_0^\infty \theta_k e^{-(\theta_k + \mu_k)x} dx = \frac{\theta_k}{\theta_k + \mu_k} = \frac{\theta_k}{\theta_1 + \ldots + \theta_n} \Bigg]$$

Exercise 6.7 Now Kolmogorov's forward equation becomes

$$p_0'(t) = -\lambda p_0(t) + \mu p_1(t)$$

and
$$p_j'(t) = \lambda p_{j-1}(t) - (\lambda + j\mu) p_j(t) + \mu(1 + j) p_{j+1}(t)$$

Consider the P.G.F. of $\{p_j(t)\}$ $g(t, u) = \sum_{j=0}^\infty p_j(t) u^j$. We get

$$\begin{aligned} \partial g(t, u)/\partial t &= \sum_{j=0}^\infty p_j'(t) u^j \\ &= -\lambda p_0(t)(1 - u) + \mu p_1(t)(1 - u) \\ &\quad - \lambda p_1(t)(1 - u)u + 2\mu p_2(t) u(1 - u) + \ldots \\ &= -\lambda(1 - u) \sum_{j=0}^\infty p_j(t) u^j + \mu(1 - u) \sum_{j=0}^\infty j p_j(t) u^{j-1} \\ &= -\lambda(1 - u) g(t, u) + \mu(1 - u) \frac{\partial g}{\partial u} \end{aligned}$$

So $g(t, u)$ satisfies linear partial differential equation

$$\frac{\partial g(t, u)}{\partial t} - \mu(1 - u) \frac{\partial g(t, u)}{\partial u} = -\lambda(1 - u) g(t, u) \tag{1}$$

Suppose that $Y(0) = N$, then $g(0, u) = u^N$ (2)

is the initial condition. Solving this by Lagrange method (Recatti) and using the initial condition, we obtain

$$g(t, u) = \{1 - (1 - u)e^{-\mu t}\}^N \exp\left\{-\frac{\lambda}{\mu}(1 - u)(1 - e^{-\mu t})\right\} \tag{3}$$

Note that the first term on the right of (3) is the P.G.F. of Binomial distribution with $p = \exp(-\mu t)$, whereas the second term is the P.G.F. of the Poisson distribution with mean $\wedge(t) = (1 - e^{-\mu t})/\mu$. Therefore $Y(t) = Y_0(t) + Y_1(t)$, where $Y_0(t)$ is a Binomial component and $Y_1(t)$ is the Poisson component and they are also independent for all $t \geq 0$.

Expanding the P.G.F., we get

$$p_j(t) = \exp\{-\lambda/\mu(1 - e^{-\mu t})\} \sum_{k=0}^{\min(N,t)} \binom{N}{k} \left(\frac{\lambda}{\mu}\right)^{j-k} \frac{e^{-\mu t k}(1 - e^{-\mu t})^{N+j-2k}}{(j-k)!}$$

for $j \geq 0$ and

$$E(Y(t)) = Ne^{-\mu t} + \frac{\lambda}{\mu}(1 - e^{-\mu t}).$$

This formula was first time obtained by Palm.

Exercise 6.8

(a) The system of difference-differential equation in this case is

$$p'_n(t) = -n(\lambda + \mu)p_n(t) + \lambda(n - 1)p_{n-1}(t) + \mu(n + 1)p_{n+1}(t), n \geq 1 \tag{1}$$

$$p'_0(t) = \mu p_1(t), p_n(0) = \delta_{n1} \tag{2}$$

Let $$G(z, t) = \sum_{j=0}^{\infty} p_n(t)z^n.$$

There $$\frac{\partial G}{\partial z} = \sum_{n=1}^{\infty} np_n(t)z^{n-1}$$

and $$\frac{\partial G}{\partial t} = \sum_{n=0}^{\infty} p'_n(t)z^n.$$

Multiplying (1) by z^n and summing over $n = 1, 2, 3, \ldots$ and adding (2), we get

$$\frac{\partial G}{\partial t} = -(\lambda + \mu)\sum_{n=1}^{\infty} np_n(t)z^n + \lambda \sum_{n=1}^{\infty}(n - 1)p_{n-1}(t)z^n$$

$$+ \mu\left[\sum_{n=1}^{\infty}(n + 1)p_{n+1}(t)z^n + p_1(t)\right]$$

$$- (\lambda + \mu)z\frac{\partial G}{\partial z} + \lambda z^2\frac{\partial G}{\partial z} + \mu\frac{\partial G}{\partial z}$$

$$= [\mu - (\lambda + \mu)z + \lambda z^2]\frac{\partial G}{\partial z}$$

$$= (z - 1)(\lambda z - \mu)\frac{\partial G}{\partial z}, G(z, 0) = z \tag{3}$$

is a partial differential equation of Lagrange type.

(b) and (c) The characteristic equations are

$$\frac{dt}{1} = \frac{dz}{(1 - z)(\lambda z - \mu)} = \frac{dG}{0}.$$

Hence $G(z, t) = c_1$ and $\frac{dz}{dt} = (1 - z)(\lambda z - \mu)$ is a Riccati differential equation.

Putting $1 - z = 1/u$ this reduces to $\frac{du}{dt} - (\mu - \lambda)u + \lambda = 0$ which is a first order linear differential equation with constant coefficient and is easy to solve. After changing back to z, we obtain

$$C_2 = e^{(\mu-\lambda)t} \frac{\lambda z - \mu}{z - 1} \text{ so that } G(z, t) = \psi\left\{e^{(\mu-\lambda)t} \frac{\lambda z - \mu}{z - 1}\right\}$$

where the function ψ is to be determined from initial condition.

$$G(z, 0) = z = \psi\left\{\frac{\lambda z - \mu}{z - 1}\right\}. \text{ Hence } \psi(u) = \frac{\mu - u}{\lambda - u}.$$

$$\text{Hence } G(z, t) = \frac{\mu(e^{(\lambda-\mu)t} - 1) + z(\lambda - \mu e^{(\lambda-\mu)t})}{(\lambda e^{(\lambda-\mu)t} - \mu) + \lambda z(1 - e^{(\lambda-\mu)t})}$$

Put
$$\xi(t) = \mu\left\{\frac{1 - e^{(\lambda-\mu)t}}{\mu - \lambda e^{(\lambda-\mu)t}}\right\}, \eta(t) = \lambda\left\{\frac{1 - e^{(\lambda-\mu)t}}{\mu - \lambda e^{(\lambda-\mu)t}}\right\} \tag{4}$$

then
$$G(z, t) = \frac{\xi(t) + z\{1 - \xi(t) - \eta(t)\}}{1 - z\eta(t)} \quad \text{if } \lambda \neq \mu \tag{5}$$

and
$$\xi(t) = \eta(t) = \frac{\lambda t}{1 + \lambda t} \text{ and } G(z, t) = \frac{\lambda t + z(1 - \lambda t)}{1 + \lambda t - z\lambda t} \quad \text{if } \lambda = \mu \tag{6}$$

By expanding $G(z, t)$ with respect to z and collecting coefficient of z^n we get

$$\left.\begin{aligned} p_0(t) &= \xi(t) \\ p_n(t) &= (1 - \xi(t))(1 - \eta(t))[\eta(t)]^{n-1}, n \geq 1 \end{aligned}\right\} \text{if } \lambda \neq \mu \tag{7}$$

and
$$\left.\begin{aligned} p_0(t) &= \frac{\lambda t}{1 + \lambda t} \\ p_n(t) &= \frac{(\lambda t)^{n-1}}{(1 + \lambda t)^{n+1}}, n \geq 1 \end{aligned}\right\} \text{if } \lambda = \mu \tag{8}$$

(d) If $\lambda = \mu$, $p_0(t) = \frac{\lambda t}{1 + \lambda t}$ and if $\mu \neq \lambda$, $p_0(t) = \xi(t)$

(e) From the definition of $\xi(t)$ and $\eta(t)$ we see that

$$\lim_{t\to\infty} \xi(t) = \begin{cases} 1 & \text{if } \mu \geq \lambda \\ \frac{\mu}{\lambda} & \text{if } \mu < \lambda \end{cases}$$

and
$$\lim_{t\to\infty} \eta(t) = \begin{cases} \dfrac{\mu}{\lambda} & \text{if } \mu < \lambda \\ 1 & \text{if } \mu \geq \lambda \end{cases}$$

Since for $\lambda \neq \mu$, the $p_n(t)$ contains either $1 - \xi(t)$ or $1 - \eta(t)$, $\lim_{t\to\infty} p_n(t) = 0$ for $n \neq 0$. Since $p_0(t) = \xi(t)$, $\lim_{t\to\infty} p_0(t) = 1$ if $\mu \geq \lambda$ and $\dfrac{\mu}{\lambda}$ if $\mu < \lambda$. Thus ultimate extinction is a certain event unless the ratio λ/μ = birth rate/death rate exceeds one.

(f) By differentiating $G(z, t)$ with respect to z, we obtain

$$E(X(t)) = \begin{cases} e^{(\lambda-\mu)t} & \text{if } \lambda \neq \mu \text{ and} \\ 1 & \text{if } \mu = \lambda \end{cases}$$

$$\text{Var }(X(t)) = \begin{cases} \dfrac{\lambda+\mu}{\lambda-\mu} e^{(\lambda-\mu)t} \{e^{(\lambda-\mu)t} - 1\} & \text{if } \lambda \neq \mu \\ 2\lambda t & \text{if } \lambda = \mu \end{cases}$$

Alternative derivation

$$m(t) = E(X(t)) = \sum_{n=1}^{\infty} np_n(t) \text{ and } m_2(t) = EX^2(t) = \sum_{n=1}^{\infty} n^2 p_n(t).$$

Multiplying both sides of (1) by n and summing over $n = 1, 2, 3, \ldots$ and adding

$$\sum_{n=0}^{\infty} np_n'(t) = -(\lambda+\mu)\sum_{n=1}^{\infty} n^2 p_n(t) + \lambda \sum_{n=1}^{\infty} n(n-1)p_{n-1}(t) + \mu \sum_{n=1}^{\infty} n(n+1)p_{n-1}(t).$$

Now $\sum_{n=1}^{\infty} n(n-1)p_{n-1}(t) = \sum_{n=1}^{\infty} (n-1)^2 p_{n-1}(t) + \sum_{n=1}^{\infty} (n-1)p_{n-1}(t)$

$$= m_2(t) + m(t).$$

Hence $\sum_{n=1}^{\infty} n(n+1)p_{n+1}(t) = \sum_{n=1}^{\infty} (n+1)^2 p_{n+1}(t) - \sum_{n=1}^{\infty} (n+1)p_{n+1}(t)$

$$= \{m_2(t) - p_1(t)\} - \{m(t) - p_1(t)\}$$
$$= m_2(t) - m(t)$$

and
$$\sum_{n=1}^{\infty} np_n'(t) = m'(t)$$

Therefore $m'(t) = -(\lambda + \mu)m_2(t) + \lambda\{m_2(t) + m(t)\} + \mu\{m_2(t) - m(t)\}$

$$= (\lambda + \mu)m(t)$$

Solving, we get
$$m(t) = Ce^{(\lambda-\mu)t}$$

Initial condition gives $m(0) = 1 = C$.

Therefore
$$m(t) = e^{(\lambda-\mu)t} \quad \text{if } \lambda \neq \mu$$

$m_2(t)$ can be found in the same way.

$$\lim_{t\to\infty} E(X(t)) = \begin{cases} 0 & \text{if } \lambda < \mu \\ 1 & \text{if } \lambda = \mu \\ \infty & \text{if } \lambda > \mu \end{cases}$$

Also
$$\text{Var}\,[X(t)] = \begin{cases} \dfrac{\lambda+\mu}{\lambda-\mu} e^{(\lambda-\mu)t}\{e^{(\lambda-\mu)t} - 1\} & \text{if } \lambda \neq \mu \\ > 0 \left(\text{since } \dfrac{e(\lambda-\mu)t - 1}{\lambda - \mu} > 0 \ \text{ for all } t \right) & \\ 2\lambda t & \text{if } \lambda = \mu \end{cases}$$

Hence
$$\lim_{t\to\infty} \text{Var}(X(t)) = \begin{cases} \infty & \text{if } \lambda > \mu \\ \infty & \text{if } \lambda = \mu \\ 0 & \text{if } \lambda < \mu \end{cases}$$

(g) In classifying the states, we observe that state 0 (since $\lambda_0 = 0$) is an absorbing state. Moreover $\{1, 2, 3 \ldots\}$ form a communicating class, so that we can consider only state 1.

Since $p_1(t)$ (= $p_{11}(t)$ essentially since $\delta_{n2} = p_n(0)$) is given above

$$\int_0^\infty p_1(t)\,dt = \int_0^\infty (1 - \xi(t))(1 - \eta(t))\,dt$$

$$= (\mu - \lambda)^2 \int_0^\infty \frac{e^{(\lambda-\mu)t}\,dt}{[\mu - \lambda e^{(\lambda-\mu)t}]^2}$$

Putting $v = \mu - \lambda e^{(\lambda-\mu)t}$, we obtain

$$\int_0^\infty p_1(t)\,dt = \begin{Bmatrix} 1/\lambda \text{ if } \lambda > \mu \\ 1/\mu \text{ if } \lambda > \mu \\ 1/\lambda \text{ if } \lambda > \mu \end{Bmatrix} < \infty.$$

The last case is obtained either by continuity from the cases $\lambda > \mu$ and $\lambda < \mu$ or by calculating $\int_0^\infty \left(1 - \frac{\lambda t}{1 + \lambda t}\right)^2 dt = 1/\lambda$ (from (8)) $< \infty$. Thus $\{1, 2, 3, \ldots\}$ are all transient states.

****Exercise 6.9** For a pure birth process

$$p_{ij}(t) = 0 \text{ for all } j < i \text{ and } t \geq 0 \tag{1}$$

The forward equation reduces to

$$p'_{ij}(t) = \lambda_{j-1} p_{ij-1}(t) - \lambda_j p_{ij}(t) \tag{2}$$

From (1) and (2), since $\lambda_0 \geq 0$

$p'_{ii}(t) = -\lambda_i p_{ii}(t)$ which with initial condition $p_{ij}(0) = \delta_{ij}$, $j \geq i$, gives $p_{ii}(t) = \exp[-\lambda_i t]$.

Since (2) is a linear differential equation, it is also easy to see that

$$p_{ij}(t) = p_{ij}(0)e^{-\lambda_j t} + \int_0^t \lambda_{j-1} e^{-\lambda_j(t-s)} p_{ij-1}(s)ds$$

$$\lambda_{j-1} \int_0^t e^{-\lambda_j(t-s)} p_{ij-1}(s)ds \quad \text{if } j > i$$

Putting $j = i + 1$

$$p_{ii+1}(t) = \lambda_i \int_0^t e^{-\lambda_{i+1}(t-s)} p_{ii}(s)ds.$$

(b) Let $\pi(\theta, i, j) = \int_0^t e^{-\theta t} p_{ij}(t)\,dt$, $\theta > 0$. Integrating by parts

$$\int_0^t e^{-\theta t} p'_{ij}(t)dt = e^{-\theta t} p_{ij}(t)\Big|_0^\infty + \theta \int_0^\infty e^{-\theta t} p_{ij}(t)\,dt$$

$$= \delta_{ij} + \theta\pi(\theta, i, j)$$

Using this in forward equation (2), we obtain

$$-\delta_{ij} + \theta\pi(\theta, i, j) = \lambda_{j-1}\pi(\theta, i, j-1) - \lambda_j\pi(\theta, i, j)$$

So that

$$[\theta + \lambda_j]\pi(\theta, i, j) = \delta_{ij} + \lambda_{j-1}\pi(\theta, i, j-1), j \geq 0 \tag{3}$$

If $j < i$, $\pi(\theta, i, j) = 0$ (being birth process).

Let $i = j$, then we get from (3)

$$\pi(\theta, i, i) = 1/(\theta + \lambda_i) \tag{4}$$

From (3) and (4)

$$\pi(\theta, i, j+1) = \frac{\lambda_i}{(\theta + \lambda_i)(\theta + \lambda_{i+1})}$$

Proceeding successively, we obtain

$$\pi(\theta, i, j) = \frac{\lambda_i \lambda_{i+1} \ldots \lambda_{j-1}}{(\theta + \lambda_i)(\theta + \lambda_{i+1}) \ldots (\theta + \lambda_j)}, j > i \tag{5}$$

Resolving the RHS of (5) into partial fraction, we get

$$\pi(\theta, i, j) = \sum_{k=i}^{j} \frac{B(k, i, j)}{\theta + \lambda_k}, B(k, i, j) = \lim_{\theta \to -\lambda_k} [\theta + \lambda_k]\pi(\theta, i, j).$$

This gives the expression of $B(k, i, j)$. But the inverse Laplace transform of $[\theta + \lambda_k]^{-1}$ is given by $\exp[-t\lambda_k]$, $t \geq 0$ and hence we obtain the desire $p_{ij}(t)$.

(c) $p_{ij}(t)$ can also be written as

$$p_{ij}(t) = \left(\prod_{k=i}^{j-i} \lambda_k\right) \sum_{k=i}^{j} \exp(-\lambda_k t)/\psi'(-\lambda_k) \quad \text{if } j > i$$

$$= \exp(-\lambda_i t) \quad \text{if } j = i$$

where $\psi'(\lambda_k) = \prod_{\substack{r=i \\ r\neq k}}^{j} (\lambda_r - \lambda_k)$. Putting $\lambda_k = k\lambda$, we get

$$p_{ij}(t) = i(i+1)\ldots(j-1)\lambda^{j-i}$$

$$\times \sum_{k=i}^{j} \frac{e^{-\lambda_k t}}{\lambda^{j-i}(i-k)(i+1-k)\ldots(-1)1\ldots(j-k)}$$

$$= e^{-i\lambda t}\binom{j-i}{j-1}(1-e^{-\lambda t})^{j-i}$$

****Exercise 6.10** (Feller and Lundberg)

Define $S(t, i, j) = \sum_{k=0}^{j} p_{ik}(t)$

$$= \sum_{k=i}^{j} p_{ik}(t) \quad (\text{since } p_{ij}(t) = 0 \text{ if } j < i \text{ and } t \geq 0)$$

Since $p_{ik}(t)$'s are transition functions, $S(t, i, \infty) = \sum_{k=i}^{\infty} p_{ik}(t) \leq 1$. Then either $S(t, i, j) = 1$ or < 1 for all t.

Using forward equation we obtain

$$S'(t, i, j) = \sum_{k=i}^{j} p'_{ik}(t) = \sum_{k=i}^{j} \{\lambda_{k-1} p_{ik-1}(t) - \lambda_k p_{ik}(t)\}$$

$$= -\lambda_j p_{ij}(t) \text{ for } j \geq i \tag{1}$$

Let $\sigma(\theta, i, j)$ be the Laplace transform of $S(t, i, j)$. Noting that $S(0, i, j) = 1$, it follows from (1) and integrating by parts that

$$\theta\sigma(\theta, i, j) - 1 = -\lambda_j \pi(\theta, i, j).$$

By previous exercise $\pi(\theta, i, j) = \prod_{k=i}^{j} \frac{\lambda_k}{\theta + \lambda_k} \cdot \frac{1}{\lambda_j}$. We see that

$$\lim_{j\to\infty} \lambda_j \pi(\theta, i, j) = \begin{cases} \prod_{k\geq i}\left[1 + \dfrac{\theta}{\lambda_k}\right]^{-1} & \text{if } \sum_{k=i}^{\infty} 1/\lambda_k < \infty \\ 0 & \text{if } \sum_{k=i}^{\infty} 1/\lambda_k = \infty \end{cases}$$

Hence

(i) $\lim_{j\to\infty} \theta\sigma(\theta, i, j) < 1$ if the series $\sum_{k\geq i} 1/\lambda_k < \infty$

(ii) $\lim_{j\to\infty} \theta\sigma(\theta, i, j) = 1$ if $\sum_{k\geq i} 1/\lambda_k = \infty$.

Consequently by monotone convergence theorem $\lim_{j\to\infty} S(t, i, j) = 1$ or < 1 for all t implies $\lim_{j\to\infty} \theta \int_0^\infty e^{-\theta t} S(t, i, j)\, dt = 1$ or < 1. Hence $\lim_{j\to\infty} S(t, i, j) = \sum_{k=i}^{\infty} p_{ik}(t) = \lim_{j\to\infty} \theta\sigma(\theta, i, j) < 1$ or $= 1$ according as the series $\sum_{k=i} 1/\lambda_k$ converges or diverges.

Let $X(0) = x$ and W_n be the nth jump time $W_n = \tau_i + \ldots + \tau_n$ where τ_k is the interarrival time between $(i + k - 1)$ and $(i + k)$th births. Since the interarrival times τ_k are independent and exponentially distributed, the Ch.f. $\psi_n(\theta)$ of W_n is $\psi_n(\theta) = \prod_{k=i}^{n-1} \left[1 - \frac{i\theta}{\lambda_k}\right]^{-1}$ and clearly (from the theory of infinite product) $\lim_{n\to\infty} \psi_n(\theta) = 0$ if $\sum_k \lambda_k^{-1} = \infty$. Therefore, as $n \to \infty$, $W_n \to \infty$ a.s.

****Exercise 6.11** Let τ_k be the duration of occupancy of the kth state. If τ_k are positive, by definition of Markov process they are independent with distribution function $P(\tau_k \leq x) = 1 - \exp[-\lambda_k x]$ and $E\tau_k = 1/\lambda_k$. Define $\tau'_k = \tau_k$ if $\tau_k \leq 1$ and $\tau'_k = 0$ for $\tau_k > 1$. Then $E\tau'_k \leq \sum_{k=1}^{\infty} E\tau_{k'}$ $\sum_{k=1}^{\infty} P(\tau_k > 1) \leq \sum_{k=1}^{\infty} E(\tau_k)$. Since τ_k's are nonnegative r.v.s, $E\tau_k'^2 \leq E(\tau'_k)$ and therefore convergence of the series $\sum_{k=i}^{\infty} E(\tau'_k)$ implies convergence of the series $\sum_{k=i}^{\infty} E\tau_k'^2$ and hence of the series $\sum_{k=i}^{\infty} \mathrm{Var}(\tau'_k)$. Therefore by Kolmogorov's three series theorem for independent r.v.s $\sum_{k=i}^{\infty} \tau_k$ converges a.s. if $\sum_{k=0}^{\infty} 1/\lambda_k < \infty$. On the other hand,

$$\sum_{k=i}^{\infty} P(\tau_k > 1) = \sum_{k=0}^{\infty} e^{-\lambda_k} \tag{1}$$

and

$$\sum_{k=1}^{\infty} E(\tau'_k) = \sum_{k=0}^{\infty} e^{-\lambda_k} + \sum_{k=0}^{\infty} (1 - e^{-\lambda_k})/\lambda_k \tag{2}$$

Hence $\sum_{k=0}^{\infty} 1/\lambda_k = \infty$ but the series (1) converges, then the series (2) diverges. Thus in the case of divergence of the series $\sum_{k=0}^{\infty} 1/\lambda_k$, one of the series (1) or (2) diverges and consequently $\sum_{k=0}^{\infty} \tau_k$ diverges with probability one. For linear process $\sum_{k=0}^{\infty} 1/\lambda_k = \frac{1}{\lambda} \sum_{1}^{\infty} 1/k = \infty$. Hence after a finite interval of time linear process does not move out to infinity.

Exercise 6.12 (a) From Kolmogorov's forward equations for birth and death process we get for the unconditional probabilities $p_i(t)$ the system of equations

$$p'_i(t) = -p_i(t)(\lambda_i + \mu_i) + p_{i+1}(t)\lambda_{i-1} + p_{i+1}(t)\mu_{i+1},\ i \geq 0 \text{ and } p_{-1}(t) = 0.$$

We assume that $p_i(t)$'s are independent of t.

Then the above system of differential equations degenerates into homogeneous difference equations

$$-(\lambda_i + \mu_i)p_i + \lambda_{i+1} + \mu_i p_{i+1} = 0,\ i \geq 1$$

$$-\lambda_0 p_0 + \mu_1 p_1 = 0.$$

Suppose that $\mu_k > 0$ for $k \geq 1$. Then $p_1 = (\lambda_0/\mu_1)p_0$ and by induction, we find

$$p_k = \frac{\lambda_0\lambda_1 \ldots \lambda_{k-1}}{\mu_1\mu_2 \ldots \mu_k} p_0, \text{ and } \sum_{k=0}^{\infty} p_k = p_0\left(1 + \sum_{k=1}^{\infty} \frac{\lambda_0\lambda_1 \ldots \lambda_{k-1}}{\mu_1\mu_2 \ldots \mu_k}\right)$$

(b) It follows from our assumptions that

$$p_{k,k+1}(\Delta t) = (m-k)(\lambda\Delta t + o(\Delta t),\ k = 0, 1, \ldots, m-1.$$

$$p_{k,k-1}(\Delta t) = \begin{cases} k\mu\,\Delta t + o(\Delta t), & 1 \leq k \leq s \\ s\mu\,\Delta t + o(\Delta t), & m \geq k \geq s \end{cases}$$

$$p_{k,k+r}(\Delta t) = o(\Delta t),\ r \geq 2;$$

that is we have birth and death process with finely many states $E_0, \ldots E_m$ with $\lambda_k = (m-k)\lambda$ for $k - 0, 1, \ldots, m$;

$$\mu_k = k\mu \text{ for } 0 \leq k \leq s,\ \mu_k = s\mu \text{ for } s \leq k \leq m.$$

From part (a) by virtue of equations $p_k = \dfrac{\lambda_0\lambda_1 \ldots \lambda_{k-1}}{\mu_1\mu_2 \ldots \mu_k} p_0,\ k = 1, \ldots, m$

and $\sum_{k=0}^{m} p_k = p_0\left(1 + \sum_{k=1}^{m} \dfrac{\lambda_0 \ldots \lambda_{k-1}}{\mu_1 \ldots \mu_k}\right)$ the stationary distribution is given by

the formulas $p_k = \binom{m}{k}\left(\dfrac{\lambda}{\mu}\right)^k p_0$ for $k \leq s$

$$p_k = \binom{m}{k}\frac{k(k-1)\ldots(s+1)}{s^{k-s}}\left(\frac{\lambda}{\mu}\right)^k p_0 \quad \text{for } s \leq k \leq m$$

$$p_0 = \left[1 + \sum_{k=0}^{s}\binom{m}{k}\left(\frac{\lambda}{\mu}\right)^k + \sum_{k=s+1}^{m}\frac{k(k-1)\ldots(s+1)}{s^{k-s}}\left(\frac{\lambda}{\mu}\right)^k\right]^{-1}$$

****Exercise 6.13**

1 (a) $1 \geq p_{ij}(s) = \sum_{n=0}^{\infty} e^{-\lambda s}\dfrac{(\lambda s)^n}{n!}p_{ij}^{(n)} \geq 0,\ s > 0$ is obvious

(b) $\sum_{j=0}^{\infty} p_{ij}(s) = \sum_{n=0}^{\infty} \sum_{j=0}^{\infty} e^{-\lambda s}(\lambda s)^n / n!\, p_{ij}^{(n)}$ (by Fubini Theorem)

$$= \sum_{n=0}^{\infty} e^{-\lambda s}(\lambda s)^n / n! = 1 \text{ for all } s > 0.$$

(c) $\sum_{j=0}^{\infty} \tilde{p}_{ij}(s)\tilde{p}_{ik}(t) = \sum_{j=0}^{\infty}\sum_{n=0}^{\infty}\sum_{m=0}^{\infty} e^{-\lambda(s+t)} \frac{(\lambda s)^n}{n!}\frac{(\lambda t)^m}{m!} p_{ij}^{(n)} p_{jk}^{(m)}$

$$= \sum_n \sum_m e^{-\lambda(s+t)} \frac{(\lambda s)^n}{n!}\frac{(\lambda t)^m}{m!} p_{ik}^{(n+m)}$$

(by Fubini and Chapman Kolmogorov)

$$= e^{-\lambda(s+t)} \sum_{n=0}^{\infty}\sum_{l=n}^{\infty} \frac{(ls)^n (lt)^{l-n}}{n!(l-n)!} p_{ik}^{(l)} \quad \text{(putting } l = m + n\text{)}$$

$$= e^{-\lambda(s+t)} \sum_{l=0}^{\infty} \frac{p_{ik}^{(l)}}{l!}\left[\sum_{n=0}^{l} \binom{l}{n} (\lambda s)^n (\lambda t)^{l-n}\right]$$

$$= e^{-\lambda(s+t)} \sum_{l=0}^{\infty} \frac{p_{ik}^{(l)}}{l!}[\lambda(s+t)]^l = \tilde{p}_{ik}(s+t).$$

(d) $\tilde{p}_{ij}(s) - \partial_{ij} = (e^{-\lambda s} - 1) + \lambda s \sum_{n=1}^{\infty} e^{-\lambda s}(\lambda s)^{n-1}/n!$ (δ_{ij}'s are Kroncker's delta)

$$\leq -\lambda s + (\lambda s)^2 . c + \lambda s k \text{ as } 0 \leq s \leq s_0.$$

So $\lim_{s\to 0} |\tilde{p}_{ik}(s) - \partial_{ij}| = 0$ (by Dominated Convergence Theorem).

2. Same argument as in d applies to prove that $\tilde{A} - \lambda[P - I]$

or
$$\begin{cases} a_{ij} = q_{ij = \lambda p_{ij}}, i \neq j \\ a_{ii} = -q_{ii = \lambda p_{ii}}, i = j \end{cases}$$

Therefore $$\sum_{j=0}^{\infty} a_{ij} = \lambda\left(\sum_{j=0}^{\infty} (p_{ij} - \partial_{ij})\right) = \lambda(1 - 1) = 0.$$

So that the M.P. is conservative.

3 (a) Suppose $i \to j$ in M then there is an $n \geq 1$ such that $p_{ij}^{(n)} > 0$.

Now $e^{-\lambda s}\frac{(\lambda s)^n}{n!} p_{ij}^{(n)} \geq 0$ for all $n \geq 0$, $\lambda > 0$ and $s > 0$, so that

$$\tilde{p}_{ik}(s) = \sum_{n=0}^{\infty} e^{-\lambda s}\frac{(\lambda s)^n}{n!} p_{ij}^{(n)} > 0.$$

Therefore $i \to j$ in $\tilde{M}$.

Conversely, if $i \to j$ in $\tilde{M}$ then there is some $s > 0$ such that $\tilde{p}_{ij}(s) > 0$. Now

$$\tilde{p}_{ij}(s) = \sum_{n=0}^{\infty} e^{-\lambda s} \frac{(\lambda s)^n}{n!} p_{ij}^{(n)} > 0$$

and $\quad \dfrac{e^{-\lambda s}(\lambda s)^n}{n!} p_{ij}^{(n)} \geq 0$ for all $n \geq 0$, some $s > 0$ and $\lambda > 0$

and there will be at least one $n \geq 1$ such that

$\dfrac{e^{-\lambda s}(\lambda s)^n}{n!} p_{ij}^{(n)} > 0 \Rightarrow p_{ij}^{(n)} > 0$ for some $n \geq 1$. So $i \to j$ in M.

(b) and (c) We have to prove

(i) $\lim_{s\to\infty} \tilde{p}_{ij}(s) = \lim_{s\to\infty} p_{ij}^{(n)}$ and (ii) $\displaystyle\int_0^{\infty} \tilde{p}_{ij}(s)\, ds = \frac{1}{\lambda} \sum_{n=0}^{\infty} p_{ij}^{(n)}$.

Proof of (i) Let $p_{ij}^{(n)} \to \alpha_j$ (exists since the M.C. is aperiodic), then $|\tilde{p}_{ij}(s) - \alpha_j| \leq \sum_{n=0}^{\infty} \dfrac{e^{-\lambda s}(\lambda s)^n}{n!} |p_{ij}^{(n)} - \alpha_j|$ and for every $\varepsilon > 0$, there exists $N(\varepsilon)$ such that for $n \geq N$ $|p_{ij}^{(n)} - \alpha_j| \leq \varepsilon/2$. Since $\dfrac{e^{-\lambda s}(\lambda s)^n}{n!} \to 0$ as $s \to \infty$, we can choose s so large that $e^{-\lambda s}\dfrac{(\lambda s)^n}{n!} \leq \dfrac{\varepsilon}{4N}$ for $n \geq N$. Also $|p_{ij}^{(n)} - \alpha_j| \leq 2$ for all $n \geq 1$. Hence $|\tilde{p}_{ij}(s) - \alpha_j| \leq \left(\sum_{n=0}^{N} + \sum_{n=N+1}^{\infty}\right)\left(\dfrac{e^{-\lambda s}(\lambda s)^n}{n!}\right) |p_{ij}^{(n)} - \alpha_j \leq \dfrac{\varepsilon}{4N} N \max_{1\leq n\leq N} |p_{ij}^{(n)} - \alpha_j| + \varepsilon/2 \sum_{n=N+1}^{\infty} e^{-\lambda s}\dfrac{(\lambda s)^n}{n!} \leq \varepsilon$. Hence $\tilde{p}_{ij}(s) \to \alpha_j$ as $s \to \infty$.

Proof of (ii) $\displaystyle\int_0^{\infty} p_{ij}(s)\, ds = \sum_{n=0}^{\infty} \frac{p_{ij}^{(n)}}{n!} \int_0^{\infty} e^{-\lambda s}(\lambda s)^n\, ds$ (by Fubini Theorem)

$$= \frac{1}{\lambda} \sum_{n=0}^{\infty} p_{ij}^{(n)}.$$

From (i) and (ii) we get (b) and (c) since j is transient in M iff

$$\sum_n p_{ij}^{(n)} < \infty \text{ iff } \int_0^{\infty} p_{ij}(s)\, ds < \infty \text{ iff } j \text{ is transient in } M.$$

Similarly let M be positive and $\lim_{s\to\infty} p_{ij}(s) = \pi_j = \lim_{n\to\infty} p_{jj}^{(n)} > 0$.

****Exercise 6.14**

4. Let M be irreducible and positive with stationary distribution $\{\pi_j, k \in S\}$.

Then $|\tilde{p}_{ij}(s) - \pi_j| \leq \sum_{n=0}^{\infty} \dfrac{e^{-\lambda s}(\lambda s)^n}{n!} \rho^n = e^{-\lambda s(1-\rho)}$.

Let M be transient. Then $\lim_{s\to\infty} \tilde{p}_{ij}(s) = 0 = \lim_{s\to\infty} p_{jj}^{(n)}$ since $p_{ij}^{(n)} \leq \rho^n, 0 \leq \rho < 1$ and also

$$\tilde{p}_{ij}(s) \leq \sum_{n=0}^{\infty} \frac{e^{-\lambda s}(\lambda s)^n}{n!} \rho^n = e^{-\lambda s(1-\rho)}.$$

Let M be null. Then $p_{ij}^{(n)} \leq \rho^n$ even though $p_{ij}^{(n)} \to 0$. For $\infty = \sum_n p_{jj}^{(n)} \leq \sum_n \rho^n = \dfrac{1}{1-\rho} < \infty$.

5. Let $\tilde{A}$ be infinitesimal generator, $\sum_j a_{ij} = 0$. Define a matrix $\tilde{P} = \dfrac{\tilde{A}}{K} + I$ where $K \geq \sup_i |a_{ii}| < \infty$ and $\tilde{P}$ a stochastic matrix since $\tilde{P} > 0$, $\sum_j \tilde{p}_{ij} = \dfrac{1}{K}\sum_j a_{ij} + 1 = 1$. So P determines a discrete M.C. $\tilde{M}$ (with some initial distribution). Define $\tilde{p}_{ij}(s) = \sum_{n=0}^{\infty} e^{-Ks}\dfrac{(Ks)^n}{n!}\tilde{p}_{ij}^{(n)}$. Then $\tilde{\tilde{A}} = K[\tilde{P} - I] = \tilde{A}$ by part (2) of Exercise 6.13, and the definition of $\tilde{P}$. Hence $\tilde{\tilde{M}}$ and $\tilde{M}$ are the same by Chung's theorem. Hence $\tilde{M}$ is positive $\Leftrightarrow$ $\tilde{\tilde{M}}$ is positive $\Leftrightarrow$ $\tilde{M}$ is positive by Foster's Theorem (By Exercise 2.36) ($\sum_i x_i \tilde{p}_{ij} = x_j$ has a solution such that $0 < \sum_i |x_i| < \infty$ (by definition of $\tilde{P}$) $\Leftrightarrow \sum_i x_i \left[\dfrac{a_{ij}}{K} + \partial_{ij}\right] = x_j$ has a solution such that $0 < \sum_i |x_i| < \infty \Leftrightarrow \sum_i x_i a_{ij} = 0$ has a solution such that $0 < \sum_i |x_i| < \infty$). This proves 5 of Exercise 6.14.

Exercise 6.15

1. $$|A - \lambda I| = \begin{vmatrix} -(\alpha+\lambda) & \beta & \gamma \\ \gamma & -(\alpha+\lambda) & \beta \\ \beta & \gamma & -(\alpha+\lambda) \end{vmatrix} = 0 \text{ or}$$

$$-(\alpha+\lambda)[(\alpha+\lambda)^2 - \gamma\beta] - \beta[\alpha\gamma - \gamma\lambda - \beta^2] + \gamma\{\gamma^2 + \alpha\beta + \beta\gamma] = 0$$

$$\text{or } -\lambda^3 - 3\lambda^2\alpha - \lambda(3\alpha^2 - 3\beta\gamma) - \alpha^3 + \beta^3 + \gamma^3 + 3\alpha\beta\gamma = 0$$

$$\text{or } \lambda^3 + 3\lambda^2\alpha + \lambda(3\alpha^2 - 3\beta\gamma) = 0 \ (\therefore \beta + \gamma = \alpha)$$

$$\text{or } \lambda\left(\lambda + \frac{3\alpha}{2} - 1/2(\beta - \gamma)\sqrt{3}\right)\left[\lambda + \frac{3\alpha}{2} + 1/2(\beta-\gamma)\sqrt{3}\right] = 0$$

So $\lambda = 0, -\rho + i\sigma$ or $-\rho - i\sigma$ where $\rho = \dfrac{3\alpha}{2}$ and $\sigma = \dfrac{1}{2}(\beta - \gamma)\sqrt{3}$.

2. Let $P(t) = R + Se^{-(\rho+i\sigma)t} + Te^{-(\rho-i\sigma)t}$. Symmetries of A tell us that

$$p_{11}(t) = p_{22}(t) = p_{33}(t)$$

$$p_{12}(t) = p_{23}(t) = p_{31}(t)$$

$$p_{13}(t) = p_{21}(t) = p_{32}(t)$$

Also A^2 is of the same form (has same symmetries) as A and so on.

Hence $p_{11}(t) = \sum_n A_{11}^{(n)} t^n/n! = \sum_n A_{22}^{(n)} t^n/n! = p_{22}(t)$, and so on. The given M.C. is finite and clearly irreducible (since $\rho > 0$ or $\beta > 0$) and hence positive recurrent (If not obvious, use skeleton chains to reduce it to discrete time). Hence $\lim_{t\to\infty} P(t) = R$. By Exercise 2.14 every row of R satisfies $\sum_i x_i a_{ij} = 0$.

Since $x_1 = x_2 = x_3 = 1/3$ is a positive solution, $\sum_{i=1}^{3} x_1 = 1$, it is the stationary distribution of $P(t)$. Hence every element of R is 1/3. In particular R is real.

Note Actual computation also shows this.
To determine P, let $S = S' + iS''$, $T = T' + iT''$ where S', S'', T'.
T'' have real entries. Then $P(t)$ becomes $P(t) = R + e^{-\rho t}[(S' + T')$

$\cos(\sigma t) + [S'' - T''] \sin(\sigma t) + i[S'' + T''] \cos(\sigma t) + i[T' - S'] \sin(\sigma t)$
(0)

Now $P(t)$ is real, hence $\quad S'' = -T''$ (1)

and $\quad S' = T'$ (2)

But $P(0) = I$ and hence $\quad I = R + S' + T'$ (3)

Also $P(0) = A$, and hence $\quad A = -\rho[s' + T'] + \sigma[S'' - T'']$ (4)

From (2) and (3), $\quad S' = T' = 1/2(I - R)$ (5)

From (1), (4) and (5) $\quad S'' = -T'' = \dfrac{A}{2\sigma} + \dfrac{\rho}{2\sigma}(I - R)$

This yields immediately (with convention above)

$$S'_{ii} = T'_{ii} = 1/3$$

$$S''_{ii} = -T''_{ii} = -\frac{\alpha}{2\sigma} + \frac{\rho}{3\sigma} = \frac{2\rho - 3\alpha}{6\sigma} = 0, \text{ since } \rho = \frac{3\alpha}{2}.$$

$$S'_{ij} = T'_{ij} = -1/6 \text{ if } i \neq j$$

$$S''_{ij+1} = -T''_{ii+1} = \frac{\beta}{2\sigma} - \frac{\rho}{3\sigma} = \frac{\beta - \gamma}{4\sigma} = 1/3 \cdot \sqrt{3/2} \quad \left(\sigma = \frac{1}{2}(\beta - \gamma)\sqrt{3}\right)$$

$$S''_{i,i+2} = -T''_{i,i+2} = \frac{\gamma}{2\sigma} - \frac{\rho}{6\sigma} = \frac{\gamma - \beta}{4\sigma} = -1/3\sqrt{3/2}$$

Using these values in (0) the desired result follows.

3. Let $h = \dfrac{4\pi}{(\rho - \gamma)\sqrt{3}}$ then $\sin(\sigma h) = 0$, $\cos \sigma h = 1$

Let $p_{ii}(h) = 1/3 + 2/3e^{-ph}$

$$p_{ij}(h) = 1/3 - 1/3e^{-\rho h} \quad \text{if } i \neq j$$

which depends only on

$$\rho h = 2\sqrt{3\pi} \cdot \frac{\beta + \gamma}{\beta - \gamma} \quad \left(\rho = \frac{3\alpha}{2} = \frac{3}{2}(\beta + \gamma)\right)$$

All continuous time Markov chain with infinitesimal generator of the given form and with $\dfrac{\beta+\gamma}{\beta-\gamma}$ constant have the same skeleton $\{X(h), X(2h), X(3h), \ldots\}$, where h is defined above.

Since we look for solution of Elfving's problem, we need that

$$P(1) = \begin{bmatrix} p & q & r \\ r & p & q \\ q & r & p \end{bmatrix}, p \neq 1 \tag{6}$$

All properties of transition function matrix $P(t)$ (namely semigroup, non negative, stochastic) are obviously satisfied if (6) has those properties. So

$$p = 1/3 + 2/3e^{-\rho} \cos \sigma$$

and
$$q = 1/3 - 1/3e^{-\rho} [\cos \sigma - \sqrt{3} \sin \sigma]$$

(Note if $p = q = r = 1/3$ then $\cos \sigma = \sin \sigma = 0$. Hence there is no solution for the problem).

Hence
$$\left.\begin{aligned} e^{-\rho} \cos \sigma &= \frac{3p-1}{2} \\ e^{-\rho} \sin \sigma &= \frac{\sqrt{3}}{2}(p+2q-1) \end{aligned}\right\} \tag{7}$$

so that $\sigma = \sqrt{3}\,\dfrac{p+2q-1}{3p-1}$.

This determines infinitely many σ's if we can find ρ also. If ρ can be determined then every continuous time Markov chain with that ρ and one of the possible σ values is a solution to Elfving's problem. Now from (7)

$$\begin{aligned} e^{-2\rho} &= e^{-2\rho} (\cos^2 \sigma + \sin^2 \sigma) \\ &= 3(p^2 + q^2 + pq - p - q) + 1 \end{aligned}$$

The right hand side is minimal for $p = q = r = 1/3$. This is excluded. Hence $e^{-2\rho} > 0$. But $q \geq 0$, so $p \leq p + q$; $r \geq 0$ so $p \leq 1 - q$. Moreover one of these inequalities is strict since otherwise $q = 0.p = 1$.

So
$$p^2 < (p+q)(1-q) \text{ or } 3[p^2 + q^2 + pq - p - q] < 0$$

or $e^{-2\rho} < 1$. Hence ρ is well determined.

Exercise 6.16

(a) is completely similar to Feller-Arley process.

(b) Following Feller-Arley the P.G.F. of $\{p_n(t)\}_0^\infty$ is

$$G(t, z) = [\xi(t) + z\{1 - \xi(t) - \eta(t)\}]\{1 - z\eta(t)\}.$$

By expanding with respect to z, we get

$$p_0(t) = \xi(t),\ p_n(t) = (1 - \xi(t))(1 - \eta(t)[\eta(t)]^{n-1},\ n \geq 1$$

where $\xi(t) = \dfrac{1 - e^{-\rho}}{W},\ \eta(t) = 1 - 1/W,\ W = e^{-\rho}\left\{1 + \int_0^t e^{\rho}\mu t\,dt\right\}$

and $\rho = \dfrac{\mu t^2}{2} - \lambda t.\ E(X(t)) = \sum_{n=1}^{\infty} np_n(t) = G_2'(t, z)\Big|\ z = 1$

$$= \frac{\{1 - \xi(t) - \eta(t)\}\{1 - \eta(t)\} + \eta(t)\{1 - \eta(t)\}}{\{1 - \eta(t)\}^2}$$

$$= \frac{1 - \xi(t)}{1 - \eta(t)} = e^{\rho} = e^{\lambda t - \mu t^2/2}$$

Hence $E(X(t)) \to \begin{cases} 0 & \text{if } \mu > 0 \\ 1 & \text{if } \lambda = \mu \\ \infty & \text{if } x > 0, \mu = 0 \end{cases}$

(c) The zero state, corresponding to extinction, is obviously an absorbing state. The probability that extinction has occurred at or before time t is the coefficient of z^0 in $G(t, z)$. Since T is the time till absorption,

$$P(T \leq t) = P[X(t) = 0] = p_0(t) = \xi(t)$$

When $\lambda = 0,\ p_0(t)\big|_{\lambda=0} = \dfrac{1 - e^{-\mu t^2/2}}{e^{-\mu t^2/2}\left\{1 + \int_0^t e^{\mu\tau^2/2}\,d\tau\right\}} = 1 - e^{-\mu t^2/2}.$

$$E(T) = \int_0^t [1 - p_0(t)]\,dt = \int_0^{\infty} e^{-\mu t^2/2}\,dt = \sqrt{\pi/2\mu} \quad \text{if } \lambda = 0$$

Chapter 7

Exercise 7.1

1st Solution Let $X(t)$ be the number of shocks upto time t. Hence

$$P[T \leq t] = \sum_{n=k}^{\infty} e^{-\lambda t}(\lambda t)^n/n! = \int_0^t \frac{\lambda x^{k-1} e^{-\lambda x}}{\Gamma k}\,dx \tag{1}$$

(integrating by parts).

Differentiating (1) gives $f(t) = \dfrac{\lambda^k t^{k-1} e^{-\lambda t}}{\Gamma k}, t > 0.$

2nd Solution

Let X_i be the interarrival time till the ith shock, then $P[T \le t] = P\left[\sum_{t=1}^{k} X_i \le t\right]$.

So the Laplace transform is $\phi_T(s) = [\phi_{x_i}(s)]^k = \left(\frac{\lambda}{\lambda + s}\right)^k$. By uniqueness theorem of Laplace transform the result follows.

Third Solution
Let $T = T_k$. Then by induction on k,

$$P[t < T_k \le t + dt] = \lambda\, dt \int_0^t e^{-\lambda(t-u)} \frac{\lambda^{k-1} u^{k-2} e^{-\lambda u}}{\Gamma(k-1)} du = \frac{\lambda^k e^{-\lambda t} t^{k-1}}{\Gamma k} dt$$

Exercise 7.2

$$P[Y(t) = m] = \sum_{k=m}^{\infty} P[Y(t) = m \mid X(t) = k]\, P[X(t) = k]$$

$$= \sum_{k=m}^{\infty} \binom{k}{m} p^m q^{k-m} e^{-\lambda t} (\lambda t)^k / k! = e^{-\lambda pt} (\lambda pt)^k / k!$$

since each arrival is registered with probability p independent of other arrivals, and probability of m registered arrivals among k arrivals is $\binom{k}{m} p^m q^{k-m}$, $q = 1 - p$.

Exercise 7.3 $F(t + h) - F(t) = P$ {failure in $(t, t + h)$ and no failure in $(0, t)$} = P {failure in $(t, t + h) \mid$ no failure in $(0, t)$}. P [no failure in $(0, t)$] = $\{g(t)h + o(h)\}$ $[1 - F(t)]$. Dividing by h and letting $h \downarrow 0$, we obtain

$$\frac{F'(t)}{1 - F(t)} = g(t) \Rightarrow -\log[1 - F(t)] = \log c + \int_0^t g(\tau)\, d\tau$$

From the initial condition $F(0) = 0 \Rightarrow c = 1$.

Hence $1 - F(t) = \exp\left[-\int_0^t f(x)\, dx\right]$.

Note For $g(x) = 1 \Rightarrow 1 - F(t) = e^{-\lambda t}$. If $1 - F(t) = e^{-t^\alpha}$ then

$$F'(t) = e^{-t^\alpha} \alpha t^{\alpha-1}. \text{ So } g(t) = \frac{F'(t)}{1 - F(t)} = \alpha t^{\alpha-1}.$$

Exercise 7.4 Let $\tau = W^* - W$. Then τ is the interval time and $P(\tau \le s) = 1 - e^{-\lambda s}$. Now $P(N = k) = P[Y(W^*) - Y(W) = k]$

$$= E\{P[Y(W^*) - Y(W) = k \mid \tau = s]\}$$

$$= \int_0^\infty e^{-\mu s}(\mu s)^k / k!\, \lambda e^{-\lambda s}\, ds = \frac{\lambda \mu^k}{k!}\int_0^\infty s^k e^{-(\lambda+\mu)s}\, ds$$

$$= \lambda/(\lambda + \mu)(\mu/(\lambda + \mu))^k.$$

Exercise 7.5

(a) Let $0 \le t_0 < t_1 < \ldots < t_m$. Then

$$Y(t_0) = \sum_{n=1}^{X(t)} \xi_n,\ Y(t_k) - Y(t_{k-1}) = \sum_{n=X(t_{k-1})}^{X(t_k)} \xi_n,\ k = 1, 2, \ldots, m$$

Now, $\{\xi_i\}$ are the i.i.d., $X(t)$ being a Poisson process, $X(t_k) - X(t_{k-1})$ and $X(t_{k-1}) - X(t_{k-2})$ are independent and $Y(t_k) - Y(t_{k-1})$ and $Y(t_{k-1}) - Y(t_{k-2})$ are also independent.

(b) $$\phi_{Y(t)}(s) = \sum_{n=0}^{\infty} E\exp[iu\, Y(t) \mid X(t) = n].\, P[X(t) = n]$$

$$= \sum_{n=0}^{\infty} [\phi_\xi(u)]^n e^{-\lambda t}(\lambda t)^n / n! = e^{-\lambda t}\exp[\lambda t \phi_\xi(u)]$$

(c)
$$\begin{aligned} EY(t) &= E[E(Y(t) \mid X(t))] = E[X(t)E(\xi)] = \lambda t E(\xi) \\ \text{Var}(Y(t)) &= E[\text{Var}(Y(t) \mid X(t))] + \text{Var}(E[Y(t) \mid X(t)]) \\ &= E(X(t)\text{Var}(\xi)) + \text{Var}[X(t)E(\xi)] = \lambda t\, \text{Var}(\xi) + \lambda t (E(\xi)^2) \\ &= \lambda t[\text{Var}(\xi) + (E(\xi))^2] = \lambda t E\xi^2. \end{aligned}$$

Exercise 7.6 Since the interarrival times $T_1, T_2, \ldots$ in a Poisson process are i.i.d. with common distribution exponential with parameter λ, $E(N(T_1) \mid T_1 = t) = \lambda t$. Hence

$$\begin{aligned} E(N(T_1)T_1) &= E[T_1 E(N(T_1) \mid T_1)] \\ &= \int_0^\infty t\lambda t \lambda e^{-\lambda t}\, dt = \frac{1}{\lambda}\int_0^\infty x^2 e^{-x}\, dx \\ &= 2/\lambda. \\ \text{Var}(N(T_1)T_1) &= E(N(T_1)T_1)^2 - E^2(N(T_1)T_1) \\ &= \int_0^\infty t^2 E(N^2(t) \mid T_1 = t)\, \lambda e^{-\lambda t}\, dt - \frac{4}{\lambda^2} \\ &= \int_0^\infty t^2[(\lambda t)^2 + \lambda t]\, \lambda e^{-\lambda t}\, dt - \frac{4}{\lambda^2} \\ &= \int_0^\infty t^4 \lambda^3 e^{-\lambda t}\, dt + \int_0^\infty t^3 \lambda^2 e^{-\lambda t}\, dt - \frac{4}{\lambda^2} \\ &= \frac{\Gamma^5}{\lambda^2} + \frac{\Gamma^4}{\lambda^2} - \frac{4}{\lambda^2} = \frac{24}{\lambda^2} + \frac{6}{\lambda^2} - \frac{4}{\lambda^2} = \frac{26}{\lambda^2}. \end{aligned}$$

Exercise 7.7 Since $EX(t) = \lambda t$, it follows that

$$EY(t) = E(X(t+1) - X(t)) = \lambda(t+1) - \lambda t = \lambda.$$

Due to independent increment of Poisson process $X(t)$ the r.v.s $X(s+1) - X(s)$ and $X(t+1) - X(t)$ are independent if $|t-s| \geq 1$. Consequently $\rho_Y(s, t) = 0$ for $|t-s| \geq 1$. Suppose $s \leq t < s+1$. Then

$$\begin{aligned}\text{Cov}(Y(s), Y(t)) &= \text{Cov}(X(s+1) - X(s), X(t+1) - X(t)) \\ &= \text{Cov}(X(t) - X(s) + X(s+1) - X(t), X(s+1) \\ &\quad - X(t) + X(t+1) - X(s+1)).\end{aligned}$$

Again by independent increment property

$$\begin{aligned}&\text{Cov}(X(t) - X(s), X(s+1) - X(t)) = 0 \\ &= \text{Cov}(X(t) - X(s), X(t+1) - X(s+1)) \\ &= \text{Cov}(X(s+1) - X(t), X(t+1) - X(s+1)).\end{aligned}$$

Also Cov $(X(s+1) - X(t), X(s+1) - X(t)) = \text{Var}(X(s+1) - X(t)) = \lambda(s+1-t)$. Therefore Cov $(Y(s), Y(t)) = \lambda(s+1-t)$. By using symmetry

$$\rho_Y(s,t) = \begin{cases} \lambda(1 - |t-s|) & \text{if } |t-s| < 1 \\ 0 & \text{if } |t-s| \geq 1 \end{cases}$$

Therefore $Y(t)$, $-\infty < t < \infty$ is a second order stationary process having mean λ and covariance function

$$\rho_Y(s,t) = \begin{cases} \lambda(1 - |t|) & \text{if } |t| < 1 \\ 0 & \text{if } |t| > 1 \end{cases}$$

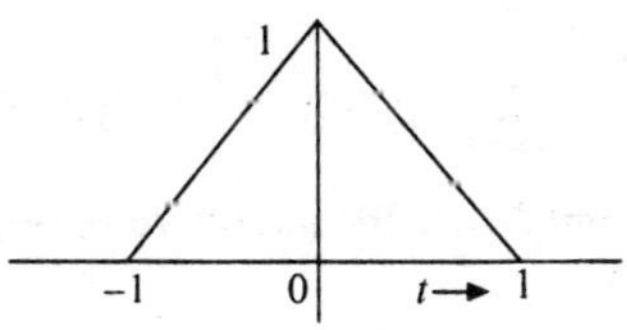

$\rho_Y(t)$ decreases linearly to zero as $|t|$ increases from 0 to 1 and remains zero for all larger values of $|t|$.

Exercise 7.8 Now $Y(t) = X(t) - \lambda t$ is a martingale.

Set $X(0) = 0$ a.s. Then $X(T_a) = a \Rightarrow E(Y(T_a)) = F(X(T_a)) - \lambda ET_a = E(Y(0))$ by the optional sampling theorem (see Appendix IV). Also, $Y(0) = X(0) - \lambda.\ 0 = 0$ a.s.

$$\Rightarrow \qquad E(X(T_a)) = \lambda E(T_a) \Rightarrow a = \lambda E(T_a) \Rightarrow \frac{a}{\lambda} = E(T_a) \qquad (1)$$

Also $U(t) = Y^2(t) - \lambda t$ is a Martingale. Since T_a is a stopping time by optional sampling theorem (see Basu [1998], Chapter 11).

$$E(U(T_a)) = E(Y^2(T_a)) - \lambda E(T_a) = EU(0) = 0$$

$$\Rightarrow \quad E[Y^2(T_a)] = \lambda E(T_a) = a \text{ (by (1))} \tag{2}$$

Also

$Y(T_a) = X(T_a) - \lambda T_a$ and $E[Y^2(T_a)] = E(X(T_a) - \lambda T_a)^2 = E[\lambda E(T_a) - \lambda T_a]^2$

$(\because\ X(T_a) = a,\ E(T_a) = a/\lambda \Rightarrow X(T_a) = \lambda E(T_a)$. Therefore

$$E[Y^2(T_a)] = \lambda^2 E[T_a - E(T_a)]^2$$

$$\Rightarrow \quad a = \lambda^2 \text{ Var } (T_a) \Rightarrow \text{Var } (T_a) = a/\lambda^2 \text{ (by (2))}$$

Again $V(t) = \exp[-\theta X(t) + \lambda t(1 - e^{-\theta})]$, $t \geq 0$ is a Martingale.

Hence

$$E[V(T_a)] = E\{\exp[-\theta X(T_a) + \lambda T_a(1 - e^{-\theta})]\} = E(V(0)) = 1$$

$$\Rightarrow \quad 1 = E\{\exp[-\theta X(T_a)] \exp(-\alpha T_a)\} \text{ where } \alpha = -\lambda(1 - e^{-\theta})$$

$$\Rightarrow \quad 1 = e^{-\theta a} E[\exp(-\alpha T_a)] \Rightarrow E[\exp(-\alpha T_a)] = e^{a\theta}.$$

Again $\quad \dfrac{\lambda + \alpha}{\lambda} = e^{-\theta} \Rightarrow \dfrac{\lambda}{\alpha + \lambda} = e^{\theta} \Rightarrow \left(\dfrac{\lambda}{\alpha + \lambda}\right)^a = e^{a\theta}$

$$\Rightarrow \quad E[\exp(-\alpha T_a)] = \left(\frac{\lambda}{\alpha + \lambda}\right)^a$$

$\Rightarrow T_a$ has a gamma distribution with parameters a and λ.

Exercise 7.9 Let $N(t) = N_1(t) + N_2(t)$, $\lambda = \lambda_1 + \lambda_2$, $\lambda_1 > 0$, $\lambda_2 > 0$ be a Poisson process, where $\{N_1(t)\}$ and $\{N_2(t)\}$ are independent renewal processes.

Then $\quad e^{-\lambda t}(\lambda t)^k/k! = e^{-\lambda t} t^k/k! \sum_{m=0}^{k} \binom{k}{m} \lambda_1^m \lambda_2^{k-m}$

$$= P(N(t) = k) = \sum_{m=0}^{k} P[N_1(t) = m]\, P[N_2(t) = k - m]$$

by independence of $N_1(t)$ and $N_2(t)$. By inductive reasoning, since this is valid for $k = 0, 1, 2, \ldots$ and all k,

$$e^{-\lambda t} \frac{t^k}{k!} \binom{k}{m} \lambda_1^m \lambda_2^{k-m} = P[N_1(t) = m]\, P[N_2(t) = k - m].$$

Putting $k - m = n$ we get

$$\frac{e^{-\lambda_1 t}(\lambda_1 t)^m}{m!} e^{-\lambda_2 t} \frac{(\lambda_2 t)^n}{n!} = P[N_1(t) = m]\, P[N_2 = n].$$

Summing over n we obtain

$$\frac{e^{-\lambda_1 t}(\lambda_1 t)^m}{m!} \cdot 1 = P[N_1(t) = m] \sum_{n=0}^{\infty} P[N_2(t) = n] = P[N_1(t) = m].$$

Exercise 7.10 This problem can be looked upon as a birth and death process with birth rates and death rates

$$\lambda_n = \begin{cases} \lambda & \text{if } 0 \le n \le m-1 \\ 0 & \text{if } n \ge m \end{cases}, \quad \mu_n = \begin{cases} n\mu & \text{if } 0 \le n \le m \\ 0 & \text{if } n > m \end{cases}$$

respectively. The arrangement is as follows:

Let I be an interval of length h. Let $\{X(t), t \ge 0\}$ be the number of busy channels at time t with $p_{ij}(h) = P[X(t + h) = j \mid X(t) = i]$.

Then

$$p_{n,n+1}(h) = P\,[\text{a new call arrives during 1}] + o(h) = \lambda h + o(h).$$

$p_{n,n-1}\,(h) = P$ [exactly one of n conversation in progress terminates during I] + $o(h)$

$$= \binom{n}{1}\left[\int_0^h \mu e^{-\mu x}\,dx\right]\left[\int_h^\infty \mu e^{-\mu x}\,dx\right]^{n-1} + o(h)$$

$$= n(1 - e^{-\mu h})\,e^{-(n-1)\mu h} + o(h)$$

$$= n\mu h + o(h).$$

Taking $h \to 0$ we get $\lambda_n = \lambda$ and $\mu_n = n\lambda$.

Let $p_n(t) = P[X(t) = n] = P[n$ channels are busy at time $t]$, $n = 0, 1, \ldots, m$. Then for a Birth and death process the basic differential equations become

$$p'_o(t) = -\lambda p_0(t) + \mu p_1(t) \tag{1}$$

$$p'_n(t) = (\lambda + n\mu)\,p_n(t) + \lambda p_{n-1}(t) + (n+1)\,\mu p_{n+1}(t) \qquad n = 1, \ldots, m. \tag{2}$$

We know that $\lim_{t\to\infty} p_n(t) = \pi_n$ exists and are independent of initial conditions

$p_{io}(0) = 1$ and $p_n(0) = 0$, $n \ne i$. Also $\lim_{t\to\infty} p'_n(t) = 0$.

Either we can derive explicit solutions of the differential equations (1) and (2) which is known from D.G. Kendall's result or just taking limits $t \to \infty$ in (1) and (2) and putting $p'_n(t) = 0$. We get

$$\lambda\pi_0 = \mu\pi_1$$

$$(\lambda + n\mu)\pi_n + \lambda\pi_{n-1} + (n+1)\mu\pi_{n+1}$$

Let $\alpha_n = n\mu\pi_n - \lambda\pi_{n-1}$. Then from (4) we get

$$\alpha_{n+1} = \alpha \text{ for all } n \ge 1. \text{ So } \alpha_1 = \alpha_2 = \ldots$$

But from (3) we get $\alpha_1 = 0$ and $\alpha_n = 0$ for all $n \ge 1$.

Therefore, from (4) we get $n\mu\pi_n = \lambda\pi_{n-1}$ for $n \ge 1$ or $\pi_n = \dfrac{\lambda}{\mu}\cdot\dfrac{1}{n}\pi_{n-1}\ (\mu \ne 0)$.

By iteration we get

$$\pi_n = \left(\frac{\lambda}{\mu}\right)^n \Big/ n!\,\pi_0 \text{ for all } n \ge 1. \text{ But } 1 = \sum_{k=0}^{m} \pi_k = \pi_0 \sum_{k=0}^{m} \left(\frac{\lambda}{\mu}\right)^k \Big/ k!$$

Therefore, $$\pi_0 = \frac{1}{\sum_{k=0}^{m} (\lambda/m)^k \frac{1}{k!}}.$$

So the stationary probabilities are given by

$$\pi_n = \frac{\left(\frac{\lambda}{\mu}\right)^n \frac{1}{n!}}{\sum_{k=0}^{m} \left(\frac{\lambda}{\mu}\right)^k \frac{1}{k!}}, n = 0, 1, \ldots, m.$$

Exercise 7.11

Let $$EY(t) = \int_0^\infty P(Y(t) \geq x)\, dx = I < \infty \tag{1}$$

Note that $$P(Y(t) \geq x) = \int_0^\infty P(Y(t) \geq x \mid S_1 = y)\, dF(y)$$

$$= \int_0^t P(Y(t-y) \geq x)\, dF(y) + \int_{t+x}^\infty dF(y)$$

where for $y \leq t$. Since the origin is shifted to y, so that point t becomes $t - y$, while for $y > t$, we observe that

$$P(Y(t) \geq x \mid s_1 = y) = 0 \text{ if } t < y \leq t + x \text{ and } = 1 \text{ if } y > t + x.$$

Substituting this recursive formula in (1) we obtain

$$I[1 - F(t)] = \int_t^\infty [1 - F(y)]\, dy \tag{2}$$

Since $F(x)$ is continuous, the right hand side of (2) is differentiable; differentiation and the solution of the resulting differential equation give the exponential distribution, since $F(0) = 0$. We know that $N(t)$ is Poisson iff $F(x)$ is exponential. In fact,

$$\infty > I = \int_0^\infty \int_0^t P(Y(y-t) \geq x)\, dF(y)\, dx + \int_0^\infty \int_{t+x}^\infty dF(y)\, dx$$

$$= \int_0^t \left[\int_0^\infty P(Y(y-t) \geq x)\, dx\, dF(y) + \int_0^\infty [1 - F(t+x)]\, dx\right.$$

(since I does not depend on t)

$$= \int_0^t I\, dF(y) + \int_t^\infty [1 - F(y)]\, dy \qquad (y = t + x)$$

$$= IF(t) + \int_t^\infty [1 - F(y)]\, dy \qquad (F(0) = 0)$$

Hence $I(1 - F(t)) = \int_t^\infty [1 - F(y)]\, dy$ and $I \frac{d}{dt}(1 - F(t)) = 1 - F(t)$ with initial condition $1 - F(0) = 1$ gives $1 - F(t) = e^{-t/I}$. Hence $F(t) = 1 - e^{-t/I}$.

Therefore, the Renewal Process $Y(t)$ is Poisson.

Exercise 7.12 (a) Let $p(t)$ be the probability that an injured man is still in the hospital at time t. Let T be the time spent in the hospital and S be the time up to the accident. Let N_t be the Poisson process.

$$P[X(t)=k] = \sum_{m=k}^{\infty} e^{-\lambda t}(\lambda t)^m/m! \binom{m}{k} (p(t))^k \; [1-p(t)]^{m-k}$$

$$= e^{-\lambda tp(t)} \; [\lambda(t)p(t)]^k/k! \tag{1}$$

Then $p(t) = P[T+S \geq t] = \int_0^t P[T \geq t-s \mid S=s] \, P[s \leq S \leq s+ds]$.

But since the accident has happened according to an exponential distribution we have by the theorem on order statistics and Poisson process that S is uniformly distributed over $[0, t]$. Hence, since T and S are independent and

$$p(t) = \int_0^t P[T \geq t-s] \frac{ds}{t} = \frac{1}{t}\int_0^t [1-F(t-s)]ds = \frac{1}{t}\,\alpha(t)$$

We have from (1) and (2) $P(X(t)=k) = e^{-\lambda\alpha(t)} \, [\lambda\alpha(t)]^k/k!$

$$k = 0, 1, 2, \ldots$$

(b) $E(T) < \infty \Rightarrow \int_0^{\infty} [1-F(t)]\,dt < \infty$.

Then $$P[\lim_{t\to\infty} X(t) = k] = e^{-\lambda E(T)} \, [\lambda E(T)]^k/k!$$

because $\lim_{t\to\infty} \alpha(t) = E(T) < \infty$ (integrating by parts).

(c) $F(t) = \begin{cases} 0 & \text{if } t < L \\ 1 & \text{if } t \geq L \end{cases}$

Therefore $$\alpha(t) = \int_0^t [1-F(x)]\,dx = \begin{cases} t & \text{if } t < L \\ L & \text{if } t \geq L \end{cases}$$

So $$P[X(t)=k] = \begin{cases} e^{-\lambda t}\,(\lambda t)^k/k! & \text{if } t < L \\ e^{-\lambda t}\,(\lambda L)^k/k! & \text{if } t \geq L \end{cases}$$

Chapter 8

Exercise 8.1 Since Brownian motion $X(t)$ is an independent stationary increment process to derive the joint density of $X(t_1), X(t_2), \ldots, X(t_n)$ $(t_1 < t_2 < \ldots < t_n)$ subject to the condition $X(0) = 0$ a.s., we only have to know the probability density of $Z_1 = X(t_1) = x_1$, of $Z_2 = X(t_2) - X(t_1) = x_2 - x_1$, etc. and finally that of $Z_n = X(t_n) - X(t_{n-1})$. Since the Jacobian of the linear transformation is unity, making change of variables the joint density is given by

$$f(x_1, \ldots, x_n) = p(x_1, t_1)p(x_2 - x_1, t_2 - t_1) \ldots p(x_n - x_{n-1}, t_n - t_{n-1}) \quad (1)$$

where $\quad p(x, t) = \dfrac{1}{\sqrt{2\pi t}} \exp(-x^2/2t)$

Now consider the case $n = 3$ with the condition $X(t_1) = X(t_3) = 0$ and more specifically suppose that $t_1 = 0$, $t_3 = 1$, $t_2 = t (0 < t < 1)$. By (1) the joint density of $X(t)$ and $X(1)$ is

$$f(x, y) = \frac{1}{2\pi\sqrt{t(1-t)}} \exp\left[-1/2\left\{\frac{x^2}{t} + (y-x)^2/(1-t)\right\}\right]$$

$$f(x, 0) = \frac{1}{2\pi\sqrt{t(1-t)}} \exp\left(\frac{x^2}{2t(1-t)}\right).$$

The marginal density of $X(1) = Y$ at the point 0 is

$$f_y(0) = \int_{-\infty}^{\infty} \frac{1}{2\pi\sqrt{t(1-t)}} \exp\left(\frac{x^2}{2t(1-t)}\right) dx = \frac{1}{\sqrt{2\pi}}$$

Therefore, the conditional density of $X(t)$ given $X(1) = X(0) = 0$ is given by

$$f(x \mid X(0) = X(1) = 0) = f(x, 0)/f_Y(0) = \frac{1}{\sqrt{2\pi t(1-t)}} \exp(x^2/[2t(1-t)]).$$

Exercise 8.2 This can be reduced to the preceding exercise 8.1 as follows. Since $X(t)$ has stationary increments, the r.v. $X(t)$ subjected to the conditions $X(t_1) = A$ and $X(t_2) = B$, has the same density as the r.v. $A + X(t - t_1)$ under the conditions $X(0) = 0$ and $X(t_2 - t_1) = B - A$. This clearly has the same density as the r.v. $A + X(t - t_1) + \dfrac{t - t_1}{t_2 - t_1}(B - A)$ under the conditions $X(0) = 0$ and $X(t_2 - t_1) = 0$.

Exercise 8.3 (a) We must compute the probability p that the particle will cross zero level at some point in the time interval (a, b). Let p_λ denote the probability that the particle located at λ at time a will cross the zero level in the time interval (a, b).

Then $$p = \int_0^{\infty} p_\lambda p_0[\,|X(a)| = \lambda]\, d\lambda$$

where $P_0(A)$ is the conditional probability of A given $X(0) = 0$.

Now $$p_\lambda = P\{\min_{a \le t \le b} X(t) \le 0 \mid X(a) = \lambda\}$$

$$= P\{\min_{a \le t \le b} X(t) \le -\lambda \mid X(a) = 0\}$$

(by spatial homogeneity of Brownian motion)

$$= P\{\min_{a \le t \le b} X(t) \le \lambda \mid X(a) = 0)\}$$

(by symmetry)

$$P_0\{\max_{a \le t \le b-a} X(t) \le \lambda\}$$ (by temporal homogeneity of $X(t)$)

$$= P_0[T_\lambda \le b - a]$$

where T_λ is the first passage time across the boundary λ of Brownian motion. Hence

$$p = \int_0^\infty P_0[T_\lambda \le b - a]\, P_0[\,|X(a)| = \lambda]\, d\lambda$$

$$= 2\int_0^\infty (2\pi a)^{-1/2} e^{-\lambda^2/2a} P_0[T_\lambda \le b - a]\, d\lambda$$

$$= (2/\pi a)^{1/2} \int_0^\infty (2\pi)^{-1/2} \lambda \int_0^{b-a} t^{-3/2} e^{-\lambda^2/2a} e^{-\lambda^2/2t}\, dt\, d\lambda$$

$$= \lambda^{-1} a^{-1/2} \int_0^{b-a} t^{-3/2} \left\{ \int_0^\infty \lambda \exp[-\lambda^2(t^{-1} + a^{-1})/2]\, d\lambda \right\} dt$$

$$= \pi^{-1} a^{-1/2} \int_0^{b-a} t^{-3/2} \{at/(t + a)\}\, dt \qquad \left(\text{putting } u^2 = \lambda^2 \frac{t + a}{2at}\right)$$

$$= \pi^{-1} a^{1/2} \int_0^{b-a} t^{-1/2} (t + a)^{-1}\, dt$$

$$= 2\pi^{-1} \int_0^{((b-a)/a)^{1/2}} ds/(1 + s^2) \qquad (\text{putting } t = as^2)$$

$$= 2\pi^{-1} \tan^{-1} \left(\frac{b - a}{a}\right)^{1/2}$$

Setting $\alpha = \tan^{-1} \left(\frac{b - a}{a}\right)^{1/2}$ gives $\cos \alpha = (a/b)^{1/2}$.

Hence, $p = 2/\pi \cos^{-1} (a/b)^{1/2}$.

(b) Note that if $T < a$, there is no zero in the interval (a, t) and conversely. Therefore

$$P[T < t] = 1 - p = 1 - 2/\pi \cos^{-1}(a/b) = 2/\pi \sin^{-1} (a/b)^{1/2},$$

since $\pi/2 = \cos^{-1} \theta + \sin^{-1} \theta$.

****Exercise 8.4** Since $X(t)$ is stationary and Gaussian,

$$E(X(s)\, X(t)) = E(X(s + \tau)X(t + \tau)) \text{ or } \Gamma(s, t) = \Gamma(s + \tau, t + \tau).$$

Taking

$$\tau = -s.\ \Gamma(s, t) = \Gamma(0, t - s) = R(t - s), \text{ say if } t \ge s \ge 0.$$

Since $\qquad E(X(t)) = 0$ and $\mathrm{Var}\,(X(t)) = 1$ for all $t \ge 0$,

$$\rho(s) = \Gamma(0, s) = R(s).$$

By the condition of Gaussian Process being Markov (see Theorem 8.4)

$$\Gamma(0, s+t) = \Gamma(0, s)\,\Gamma(s, s+t)/\Gamma(s, s)$$

$$= \Gamma(0, s)\Gamma(0, t)$$

i.e.
$$\rho(s+t) = \rho(s)\,\rho(t). \tag{1}$$

Since $X(t)$ is stochastically continuous, $\rho(s)$ is continuous at $s = 0$ (see Appendix II), then because $\rho(0) = 1$, we have

$$\rho(t+h) - \rho(t) = \rho(t)[\rho(h) - 1] \to 0 \text{ as } h \to 0.$$

Therefore $\rho(t)$ is continuous at every real number (see Appendix II). Then the unique solution of the Cauchy equation (1)

$$\rho(t) = e^{-ct},\ c > 0 \text{ and } t > 0.$$

Since
$$\rho(t) = \rho(-t),\ \rho(t) = e^{-c|t|},\ c > 0.$$

Exercise 8.5

$$X(t) = Z\cos(2\pi t + \theta) = Z\cos\theta\cos 2\pi t - Z\sin\theta\sin 2\pi t$$

$$= A\cos(2\pi t) - B\sin(2\pi t)$$

where $A = Z\cos\theta$ and $B = Z\sin\theta$.

The joint density of A and B is given by

$$f_{A,B}(Z\cos\theta, Z\sin\theta) = 1/2\,\pi Z e^{-1/2(z^2)} \begin{vmatrix} \cos\theta & -Z\sin\theta \\ \sin\theta & Z\cos\theta \end{vmatrix}^{-1}$$

$$= 1/2\,\pi z e^{-(1/2)z^2}\,1/z = 1/2\pi\, e^{z^2/2}$$

Therefore $f_{A,B}(a, b) = 1/2\pi\ e^{-1/2(a^2+b^2)}$. So A and B are independent normal variables.

Therefore every linear combination of $X(t_j)$'s

$$\sum_{j=1}^{n} c_j X(t_j) = \left(\sum_{j=1}^{n} c_j \cos(2\pi t_j)\right) A - \left(\sum_{j=1}^{n} c_j \sin(2\pi t_j)\right) B$$

is a Gaussian r.v. By definition $\{X(t), -\infty < t < \infty\}$ is a Gaussian Process.

Exercise 8.6 (a) Consider the characteristic function

$$E\left\{\exp\left(i\sum_{k=1}^{n} u_k X(t_k+\tau)\right)\right\}$$

$$= E\left\{1/2\pi \int_0^{2\pi} \exp\left[iA\sum_{k=1}^{n} u_k \cos(2\pi t_k + 2\pi\tau + \theta)\right] d\theta\right\}$$

$$= E\left\{\frac{1}{2\pi}\int_{2\pi\tau}^{2\pi(\tau+1)} \exp\left[iA\sum_{k=1}^{n} u_k\cos(2\pi t_k + \psi)\right] d\psi\right\} \qquad (\psi = 2\pi t + \theta)$$

$$= E\left\{\frac{1}{2\pi}\int_0^{2\pi} \exp\left[iA\sum_{k=1}^{n} u_k \cos(2\pi t_k + \psi)\right] d\psi\right\}$$

$$= E\left(i\sum_{k=1}^{n} u_k X(t_k)\right) \qquad (\because X(t) = A\cos(2\pi t + \psi))$$

So by uniqueness theorem of characteristic function $\{X(t)\}$ is stationary.

(b) $E(X(t)) = E\left\{A(1/2\pi\int_0^{2\pi} \cos(2\pi t + \theta)\, d\theta\right\} = E(A).$

$$E\left(1/2\int_0^{2\pi} \cos(2\pi t+\theta)\, d\theta\right) = 0 \text{ for all } t \geq 0.$$

$$M_{T(\omega)} = 1/2T\int_{-T}^{T} X_T(\omega)\, dt = A(\omega)\frac{1}{2T}\int_{-T}^{T} \cos[2\pi t + \theta(\omega)]\, dt \to 0$$

a.s. as $T \to \infty$.

Therefore $M_{T(\omega)} \to E(X(t))$ a.s. as $T \to \infty$.

(c) Let $Y = A\cos\theta$ and $Z = A\sin\theta$, then

$$X(t) = Y\cos 2\pi t - Z\sin 2\pi t. \text{ So } \{X(t), -\infty < t < \infty\}$$

is Gaussian iff (Y, Z) are Gaussian. Since

$$E(YZ) = E(A^2)1/2\pi\int_0^{2\pi} \sin\theta\cos\theta\, d\theta = 0$$

$$E(Y^2) = EZ^2 = 1/2\, E(A^2) = \sigma^2.$$

Therefore (Y, Z) is Gaussian iff

$$f_{Y,Z}(y, z) - \frac{1}{2\pi\sigma^2}\exp\left[\frac{1}{2\sigma^2}(y^2 + z^2)\right].$$

Hence, by the transformation rule for r.v.s (c.f. exercise 8.5) we have,

$$1/2\pi f_A(a) = f_{Y,Z}(a\cos\theta, a\sin\theta)\begin{vmatrix}\cos\theta & a\sin\theta \\ \sin\theta & a\cos\theta\end{vmatrix}$$

$$= \frac{a}{2\pi\sigma^2}\exp(-a^2/2\sigma^2),\ a \geq 0,$$

which is the Raleigh density.

$\{X(t), -\infty < t < \infty\}$ is not a Markov Process because for $t \geq 1/4$

$$E(X(t) \mid X(0) = y, X(1/4) = -z) = y\cos(2\pi t) - z\sin(2\pi t)$$

which depends on y, contrary to the Markov property.

Exercise 8.7

$$\begin{aligned} E(U(t)U(s)) &= e^{-t}e^{-s}E(X(e^{2t})X(e^{2s})),\ t > s \\ &= e^{-t}e^{-s}\min(e^{2t}, e^{2s}) = e^{-t}e^{-s}e^{2s} = e^{-t+s} = e^{-(t-s)},\ t > s \end{aligned}$$

Therefore $E(U(t)\, U(s)) = e^{-|t-s|}$.

Exercise 8.8

$$EX^2(t) = E\left[\left\{\int_0^t W(s)ds\right\}^2\right] = E\left[\left\{\int_0^1 W(u)du\right\}\left\{\int_0^t W(v)dv\right\}\right]$$

$$= \int_0^t \int_0^t E[W(u)W(v)]\, dudv \quad \text{(by Fubini's Theorem)}$$

$$= \int_0^1 \left\{\int_0^v E[W(u)W(v)]du + \int_v^t E[W(u)W(v)]\, du\right\} dv$$

$$= 2\int_0^t \int_0^v \{W(u)W(v)du\}\, dv = 2\int_0^t \left\{\int_0^v u\, du\right\} dv$$

$$= 2\int_0^t v^2/2\, dv = t^3/3.$$

Since $X(t)$ is the limit of Riemann sums of normal random variables, $X(t)$ is normal. Also W and $X(t)$ have jointly normal distribution. Given $W(t) = w$, $X(t)$ is normally distributed with mean

$$E[X(t) \mid W(t) = w] = 1/2tw \text{ and variance } E[(X(t) - 1/2tw)^2 \mid W(t) = w]$$
$$= t^3/12.$$

****Exercise 8.9** Since Brownian motion is separable

$$P(\sup_{0 \le t \le t} \mid W(t) \mid \ge \varepsilon) = \lim_{n \to \infty} P(\max_{t \in s_n} \mid W(t) \mid \ge \varepsilon),\ \varepsilon > 0$$

where $S_n = \{k/2^n, k = 0, 1, \ldots, 2^n - 1\}$. For fixed W, let $t_1, t_2, \ldots t_N$ be the points of S_n with $t_1 < t_2 < \ldots < t_N$. Let $T(\omega)$ be the first i (if any) such that $\mid W(t_i, \omega) \mid \ge \varepsilon$, and write

$$EW^2(1) = \sum_{k=1}^{N} P(T = k)E(W^2(1) \mid T = k) + P(\max_{t \in s_n} \mid W(t) \ge \varepsilon)$$

$$E(W^2(1) \mid \max_{t \in s_n} \mid X(t) \mid < \varepsilon) \ge \sum_{k=1}^{N} P(T= k)E(W^2(1) \mid T = k)$$

$$E(W^2(1) \mid T = k) = E[(W(1) - W(t_k))^2 \mid T = k] + E(W^2(t_k) \mid T = k)$$
$$+ 2E[W(t_k)(W(1) - W(t_k))^2 \mid T = k] \ge \varepsilon.$$

Because the event $T = k$ depends only on $W(t_1), W(t_2), \ldots, W(t_k)$, we have

$$W[W(t_k)(W(1) - W(t_k))T= k]$$
$$= E[W(t_k)E[(W_1 - W(t_k) \mid W(t_1), \ldots, W(t_k)] \mid T = k] = 0$$

(by martingale property of Brownian motion, see Appendix I).

It follows that $E(W^2(1) \mid T = k) \geq \varepsilon^2$ and

$$EW^2(1) \geq \varepsilon^2 \sum_{k=1}^{N} P(T = k) = \varepsilon^2 P(\max_{t \in S_n} \mid W(t) \mid \geq \varepsilon)$$

$$\rightarrow \varepsilon^2 P(\sup_{0 \leq t \leq 1} \mid W(t) \mid \geq \varepsilon) \text{ as } n \rightarrow \infty$$

proving Kolmogorov's inequality for Brownian motion.

Exercise 8.10 Let $Y_k = W(k) - W(k-1)$. Then Y_k are independent $N(0, 1)$ variables, $X(n) = Y_1 + \ldots + Y_n$ is the sum of i.i.d. r.vs. Since $EY_k = 0$, by strong law of large numbers for i.i.d. r.v.s, $X(n)/n \rightarrow 0$ a.s. as $n \rightarrow \infty$.

$$\text{Define } Z_k = \max_{0 \leq t \leq 1} \mid W(k + t) - W(k) \mid$$

For $t \in [k, k+1]$, $\mid X(t)/t - X(k)/k \mid \leq \dfrac{1}{k(k+1)} \mid X(k) \mid + \dfrac{1}{k+1} Z_k$.

The first term $\rightarrow 0$ a.s. Since Z_k has the same distribution as

$$\max_{0 \leq t \leq 1} \mid W(t) \mid \text{ and } P(\sup_{0 \leq t \leq 1} \mid W(t) \mid \geq \varepsilon) = 2P(\mid W(t) \mid \geq \varepsilon)$$

(or by Kolmogorov's inequality for martingle; see exercise 8.9), $EZ_k < \infty$. Now Z_k's are i.i.d. r.v.s with $EZ_k < \infty$ implies $Z_k/k \rightarrow 0$ a.s. as $k \rightarrow \infty$.

Therefore $\lim \dfrac{W(t)}{t} = 0$ a.s.

Exercise 8.11 Fix $x > 0$. The characteristic function of $Y(x)$ is given by

$$\phi(u) = E[\exp(iu\, Y(x))] = \int_0^{\infty} E[\exp(iu\, W_2(t))]\, dP[T(x) \leq t],$$

since $T(x)$ is determined by the W_1 process, $T(x)$ and W_2 are independent.
Since $W_2(t)$ is $N(0, t)$, $E[\exp(iu\, W_2(t))] = \exp(-1/2\, u^2 t)$.
Also Laplace transform of $T(x)$ is given by

$$E[e^{-\theta T(x)}] = \int_0^{\infty} e^{-\theta t}\, dP[T(x) \leq t] = \exp[-x\sqrt{2\theta}]$$

$$\text{Therefore } \phi(u) = \int_0^{\infty} \exp(-1/2\, u^2 t)\, dP[T(x) \leq t]$$

$$= E[\exp(-1/2\, u^2 T(x))] = \exp(-\mid u \mid x), \; -\infty < u < \infty.$$

Therefore $Y(x)$ is a Cauchy process.
The characteristic function of $T(x_1 + x_2)$ is given by

$$\phi_{x_1+x_2}(\theta) = E(e^{i\theta T(x_1+x_2)}) = \exp[-(x_1 + x_2)\sqrt{2i\theta}]$$

$$= \exp[-x_1\sqrt{2i\theta}]\exp[-x_2\sqrt{2i\theta}] = \phi_{x_1}(\theta) \cdot \phi_{x_2}(\theta).$$

This implies that the first passage time r.v. $T(x_1 + x_2)$ for reaching $x_1 + x_2 > 0$ when $W(0) = 0$ a.s. is the same as the distribution of the sum of $T(x_1)$ and $T(x_2)$, where $T(x_1)$ and $T(x_2)$ are independent, $x_1 > 0$, $x_2 > 0$, i.e. $\{T(x), x \geq 0\}$ has stationary independent increaments and it is nondecreasing. From this it follows that

$$Y(x) = W_2(T(x))$$

also has stationary independent increments.

Exercise 8.12 Since $Y(t) = Y(0)e^{X(t)-X(0)}$, using the characteristic function of normal distribution, we get

$$E[Y(t) \mid Y(0) = y] = yE[e^{X(t)-X(0)}] = y \exp \{t(\mu + (1/2)\sigma^2)\}$$

and

$$E[Y^2(t) \mid Y(0) = y] = y^2 E[e^{2[x(t)-x(0)]}]$$
$$= y^2 \exp \{t[2\mu + (1/2)\, 4\sigma^2]\},$$

so that the variance of $Y(t)$ is

$$\text{Var}\,[Y(t) \mid Y(0) = y] = E[Y^2(t) \mid Y(0) = y] - \{E[Y(t) \mid Y(0) = y]\}^2$$
$$= y^2\{\exp\,[2t(\mu + \sigma^2)] - \exp\,[2t(\mu + 1/2\sigma^2)]\}$$
$$= y^2 \exp\,[2t(\mu + 1/2\;\sigma^2)][\exp\,(t\sigma^2) - 1].$$

$Y(t + h) - Y(t) = Y(t)[e^{X(t+h)-X(t)-1} - 1]$, again using characteristic function, we get

$$E[Y(t + h) - Y(t) \mid Y(t) = y] = yE[e^{X(t+h)-X(t)} - 1]$$
$$= y(\exp\,\{h(\mu + (1/2)\sigma^2)\} - 1). \text{ Therefore } b(y) = y(\mu + 1/2\sigma^2).$$

Similarly

$$\frac{E[Y(t + h) - Y(t)\}^2 \mid Y(t) = y]}{h} = y^2 E[e^{X(t+h)-X(t)} - 1]^2/h$$

Hence $\lim\limits_{h\downarrow 0} E[\{Y(t + h) - Y(t)\}^2 \mid Y(t) = Y]/h = y^2\sigma^2$.

Exercise 8.13 By Reflection principle

$$P[M(t) \geq m,\, W(t) \leq x] = P(W(t) \geq 2m - x]$$
$$= 1 - \phi\left(\frac{2m - x}{\sqrt{t}}\right), m \geq 0, m \geq x,$$

where $\phi(x)$ is the c.d.f. of $N(0, 1)$. Differentiating with respect to x and then with respect to m, changing the sign and using the relation $d/dx\phi(x) = -\phi(x)$, we get the joint p.d.f. of $M(t)$ and $W(t)$

$$= \frac{d}{dm}\frac{d}{dx}\left[1 - \phi\left(\frac{2m - x}{\sqrt{t}}\right)\right] = -\frac{d}{dm}\left[\frac{1}{\sqrt{t}}\phi\left(\frac{2m - x}{\sqrt{t}}\right)\right]$$
$$= \frac{2m - x}{t}\frac{2}{\sqrt{t}}\phi\left(\frac{2m - x}{\sqrt{t}}\right),$$

where $\phi(x)$ is the p.d.f. of $N(0, 1)$. Joint density of M and W is given by

$$f(m, x) = \sqrt{\frac{2}{\pi t^3}}\ (2m - x) \exp [-(2m - x)^2/2t] \text{ if } 0 \leq m \text{ and } x \leq m.$$

Hence the Joint p.d.f. of $M(t)$ and $Y(t) = M(t) - W(t)$ is given by

$$g(m, y) = f(m, m - y)$$

$$= \sqrt{\frac{2}{\pi t^3}}\ (m + y) \exp [-(m + y)^2/2t],\ m \leq 0 \text{ and } y \geq 0.$$

****Exercise 8.14** Let $T_{ab} = \inf \{t \geq 0 : W(t) = a \text{ or } W(t) = b\}$

and
$$T \wedge n = \min (T, n).$$

Now $Y(t) = W^2(t) - t$ is a continuous parameter martingale,

$$E[Y(T \wedge n)] = E[Y(0)] = 0.$$

Therefore

$$E[T \wedge n] = E[W^2(T \wedge n)] \leq (|a|^2 + b)^2$$

Thus $\quad E(T) = \lim_{n\to\infty} \int_0^n P[T > t]\, dt = \lim_{n\to\infty} E[T \wedge n] \leq (|a| + b)^2 < \infty$

Let $p = P(W(T_{ab}) = b)$.

Now $\{W(t)\}$ is a martingale, $T = T_{ab} < \infty$ and $\{X(T \wedge t)\}$ is bounded. By optional stopping theorem (see Appendix IV)

$$0 = E(W(T))\ \ aP[W(T) = a] + bP[W(T) = b]$$

$$= a(1 - p) + bp,$$

Therefore $\quad p = \dfrac{|a|}{|a| + b}$

$$E(Y(T)) = 0 \text{ implies } E(T_{ab}] = E[W^2(T_{ab})] = a^2(1 - p) + b^2 p$$

$$= a^2 \frac{b}{|a| + b} + b^2 \frac{|a|}{|a| + b} = |a|\, b.$$

By changing the origin and scale we get the result for Wiener process with variance σ^2 and starting position $W(0) = x$.

Exercise 8.15 Let $Z(t) = \exp \{\lambda(W(t)) - (\lambda\mu + 1/2\ \lambda^2\sigma^2)t\}$ is a martingale, where $W(t)$ is a wiener process with drift $\mu \neq 0$ and variance parameter σ^2. If $\lambda_0 = -2\mu/\sigma^2$. Then $Z_0(t) = \exp \{\lambda^2(W(t))\}$ is a martingale. Let T_{ab} be the first time the process reaches $a < 0$ or $b < 0$. Then by optional stopping theorem (see Appendix IV) $1 = E(Z_0(T_{ab})) = P(W(T_{ab}) = a) \exp (\lambda_0 a) + P(W(T_{ab}) = b) \exp (\lambda_0 b)$, and

$$P(W(T_{ab}) = b) = \frac{1 - \exp (\lambda_0 a)}{\exp (\lambda_0 b) - \exp (\lambda_0 a)}, \text{ where } \lambda_0 = -2\mu/\sigma^2,$$

changing the origin to the case $W(0) = x$. We get the result

$$P(W(T_{ab}) = b \mid W(0) = x) = \frac{\exp(-2\mu x/\sigma^2) - \exp(-2\mu a/\sigma^2)}{\exp(-\mu a/\sigma^2) - \exp(-2\mu a/\sigma^2)}$$

Since $\mu < 0$, letting $a \to -\infty$, $\exp(-\mu a/\sigma^2) \to 0$ and

$$\lim_{a \to -\infty} P(W(T_{ab}) = b \mid W(0) = x) = \exp\{2\mu(b - x)/\sigma^2\}.$$

Now left hand side becomes the probability that the process ever reaches b, i.e. the probability that the maximum of the process ever exceeds b. Thus, for $m = b - x$,

$$P(M \leq m) = P(\max_{0 \leq t < \infty} W(t) > m \mid W(0) = x) = \exp\left(-\frac{2\,|\,\mu\,|\,m}{\sigma^2}\right),\ m \geq 0.$$

****Exercise 8.16** Now $Y(t) = \exp\{\lambda W(t) - 1/2\lambda^2 t\}$ is a martingale for a standard Brownian motion $W(t)$. Also $W(T) = a + bT$, $a > 0$, $b > 0$. By optional stopping theorem $E[\exp\{\lambda W(T) - 1/2\,\lambda^2 T\}] = 1$. Therefore

$$E[\exp\{\lambda(a + bT) - 1/2\,\lambda^2 T\}] = 1.$$

Putting $\theta = \lambda - 1/2\,\lambda^2$, yields $E[\exp\{\theta T + \lambda a\}] = 1$.

Hence $E[\exp(\theta T)] = e^{-\lambda a}$.

Now to get the value of λ, we have $\lambda^2 + 2\theta - 2\lambda b = 0$ and

$$\lambda = b \pm \sqrt{b^2 - 2\theta}.$$

We require $\lambda \geq 0$, which implies

$$\lambda = b + \sqrt{b^2 - 2\theta} \quad \text{if} \quad \theta = b^2/2.$$

Therefore $E[\exp(\theta T)] = \exp\{-a(b + \sqrt{b^2 - 2\theta})\}$, $0 < \theta \leq b^2/2$.

Chapter 9

Exercise 9.1

The autocovariance generating function for ARMA (1, 1) process is given by

$$C(z) = \sigma^2 \frac{(1 + \beta z)(1 + \beta z^{-1})}{(1 - \alpha z)(1 - \alpha z^{-1})} = \sigma^2 \frac{(1 + \beta^2 + \beta z + \beta z^{-1})}{[1 - (\alpha^2 + \alpha z + \alpha z^{-1})}$$

$$= \sigma^2 (1 + \beta^2 + \beta z + \beta z^{-1})[1 + (\alpha^2 + \alpha z + \alpha z^{-1}) + (\alpha^2 + \alpha z + \alpha z^{-1})^2 + \ldots + (\alpha^2 + \alpha z + \alpha z^{-1})^k + \ldots]$$

$$= \sigma^2[(1 + 2\alpha\beta + \beta^2)(1 + \alpha^2 + \alpha^4 + \ldots) + (1 + \alpha)(1 + \beta)(1 + \alpha^2 + \alpha^4 + \ldots) + \alpha(1 + \alpha + \beta + \alpha\beta)(1 + \alpha^2 + \alpha^4 + \ldots) + \ldots + \alpha^{k-1}(1 + \alpha + \beta + \alpha\beta)(1 + \alpha^2 + \alpha^4 + \ldots) + \ldots]$$

Therefore coefficients of z^k give

$$\gamma(0) = \sigma^2 \frac{1 + 2\alpha\beta + \beta^2}{1 - \alpha^2}, \gamma(1) = \sigma^2 \frac{(1 + \alpha)(1 + \beta)}{1 - \alpha^2}$$

$$\gamma(k) = \alpha\gamma(k - 1), k \geq 2.$$

The auto-correlations are given by

$$\rho(0) = 1, \rho(1) = \gamma(1)/\gamma(0) = (1 + \alpha)(1 + \beta)/(1 + 2\alpha\beta + \beta^2)$$

and $\rho(k) = \gamma(k)/\gamma(0) = \alpha^{k-1}\rho(1), k \geq 2.$

Exercise 9.2 The autocovariance of Z_t is given by

$$\gamma_z(\tau) = E(Z_t Z_{t-\tau}) = E(X_t + Y_t)(X_{t-\tau} + Y_{t-\tau})$$

$$= \gamma_X(\tau) + 0 + 0 + \gamma_Y(\tau) = \gamma_X(\tau) + \gamma_Y(\tau)$$

Since X_t is a $MA(q_1)$, $\gamma_X(\tau)$ 0 for $\tau > q_1$ and Y_t is a $MA(q_2) \Rightarrow \gamma_Y(\tau) = 0$ for $\tau > q_2$. Therefore $\gamma_Z(\tau) = \gamma_X(\tau) + \gamma_Y(\tau) = 0$ for $\tau >$ Max (q_1, q_2).

From a result of Wold (1954) a necessary and sufficient condition for a $MA(q)$ process $\{X_t\}$ is that it has spectral density $f(\lambda) \geq 0, -\pi \leq \lambda \leq \pi$ given by

$$f(\lambda) = C(z) = \sum_{j=-q}^{q} \gamma_2(j) z^j, z = e^{i\lambda}.$$

Therefore the process $Z_t = X_t + Y_t$ is a $MA(q)$, where $q \leq \max(q_1, q_2)$.

Exercise 9.3 Let B be the Backshift operator. Then $\phi_{1_{p1}}(B)X_t = \theta_{1_{q1}}(B)\varepsilon_{1t}$ and $\phi_{2_{q2}}(B)Y_t = \theta_{2_{q2}}(B)\varepsilon_{2t}$, where $\phi_{1_{q1}}(z)$, $\phi_{2_{q2}}(z)$, $\phi_{1_{q1}}(z)$ and $\phi_{2_{q2}}(z)$ are polynomials in z of degree p_1, p_2, q_1, q_2 respectively and ε_{1t} and ε_{2t} are independent, Zero means white noise processes.

Since $Z_t = X_t + Y_t$, it follows that

$$\phi_{1_{p1}}(B), \phi_{2_{p2}}(B)\, Z_t = \phi_{2_{p2}}(B)\, \phi_{1_{p1}}(B) = X_t + \phi_{1_{p1}}(B)\, \phi_{2_{p2}}(B) Y_t$$

$$= \phi_{2_{p2}}(B)\, \theta_{1_{q1}}(B)\, \varepsilon_{1t} + \theta_{1_{p1}}(B)\, \phi_{2q2}(B)\varepsilon_{2t}.$$

The first term on the right hand side is $MA(p_2 + q_1)$ and the second term is $MA(p_1 + q_2)$ so the whole of the expression on the right-hand side is $MA(q)$ where $q \leq \max(p_1 + q_2, p_2 + q_1)$ (by exercise 9.2). Now the polynomials $\phi_{1_{p1}}(z)$ and $\phi_{2_{p2}}(z)$ may contain common root and so the part of the operator need not be applied twice and hence the order p of the polynomial $\phi_{1_{p1}}(B)\, \phi_{2_{p2}}(B)$ is less than equal to $p_1 + p_2$.

Exercise 9.4 For ARMA (1, 1) model equate the coefficients of the right and left sides of $(1 - \phi - B)\, \psi(B) = 1 + \theta B$, where $\psi(B)$ is a function of back shift operator B given by $\psi(B) = \sum_{j=0}^{\infty} \psi_j B^j$. From the constant term $\psi_0 = 1$, the coefficient of B gives $-\phi + \psi_1 = \theta$, so $\psi_1 = \theta + \phi$; equating the coefficients of B^2 gives $-\phi\psi_1 + \psi_2 = 0$, so $\psi_2 = \phi\psi_1 = \phi(\theta + \phi)$, and so on, to give

$$X_t = \varepsilon_t - (\phi + \theta)\varepsilon_{t-1} + \phi(\phi + \theta)\varepsilon_{t-2} + \phi^2(\phi + \theta)\varepsilon_{t-3} + \dots$$
$$(-\phi)^{k-1}(\phi + \theta)\varepsilon_{t-k} + \dots$$

Therefore $X_{t-k} = \varepsilon_{t-k} + \psi_1\varepsilon_{t-k-1} + \psi_2\varepsilon_{t-k-2} + \dots$, so X_{t-k} is a linear function of $\varepsilon_{t-k}, \varepsilon_{t-k-1}, \dots$ etc. but not of ε_t. As the ε_t's are independent by assumption, ε_t must therefore be independent of $\varepsilon_{t-k}, \varepsilon_{t-k-1}, \dots$ and so of X_{t-k}. This is true for any $k > 0$ and any stationary ARMA process.

(b) In the ARMA (1, 1) case the ARMA equation gives

$$\varepsilon_t = X_t - \phi X_{t-1} - \theta\varepsilon_{t-1} = X_t - \phi X_{t-1} - \theta(X_{t-1} - \phi(X_{t-2} - \theta\varepsilon_{t-2}) = \dots$$
$$= X_t - (\phi + \theta)X_{t-1} + \theta(\phi + \theta)X_{t-2} - \theta^2(\phi + \theta)X_{t-3} + \dots$$
$$+ (-\theta)^{k-1}(\phi + \theta)X_{t-k} + \dots$$

Exercise 9.5 (a) If $\{X_t\}$ is zero mean second order stationary non-deterministic time series then by Wold's Theorem X_t has a representation of the form

$$X_t = \sum_{j=0}^{\infty} \psi_j \varepsilon_{t-j} \tag{1}$$

Since we have the knowledge of the entire past $X_t, X_{t-1}, \dots$ of the series at time t, the linear predictor of X_{t+k}, $k > 0$ $X_{t+k|t}$ is a linear combination of the known past values, viz.

$$X_{t+k|t} = \sum_{i=0}^{\infty} h_i X_{t-i} \tag{2}$$

So we can rewrite (1) in the form

$$X_{t+k|t} = \sum_{i=0}^{\infty} g_i\varepsilon_{t-i} \text{ for some sequence } \{g_i\}.$$

Therefore $X_{t+k} - X_{t+k|t} = \sum_{i=0}^{\infty} \psi_i\varepsilon_{t+k-i} - \sum_{i=0}^{\infty} g_i\varepsilon_{t-i}$

$$= \sum_{i=0}^{k-1} \psi_i\varepsilon_{t+k-i} + \sum_{i=k}^{\infty} \psi_i\varepsilon_{t-k-i} - \sum_{j=0}^{\infty} g_i\varepsilon_{t-i}$$

$$= \sum_{i=0}^{k-1} \psi_i\varepsilon_{t+k-i} + \sum_{i=k}^{k-1} (\psi_{i+k} - g_i)\varepsilon_{t-i}$$

Since ε_j's are uncorrelated, expectations of the cross terms vanish and we have

$$E[(X_{t+k} - X_{t+k|t})^2] = \sigma^2 \sum_{i=0}^{k-1} \psi_i^2 + \sigma^2 \sum_{i=0}^{\infty} (\psi_{i+k} - g_i)^2$$

which is minimized by choosing

$$g_i = \psi_{i+k},$$

so the optimal mean square predictor is

$$X_{t+k|t} = \sum_{i=0}^{\infty} \psi_{i+k}\ \varepsilon_{t-i}$$

and the minimum value of the mean square error is

$$E[(X_{t+k} - X_{t+k|t})^2] = \sigma^2 \sum_{i=0}^{k-1} \psi_i^2.$$

(b) For the AR(1) process $X_t = \alpha X_{t-1} + \varepsilon_1$, $|\alpha| < 1$

$$X_t = \sum_{i=0}^{\infty} \alpha^j \varepsilon_{t-j} \text{ so that } \psi_j = \alpha^j$$

We find the k-step predictor has

$$g_j = \alpha^{j+k}, \text{ given } X_{t+k|t} = \sum_{j=0}^{\infty} \alpha^{j+k} \varepsilon_{t-j} = \alpha^k \sum_{j=0}^{\infty} \alpha^j \varepsilon_{t-j} = \alpha^k X_t$$

For $MA(1)$ process $X_t = \varepsilon_t + \beta\varepsilon_{t-1}$, $|\beta| < 1$, the optimum one step ahead predictor is

$$X_{t+1|t} = \beta\varepsilon_t \tag{3}$$

However, we do not observe that the $\{\varepsilon_t\}$ series, but just the $\{X_t\}$ series. To get a useful predictor we express ε_t interms of X_t, X_{t-1} X_{t-2}, ...

$MA(1)$ can be written as

$$\varepsilon_t = X_t - \beta\varepsilon_{t-1} = X_t - \beta(X_{t-1} - \beta\varepsilon_{t-2}) = \ldots$$

$$= X_t - \beta X_{t-1} + \beta^2 X_{t-2} - \beta^3 X_{t-3} + \ldots$$

Substituting this in (3) we get one-step predictor as

$$X_{t+1|t} = \beta X_t - \beta^2 X_{t-1} + \beta^3 X_{t-2} - \beta^4 X_{t-3} + \ldots$$

Exercise 9.6 Let $X_t = \theta\,\varepsilon_{t-1} + \varepsilon_t$ be a $MA(1)$ process where $\{\varepsilon_t\}$ are i.i.d. r.v.s with zero mean and finite variance σ^2.

Then $$\text{Cov}(X_{t+h}, X_t) = \begin{cases} (1+\theta^2) & \text{if } h = 0 \\ \theta\sigma^2 & \text{if } h = \pm 1 \\ 0 & \text{if } |h| > 1 \end{cases}$$

If $|\rho| > 1/2$ then $k(h)$ is the autocovariance function of the above $MA(1)$ process with $\sigma^2 = (1+\theta^2)^{-1}$ and $\theta = (2\rho)^{-1}(1 \pm \sqrt{(1-4\rho^2)})$.

If $\rho > 1/2$, $K = [k(i-j)]_{i,j=1}^{n}$ and $a = (1, -1, 1, -1, \ldots)$

with n-component, then

$$a'\,K\,a = n - 2(n-1)\rho < 0 \text{ for } n > 2\rho/(2\rho - 1),$$

which shows that $k(h)$ is not non-negative definite and therefore by Bochner's Theorem it is not an autocovariance function. If $\rho < -(1/2)$, the same argument using n-component vector $a = (1, 1, 1, \ldots)'$ again shows that $k(h)$ is not non-negative definite.

Exercise 9.7

$$S_1 = \mu + S_0 + X_1 = \mu + X_1$$

$$E(S_1) = \mu,\ E(S_t) = \mu + E(S_{t-1}) = t\mu$$

$$\text{Cov}(S_t, S_{t+h}) = E(S_t - t\mu)(S_{t+h} - \overline{t+h}\,\mu) = \text{Var}\,(S_t - t\mu) = t\sigma^2$$

Thus, S_t is not second order stationary,

Let $Y_t = \nabla S_t$. Then $\nabla S_t = (1 - B)S_t - S_{t-1} = \mu + X_t$

$E(\nabla S_t) = \mu$, $E(Y_{t+h}) = \mu$ for all $h = \ldots, -1, 0, 1, 2, \ldots$

$$\text{Cov}(Y_t, Y_{t+h}) = E(X_t, X_{t+h}) = \begin{cases} \sigma^2 & \text{if} \quad h = 0 \\ 0 & \text{if} \quad |h| \neq 0 \end{cases}$$

Obviously $Y_t = \mu + X_t$ is a (i.i.d.) stationary process and also a covariance stationary process.

Exercise 9.8 For one-sided infinite moving average representation of $AR(2)$, the roots of $\phi(B) = 1 - \phi_1 B - \phi_2 B^2 = 0$ must lie outside the unit circle, from the properties of quadratic equation this implies that the AR(2) parameters ϕ_1 and ϕ_2 must lie in the triangular region

$$\left.\begin{aligned} \phi_2 + \phi_1 &< 1 \\ \phi_2 - \phi_1 &< 1 \\ -1 < \phi_2 &< 1 \end{aligned}\right\} \tag{1}$$

From Yule-Walker equations in $AR(2)$

$$\rho(1) = \phi_1 + \phi_2\,\rho(1) \text{ and } \rho(2) = \phi_1\,\rho(1) + \phi_2.$$

Solving we get

$$\phi_1 = \frac{\rho(1)(1 - \rho(2))}{1 - \rho^2(1)},\ \phi_2 = \frac{\rho(2) - \rho^2(1)}{1 - \rho^2(1)}$$

which imply

$$\rho(1) = \frac{\phi_1}{1 - \phi_2} \text{ and } \rho(2) = \phi_2 + \frac{\phi_1^2}{1 - \phi_2} \tag{2}$$

From (1) and (2), we get the admissible values of $\rho(1)$ and $\rho(2)$ given by $\rho^2(1) < (1/2)(\rho(2) + 1)$, $|\rho(1)| < 1$, $|\rho(2)| < 1$.

Exercise 9.9 Now since $|\theta| < 1$, $X_t = \theta X_{t-1} + \varepsilon_t$ can be written as a linear process as

$$X_t = \sum_{i=0}^{\infty} \theta^i \varepsilon_{t-i} \text{ and } X_{t+k} = \sum_{i=0}^{\infty} \theta^i\, \varepsilon_{t+k-i}.$$

Hence $\gamma(k) = E\left(\sum_{i=0}^{\infty} \theta^i \varepsilon_{t-i}\right)\left(\sum_{j=0}^{\infty} \theta^j\, \varepsilon_{t+k-j}\right) = \sigma^2\theta^k \sum_{i=0}^{\infty} \theta^{2i}$ if $k \geq 0$.

Therefore $\gamma(k) = \dfrac{\sigma^2}{1 - \theta^2}\, \theta^{|k|}$.

Since spectral density of $\{\varepsilon_t\}$ is $\sigma^2/2\pi$, by calculating covariance generating function we easily get

$$f(\lambda) = \frac{\sigma^2}{2\pi} \mid 1 - \theta e^{-i\lambda} \mid^2.$$

Since $\{\varepsilon_1, \varepsilon_2, \ldots, \varepsilon_n\}$ is distributed as multivariate normal with zero mean vector and covariance matrix

$$\Sigma = \begin{bmatrix} (1-\theta^2)^{-1} & \sigma^2 & 0\ldots0 \\ 0 & \sigma^2 & 0\ldots0 \\ 0 & 0 & \sigma^2\ldots0 \\ 0 & 0\ldots\ldots\sigma^2 & \end{bmatrix}$$

$$\mid \Sigma \mid = \frac{\sigma^{2n}}{1-\theta^2}$$

and $$X'\Sigma^{-1} X = \sigma^{-2}\left[(1-\theta^2) X_1^2 + \sum_{j=2}^{n} (X_j - \theta X_{j-1})^2\right]$$

Therefore

$$L_n = -1/2 \; \{n \log(2\pi\sigma^2) - \log(1-\theta)^2$$

$$+ \sigma^{-2}\left[(1-\theta^2) X_1^2 + \sum_{j=2}^{n} (X_j - \theta X_{j-1})^2\right]$$

****Exercise 9.10** Let $\{X_t\}$ be a ARMA (p, q) process $\sum_{r=0}^{p} \alpha_r X_{t-r} = \sum_{r=0}^{q} \beta_s \varepsilon_{t-s}$, $\alpha_0, \beta_0 \neq 0$ and $\{\varepsilon_t\}$ is a white noise with $E(\varepsilon_t) = 0$ and $\{\varepsilon_t\}$ are i.i.d. r.v.s with $E\varepsilon_t^2 = \sigma^2$. Then the spectral density of X_t is given by $\dfrac{\sigma^2}{2\pi}\left|\dfrac{\sum_{s=0}^{q} \beta_s z^s}{\sum_{r=0}^{p} \alpha_r z^r}\right|^2$,

$z = e^{-i\lambda}$. Let $\mathcal{M}_n = \mathcal{B}(X_t, t \leq n)$ denote the closed linear space (closed in the sense of mean square convergence) generated by linear forms in the X_t's, $t \leq n$. Now $\varepsilon_t \in \mathcal{M}_t$. X_t has a linear process (one sided) representation $X_t = \sum_{j=0}^{\infty} \gamma_j \in_{t-j}$, since the zeros of $a(z) = \sum_{r=0}^{p} \alpha_r z^r$ have modulus greater than one. Therefore ε_t is orthogonal to (independent of) $\mathcal{M}_{t-1}$. It follows that

$\hat{X}_t = -\sum_{k=1}^{p} \alpha_k/\alpha_0 X_{t-k} + \sum_{k=1}^{q} \beta_j/\alpha_0 \varepsilon_{t-j}$ is in $\mathcal{M}_{t-1}$ and $X_t - \hat{X}_t = \beta_0/\alpha_0 \varepsilon_t$ is orthogonal (independent of) to $\mathcal{M}_{t-1}$. Thus $\hat{X}_t = E(X_t \mid \mathcal{M}_{t-1})$, i.e. $\hat{X}_t$ is the projection of X_t on $\mathcal{M}_{t-1}$ and consequently the best linear predictor of X_t in terms of the past

$X_j, j \leq t-1$. The variance of the prediction error is $E \mid X_t - \hat{X}_t \mid^2 = \left| \frac{\beta_0}{\alpha_0} \right|^2 \sigma^2$.

If Z_μ's are the zeros of $\alpha(z) = \sum_{r=0}^{p} \alpha_r z^r$

$$| a(e^{-i\lambda}) |^2 = \left| \prod_{\mu=1}^{p} (e^{-i\lambda} - z_\mu) \right|^2 | \alpha_p |^2$$

$$= \prod_{\mu=1}^{p} | z_\mu |^2 \prod_{\mu=1}^{p} | 1 - e^{-i\lambda} z_\mu^{-1} |^2 \, | \alpha_p |^2$$

Now $\log | a(e^{-i\lambda}) |^2 = \log | \alpha_0 |^2 + \prod_{\mu=1}^{p} \log | 1 - e^{-i\lambda} z_\mu^{-1} |^2$.

Since for $| z_\mu | > 1$, the $\log(1 - z_\mu^{-1})$ expansion is valid and hence

$$\int_{-\pi}^{\pi} \log(1 - e^{-i\lambda} z_\mu^{-1}) d\lambda = -\sum_{j=1}^{\infty} z_\mu^{-1} / j! \int_{-\pi}^{\pi} e^{-i\lambda j} d\lambda = 0.$$

By assumption all the zeros of $a(z)$ and $b(z)$ are outside the unit disc. Therefore

$$1/2\pi \int_{-\pi}^{\pi} \log f(\lambda) d\lambda = \log \frac{\sigma^2}{2\pi} + \log | \beta_0 |^2 - \log | \alpha_0 |^2$$

and hence
$$\sigma^2 \left| \frac{\beta_0}{\alpha_0} \right|^2 = 2\pi \exp \left\{ \frac{1}{2\pi} \int_{-\pi}^{\pi} \log f(\lambda) d\lambda \right\}.$$

Exercise 9.14 (*Explosive AR*(1) *process*)
Using backshift operator $X_t = \alpha X_{t-1} + \varepsilon_t$ can be written as $(1 - \alpha B)X_{t-1} = \varepsilon_t$ or as $\alpha B(\alpha^{-1}B^{-1} - 1) X_t = \varepsilon_t$ and hence that

$$X_t = -\alpha^{-1}B^{-1}[1 + \alpha^{-1}B^{-1} + \alpha^{-2}B^{-2} + \ldots + \alpha^{-j} B^{-j} + \ldots]\varepsilon_t \text{ (since } | \alpha^{-1} | < 1)$$

$$= -\sum_{j=1}^{\infty} \alpha^{-j} \varepsilon_{t+j}, \text{ where } B^{-j} \varepsilon_t = \varepsilon_{t+j}.$$

Exercise 9.15

$$X_n = \alpha X_{n-1} + \varepsilon_n,\ n \geq 2 \text{ gives } X_n = \sum_{i=1}^{n} \varepsilon_i \alpha^{n-i},$$

and
$$\overline{X}_n = n^{-1} \sum_{j=1}^{n} \sum_{j=1}^{j} \varepsilon_i \alpha^{j-i} = n^{-1} \sum_{j=1}^{n} \varepsilon_j (1 - \alpha^{n-j+1})/(1 - \alpha).$$

Hence

$$E\overline{X}_n = n^{-1} (\mu/(1 - \alpha)) \sum_{j=1}^{n} (1 - \alpha^j)$$

$$= n^{-1}(\mu/(1 - \alpha))(n - \alpha (1 - \alpha^n)/(1 - \alpha))$$

$$= \mu/(1 - \alpha) - (\alpha\mu/n)((1 - \alpha^n)/(1 - \alpha^2)^2) \to \mu/(1 - \alpha),$$

$$\text{Var} (\overline{X}_n) = n^{-2} \sum_{j=1}^{n} \sigma^2(1 - \alpha^{n-j+1})^2/(1 - \alpha)^2 \leq \sigma^2/n \to 0,$$

and $$(E(\overline{X}_n) - \mu/(1-\alpha))^2 = Var(\overline{X}_n) + (E\overline{X}_n - \mu/(1-\alpha))^2 + 2E(\overline{X}_n)(E\overline{X}_n - \mu/(1-\alpha)) \to 0.$$

Exercise 9.16 Let $|\alpha| < 1$. Since $X_0 = 0$ and $X_j = \beta X_{j-1} + \varepsilon_j$, we have

$$\sum_{j=1}^{n} X_j = \alpha \sum_{j=1}^{n-1} X_j + \sum_{j=1}^{n} \varepsilon_j.$$

This implies that $(1-\alpha) \sum_{j=1}^{n} X_j = \sum_{j=1}^{n} \varepsilon_j - \alpha X_n$

or $\sqrt{n}((1-\alpha)\overline{X}_n - \mu) = \sqrt{n}(\overline{\varepsilon}_n - \mu) - \alpha X_n/\sqrt{n}$.

By C.L.T for i.i.d. r.v.s, $\sqrt{n}(\overline{\varepsilon}_n - \mu) \xrightarrow{\mathscr{L}} N(0, \sigma^2)$.

By Exercise 9.15, $X_n = \sum_{i=1}^{n} \varepsilon_j\, \alpha^{n-j}$ and hence

$$E(X_n/\sqrt{n}) = \mu \sum_{j=1}^{n} \alpha^{n-j}/\sqrt{n} \to 0$$

and $$\text{Var}\ (X_n/\sqrt{n}) = \sigma^2 \sum_{j=1}^{n} \alpha^{2(n-j)}/n \to 0.$$

This implies $X_n/\sqrt{n} \xrightarrow{L_2} 0$ and hence $X_n/\sqrt{n} \xrightarrow{P} 0$.
By Slutsky's Theorem $\sqrt{n}((1-\alpha)\overline{X}_n - \mu) \xrightarrow{\mathscr{L}} N(0, \sigma^2)$

or $$\sqrt{n}\left(\overline{X}_n - \frac{\mu}{1-\alpha}\right) \xrightarrow{\mathscr{L}} N(0, \sigma^2/(1-\alpha)).$$

For $\alpha = 1$, $\overline{X}_n = n^{-1}(\varepsilon_2 + \varepsilon_4 + \ldots + \varepsilon_n)$ for even n and $\overline{X}_n = n^{-1}(\varepsilon_1 + \varepsilon_3 + \ldots + \varepsilon_n)$ for n odd. In either case,

$$\sqrt{n}(\overline{X}_n - \mu/2) \xrightarrow{\mathscr{L}} N(0, \sigma^2/2).$$

For $\alpha = 1$, we have $X_n = \sum_{i=1}^{n} \varepsilon_j$ and $\overline{X}_n = n^{-1} \sum_{j=1}^{n} (n+1-j)\varepsilon_j$, If $\mu > 0$ $\overline{X}_n \xrightarrow{P} \infty$ (by Weak law of large members).

Since ε_i's are i.i.d., $\overline{X}_n$ has same distribution as $n^{-1}Z_n$, where $Z_n = \sum_{i=1}^{n} j\varepsilon_j$, so we consider $Z_n - EZ_n = \sum_{i=1}^{n} j(\varepsilon_j - \mu)$. Define for $1 \le j \le n$

$$X_{nj} = j(\varepsilon_j - \mu) \text{ and } B_n^2 = E(\Sigma X_n)^2 = \sum_{i=1}^{n} j^2\sigma^2 = \sigma^2 n(n+1)(2n+1)/6,$$

since we have $E(X_{nj}) = 0$, Var $(X_{nj}) = j^2\sigma^2$. Then,

$$B_n^{-2} \sum_{i=1}^{n} E\{X_{nj}^2 I(|X_{nj}| \ge \varepsilon B_n)\}$$

$$= B_n^{-2} \sum_{i=1}^{n} j^2 \{(\varepsilon_j - \mu)^2 \; I[(\varepsilon_j - \mu)^2 \; \varepsilon B_n^2 / j^2]\}$$

$$\leq B_n^{-2} \sum_{i=1}^{n} j^2 E\{(\varepsilon_j - \mu)^2 I[(\varepsilon_j - \mu)^2 \geq \varepsilon B_n^2 n^{-2}]\}$$

$$= \sigma^{-2} E\{(\varepsilon_1 - \mu)^2 I[(\varepsilon_1 - \mu)^2 \geq \varepsilon B_n^2 n^{-2}]\}$$

(since ε_j's are i.i.d. and $B_n^2 = \sum_1^n j^2 \sigma^2$)

$\rightarrow 0$, because $B_n^2 / n^2 \rightarrow \infty$.

Hence by Lindberg's CLT $(Z_n - EZ_n)/B_n \xrightarrow{\mathscr{L}} N(0, 1)$ which implies

$$n^{-1/2} (\overline{X}_n - n\mu/2) \xrightarrow{\mathscr{L}} N(0, \sigma^2/3).$$

**** Exercise 9.17** Let $\overline{Z}_n = n^{-1} \sum_{i=1}^{n} X_i X_{i+1}$ since $a\overline{X}_n + b\overline{Z}_n = n^{-1} \sum_{i=1}^{n} X_i (a + bX_{i+1})$. We let $Y_i = X_i(a + bX_{i+1})$. Then $Y_1, Y_2, \ldots$ is a stationary 1-dependent sequence with

$$E(Y_i) = a\mu + b\mu^2,\ \eta^2 = \text{Var}(Y_1) = EX_1^2\ E(a + bX_2)^2 - \mu^2(a + b\mu)^2$$

$$= (\sigma^2 + \mu^2)(b^2\sigma^2 + b^2\mu^2 + 2ab\mu + a^2) - \mu^2(a^2 + 2ab\mu + b^2\mu^2)$$

$$= \sigma^2 (a\mu + b\sigma^2 + b\mu^2)(a + b\mu) - \mu^2(a + b\mu)^2$$

$$= \mu(a + b\mu)b\sigma^2.$$

Therefore,

$$\sqrt{n}(a\overline{X}_n + b\overline{Z}_n - \mu(a + b\mu)) \xrightarrow{\mathscr{L}} N(0, \eta^2 + 2\xi)$$

$$= N(0, \sigma^2 (a^2 + 4ab\mu + b^2(\sigma^2 + 4\mu^2))).$$

Now $a\sqrt{n}(\overline{X} - \mu) + b\sqrt{n}(\overline{Z}_n \mu^2) \xrightarrow{\mathscr{L}} aX + bZ$,

$$(X, Z) \sim N_2(0, \Sigma),\ \Sigma = \begin{pmatrix} \sigma^2 & 2\sigma^2\mu \\ 2\sigma^2\mu & \sigma^4 + 4\sigma^2\mu^2 \end{pmatrix}$$

which implies $\sqrt{n}(\overline{X} - \mu, \overline{Z}_n - \mu^2) \xrightarrow{\mathscr{L}} N_2(0, \Sigma)$.

Now $\hat{\gamma}(1) = \overline{Z}_n - \overline{X}_n^2$. Apply Cramer's (delta method) Theorem using the function $g(x, z) = z - x^2$. We have $\dot{g}(x, z) = \left(\frac{\partial g}{\partial x}, \frac{\partial g}{\partial z}\right) = (-2x, 1)$, $g(\mu, \mu^2) = 0$ and $\dot{g}(\mu, \mu^2) = (-2\mu, 1)$. Therefore

$$\sqrt{n}\hat{\gamma}(1) = \sqrt{n}(\bar{Z}_n - \bar{X}_n^2) \xrightarrow{\mathscr{L}} N(0, \sigma^4).$$

(b) (i) Let $U_i = X_i^2$ and $V_i = X_iX_{i+1}$. Then $W_i = aX_i + bU_i + cV_i$ is a 1-dependent sequence with mean $EW_i = b\sigma^2$. Then $EW_i^2 = a^2\sigma^2 + 2ab\mu_3 + b^2\mu_4 + c^2\sigma^4$ and $EW_iW_{i+1} = b^2\sigma^4$, so that Var $(W_i) = a^2\sigma^2 + 2ab\mu_3 + b^2\mu_4 + c^2\sigma^4 - b^2\sigma^4$.

$$\sqrt{n}(a\bar{X}_n + b\bar{U}_n + c\bar{V}_n - b\sigma^2) \xrightarrow{\mathscr{L}} N(0, a^2\sigma^2 + 2ab\mu_3 + b^2(\mu_4 - \sigma^4) + c^2\sigma^4).$$

and
$$\sqrt{n}(\bar{X}_n, \bar{U}_n - \sigma^2, \bar{V}_n) \xrightarrow{\mathscr{L}} N_3(0, \Sigma),$$

where
$$\Sigma = \begin{pmatrix} \sigma^2 & \mu_3 & 0 \\ \mu_3 & \mu_4 - \sigma^4 & 0 \\ 0 & 0 & \sigma^4 \end{pmatrix}, \mu_3 = EX_1^3 \text{ and } \mu_4 = EX_1^4$$

Therefore $\sqrt{n}(\bar{X}_n, \bar{U}_n - \sigma^2, \bar{V}_n) \xrightarrow{\mathscr{L}} N_3(0, \Sigma)$.

(ii) $\hat{\rho}(1) = g(\bar{X}_n, \bar{U}_n, \bar{V}_n)$, where $g(x, u, v) = (v - x)^2/(u - x^2)$. We have $g(0, \sigma^2, 0) = 0$, and hence $\dot{g}(x, u, v) = ((2x(v - u), -(v - x^2), (u - x^2))/(u - x^2)^2)$, we have $\dot{g}(0, \sigma^2, 0) = (0, 0, 1/\sigma^2)$. Hence by Cramer's theorem (delta method), $\sqrt{n}(\hat{\rho}(1) - 0) \xrightarrow{\mathscr{L}} N(0, \sigma^4/\sigma^4) = N(0, 1)$.

Chapter 10

Exercise 10.1 This is a Queueing system with mean arrival and service rates λ and μ, respectively, where no more than k units are allowed to enter the system at any time, we set

$$\begin{aligned} \lambda_n &= \lambda(0 \le n \le k) \\ &= 0(n \ge k) \\ \mu_n &= \mu(n \ge 1) \end{aligned}$$

Then π_0^{-1} in the generalized Birth and Death model in the text is given by $s = \pi_0^{-1} = 1 + \rho + \rho^2 + \ldots + \rho^k = \dfrac{1 - \rho^{k+1}}{1 - \rho}$ where $\rho = \lambda/\mu$ is the traffic intensity. Since s is finite, we conclude that the steady state exists for all values of ρ. Then $\pi_n = (1 - \rho)\rho^n/(1 - \rho^{k+1})$, $\rho \neq 1$, $n = 0, 1, \ldots, k$. When $\rho = 1$, $s = k + 1$ and therefore $\pi_n = \dfrac{1}{k + 1}$, for $\rho = 1$ and $n = 0, 1, \ldots, k$.

Then the expected number in the system is

$$L = \sum_{n=0}^{k} n\pi_n = \rho[1 - (k + 1)\rho^k + k\rho^{k+1}]/(1 - \rho)(1 - \rho^{k+1}) \quad \text{if } \rho \neq 1$$

$$= k/2 \text{ if } \rho = 1.$$

The expected number in the Queue is

$$L_q = \sum_{n=1}^{k} (n-1)\pi_n = \rho^2[1 - k\rho^{k-1} + (k-1)\rho^k]/(1-\rho)(1-\rho^{k+1}) \quad \text{if } \rho \neq 1$$

$$= k(k-1)/\{2(k+1)\} \quad \text{if } \rho = 1.$$

Exercise 10.2

Here

$$\lambda_n = \lambda \text{ if } 0 \le n < k$$
$$= 0 \text{ if } n \ge k$$
$$\mu_n = n\mu \text{ if } n < c$$
$$= c\mu \text{ if } c \le n \le k$$

Writing $\lambda/c\mu = \rho$, we have

$$\pi_n = \pi_0\, c^n\rho^n/n! \text{ if } 0 \le n < c$$
$$= \pi_0 c^c\rho^n/n! \text{ if } c \le n \le k$$

where π_0 will be found from the boundary condition $\sum_{n=0}^{k} \pi_n = 1$ i.e.,

$$\sum_{n=0}^{c-1} \frac{1}{n!}\left(\frac{\lambda}{\mu}\right)^n \pi_0 + \sum_{n=c}^{k} \frac{1}{c^{n-c}c!}\left(\frac{\lambda}{\mu}\right)^n \pi_0 = 1$$

and hence

$$S = \pi_0^{-1} = \sum_{n=0}^{c-1} (c\rho)^n/n! + \sum_{n=c}^{k} \frac{1}{c^{n-c}c!}\left(\frac{\lambda}{\mu}\right)^n$$

$$= \sum_{n=0}^{c-1} (c\rho)^n/n! + \frac{(c\rho)^c}{c!}\sum_{n=0}^{k-c} \rho^n$$

$$= \sum_{n=0}^{c-1} (c\rho)^n/n! + \frac{(c\rho)^c}{c!}\,\frac{1-\rho^{k-c+1}}{1-\rho} \quad \text{if } \rho \neq 1$$

$$= \sum_{n=0}^{c-1} \frac{c^n}{n!} + \frac{c^c}{c!}(k-c+1) \quad \text{if } \rho = 1$$

Note that the steady state exists for all values of ρ. Now, the expected number in the queue (waiting) is

$$L_q = \sum_{n=c+1}^{k} (n-c)\,\pi_n$$

$$= \{\pi_0(cp)^c/c!\} \sum_{n=c+1}^{k} (n-c)\,\rho^{n-c}$$

$$= \{\pi_0\,(c\rho)^c\,\rho/c!\} \sum_{n=1}^{k-c} n\rho^{n-1}$$

$$= \{\pi_0 (c\rho)^c \rho / c!\} \frac{d}{d\rho} [(1 - \rho^{k-c+1})/1 - \rho] \quad \text{if } \rho \neq 1$$

$$\frac{\pi_0 (c\rho)^c \rho}{c! (1 - \rho)^2} [1 - \rho^{k-c+1} - (1 - \rho)(k - c + 1) \rho^{k-c}] \quad \text{if } \rho \neq 1$$

The average number being served or the mean number of busy channels is

$$L_s = \sum_{n=0}^{c-1} n \pi_n + \sum_{n=c}^{k} c \pi_n = \sum_{n=0}^{c-1} n \pi_0 (c\rho)^n / n! + \sum_{n=c+1}^{k} c \pi_0 c^c \rho^n / c!$$

$$= c\rho \pi_0 \left[\sum_{n=1}^{c} \{(c\rho)^{n-1} / (n - 1)!\} + \frac{c^c}{c!} \sum_{n=c+1}^{k} \rho^{n-1} \right]$$

$$= c \rho \pi_0 \left[\sum_{j=0}^{c-1} (c\rho)^j / j! + (c\rho)^c / c! \sum_{n=0}^{k-c+1} \rho^n \right]$$

$$= c \rho \pi_0 [\pi_0^{-1} - (c \rho)^c / c! \rho^{k-c}]$$

$$= c\rho (1 - \pi_k) = \frac{\lambda}{\mu} (1 - \lambda_k)$$

Therefore expected number in the system is

$$L - L_q + L_s = L_q + \frac{\lambda}{\mu} (1 - \pi_k)$$

The effective arrival rate at the system is $\lambda' = \lambda(1 - \pi_k)$.

Therefore from Little's formula $E(W) = L/\lambda'$

$$= L_q / \lambda' + \frac{1}{\mu} = W_q + \frac{1}{\mu}, \quad \text{where} \quad W_q = L_q / \lambda'.$$

Exercise 10.3 By Bayes theorem $q_n = P$ {n is system | arrival about to occur}

$$= \frac{P \{\text{arrival about to occur} \mid n \text{ in system}\} \pi_n}{\sum_{n=0}^{k} P \{\text{arrival about to occur} \mid n \text{ in system}\} \pi_n}$$

$$= \lim_{\Delta t \to 0} \left\{ \frac{[\lambda_n \Delta t + o(\Delta t)] \pi_n}{\sum_{n=0}^{k-1} [\lambda_n \Delta t + o(\Delta t)] \pi_n + 0.\pi_k} \right\}$$

$$= \lim_{\Delta t \to 0} \left\{ \frac{[\lambda_n + o(\Delta t)/(\Delta t)] \pi_n}{\sum_{n=0}^{k-1} [\lambda_n \Delta t + o(\Delta t)/\Delta t] \pi_n} \right\}$$

$$= \frac{\lambda_n \pi_n}{\sum_{n=0}^{k-1} \lambda_n \pi_n} = \frac{\lambda \pi_n}{\lambda \sum_{n=0}^{k-1} \pi_n} = \frac{\pi_n}{1 - \pi_k} \, (n \leq k - 1) \qquad (1)$$

The waiting time c.d.f.

$$W_q(t) = \sum_{n=c}^{k-1} P\{n - c + 1 \text{ completions in} \le t \mid \text{arrival founds } n \text{ in system}\}$$

$$q_n + W_q(0)$$

$$= \sum_{n=c}^{k-j} q_n \int_0^t \frac{\mu c(\mu cx)^{n-c}}{(n-c)!} e^{-\mu cx} dx + W_q(0)$$

$$= \sum_{n=c}^{k-j} q_n \left[1 - \int_t^{\infty} \frac{\mu c(\mu cx)^{n-c}}{(n-c)!} e^{-\mu cx} dx\right] + W_q(0)$$

$$= \sum_{n=c}^{k-1} q_n - \sum_{n=c}^{k-1} q_n \sum_{j=0}^{n-c} \frac{(\mu ct)^j e^{-\mu ct}}{j!} + W_q(0)$$

$$= \sum_{n=c}^{k} q_n - \sum_{n=c}^{k-1} q_n \sum_{j=0}^{n-c} \frac{(\mu ct)^j e^{-\mu ct}}{j!}$$

$$= 1 - \sum_{n=c}^{k-1} q_n \sum_{j=c}^{n-c} \frac{(\mu ct)^j \exp(-\mu ct)}{j!}$$

In $M/M/C/\infty$, $q_n = \dfrac{\lambda_n \pi_n}{\sum_{n=0}^{\infty} \lambda_n \pi_n}$, by leting $k \to \infty$ in (1)

$$= \frac{\lambda \pi_n}{\lambda \sum_{n=0}^{\infty} \pi_n} = \pi_n (n \ge 1)$$

since $\lambda_n = \lambda$ for all $n \ge 1$ in $M/M/c/\infty$.

Exercise 10.4 Taking λ as the mean arrival rate and μ as the mean service rate for each channel, we have

$$\lambda_n = \lambda \text{ for } n < c$$

$$= 0 \text{ for } n \ge c$$

$$\mu_n = n\mu \text{ for } n \le c$$

Therefore $s = \pi_0^{-1} = 1 + \lambda/\mu + (\lambda/\mu)^2/2! + \ldots + \left(\dfrac{\lambda}{\mu}\right)^c \Big/ c!$

Thus, the steady state exists for all values of λ/μ and

$$\pi_n = (\lambda/\mu)^k/n!\pi_0 \; (0 \le n \le c).$$

The probability that all the channels are busy in the steady state is given by

$$\pi_c = \frac{(\lambda/\mu)^c/c!}{1 + (\lambda/\mu) + (\lambda/\mu)^2/2! + \ldots + (\lambda/\mu)^c/c!}$$

π_c is called Erlang's first loss formula. An application of this model is a telephone exchange with a trunk lines with no facility for holding subcribers who fail to get a free line.

Exercise 10.5 Using $M/M/C$ system's result

$$\pi_n = \begin{cases} \dfrac{N!/(N-n)!}{n!}\left(\dfrac{\lambda}{\mu}\right)^n \pi_0 & (0 \le n < c) \\[2ex] \dfrac{N!/(N-n)!}{c^{n-c}c!}\left(\dfrac{\lambda}{\mu}\right)^n \pi_0 & (c \le n \le N) \end{cases}$$

or equivalently
$$\pi_n = \begin{cases} \dbinom{N}{n}\left(\dfrac{\lambda}{\mu}\right)^n \pi_0 & (0 \le n < c) \\[2ex] \dbinom{N}{n}\dfrac{n!}{c^{n-c}c!}\left(\dfrac{\lambda}{\mu}\right)^n \pi_0 & (c \le n \le N) \end{cases}$$

where π_0 is found in the usual way from $\sum_{n=0}^{N} \pi_n = 1$ as

$$\pi_0^{-1} = \left[\sum_{n=0}^{c-1}\binom{N}{n}\left(\frac{\lambda}{\mu}\right)^n + \sum_{n=c}^{N}\binom{N}{n}\frac{n!}{c^{n-c}c!}\left(\frac{\lambda}{\mu}\right)^n\right].$$

This is a finite source queue. A typical application of this model is that of servicing of N machines, where arrivals corresponds to breakdown of machines and the repairmen are the servers. The service times are assumed to be exponentially distributed with mean $1/\mu$.

Exercise 10.6

$$L = \sum_{n=0}^{N} n\pi_n = \pi_0\left[\sum_{n=0}^{c-1} n\binom{N}{n}\left(\frac{\lambda}{\mu}\right)^n + \sum_{n=c}^{N} n\binom{N}{n}\frac{n!}{c^{n-c}c!}\left(\frac{\lambda}{\mu}\right)^n\right]$$

$$= \pi_0\left[\sum_{n=0}^{c-1} n\binom{N}{n}\left(\frac{\lambda}{\mu}\right)^n + \frac{1}{c!}\sum_{n=c}^{N} n\binom{N}{n}\frac{n!}{c^{n-c}c!}\left(\frac{\lambda}{\mu}\right)^n\right]$$

where
$$\pi_0^{-1} = \left[\sum_{n=c}^{c-1}\binom{N}{n}\left(\frac{\lambda}{\mu}\right)^n + \frac{1}{c!}\sum_{n=0}^{N}\binom{N}{n}\frac{n!}{c^{n-c}c!}\left(\frac{\lambda}{\mu}\right)^n\right]$$

The average number of operational machines is $N–L$.

$$L_q = \sum_{n=0}^{N}(n-c)\pi_n = \sum_{n=0}^{N} n\pi_n - c\sum_{n=0}^{N}\pi_n = L - \sum_{n=0}^{c-1} n\pi_n - c\left[1 - \sum_{n=0}^{c-1}\pi_n\right]$$

$$= L - c + \sum_{n=0}^{c-1}(c-n)\pi_n$$

Hence $L_q = L - c + \pi_0 \sum_{n=0}^{c-1} (c-n)\binom{N}{n}\left(\frac{\lambda}{\mu}\right)^n$.

If the system has n units in it, there are $N - n$ possible arrival units outside the system, each with a mean arrival rate λ, so the mean arrival to the system is $(N-n)\,\lambda$.

Hence the effective mean arrival rate,

$$\lambda' = \sum_{n=0}^{N} (N-n)\lambda\pi_n = \lambda M \sum_{n=0}^{N} \pi_n - \lambda \sum_{n=0}^{N} n\pi_n$$

$$= \lambda M - \lambda L = \lambda(M - L).$$

Hence by Little's formula we have

$$E(W) = \frac{L}{\lambda(M-L)} \quad \text{and} \quad W_q = L_q/\lambda(M-L).$$

Hence the expected downtime is $L/\lambda(M-L)$. The average percent idle time of each server is

$$\pi_0 + \frac{1}{c}(\pi_1 + \pi_2 + \ldots + \pi_{c-1}).$$

So the expected percent idle time of each repairman is

$$100\left(\pi_0\left[1 + 1/c\left\{\sum_{n=0}^{c-1}\binom{N}{n}\left(\frac{\lambda}{\mu}\right)^n\right\}\right]\right).$$

Exercise 10.7 From Exercise 10.1

$$\mu(L - L_q) = \mu\frac{\rho - \rho^{k+1} - \rho^2 + \rho^{k+2}}{(1-\rho)(1-\rho^{k+1})} = \mu\frac{\rho(1-\rho) - \rho^{k+1}(1-\rho)}{(1-\rho)(1-\rho^{k+1})}$$

$$= \mu\left[\frac{\rho(1-\rho^k)}{1-\rho^{k+1}}\right]$$

Also $\lambda(1-\pi_k) = \lambda\dfrac{[1-\rho^k]}{1-\rho^{k+1}} = \rho\mu\dfrac{(1-\rho^k)}{1-\rho^{k+1}}$.

Therefore $\mu(L - L_q) = \lambda(1 - \pi_k)$.

Exercise 10.8 Let X_n be the number of arrivals during the service time of the nth customer. Then in the $M/G/1$ system we have $Q_{n+1} = Q_n - H(Q_n) + X_{n+1}$, where $H(x)$ is the heavyside's function, given by

$$H(x) = 1 \text{ if } x > 0$$

$$= 0 \text{ if } x \le 0.$$

Equating expectations Q_{n+1} and Q_n gives $E(H(Q_n)) = E(Q_{n+1})$.

Now $E(X_{n+1}) = \int_0^{\infty} \sum_{j=1}^{\infty} j[\lambda v)^j e^{-\lambda\mu} / j!] dG(v) = \rho.$

Exercise 10.9 (**Takacs integral equation**) Regard the busy period as made up of the service time of the first customer plus the busy period initiated by customers in the system.

Now $G(x) = P(T \le x)$, where T is the busy period with

$$G^*(s) = \int_0^{\infty} e^{-sx} dG(x).$$

Let V be the service time with d.f. $F(v) = P(V \le v)$,

$$F^*(s) = \int_0^{\infty} e^{-sv} dF(v).$$

Let $G^{(n)}$ be the n-fold convolution of G with itself and $A(t)$ denotes the number of arrivals in $[0, t]$. By conditioning on the duration of the service time V of the initial customer initiating the busy period and on the number A of arrivals during his service time, we get

$$G(x) = e^{-\lambda x} V(x) + \sum_{n=1}^{\infty} \int_0^x P\{T \le x \mid V = v, A(v) = n\} dV(v) P\{A(v) = n\}.$$

The first term corresponds to the case that none arrives during the service time of the initial customer (and then $v = x$), let T_i be the total service time of the ith "descendent" and all its other customer initied by him. Then T_i's are i.i.d. r.v.s with same distribution as T. We have

$$(T \le t \mid V = v, A(v) = n) = P(T_1 + \ldots T_n \le t - v) = G^{(n)}(t - v),\ n \ge 1.$$

Therefore $G(x) = \sum_{n=0}^{\infty} \int_0^x \frac{e^{-\lambda v}(\lambda v)^n}{n!} G^{(n)}(x - v) dV(v).$

Pluging the above expression of $G(x)$ on the r.h.s. of

$$G^*(s) = \int_0^{\infty} e^{sx} dG(x)$$

and changing the order of integration (on the r.h.s.) and after simplification, we get

$$G^*(s) = V^* (s + \lambda - \lambda G^*(s)).$$

References

1. Anderson, T.W. The Statistical Analysis of Time Series, John Wiley, New York (1971).
2. Andre, D. (1887).
3. Arthey, K.B. and Ney, P. Branching Processes, Springer-Verlag, Berlin (1972).
4. Bachliar, L. Theorie de la speculation. Ann. Sci. Ecole Norm Sup (3), 21–86 (1900).

5a. Bailey, N.T.J. The Elements of Stochastic Proceses. John Wiley, New York (1964).

5b. Bailey, N.T.J Mathematical Theory of Epidemics, Giriffin and Co., London.

6. Barlow, R.E. and Proschan, F. Statistical Theory of Relability and Life Testing, Holt, Rinehart, New York (1975).
7. Bartlett, M.S. An introduction to stochastic process, Camb. Univ. Press, London (1954).
8. Basu, A.K. An introduction to measure theoretic probability, Prentice Hall, New Delhi (1999).
9. Bharucha-Ried, A.T. Elements of the theory of Markov processes and their applications, McGraw-Hill Book Co, Inc, New York (1960).
10. Bhatt, U.N. Elements of Applied Stochastic Processes. John Wiley, New York (1984).
11. Bhattacharyya, R.N. and Waymire, E. Stochastic Processes with Application, John Wiley, New York (1990).
12. Billingsley, P. Convergence of Probability measure, 2nd ed, John Wiley New York (1999).
13. Blackwell, D. Extension of renewal theorem, Pacific J. Math, 3, 315–320 (1953).
14. Blumenthal, R.M. and Gatoor, R.K. Markov Processes and Potential Theory, Academic Press, New York (1960).
15. Bloomfield, P. Fourier Analysis of Time Series: An introduction, John Wiley, New York (1976).
16. Bochner, S. Stochastic Processes, Ann. Math, 48, 1014–1061 (1947).
17. Box, G.E.P. and Jenkins, G.M. time series Analysis: Forecasting and control, Holdey Dey, Sanfrancisco (1996).
18. Blanc - Lapierre, A. and Fortet, R. Theorie des Fonetions Aleatories, Masson et. Cie, Paris (1953).
19. Brieman, L. Probability, Addison-Wesley, Reading. Massachusetts (1998).
20. Champernowne, D.C. An elementary method of solution of the Queueing problem with a single server and a constant parameter. J.R.S.S. B, 18, 125–128(1956).
21. Chatfield, C. The analysis of Time Series: An Introduction, 5th ed. Chapman and Hall, London (1989).
22. Christer, A.H. Refined Asymptotic costs for renewal reward processes. J. Opns. Res. Soc., 29, 577–583 (1978).
23. Chow, Y.S. and Tiecher, H. Probability Theory, Springer Verlag, 3rd ed. (1998).
24. Chung, K.L. Markov Chains with Stationary Tranisition Probabilities, Springer Verlag, 2nd ed. Berlin (1967).

25. Chung, K.L. and Ornstein, D. On the Recurrence of Sums of Random Variables, Bull. Am. Math. Soc., 5, 332–334 (1954).
26. Cox, D.R. Renewal Theory. Methuen, London (1962).
27. Cox, D.R. and Miller, H.D. The Theory of Stochastic Processes, Methuen, London (1977).
28. Cramer, H. and Leadbetter, M.R. Stationary and Related Stochastic Processes, John Wiley, New York (1967).
29. Doeblin, W. Sur les proprietis asymptotiques de movement rigis par certains types des chaines simples. Bull. Math. Soc. Roum. Sci, 39, No. 1, 57–115, No. 2, 3–61 (1936).
30. Doob, J.L. Stochastic processes, John Wiley, New York (1953).
31. Doob, J.L. Renewal theory from the point of view of the theory of probability. Trans. Amer. Math. Soc. 63, 422–438 (1948).
32. Dvoretzky, A, Erdos, P. and Katutani, S. Nonincrease everywhere of the Brownion motion process. Proc. 4th Berkley. Symp. On Math. Stat. and Prob. vol. II, 103–116 (.961).
33. Erdos, P., Feller, W., and Pollard, H. A Property of power series with positive coefficients, Bull. Amer. Math. Soc. Vol–55, 201–204 (1949).
34. Fabens, A.T. The solution of Queueing and Inventory models by Semi-Markov processes, J.R.S.S. B, 23, 113–127 (1961).
35. Feller, W. On the integral equation of renewal theory, Ann. Math. Statistics, 12, 243–267 (1941).
36. Feller, W. An introduction to probability theory and its applications. vol. I & II, John Wiley, New York (1968, 71).
37. Fisher, R.A. Tests of significance in harmonic Analysis. Proc. Roy. Soc, Ser. A, 125, 54–59.
38. Folkes, J.L. and Chikara, R.S. The Inverse Gaussian Distribution and its Statistical Application—a review, J.R.S.S.B. 40, 263–275(1978).
39. Foster, F.G. On Stochastic Matrices Associated with Certin Queueing Processes. Ann. Math. Statist., 24, 355–360 (1953).
40. Fuchs, W.H.J. and Chung, K.L. On the distribution of values of sums of random variables. Mem. Am. Math. soc. No. 6, 1–12 (1951).
41. Fuller, W.A. Introduction to Statistical Time Series, John Wiley, 2nd ed, New York (1996).
42. Galton, F. and Watson, H.W. On the Probability of extinction of families. J. of Anthr. Inst., 6, 138–144 (1874).
43. Gikhman, I.I. and Skorohod, A.V. Introduction to the theory of random functions, Saunders, Philadelphia (1969).
44. Granger, C.W.J. and Newbold, P. Forecasting Economic Time Series. Academic Press, New York (1977).
45. Grenander, V. and Rosenblatt, M. Statistical Analysis of Stotionary Time Series, John Wiley, New York (1957).
46. Grimmet, G. and Stirzakar, D. Probability and Random Processes, Oxford University Press, 3rd. ed. (2001).
47. Good, I.J. The Number of Individuals in a cascade process. proc. Camb. Phil. Soc., 45, 360–363 (1948).
48. Gross, D. and Harris, C.M. Fundamentals of Queueing theory. 2nd ed. John Wiley, New York (1985).
49. Harris, T.E. The theory of Branching Processes, Springer-Verlag, Berlin (1963).
50. Hannan, E.J. Time Series Analysis. Methuen, London (1960).
51. Hannan, E.J. Multiple Time Series. John Wiley, New York (1970).

52. Isifescu, M. and Tautu, P. Stochastic Processes and Applications in Biology and Medicine, I, Springer-Verlag, Berlin (1973).
53. Ito. K. Stochastic integral. Proc. Imp. Acad. Tokyo 20, 519–524 (1944).
54. Jagers, P. Branching Process with Biological Applications, John Wiley, New York (1975).
55. Kalkutani, S. Ergodic theorems and the Markov Process with a stable distribution. proc. Acad. Tokyo, 16, 49–54 (1940).
56. Karakowski, M. Conservation methods in queueing Theory. Rev.Fran d'auto. Infor. Rech. Oper (R.A.I.R.O) 7e annee, vol. I, 3–20 (1973).
57. Karhunen, K. Uber Lineare Method Methoden in der Wahrschein lich Keit srechnung. Ann. Acad. Sci. Fenn. A. 37 (1947).
58. Karlin, S. and Taylor H.M. A first course in stochastic processes, Academic Press, New York, 2nd ed. (1975).
59. Keilson, J. and Kooharian, A. On time dependent Queueing processes. Ann. Math. Statist, 31, 104–112 (1960).
60. Kemeny, J.G. and Snell, J.L. Finite Markov Chains, Academic Press, New York (1966).
61. Kendall, D.G. Some Problems in the Theory of Queues, J.R.S.S. B, 13, 151–185 (1951).
62. Kendall, D.G. On the generalized birth and death processes, Ann Math, Statist., 19, 1–15.
63. Khintchine, A.Y. Correlation theory of stationary random processes. Usp. Math. Nauk, 5, 42–51 (1938).
64. Kingman, J.F.C. The single server queue in heavy traffic. proc. Camb. Phil. Soc., 57, 902–904 (1961).
65. Kleinrock, L. Queueing Systems, Vol. I & II, John Wiley, New York (1975).
66. Kolmogorov, A.N. Interpolation and extrapolation of Stationary random sequences, IZV. Akad. Nauk. SSSR Ser. Math. 5, 3–14 (1941).
67. Kolmogorov, A.N. On the problem of differencetiability of transition probabilities of time-homogenous Markov Processes with Countable number of states. Uchenye Zapiski Markovskovo Gos. 148 (Mathematika 4), 53–59 (in Russian).
68. Loeve, M, Probability Theory. Vol I & II, Springer-verlag New York (1982).
69. Levy, P. Theorie de l'addition des variables aleatories, Paris (1937).
70. Levy, P. processus stochastiques et mouvement Brownien, Paris (1948).
71. Lotka, J.A. Stochastic convergence. Academic Press (1937).
72. Markov, A.A. Calclus of Probability (Russian) 4th ed. Moscow (1924).
73. Medhi, J. Stochastic Processes, New Age International Pub. New Delhi, 2nd ed (1994).
74. Miller, H.D. A Generalization of Wald's identity with Applications to Random Walks. Ann. Math. Statist., 32, 549–560 (1961).
75. Morse, P.M. Queues, Inventories and Maintenance. John Wiley, New York (1958).
76. Neuts, M.F. A general class of bulk queues with Poisson input. Ann. Math. Statist., 38, 759–770 (1967).
77. Ornstein, L.S. and Uhlenbeck, G.E. on the theory of Brownian motion. Physical Review, 36, 93–112 (1930). [Reprinted in selected papers on Noise and Stochastic Processes, (Ed. Wase, N.) Dover, New York (1954)].
78. Palm, C. (1943) Intensitatschwarkungen in Fernsprechovorkehr Eriesson technics, 44, 1–189, Stocholm: Tele fonaktiebolaget I.M. Ericeson. I.M. Ericeson.
79. Parzen, E. Stochastic Processes, SIAM, Philadelphia (1999).
80. Prabhu, N.U. Queues and Inventories, John Wiley, New York (1965).
81. Prabhu, N.U. Stochastic Processes, Macmillan, New York (1965).

82. Priestley, M.B. Spectral analysis and time Series, Vol. I & II, Academic Press, New York (1981).
83. Quenouille, M.H. Approximate tests for correlation in time Series. J. Roy. Stat. Soc. B, 11, 68–84 (1957).
84. Resnik, S.I. Adventures in stochastic Processes. Birkhairser, Boston (1999).
85. Rolski, T., schmidli, H., Schmidt, V., Tengels, J. Stochastic Processes for Insurance and Finance, Wiley, New York (1998).
86. Rosenblatt, M. Some regression problems in time series Analysis. Proc. 3rd. Berkeley. Symp. On Math. Statist. Prob., Univ. of California Press, Berkeley 165–186 (1956).
87. Ross, S. Stochastic Processes. 2nd ed. John Wiley, New York (2000).
88. Slutsky, E. Sur Les Fonctions Eventuelles Continues, Integrables et Drivables Dans le sens stochastique. C.R. Acad. Sci. Paris, 187, 878–880 (1928).
89. Smith, W.L. Renewal Theory and its Ramifications, J. Roy. Statist. Soc., B 20, 243–302 (1958).
90. Smoluchowlski, M.V. Molekularthearetische studen uber Umkehn thermodynamisch irreversilder vorgange and uber wideskehn abnormale zustance, S.B. Akad. Wiss Wien (2a): 124–339 (1915).
91. Spitzer, F. Principles of random walk. D. Van Nonstrand, Co. Princeton (1964).
92. Stidham, S. A last work on $L = W$, Opns. Res. 22, 417–421 (1974).
93. Stratonovich, R.L. A new representation for stochastic Integrals and equation, Bull. Moscow Univ. Math. and Mech. Ser. No: 1, 3–12 (in Russian) (1964) Translation in J. SIAM CONTROL, 4, 362–371 (1965).
94. Takacs, L. Investigation of wating time problems by reduction to Markov processes, Act. Math. Hung., 6, 101–129 (1955).
95. Tichmmarsh, E.C. Theory of Functions, Griffin, London (1936).
96. Wald, A. Sequential Analysis, John wiley, New York (1945).
97. Wiener, N. Extrapolation, interpolation and smoothing of stationary time series; with Engineering Applications. Cambridge - New York (1949).
98. Wold, H. A study in the Analysis of stationary Time series, 2nd ed. Almqvist & Wiksells, Stockholm (1954).
99. Wong, E. Stochastic Processes in Information and Dynamical systems. McGraw-Hill, New York (1971).
100. Yule, G.D. A Mathematical Theory of Evolution, Phil. Trans. Roy. Soc., London B, 213, 21–87.

Subject Index

Abel's Lemma, 16
Absorbing state, 13
Accessible state, 12
Amplitudes, 185, 186, 187
Age replacement, 82
Aperiodic Markov Chains, 23
Arley Process (Feller), 128, 129
(Lattice) Arithmatic distribution, 70, 85
Autocorrelation functions, 178, 187
Autocovariance function, 178
Autocorrelation Generating function (AGF), 201
Autoregressive process, 180, 188, 190
Autoregressive moving averages, 181
  ARMA (p, q), 181, 203
  ARIMA (p, d, q), 181, 193
Average Power of random process, 185

Backward Kolmogorov differential equation, 155
Bessel process, 170
Birth and Death Processes, 117
  with linear growth, 125
  linear growth with immigration, 126
  nonhomogeneous, 121
  probability of absorption, 126
  pure birth processes, 134
Block replacement, 82
Bochner-Khintchine's Theorem, 195, 197
Borel Zero-one law, 267
Branching processess, 85
  in continuous time, 98
  extinction probability, 91
  generating functions, 86
  with linear birth and death process, 102
  factorial moments, 103
  Kolmogorv equations for branching process, 100, 129
  with multiple types,
Brownian motion, 156
  with drift, 158, 161
  geometric, 174
  martingales related to, 260
  multidimensional Brownian motion, 170
  radial, see Bessel process, 170
  reflected at the origin, 173
  quadratic variation of, 262
  (bounded)
  total variation of, 263

Caley-Hamilton Theorem, 42
Cauchy - Schwartz inequality,
Central limit theorem
  in a renewal process, 74
Chapman-Kolmogorov equation, connected with branching processes, 99
Closed (Transient) class, 13
Communicating class of states, 12
Conservative (nonexplosive) Markov process, 112
Continuity of sample paths, 151, 261
Correlation function, 178
Correlogram analysis, 184, 187
Covariance function, 178
Covariance matrix, 4, 167, 177
Covariance stationary process, 4, 177
Current life,
  limiting distribution, 80
  in a Poisson process, 137
  in a renewal process, 78

Deterministic process, 194
Differential difference equation, 112
  of conservative process, 112
  of birth and death processes, 117
Diffusion equation, 155
Diffusion process, 151
Dominated Convergence theorem, 17
  Doob's renewal theorem, 71

Elementary renewal theorem, 70, 71
Erlang's models, 230, 247
Erlang loss system, 247
Entry time, 14
Ergodicity, 3, 182
Ergodic states in a Mrkov Chain, 28
Ergodic Theorem, 182
Ehrenfest chain, 34, 35
embedded markov chain, 243
Essential state, 12
Inessential state, 12
Excess life, 78
Exponential Smoothing, 212
Extinction in branching processes
in continuous time, 104
in discrete time, 91

Fair game, 2
FCFS, 227
First come, first serve discipline (FCFS), 227
First entrance time, 14
First Fourier Transform, 210
Fisher's Information, 216
Fisher's test for Hidden periodicties, 210
Forward Kolmogorov differential equations, 155
Fourier Analysis, 196
Frequency, 185, 195

Gambler's ruin, 52
Gaussian process, 4, 167
Generating functions
Generating functions relations for branching processes, 88, 96
Fokker - Planck equation, 155
Foster type theorems, 136
Genetic models,
General input, 228
Geometric Brownian motion, 174

Hamming, 216
Hanning, 210, 212
Heat equation, 155
Hilbert space, 275

Independent increments, 2, 133, 156
Index parameter, 1
Inequalities for Brownian motion, 174
Infinitesimal generator, 114
Infinitesimal matrix, 114
Instantaneous state, 112
Inventory (Storage) models,
Irreducible (ergodic) Markov chains, 13

Joint distribution, 4, 154

Kemperman and Miller's indentity, 60
Karhunen-Loe've expansion, 274, 282
Kolmogorov's formula, 261
Kronecker's lemma, 183, 198

Lengevian equation, 163
Laplace transform
of a renewal equation, 67, 76
Limit theorems
for renewal process, 70-71
for branching processes,
for Markov chains, 23, 26, 28, 30
Limiting distribution
of current life in a renewal process, 80
of excess life in a renewal process, 79
in a Markov renewal process,
Linear fractional transformation, 98
Theory of Linear predictor,
Little's formula, 97
Lotka model, 97

Markov branching process, 98
Markov chain,
Birth and Death, 32
Markov process
basic limit theorems, 23, 26
Classifications of states, 14
Periodicity, 23
recurrence of , 16
Markov renewal process,
Markov time, 301
Martingales
Continuous parameter, 260, 301
Maxwell-Boltzmn distribution, 164, 165
Mean recurrence time, 16
Mean square convergence, 182, 271
Mean square distance, 275
Mean square error, 182, 270
Measurable progressively, 260
Mercer's theorem, 282
Moving average processes, 179
Multivariate normal distribution, 4, 167

NBU distribution, 78
NBUE, 78

Negative binomial distribution, 6
Nonanticipating, 289
Nonhomogeneous Poisson process, 143
null recurrent states, 16

Offspring distribution, 87-88
Optional sampling theorem, 300, 302
Optional stopping theorem, 300
Order statistics, 146
Ornstain-Uhlenbeck process, 162

Periodic chain, 23
Periodicity of Markov Chains, 23
Periodogram, 184, 209
Point processes, 2
Poisson distribution, 133
Poisson process, 133
Poisson increment process, 145
Characterization of, 140
Compound (cumulative), 144
Interrival time of, 139
Nonhomogeneous (time dependent), 143
Postulates, for, 134
Properties of, 138
Positive recurrent, state, 16
Prediction theorem, 195
Pollaczek - Khintchine formula, 229
- mean-value formula, 249
- transform formula, 250
Polya's urn model, 45
Power spectrum, 185
Probability generating function, 16
Probability vector, 9
Process-Kendall, 126
Process with independent increments, 2
Separable, 259
Pure birth process, 118
Purely random process, 179

Queueing models
M/*M*/*C*, 239
M/*D*/1, 249
E_k/*M*/1, 251
GI/*G*/1, 246
M/E_k/1, 249
M/*G*/1, 246
M/*M*/1, 230
$M^{(x)}$/*M*/1, 256
Queing chain, 23
Queing processes, 238

Queueing sheory, 237
bulk service rule, 257
Queueing time, 232
Steady state distribution, 228
non-Markovian, 243
utilization (load) factor, 231
Queues
bulk, 256
group arrival, 256
multichannel, 240
non-Markovian, 243
Zero avoiding process, 238

Random function (process), 1
Random walk,
Correlated, 49
Realization of aprocess,
Recurrence time (forward/backward), 78
Recurrent events, 1
Recurrent states, 16
Reducible chain, 12
closed class, 12
Reflection principle, 159
Regeneration point (epoch), 243
Regenerative stochastic process, 243
Regular (standard) Markov process,
Reliability theory, 77
Renewal argument, 74
Renewal counting process, 65
Renewal function, 65
Renewal density, 74
Renewal equation, 66
Renewal process, 65
Delayed, 75
equilibrium (stationary), 77, 75
in continuous time, 65
in discrete time, 65
Renewal reward process, 80
Renewal theorems,
Blackwell's, 70
Doob's, 71
elementary, 70
Key (Smith's), 70
Renewal theory,
Replacement policies,
age, 82
block, 82
Residual and excess (or spent life) life times, 79, 80
asymptotic distributions of - 79, 80
Rouche's theorem, 234

Sample function, 1
Sample correlogram, 191
SARIMA (p, d, q) × (P, D.Q), 213
Seasonal components, 209
Second order stationary process, 276
Semi-markov process, 246
  embedded markov chain, 246
Slutsky's theorem, 217
Sojurn time, 232
Spectral analysis, 184
Spectral decomposition, 195
Spectral density function, 184
Spectral distribution function, 186, 196
Spectral representation, 186, 196
Spectral theorem, 195
Spectrum, 185
Spectrogram, 207
Steady state,
Standrd Brownian motion, 156
State space, 1
Stationary increments, 133
Stationary distribution for a M.C., 27
Stationary process, 2, 177
Statistical equilibrium, 28
  covariance function of,
  convariance stationary (weakly or wide sense stationary), 4
  evaluationary process, 4
  Gaussian process, 4
  multi-dimensional, 4
Stictly stationary,
  with independent increments, 177
Stationary transition functions (probabilities), 8
Stochastic matrix, 8
Stochastic process, 1
  regular, 5
  orthogonal, 5
Stratonovich integral, 289
Stochastic sequence, 1
  independent, 2
Stochastic Integral, 273, 287
Stopping time (Markov time), 59, 299, 301
Submatringales, 280
  martingales, 2, 95, 260
Subordinator, 174
Success runs, 21
Sums of independent random variables
  associated with renewal processes, 64
Sum of a random number of r.vs., 59
  Supermartingales, 299
Supplementary variable technique, 246

Tackcs integral equation, 258
Tauberian theorems, 16
Telegraph signals, 7
  random - , 7
Time domian, 184
Time series
  autoregressive process, 180
  first order Markov process, 188
  moving average process, 181
Total life
  in a renewal process, 78
Transient state, 16
Traffic intensity, 231
Transient behaviour
  of M/M/1 queue, 233
Transition density
  of continuous time Markov chain, 152, 157
Transition function, 151
Transition distribution of waiting time, 233
Transition (Markov) matrix, 8
Two state process, 30

Uniformly integrable random variables, 72
  Urn models, 45, 47
  Ehrenfest model, 34
  modified Ehrenfest model, 35

Waiting times
  in a Poisson process, 138
Waiting time distribution
  - in M/M/1 queue, 231
  - M/M/C queue, 242
Wald's equation, 59
Wald's indentity, 60
Wald's martingale, 393
Weakly stationary process, 2, 117
White noise process (Purely random process), 117
Wiener process (Wiener Einstein), 156
Wiener-Khinchine's theorem, 195
  backward diffusion equation, 155
  definition of, 155
  differential equation of, 155
  distribution of maximum of, 160
  drift of, 161

first passage time distribution of, 161
forward diffusion equation of, 155, 165
Kolmogorov equation of, 155
sample path of Brownian motion, 260
Windows
Parzen, 212
- Tukey, 212
Wold Decomposition Theorem, 195, 284

Yule (Yule-Furry) process, 118
Yule-Walker equations, 190

Zero-avoiding state probability, 238